Adrenergic Mechanisms in Myocardial Ischemia

Gerd Heusch, John Ross, Jr. (Eds.)

Adrenergic Mechanisms in Myocardial Ischemia

With contributions by

M. Anliker, C. Berger, G. Berkenboom, G. Breithardt, O. E. Brodde, P. Busch, W. M. Chilian, P. B. Corr, S. D. Da Torre, A. Deussen, E. O. Feigl, D. Frehen, A. Gaglione, K. P. Gallagher, M. W. Gerdisch, K. L. Gould, H. Gülker, B. D. Guth, P. A. Gwirtz, C. Haemmerli, D. G. Harrison, W. Haverkamp, O. M. Hess, G. Heusch, G. Hindricks, H. Hirche, A. Hjalmarson, J. Holtz, H. Homburg, C. Indolfi, P. A. Insel, C. E. Jones, S. Jost, R. L. Kirkeeide, D. R. Knight, H. Knopf, H. P. Krayenbuehl, P. R. Lichtlen, A. S. Maisel, A. Malliani, R. Marquetant, M. Muser, H. P. Osenberg, J. E. Quillen, W. Rafflenbeul, L. A. Ransnäs, G. Richardt, J. Ross, Jr., J. Schipke, A. Schömig, R. Schulz, P. J. Schwartz, R. Seitelberger, F. W. Sellke, Y. T. Shen, R. H. Strasser, T. Suter, V. Thämer, E. Thaulow, J. X. Thomas, T. Tölle, P. Unger, E. Vanoli, D. E. Vatner, S. F. Vatner, R. Walser, K. A. Yamada, M. A. Young.

 Springer-Verlag Berlin Heidelberg GmbH

CIP-Titelaufnahme der Deutschen Bibliothek

Adrenergic mechanisms in myocardial ischemia / Gerd Heusch ;
John Ross, jr. (ed.) With contrib. by M. Anliker ... –
Darmstadt : Steinkopff ; New York : Springer, 1991
  (Supplement to Basic research in cardiology ; Vol. 85,1)

NE: Heusch, Gerd [Hrsg.]; Anliker, Max; Basic research in cardiology /
  Supplement

  ISBN 978-3-662-11040-9   ISBN 978-3-662-11038-6 (eBook)
  DOI 10.1007/978-3-662-11038-6

Basic Res. Cardiol. ISSN 0300-8428
Indexed in Current Contents.

Medical editor: Sabine Müller – English editor: James C. Willis – Production: Heinz J. Schäfer

Typesetting: Alden Press, Oxford
Printed on acid-free paper

Foreword

Stress-induced myocardial ischemia is a frequent manifestation of coronary heart disease, and sympathetic activation is an important precipitating and aggravating factor in such stress-induced ischemia. However, the complex interplay between the sympathetic initiation of myocardial ischemia, ischemia-induced alterations in sympathetic neurotransmission, as well as changes in adrenoceptor density and post-receptor signal transduction that can occur during ischemia remains incompletely understood. Not only the activation of myocardial β-adrenoceptors, but also the activation of coronary α-adrenoceptors can contribute to myocardial ischemia. However, the role of β-adrenoceptor-mediated increases in contractility relative to heart rate in the initiation of ischemia is not clear, and the significance of α-adrenoceptor-mediated changes in coronary vasomotor tone, as well as the responsible α-adrenoceptor subtypes are highly controversial. Malignant arrhythmias may be triggered by both α- and β-adrenergic mechanisms. Current research in this field is focussed not only on the underlying physiological and pathophysiological mechanisms, but also on clinical treatment strategies, e.g., by β-blockade, α-blockade, bradycardic agents and calcium antagonists.

Recent findings were presented and future research directions discussed during the 61st International Titisee Conference, held at the Schwarzwald-Hotel, Titisee, March 29–31, 1990 under the sponsorship of the Boehringer Ingelheim Foundation. Dr. Hasso Schroeder and Dr. Hermann Fröhlich deserve special thanks for their generous support and pleasant organization of the meeting. The publication of the proceedings has been made possible by grants from Astra Chemicals, Bayer, ICI, Dr. Karl Thomae, and Upjohn. Finally, we would like to thank the publisher, in particular Ms. Sabine Müller, for the pleasant cooperation and constructive support in publishing these proceedings.

Essen, FRG, and La Jolla, California, USA, 1990

Gerd Heusch
John Ross, Jr.

Contents

Foreword. V

I. Cardiac Sympathetic Innervation, Neurotransmission, Adrenoceptors
and Signal Transduction in Myocardial Ischemia

Topical Organization of the Cardiac Sympathetic Nervous System
Thomas JX, Gerdisch MW 3

Cardiac Sympathetic Activity in Myocardial Ischemia: Release and Effects of Noradrenaline
Schömig A, Richardt G 9

Modulation of α-Adrenergic Receptors and their Intracellular Coupling in the Ischemic Heart
Corr PB, Yamada KA, Da Torre SD 31

β-Adrenergic Receptors and the G_s Protein in Myocardial Ischemia and Injury
Maisel AS, Ransnäs LA, Insel PA 47

β- and α-Adrenoceptor-Agonists and -Antagonists in Chronic Heart Failure
Brodde OE 57

Supersensitivity of the Adenylyl Cyclase System in Acute Myocardial Ischemia: Evaluation of three Independent Mechanisms
Strasser RH, Marquetant R 67

II. Adrenergic Coronary Vasomotion

α-Adrenoceptor Subtypes in the Coronary Circulation
Holtz J 81

α- and β-Adrenergic Control of Large Coronary Arteries in Conscious Calves
Young MA, Vatner DE, Vatner SF 97

Adrenergic Vasomotion in the Coronary Microcirculation
Chilian WM 111

Neurohumoral Regulation of Coronary Collateral Vasomotor Tone
Harrison DG, Sellke FW, Quillen JE 121

α-Adrenergic Regulation of Myocardial Performance in the Exercising Dog: Evidence for Both Presynaptic $α_1$- and $α_2$-Adrenoceptors
Guth BD, Thaulow E, Heusch G, Seitelberger R, Ross J, Jr. 131

III. Adrenergic Coronary Vasoconstriction and Blood Flow Distribution in Ischemic Myocardium

Transmural Steal with Isoproterenol and Exercise in Poststenotic Myocardium
Gallagher KP 145

Mechanisms of Benefit in the Ischemic Myocardium due to Heart Rate Reduction
Guth BD, Indolfi C, Heusch G, Seitelberger R, Ross J, Jr. 157

Adrenergic Control of Transmural Coronary Blood Flow
Feigl EO. 167

α_1-Adrenergic Coronary Constriction during Exercise and Ischemia
Jones CE, Gwirtz PA 177

Contribution of Postsynaptic α_2-Adrenoceptors to Reflex Sympathetic Constriction of Stenotic Coronary Vessels
Deussen A, Busch P, Schipke J, Thämer V, Heusch G 193

α_2-Adrenergic Coronary Constriction in Ischemic Myocardium during Exercise
Seitelberger R, Guth BD, Heusch G, Ross J, Jr.. 207

Prevention of α-Adrenergic Coronary Constriction by Calcium-Antagonists
Heusch G, Deussen A, Guth BD 219

Effects of Selective Cardiac Denervation on Collateral Blood Flow after Coronary Artery Occlusion in Conscious Dogs
Shen YT, Knight DR, Thomas JX, Vatner SF 229

IV. Sympathetic Activation in Myocardial Ischemia

Cardiocardiac Excitatory Reflexes during Myocardial Ischemia
Malliani A 243

Pain and Myocardial Ischemia: the Role of Sympathetic Activation
Thämer V, Deussen A, Schipke JD, Tölle T, Heusch G. 253

No Impairment of Sympathetic Neurotransmission in Stunned Myocardium
Schulz R, Frehen D, Heusch G 267

V. Adrenergic Mechanisms Triggering Arrhythmias during Myocardial Ischemia

Does Noradrenaline Influence the Extracellular Accumulation of Potassium, Sodium, Calcium, and Hydrogen Ions ($[K^+]_e$, $[Na^+]_e$, $[Ca^{2+}]_e$, $[H^+]_e$) during Global Ischemia in Isolated Rat Heart?
Hirche H, Knopf H, Homburg H, Walser R 283

Effects of β-Blockade on the Incidence of Ventricular Tachyarrhythmias during Acute Myocardial Ischemia: Experimental Findings and Clinical Implications
Haverkamp W, Gülker G, Hindricks G, Breithardt G 293

Sympathetic–Parasympathetic Interaction and Sudden Death
Vanoli E, Schwartz PJ. 305

VI. Adrenergic Mechanisms in Clinical Coronary Heart Disease

Heart Rate and β-Adrenergic Mechanisms in Acute Myocardial Infarction
Hjalmarson A 325

Coronary Vasomotor Tone in Large Epicardial Coronary Arteries with Special Emphasis on β-Adrenergic Vasomotion, Effects of β-Blockade
Lichtlen PR, Rafflenbeul W, Jost S, Berger C 335

The Poststenotic Vessel Segment during Dynamic Exercise: Effect of Oral Isosorbid-dinitrate
Gaglione A, Hess OM, Haemmerli C, Suter T, Kirkeeide RL, Osenberg HP, Muser M,
Anliker M, Gould KL, Krayenbuehl HP 347

α-Adrenergic Coronary Constriction in Effort Angina
Berkenboom G, Unger P 359

Subject Index 371

I. Cardiac Sympathetic Innervation, Neurotransmission, Adrenoceptors and Signal Transduction in Myocardial Ischemia

Topical Organization of the Cardiac Sympathetic Nervous System

J. X. Thomas, Jr. and M. W. Gerdisch

*Departments of Physiology and Surgery, Loyola University of Chicago, Stritch School of Medicine, Maywood, USA

Summary

A thorough understanding of the innervation pathways to the heart is requisite for studying the effects of the sympathetic nervous system. Knowledge of these pathways and how to selectively activate or eliminate them allows for the production of unique animal models for further investigation about the interaction between the cardiac sympathetics and the heart during ischemia.

Introduction

The autonomic nervous system is an important regulating mechanism of cardiovascular function. The autonomic innervation of the myocardium has been discussed by numerous authors (3,27,29–31,34) for the past 65 years. Projections from the parasympathetic nervous system as well as the sympathetic nervous system are seen in each of the four cardiac chambers to some extent. Although both the parasympathetic and sympathetic nerves participate in the moment-to-moment regulation of the heart, the primary influence of the parasympathetic nerves is on heart rate and rhythm. Parasympathetic activation results in sinus bradycardia, slowed conduction through the A-V nodal region, decreased atrial contractility, and a mild coronary vasodilation. The sympathetic nerves affect rate and rhythm as well, but additionally have considerable influence on ventricular excitability, function, and coronary blood flow. The importance of the sympathetic nerves in response to pathophysiologic stresses such as myocardial ischemia is obvious. The purpose of this paper will be to briefly review the organization of the canine cardiac sympathetic efferent system. The organization of other species is similar.

The primary neurotransmitter of the sympathetic nerves in the myocardium is norepinephrine. Neuropeptide Y which has been colocalized with norepinephrine (1,10,12) in the heart may play some role in the process of sympathetic nevous control, but will not be discussed here. Likewise, specifics of the responses obtained when the sympathetics are activated, directly or reflexly, have been extensively discussed elsewhere (30,39).

Anatomic distribution of the cardiac sympathetic nerves

A schematic diagram of the cardiac autonomic nerves is shown in Fig. 1. The nomenclature used to describe these nerves is the same as that used by Mizeres (27) and by Randall and colleagues (30,31). Additions to this nomenclature have been proposed by Armour and his colleagues (2,3,6,16,20) based on extensive study of ganglionic interconnections. Other species have also been recently reviewed (3,18,28).

---

*Support: Thoracic and Cardiovascular Surgery Research Fund

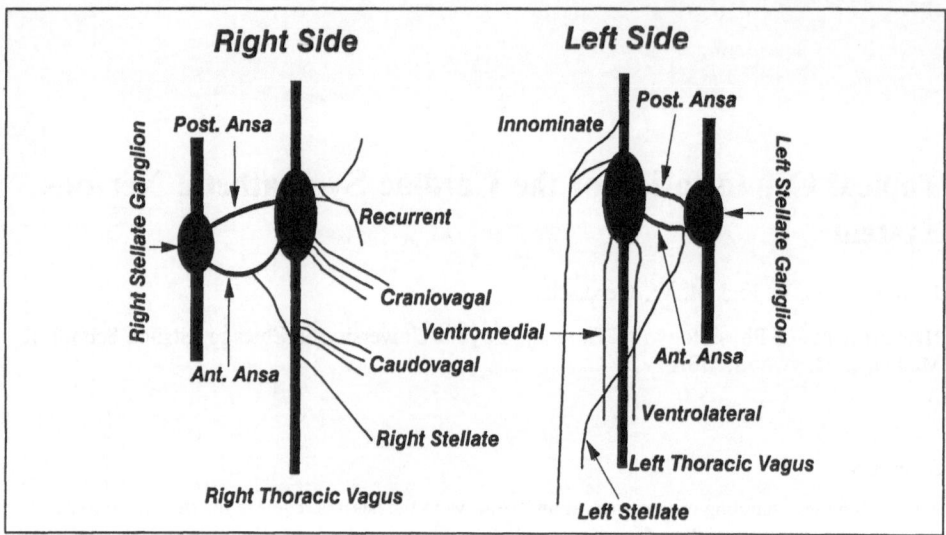

Fig. 1. Schematic diagram of the canine autonomic nerve distribution to the heart.

Anatomy of the cardiac nerves on the right side

Arising from the right stellate ganglion or the anterior ansa subclavia are several small branches which proceed caudally to the dorsal surface of the right atrium. These nerves coalesce to comprise the right stellate cardiac nerve. Stimulation of this nerve elicits a profound chronotropic response with some inotropic response seen in the right atrium. However, this nerve carries few fibers to the right or left ventricles. The right recurrent laryngeal nerve gives rise to the right recurrent cardiac nerve. This nerve contains both parasympathetic as well as sympathetic fibers, with branches which innervate all four cardiac chambers to some extent. In addition to these nerves, there are the craniovagal and caudovagal nerves arising from the vagosympathetic trunk. These nerves are mixed nerves carrying both sympathetic as well as parasympathetic fibers. However, they are predominantly parasympathetic in their effect. Thus, the innervation from the right side sympathetics to the heart is via the right stellate cardiac nerve, the recurrent cardiac nerve, and portions of the craniovagal and caudovagal cardiac nerves. Stimulation of the right stellate cardiac nerve will elicit a pure sympathetic response (not mixed) and result in profound sinus tachycardia. Stimulation of all other nerves on the right side will result in mixed (sympathetic and parasympathetic) responses, with the vagal effects predominating in the atrial chambers.

Cardiac nerves on the left side

Nerves arising from the left stellate ganglion or ansae subclaviae sometimes give rise to a left stellate cardiac nerve (approximately 20% of all dissections). This nerve innervates the atrial chambers and, like the right stellate cardiac nerve, results in profound chronotropic effects, as well as some inotropic effects in the atrial chambers. The distribution of this nerve appears to be limited primarily to the atria. The innominate cardiac nerve arises from the caudal cervical ganglion and receives input from the ansae subclaviae. Stimulation of this nerve will result in variable sympathetic or parasympathetic effects. This nerve pathway carries significant input to the right ventricular region as well as to the left coronary artery. The ventromedial cardiac nerve arises from the medial aspect of the vagus and also receives input from the caudal cervical ganglion. It also is a mixed nerve, having both sympathetic as well as parasympathetic efferent fibers. Stimulation of this nerve in the presence of atropine results in profound inotropic effects on the left side of the heart. The ventrolateral cardiac nerve comes from the lateral aspect of the caudal cervical ganglion and receives branches from the left stellate ganglion. The course of the ventrolateral cardiac nerve caudally is along the vagosympathetic trunk and it enters the pericardium along the remnant of the left superior vena cava (vein of Marshall). This nerve primarily carries sympathetic input. Stimulation of this nerve will result in profound inotropic responses, particularly in the posterior lateral wall of the left ventricle. In addition, stimulation of this nerve will result in a pacemaker shift from the sinus node to the atrioventricular region. This nerve can be easily identified inside the pericardium as it crosses the superior pulmonary vein.

Regardless of whether the nerves originate on the right or left side, there are several plexi which can be found, either in the pericardium or just outside the pericardium. The one which is most easily identifiable is referred to as the pre-tracheal cardiac plexus. This plexus gives rise to a number of intrapericardial projections, both sympathetic as well as parasympathetic. All of the nerves enter the pericardium and eventually the heart by first coursing along one of the major blood vessels. By thoroughly removing the adventitial layers at the pericardial reflections of each of the major blood vessels as they enter or leave the pericardium, it is possible to ablate the autonomic nerves. The specific patterns of the sympathetic nerve projections onto the heart also have been described (34). The specifics of these denervation techniques will be discussed in the next section.

The nerves which may be considered to be primarily sympathetic efferent pathways include the stellate cardiac and ventrolateral on the left and the stellate cardiac on the right side. All other nerves are mixed, containing both sympathetic and parasympathetic efferent fibers. Although the mixed nerves may exhibit selective activation of the sympathetic component physiologically, direct electrical stimulation of these nerves will result in complex functional responses.

Cardiac denervation

There are numerous techniques for accomplishing removal of the autonomic input to the heart, including surgical ablation, topical application of a substance that destroys neural tissue, and chemical treatment to destroy or deplete the nerve terminals of norepinephrine. A relatively simple surgical technique for accomplishing total cardiac denervation (without autotransplantation) was described (33) and has been used successfully by many groups. More selective surgical techniques have also been described (35). The technique for removing the innervation to the left ventricle, intraventricular septum, and anterior surface of the right ventricle will be described below. This technique effectively removes sympathetic input to the left ventricle. Following a left lateral thoracotomy, the succeeding steps are accomplished:

1) The pericardium is opened and the ventrolateral cardiac nerve and its branches are cut as they pass over the left superior pulmonary vein.
2) The tissue between the left pulmonary artery and the superior surface of the left atrium is removed.
3) The adventitial layers of the pulmonary artery are completely removed and stripped back, leaving a band of the artery without the adventitial layers.
4) The branches of the innominate, ventromedial, and ventrolateral nerves travelling beneath the left pulmonary artery are cut. Sometimes these nerves are together in a bundle travelling between the common pulmonary artery and the aorta.

Completeness of denervation may be assessed at surgery by recording functional parameters during direct electrical stimulation of the left stellate ganglion before and following the denervation procedure.

A less difficult technique to denervate specific regions of the heart employs application of phenol (22). This technique allows a specific anatomic area to be denervated and has been used by many investigators to study neural control of the heart.

Sympathetic imbalance

Sympathetic imbalance refers to the concept of an unbalanced activation of the cardiac sympathetics to specific areas of the heart. This may occur due to selective stimulation or activation of a sympathetic pathway. Alternatively, when an area of the myocardium is denervated leaving intact innervation to other regions, uniform sympathetic activation from the brainstem will also result in sympathetic imbalance. Randall and co-workers reported, following total denervation of the heart leaving the ventrolateral cardiac nerve intact, an animal model in which reproducible tachydysrhythmias were observed when the animals were subjected to exercise (32). Similar results have been reported when selective nerve stimulation was used in anesthetized dogs (9).

Selective activation of the myocardium by isolated infusion of adrenergic agonists has been shown to produce a dyssynchronous contraction pattern (13,17). The area receiving the agonist begins shortening during early systole, causing passive stretching of the myocardium not receiving the agonist. Regional denervation of selected areas of the myocardium followed by sympathetic activation results in a similar effect (15,23,24). In anesthetized dogs, normally innervated myocardium responds to left stellate stimulation (24) with early shortening, while the denervated myocardium shows early systolic expansion creating an abnormal pressure-segment length loop. In a series of conscious animals, we found post-systolic wall thickening in the denervated region of the dog at rest. The magnitude of the post-systolic thickening was augmented with excitement (↑ sympathetic activity) (23). Infusion of norepinephrine reversed this process, such that the normally innervated myocardium reached peak wall thickness later, presumably due to the supersensitivity of the denervated region to norepinephrine. Although regional contractile abnormalities secondary to autonomic imbalance may be considerable, the effects on global function are dependent upon the amount of myocardium involved. Thus, small areas of myocardium with disrupted sympathetic nerves may be without any functional consequence. Interruption of the sympathetic nerves may occur secondary to a neuropathy, to transmural myocardial infarction (5), or as a result of extensive dissection or instrumentation of the coronary arteries (15,36). Dissection of the circumflex artery for implantation of a Doppler flow transducer did not attenuate the functional response to reflex sympathetic activation. However, a similar instrumentation of the left anterior descending coronary artery resulted in unbalanced functional responses in the area distal to the transducer, which correlated to the morphological extent of denervation (15). The dysfunction was similar to that discussed above.

Effects of myocardial ischemia on sympathetic innervation

Much is known about sympathetic adrenergic function in the healthy heart. Since the cardiac sympathetic nerves represent the basis of the control system affecting heart rate, rhythm, function, and flow during a pathophysiological stress, another important consideration is the consequence of myocardial ischemia on the nerves themselves. Myocardial ischemia has been shown to release (37,38) and deplete tissue norepinephrine (26), inhibit neurotransmission (25), and alter the amount of norepinephrine that is released in an ischemic area (11).

It has also been demonstrated that transmural myocardial infarction interrupts sympathetic nerves travelling in the epicardium to more apical areas (4,5) from the base. In this study (5), a branch of the left anterior descending coronary artery was injected with a rapidly hardening vinyl latex which produced a limited, but transmural infarction. Subsequent measurement of the effective refractory period at multiple sites above and below the injection site, before and during sympathetic activation, demonstrated normal shortening above, but no change below the majority of sites during sympathetic stimulation. This phenomenon was observed in acutely prepared dogs and in chronically infarcted dogs. The authors concluded that the transmural infarction produced a denervation of the noninfarcted sites distal to the area of necrosis by damaging nerves coursing over the infarcted region. However, in the study of Janes et al. (19) in which the conduction velocities of compound action potentials produced in the nerves coursing over an infarct were directly measured, the conduction velocities remained normal for the 12 h of monitoring following the infarction. Furthermore, following ventricular fibrillation, compound action potentials were conducted for a minimum of 2 h, although at a slower velocity. These results strongly suggest that the mechanism that results in denervation distal to an infarction is not damage to the nerves coursing over the infarct, since they show normal velocities during the same time period that functional impairment was reported in the other study (5).

In a related issue, post-ischemic myocardial dysfunction or the stunned myocardium (7) has been linked to an abnormality in the sympathetic nerves. Ciuffo et al. (8) reported that a 25 min occlusion period reduced the responsiveness to sympathetic stimulation for up to 2 h following reperfusion. The responses to infused norepinephrine were normal. Infusion of bretylium tosylate, an agent which releases norepinephrine from the nerve terminals, resulted in a marked inotropic response in the postischemic segment, indicating that norepinephrine depletion in the affected region was not the mechanism. They concluded that disruption of the neural pathway (above the level of the terminal) was, in part, the mechanism of myocardial stunning. However, other studies using either a single 15 min coronary occlusion (14), or using a series of brief occlusions followed by reperfusion (21) have shown that the response to sympathetic stimulation remains intact. Thus, whether myocardial ischemia can damage the nerves that course through the ischemic area, either acutely or chronically, remains a controversial issue. Further study using models that correlate both regional ventricular function and neural function are required.

References

1. Allen JM, Gjorstrup P, Bjorkman JA, Ek L, Abrahamsson T, Bloom SR (1986) Studies on cardiac distribution and function of neuropeptide Y. Acta Physiol Scand 126: 405–411
2. Armour JA, Hopkins DA (1981) Localization of sympathetic postganglionic neurons of physiologically identified cardiac nerves in the dog. J Comp Neurol 202: 169–184
3. Armour JA, Hopkins DA (1984) Anatomy of the extrinsic efferent autonomic nerves and ganglia innervating the mammalian heart. In: Randall WC (ed) Nervous Control of Cardiovascular Function; Oxford University Press, New York, pp 21–45
4. Barber MJ, Mueller TM, Davies BG, Gill RM, Zipes DP (1985) Interruption of sympathetic and vagal-mediated afferent responses by transmural myocardial infarction. Circulation 72: 623–631
5. Barber MJ, Mueller TM, Henry DP, Felten SY, Zipes DP (1983) Transmural myocardial infarction in the dog produces sympathectomy in noninfarcted myocardium. Circulation 67: 787–796
6. Brandys JC, Randall WC, Armour JA (1986) Functional anatomy of the canine mediastinal cardiac nerves located at the base of the heart. Can J Physiol Pharmacol 64: 152–162
7. Braunwald E, Kloner RA (1982) The stunned myocardium: prolonged, postischemic ventricular dysfunction. Circulation 66: 1146–1153
8. Ciuffo AA, Ouyang P, Becker LC, Levin L, Weisfeldt ML (1985) Reduction of sympathetic inotropic response after ischemia in dogs. Contributor to stunned myocardium. J Clin Invest 75: 1504–1509
9. D'Agrosa LS (1977) Cardiac arrhythmias of sympathetic origin in the dog. Am J Physiol 233: H535–H540
10. Dalsgaard CJ, Franco Cereceda A, Saria A, Lundberg JM, Theodorsson Norheim E, Hokfelt T (1986) Distribution and origin of substance P- and neuropeptide Y-immunoreactive nerves in the guinea-pig heart. Cell Tissue Res 243: 477–485
11. Dart AM, Schömig A, Dietz R, Mayer E, Kübler W (1984) Release of endogenous catecholamines in the ischemic myocardium of the rat. Part B: Effect of sympathetic nerve stimulation. Circ Res 55: 702–706
12. Gu J, Polak JM, Allen JM, Huang WM, Sheppard MN, Tatemoto K, Bloom SR (1984) High concentrations of a novel peptide, Neuropeptide Y, in the innervation of mouse and rat heart. J Histochem Cytochem 32: 467–472
13. Gwirtz PA, Franklin D, Mass HJ (1986) Modulation of synchrony of left ventricular contraction by regional adrenergic stimulation in conscious dogs. Am J Physiol 251: H490–H495
14. Heusch G, Frehen D, Kröger K, Schulz R, Thämer V (1988) Integrity of sympathetic neurotransmission in stunned myocardium. J Appl Cardiol 3: 259–272

15. Heusch G, Guth B, Roth DM, Seitelberger R, Ross J Jr (1987) Contractile response to sympathetic activation after coronary instrumentation. Am J Physiol 252: H1059–H1069
16. Hopkins DA, Armour JA (1984) Localization of sympathetic postganglionic and parasympathetic preganglionic neurons which innervate different regions of the dog heart. J Comp Neurol 229: 186–198
17. Ilebekk A, Lekven J, Kiil F (1980) Left ventricular asynergy during intracoronary isoproterenol infusion in dogs. Am J Physiol 239: H594–H600
18. Janes RD, Brandys JC, Hopkins DA, Johnstone DE, Murphy DA, Armour JA (1986) Anatomy of human extrinsic cardiac nerves and ganglia. Am J Cardiol 57: 299–309
19. Janes RD, Johnstone DE, Armour JA (1987) Functional integrity of intrinsic cardiac nerves located over an acute transmural myocardial infarction. Can J Physiol Pharmacol 65: 64–69
20. Janes RD, Johnstone DE, Brandys JC, Armour JA (1986) Functional and anatomical variability of canine cardiac sympathetic efferent pathways: implications for regional denervation of the left ventricle. Can J Physiol Pharmacol 64: 958–969
21. Johnstone DE, Janes RD, Klassen GA, Armour JA (1989) Functional integrity of sympathetic efferent postganglionic axons in a region of stunned myocardium. Can J Cardiol 5: 357–364
22. Kaye MP, Brynjolfsson GG, Geis WP (1968) Chemical epicardiectomy. A method of myocardial denervation. Cardiology 53: 139–149
23. Knight DR, Shen YT, Thomas Jr JX, Randall WC, Vatner SF (1988) Sympathetic activation induces asynchronous contraction in awake dogs with regional denervation. Am J Physiol 255: H358–H365
24. Komatsu E, Yamaguchi I, Miyazawa K (1989) Effect of local cardiac sympathectomy on regional myocardial contraction. Jpn Circ J 52: 617–623
25. Martins JB, Kerber RE, Marcus ML, Laughlin DL, Levy DM (1980) Inhibition of adrenergic neurotransmission in ischaemic regions of the canine left ventricle. Cardiovasc Res 14: 116–124
26. Mathes P, Gudbjarnason S (1971) Changes in norepinephrine stores in the canine heart following experimental myocardial infarction. Am Heart J 81: 211–219
27. Mizeres NJ (1955) The anatomy of the autonomic nervous system in the dog. Am J Anat 96: 285–318
28. Phillips JG, Randall WC, Armour JA (1986) Functional anatomy of the major cardiac nerves in cats. Anat Rec 214: 365–371
29. Priola DV, Spurgeon HA, Geis WP (1977) The intrinsic innervation of the canine heart: a functional study. Circ Res 40: 50–56
30. Randall WC (1977) Sympathetic control of the heart. In: Randall WC (ed) Neural Regulation of the Heart; Oxford University Press, New York, pp 43–94
31. Randall WC, Armour JA (1977) Gross and microscopic anatomy of the cardiac innervation. In: Randall WC (ed) Neural Regulation of the Heart; Oxford University Press, New York, pp 13–42
32. Randall WC, Kaye MP, Hageman GR, Jacobs HK, Euler DE, Wehrmacher WH (1976) Cardiac dysrhythmias in the conscious dog after surgically induced autonomic imbalance. Am J Cardiol 38: 178–183
33. Randall WC, Kaye MP, Thomas JX Jr, Barber MJ (1980) Intrapericardial denervation of the heart. J Surg Res 29: 101–109
34. Randall WC, Szentivanyi M, Pace JB, Wechsler JS, Kaye MP (1968) Patterns of sympathetic nerve projections onto the canine heart. Circ Res 22: 315–323
35. Randall WC, Thomas JX Jr, Barber MJ, Rinkema LE (1983) Selective denervation of the heart. Am J Physiol 244: H607–H613
36. Roth DM, White FC, Mathieu Costello O, Guth BD, Heusch G, Bloor CM, Longhurst JC (1987) Effects of left circumflex Ameroid constrictor placement on adrenergic innervation of myocardium. Am J Physiol 253: H1425–H1434
37. Schömig A (1988) Adrenergic mechanisms in myocardial infarction: cardiac and systemic catecholamine release. J Cardiovasc Pharmacol 12: 1–7
38. Schömig A, Dart AM, Dietz R, Mayer E, Kübler W (1984) Release of endogenous catecholamines in the ischemic myocardium of the rat. Part A: Locally mediated release. Circ Res 55: 689–701
39. Szentivanyi M, Pace JB, Wechsler JS, Randall WC (1967) Localized myocardial responses to stimulation of cardiac sympathetic nerves. Circ Res 21: 691–702

Author's address:

John X. Thomas, Jr., Ph.D., Department of Physiology, Loyola University of Chicago, Stritch School of Medicine, 2160 South First Avenue, Maywood, IL 60153, USA

Cardiac Sympathetic Activity in Myocardial Ischemia: Release and Effects of Noradrenaline

A. Schömig and G. Richardt

Department of Cardiology, University of Heidelberg, Heidelberg, FRG

Summary

Sympathetic overactivity in myocardial ischemia is closely associated with the progression of myocyte injury and the incidence of malignant arrhythmias. Adrenergic stimulation of the ischemic myocardium is predominantly due to increased local noradrenaline concentrations in the heart, whereas plasma catecholamine levels are of minor relevance. During the first few minutes of ischemia, efferent sympathetic nerves are activated. Excessive accumulation of noradrenaline, however, is prevented since adenosine, formed in the ischemic myocardium, suppresses exocytotic noradrenaline release, and released noradrenaline is rapidly removed as long as catecholamine reuptake is functional.

With progression of ischemia to more than 10 min, the myocardium is no longer protected against excess catecholamine accumulation in the interstitial space, since local metabolic release mechanisms become increasingly important. This release, which is independent of central sympathetic activity and from extracellular calcium, occurs in two steps: First, noradrenaline escapes from its intracellular storage vesicles and accumulates in the cytoplasm of the neuron. In a second, rate-limiting step, noradrenaline is transported across the plasma membrane into the interstitial space, using the neuronal uptake carrier in reverse of its normal transport direction. As a consequence of local metabolic catecholamine release, extracellular noradrenaline reaches 1000 times the normal plasma concentration within 20 min of ischemia.

Studies using acute and chronic sympathetic denervation and antiadrenergic agents demonstrate that local metabolic, rather than centrally induced noradrenaline release is critically involved in the progression of ischemic cell damage within the occurrence of ventricular fibrillation in early ischemia.

Myocardial ischemia results in a temporary supersensitivity of the myocytes to catecholamines. This is due to a twofold increase of α_1- and a 30% increase of β-adrenergic receptor number at the cell surface. The sensitization of adenylate cyclase during the first 20 min of total ischemia is followed by a rapid inactivation of the enzyme. The β-adrenergic hyperresponsiveness to catecholamines is therefore limited to the first few minutes of ischemia.

The deleterious combination of extremely high noradrenaline concentrations with a temporarily enhanced responsiveness to catecholamines of the tissue is thought to accelerate the propagation of the wavefront of irreversible cell damage within the ischemic myocardium. Moreover, the inhomogenous distribution of catecholamine excess within the heart is considered to promote malignant arrhythmias by unmasking and enhancing electrophysiological disturbances in early ischemia.

The work was supported by a grant of the Deutsche Forschungsgemeinschaft (SFB 320 - Cardiac Function and its Regulation).

Introduction

The course of acute myocardial infarction is considered to be affected by cardiac sympathetic activity. The aim of this review is to outline reasons for and effects of excess sympathetic stimulation of the ischemic myocardium. The emphasis will be placed on the causes of sympathetic overactivity in the ischemic heart. In addition, the contribution of adrenergic overstimulation to myocyte injury and ischemia-induced early arrhythmias will be discussed.

Experimental studies suggest that catecholamines accelerate the progression of myocardial injury [102,144]. Arrhythmogenic effects of enhanced sympathetic activity or elevated cardiac catecholamine concentrations are well documented [15, 21, 77]. Cardiac sympathetic activity as mediated through these effects is thought to be a major determinant of the patient's outcome in acute myocardial infarction. This assumption has been substantiated in large clinical studies with early administration of β-adrenoceptor blockers in myocardial infarction [57,81]. They revealed an overall reduction of mortality of 14% in the group of patients receiving β-adrenoceptor blockers. Moreover, in randomized long-term studies including post-infarct patients, chronic β-blockade resulted in a more than 30% reduction of sudden cardiac death and a more than 20% reduction in total mortality [152].

On the other hand, in the failing heart, adrenergic stimulation of the surviving myocardium may be important for maintaining sufficient cardiac output to prevent acute circulatory failure. Both the detrimental and beneficial effects of sympathetic stimulation of the myocardium depend on increased local myocardial catecholamine concentrations. Two different mechanisms may be involved in the increase of catecholamine concentrations during myocardial ischemia: i) reflex increase in systemic and local cardiac release of catecholamines which is accompanied by a rise of plasma catecholamines, and ii) local metabolic release from cardiac sympathetic neurons – irrespective of central sympathetic activity – which is limited to the ischemic myocardium.

Reflex activation of the sympathetic system

Several lines of evidence indicate an activation of the systemic sympathetic system in myocardial ischemia. Symptoms of sympathetic activation such as tachycardia, peripheral vasoconstriction, and sweating often dominate the clinical picture of patients with acute myocardial infarction. These clincial evidences of sympathetic stimulation are supported by measurements of plasma catecholamines in myocardial infarction. There is agreement in the literature that plasma catecholamine concentrations are enhanced in early myocardial infarction [7,9,43,47,63,85,117,125,129,142]. In uncomplicated infarction, plasma noradrenaline and adrenaline concentrations rise up to five times the normal levels at rest. Dramatically elevated plasma concentrations have been observed in patients with pulmonary edema or cardiogenic shock [7], when sympathetic reflex activity is maximally increased, and the elimination of noradrenaline and adrenaline is reduced due to severely hampered tissue perfusion.

At this point one must keep in mind that plasma noradrenaline levels are significantly lower than the effective noradrenaline levels in the synaptic space, because most of the released catecholamines are subjected to the very effective local elimination processes, uptake$_1$ and uptake$_2$ [66], and only about 10% of synaptic noradrenaline is washed out into the systemic circulations. This contributes to a major gradient between synaptic and plasma catecholamine concentrations [66]. Nevertheless, the plasma concentration of noradrenaline is considered to be a reliable marker of sympathetic activation [36] reflecting the activity of the whole sympathetic system rather than the local activity in the heart, since the contribution of the heart to total noradrenaline turnover amounts to less than 3% under normal conditions [36].

Pain, anxiety, and reflex activation of the sympathetic system are the fundamental causes of an increased peripheral catecholamine release and of enhanced sympathetic nerve activity to the heart. Reflex increase in sympathetic activity is mainly determined by two different mechanisms: i) cardiovascular reflexes, induced by activation of pressor and volume receptors following a decrease in blood pressure and cardiac output, and ii) reflexes that are activated by afferents from ischemic myocardial areas [76]. For activation of these reflexes, local acidosis, accumulation of metabolites, and increased wall stretch [137] are of importance. These sympathetic and parasympathetic reflexes, which originate in the ischemic myocardium, are different in anterior and posterior wall infarctions and may lead to either an increase or a depression of cardiac sympathetic activity [134]. During experimental coronary occlusion, frequently only a moderate and short-term increase in the impulse rate of sympathetic cardiac nerves is observed.

Plasma catecholamine concentrations in early infarction have been shown to depend on the amount of damaged myocardium and the hemodynamic consequences of infarction. In experimental studies, Karlsberg et al. demonstrated a good correlation between plasma catecholamine concentrations and the anatomic size of myocardial necrosis, as well as the reduction of cardiac output following coronary occlusion [64]. In human infarction, a relationship was found between plasma concentrations of noradrenaline and adrenaline and the angiographically determined reduction of left ventricular ejection fraction [117].

Following reperfusion of the ischemic myocardium – induced by lysis of the coronary clot within the first 3 h of infarction – the plasma catecholamine concentrations briskly returned to nearly normal values [117], despite a still compromised hemodynamic performance of the heart [34]. This effect of early reperfusion on plasma catecholamine concentrations preferentially may be due to a reduced activity of cardiosystemic reflexes directly activated by metabolic consequences of ischemia in the underperfused myocardium [76].

As a consequence of systemic and cardiac sympathetic stimulation, heart rate and contractility increase and peripheral resistance is elevated. These hemodynamic changes may accelerate the damage of ischemic myocardium due to augmented mechanical work and increased energy consumption.

Direct effects of enhanced plasma catecholamines in further damaging the ischemic myocardium cannot be assumed to be of major importance since ischemic myocardial areas are cut off from systemic circulation and can therefore hardly be reached by circulating plasma catecholamines. Experimental and clinical evidence of a causal relationship between plasma catecholamines and both the progression of myocardial damage and the development of malignant arrhythmias in infarction [9,63] is rather poor. The sparse data supporting a causal connection can easily be interpreted as parallel phenomena, both caused by the activated sympathetic system.

In the normally perfused myocardium, an increase in the impulse rate of cardiac efferent nerves causes an increase in noradrenaline release and, consequently, an increase in sympathetic stimulation of the myocardial cells. During short-term myocardial ischemia no enhanced net cardiac release of noradrenaline could be demonstrated in human heart [108]. During percutaneous transluminal angioplasty with total occlusion of the left anterior descending coronary artery for 30–150s, no significantly increased noradrenaline concentrations were found in the coronary sinus, neither during occlusion nor during the subsequent reperfusion period. In these patients ischemia was accompanied by angina pectoris, ST-segment deviation, and cardiac lactate production. Similar observations were made during pacing-induced ischemia in patients with high-grade stenoses of the coronary arteries [35]. Though these patients showed severe myocardial ischemia with angina pectoris and increased systemic noradrenaline levels, no significant cardiac noradrenaline release was found in the coronary sinus [35].

These data contrast to early observations of Wollenberger and Shahab, who described a massive release of catecholamines as early as within the first 3 min of ischemia, both in isolated hearts [149] and following coronary ligation in the anesthetized dog [121]. This accelerated time-course, however, could not be reproduced, neither in the isolated organ [2,10,101,113,114] nor in the intact animal [40,52,79]. During the first minutes of ischemia no major extracellular catecholamine accumulation has been demonstrated in the ischemic myocardium of rat, guinea pig, rabbit, dog, and pig.

Several experimental studies suggest that poorly perfused myocardium is protected within

11

the first few minutes of ischemia, via various mechanisms, against excessively high local concentrations of catecholamines [22,25,78,100], although energy deficiency induced by anoxia or cyanide intoxication rather enhances exocytotic release of noradrenaline and neuropeptide Y in isolated guinea pig hearts.

Rapid elimination of released noradrenaline via neuronal uptake (uptake$_1$) is one of the protective mechanisms that prevents excessive noradrenaline accumulation [25]. This reuptake mechanism energetically depends on an intact sodium gradient across the cell membrane of the sympathetic neuron [105], and in total ischemia or anoxia it is active at least up to 10 min of ischemia [114] or anoxia [109]. In very early ischemia, the reduced perfusion flow results in an even increased efficacy of neuronal elimination of noradrenaline, because of the prolonged exposure of the released catecholamines to the uptake process.

Moreover, acute myocardial ischemia results in a reduction of stimulation-induced noradrenaline release. Various observations suggest that this is due to altered sympathetic nerve terminal function rather than structural neuronal damage, because the nerve terminals recover their ability to release noradrenaline after reinstitution of coronary flow within a few minutes [22]. The early neuronal dysfunction is predominantly not caused by direct effects of ischemia on the nerve terminal and the release process, but is rather due to indirect influences of the surrounding ischemic myocardium.

Various factors have been discussed as contributing to the inhibition of noradrenaline release in ischemic myocardium [82]. For example, the increase of extracellular potassium and the reduction of pH are known to occur within a few minutes of global ischemia [65,145], and have been shown to affect noradrenaline release [67,73,94]. Clear evidence for the relevance of these effects in ischemia has not been documented so far.

It has been demonstrated that in ischemia, stimulation-induced exocytotic release of noradrenaline is modualated via presynaptic receptors [25,100]. While the most effective modulation under conditions of normal oxygen supply is via adrenergic α_2-receptors [127], in ischemia inhibition of release by adenosine seems to be more important. Within 3 min of ischemia, adenosine concentrations are reached in the extracellular space which inhibit stimulation-induced noradrenaline release by two-thirds [100]. This suppression of noradrenaline release is transmitted through A_1-adenosine receptors and can be abolished by receptor antagonists such as 8-phenyltheophylline [100]. Myocardial energy metabolism in early ischemia thus has a considerable effect, via the formation of adenosine, on exocytotic release of noradrenaline from the sympathetic neuron (Fig. 1).Recently, the inhibitory effect of endogenous adenosine on cardiac noradrenaline has been confirmed during low-flow ischemia [28], as well as during sympathetic stimulation of rat hearts with uncompromised energy metabolism [99].

Local metabolic release of noradrenaline and responsiveness of the ischemic myocytes to catecholamines

With progression of ischemia, the myocardium is no longer protected against excess adrenergic stimulation, since local metabolic release mechanisms gain increasing importance (Fig. 2). This release is independent from central activation of sympathetic nerves and does not become relevant before the 10th min in total ischemia [113]. With longer duration of ischemia, increasing quantities of catecholamines are released, and within 40 min of ischemia, more than 30% of total noradrenaline content of the heart has been released into the interstitial space. Adrenaline and dopamine release taken together account for less than 5% of the corresponding noradrenaline release [113]. This release from sympathetic terminals is reflected histologically by reduction or disappearance of neuronal catecholamine fluorescence due to redistribution of noradrenaline from storage vesicles in sympathetic nerve terminals to other tissue compartments [53,54,84].

The concept of ischemia-induced noradrenaline release was challenged by the objection that the release might be caused by reperfusion injury. Indeed, ischemia-induced transmitter release from the heart can only be measured during reperfusion, and it is not possible, without examining release kinetics, to decide whether release from sympathetic neurons occurs during

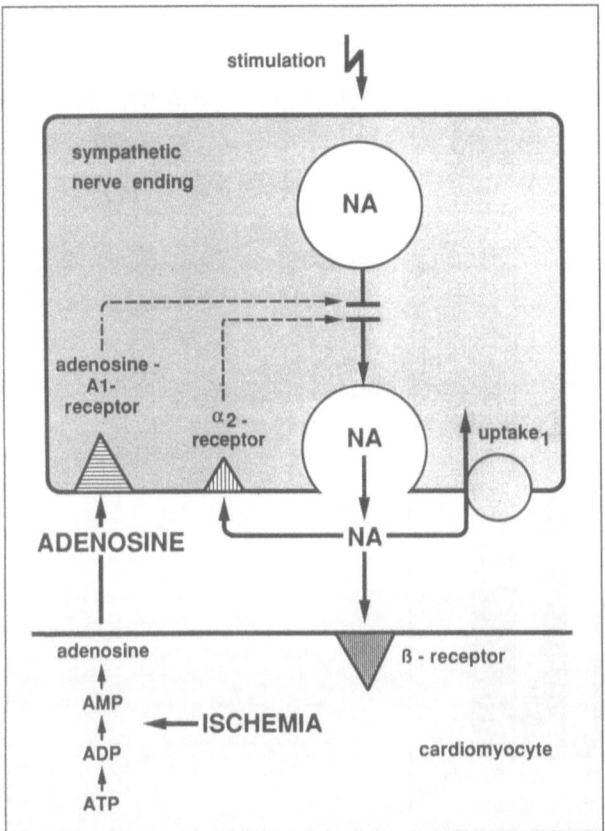

Fig. 1. Scheme of exocytotic noradrenaline release and of protective mechanisms active during early ischemia which prevent major catecholamine accumulation in the interstitial space of the ischemic myocardium.

Efferent nerve activity induces calcium-dependent exocytosis of noradrenaline (NA) and of cotransmitters such as neuropeptide Y. Further release is reduced through stimulation of presynaptic α_2-receptors which inhibit exocytosis. The second protective factor is the rapid elimination of released noradrenaline by neuronal uptake (uptake$_1$) which is active up to 10 min of ischemia. The most potent protection is provided by the extracellular accumulation of adenosine suppressing noradrenaline release through stimulation of inhibitory A_1-adenosine receptors. Via this pathway – formation of adenosine – myocardial energy metabolism in early ischemia modulates the exocytotic release of noradrenaline from the sympathetic neurons, and sympathetic overstimulation is prevented in situations of compromised myocardial energy state. (For details see [25,100].)

ischemia or in the reperfusion phase. The time-course of release can, as a good approximation, be described by means of a two-compartment model [112]. The first compartment corresponds to the washout kinetics from the extracellular space [72]. During 20 min ischemia, this compartment comprises about 90% of total release. The remaining 10% are associated to a release induced by reperfusion. As a further argument against major reperfusion-induced release and in favor of release during ischemia, it can be stated that blockade of energy metabolism combined with normal perfusion flow (e.g. anoxia or cyanide intoxication in combination with glucose depletion) also produces a noradrenaline release, which shows similar quantities and identical characteristics as ischemia-induced release. The amine release rapidly ceases after reoxygenation or after the addition of glucose.

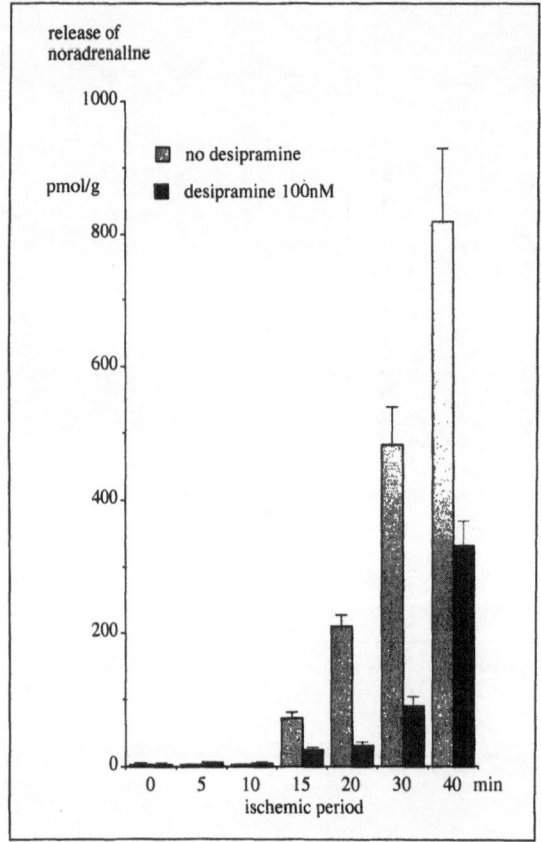

Fig. 2. Effect of global and total ischemia and desipramine on noradrenaline release from isolated perfused rat hearts. (Data from [113].)

Noradrenaline release was determined by an HPLC method from the reperfusate after various periods of isothermic perfusion stop. During normoxic control conditions and after 5 and 10 min ischemia only minor amounts of noradrenaline were collected in the perfusate, which slightly increased following application of desipramine. In contrast, during 15, 20, 30 and 40 min of ischemia noradrenaline release dramatically increased rapidly reaching micromolar concentrations in the interstitial space (calculated from noradrenaline in reperfusate). During 15, 20, 30 and 40 min ischemia desipramine effectively suppressed the ischemia-induced local metabolic noradrenaline release.

With low-flow ischemia, noradrenaline release likewise has been demonstrated to occur during the ischemic phase. In this experimental model, flow distribution is heterogeneous and noradrenaline is released into regions of particularly profound ischemia from which it is subsequently eluted during reperfusion [23].

Assuming uniform distribution of released noradrenaline in the extracellular space, during 20 to 40 min ischemia, extracellular concentrations reach the micromolar range, which is 100 to 1000 times the normal plasma concentration. Catecholamine concentrations of this order are capable of producing myocardial necrosis, even in the nonischemic heart [144]. These excessively high, local noradrenaline concentrations within the myocardium are accompanied by a temporary hyperresponsiveness of the ischemic myocyte to catecholamines. Both the α- and the β-adrenergic system exhibit a nearly twofold increased sensitivity after 15 to 30 min of ischemia.

Most authors describe an upregulation of α_1-receptor density in the myocyte plasma

Table 1. Characteristics of stimulation-induced exocytotic noradrenaline release and of local metabolic noradrenaline release caused by myocardial ischemia

	Exocytotic release	Release induced by ischemia
Stimulation by central sympathetic activity	+	−
Dependence from:		
extracellular calcium	+	−
N-type calcium channels	+	−
protein kinase C	+	−
Inhibition by verapamil and diltiazem	−	+
Co-transmission with neuropeptide Y	+	−
Modulation by presynaptic receptors	+	−
Effect of conventional sympatholytic agents	+	−
Inhibition by propranolol	−	+
Inhibition by blockade of neuronal noradrenaline uptake	−	+
Inhibition by blockade of sodium/proton exchange	−	+

membrane, resulting in a twofold increase after 30 min ischemia [16,56,104]. This augmented α_1-receptor density is associated with an increased protein kinase C activity which, in part, may be independent from receptor stimulation [74]. The α_1-adrenergic activity has been demonstrated to be intimately linked to the calcium homoeostasis of the mycoyte and may, therefore, be involved in the process of cellular calcium overload during ischemia and reperfusion [14, 122].

Concerning the β-adrenergic system, the hypersensitivity is caused by a 30% enhanced density of β-adrenergic receptors in the plasma membrane of the myocyte [75,83,132,140] and a temporary increase of adenylate cyclase activity which, in part, is independent from receptor stimulation [130,132]. Due to the decrease of high-energy phosphates during ischemia the agonist-induced β-receptor internalization is suppressed. Consequently, the balance between receptor internalization and externalization is shifted towards increased receptor numbers at the cell surface at the expense of the intracellular light membrane pool where the receptors are not accessible to stimulation by catecholamines [131]. The increased activity of the adenylate cyclase during the very early phase of ischemia turns into a rapid inactivation of the enzyme after 20 min of ischemia [139,140]. Hence, the enhanced β-adrenergic responsiveness to catecholamines is limited to the first 20 min of ischemia.

Besides the receptor-mediated actions of catecholamines, direct toxic effects of adrenaline and noradrenaline have been postulated [150]. These effects have been attributed to the formation of adrenochrome radicals and may become relevant during conditions of excessively high catecholamine concentrations within the myocardium.

Characteristics and metabolic requirements of ischemia-induced noradrenaline release

Noradrenaline release in early ischemia is completely independent of central sympathetic activity. It shows characteristics that make it possible to differentiate it from exocytotic release and to assign it to a transmembrane efflux via a specific transport system (Table 1) [113]. The transport system has been identified as the uptake₁ carrier which in ischemia reverses its transport direction and enables a carrier-mediated efflux of noradrenaline from the cytoplasm of the neuron to the extracellular space. Noradrenaline efflux via specific transport systems in the neuronal plasma membrane was first suggested by Paton [89], and several authors have described pharmacological conditions for such carrier-mediated efflux of catecholamines from sympathetic nerve cells [45,95,103].

Local metabolic release induced by ischemia is independent of extracellular calcium, and not even the addition of EGTA to the calcium-free perfusate inhibits the release [113]. The inhibitory effect of high concentrations of organic calcium antagonists such as verapamil and diltiazem on ischemia-induced amine release, which has been described by Nayler and Sturrock [86,87], is completely independent of extracellular calcium [97,110] and, therefore, cannot be attributed to the calcium channel-blocking properties of the agents. It rather depends on an interference of the drugs with the carrier-mediated noradrenaline transport itself. On the other hand, blockade of N-type calcium channels, utilizing ω-conotoxin or cadmium ions, inhibited stimulation-induced exocytotic release [50], but had no influence on nonexocytotic release induced by ischemia.

Similarly, activation of neuronal protein kinase C is a precondition for exocytotic noradrenaline release [50]. Inhibition of protein kinase C with polymyxine did not affect ischemia-induced release.

Release of the adrenergic cotransmitter neuropeptide Y (NPY) has been demonstrated to indicate exocytotic mechanisms of transmitter liberation [48,49]. While stimulation-induced noradrenaline release was accompanied by the liberation of neuropeptide Y from cardiac sympathetic nerve endings [49], total ischemia for 5 to 30 min or anoxia (up to 90 min) did not induce the release of neuropeptide Y-like immunoreactivity from the guinea pig heart [42,49].

Presynaptic receptors did not modulate the release, and the α_2-antagonist yohimbine, which more than doubled exocytotic release, was without any effect in the case of ischemia-induced release [108]. Consequently, conventional sympatholytic agents such as clonidine were not capable of suppressing nonexocytotic release in ischemia. Propranolol, however, reduced ischemia-induced release, whereas other β-blocking agents such as atenolol, metoprolol, and timolol were not effective [98]. It was found that propranolol is the only one of these agents interacting with neuronal noradrenaline uptake [98].

Extracellular catecholamine accumulation was not enhanced, as would be expected, by blockade of the neuronal uptake, but was markedly reduced [113]. Reduction of catecholamine release by uptake blockade was more than 80% up to 30 min of ischemia (Fig. 2). The release was suppressed by differently structured inhibitors of neuronal uptake (e.g. desipramine, cocaine, oxaprotiline, and nisoxetine) and the concentration-dependency of release suppression corresponded to that of uptake blockade [113].

Amiloride and its derivatives dimethylamiloride and ethylisopropylamiloride were found to suppress ischemia-induced noradrenaline release in the same concentration range that was necessary for inhibition of sodium/proton exchange [116]. Tetrodotoxin, however, inhibited stimulation-induced exocytotic release and had no effect on nonexocytotic release caused by ischemia [116].

Similar characteristics were found when catecholamine release was induced by anoxia or cyanide intoxication in the absence of glucose [24,115], indicating a common release mechanism induced by energy depletion of the heart that cannot be reconciled with the concept of exocytosis.

Release of catecholamines from the heart via leak diffusion of the transmitter from the neuron is not of major importance in ischemia of less than 40 min duration. If release occurs via passive diffusion or efflux via leaky membranes, inhibition of neuronal reuptake or sodium/proton exchange cannot result in reduced accumulation of catecholamines in the extracellular space. Only in ischemia of longer duration can release be explained by leak diffusion since, after 60 min, inhibition of neuronal reuptake no longer caused a reduction in release [113].

Stop-flow ischemia permits little variation of metabolic conditions, whereas experimental models with depleted energy (such as anoxia and cyanide intoxication) and unchanged perfusion flow provide a less complex experimental situation to study the metabolic requirements of nonexocytotic noradrenaline release. On the basis of these studies [24,115], the energy state of the sympathetic nerve terminal appears to be the main determinant of nonexocytotic amine release. Both the interruption of oxidative phosphorylation (either by oxygen deficiency or by cyanide poisoning) and the inhibition or exhaustion of anaerobic glycolysis were necessary for a significant release of noradrenaline from the nerve terminals. Activity of either energy-providing process was sufficient to strongly reduce the release. Thus, ongoing glycolysis completely prevented noradrenaline release, even during complete anoxia or cyanide intoxication. Similar protective effects of glucose were found in incomplete (low-flow) or regional ischemia [11,24].

Energy depletion of the nerve terminal is both a necessary and sufficient cause for nonexocytotic noradrenaline release to occur during ischemia. Other factors implicated in ischemia, such as acidosis, increased interstitial potassium concentrations, and accumulation of metabolites, are not likely to play a major role in causing nonexocytotic release, because in experimental models with ongoing flow they are present to a lesser extent [39]. The cumulative noradrenaline release, however, was even higher than during total ischemia [115].

Nonexocytotic catecholamine release has recently been described to occur in human cardiac tissue during energy deficiency, as well [68]. Anoxia or cyanide intoxication in combination with glucose deficiency induced calcium-independent release of noradrenaline from atrial tissue obtained during cardiac surgery. The quantities of amines released, the release characteristics, the inhibition by uptake$_1$-blockers, and the dependence from local metabolic factors were comparable to those obtained in perfused rat hearts and rat atrial tissue [68]. These results provide strong evidence for the occurrence of nonexocytotic release processes in the human heart during myocardial ischemia.

Mechanism of local metabolic catecholamine release in ischemia

Under conditions of energy depletion, as is the case in myocardial ischemia, anoxia, or cyanide intoxication, catecholamine release has been demonstrated to be a two-step process [115,116]. First, noradrenaline escapes from the storage vesicles, resulting in enhanced axoplasmic amine concentrations. Second, noradrenaline is transported across the plasma membrane into the extracellular space using the uptake$_1$ carrier in reverse of its normal transport direction.

For better understanding of these mechanisms, the processes involved in neuronal catecholamine storage and transport will be described briefly (Figs. 3,4). By accumulation of protons within the neuronal catecholamine storage vesicles, the H^+-ATPase located in the vesicular membrane generates a transmembrane H^+ electrochemical potential of close to 200 mV that is composed of the vesicular membrane potential (inside positive) and a proton gradient (inside pH 5.5) [6,92,148]. The proton potential is the driving force of vesicular noradrenaline uptake, and amine inward transport is coupled with proton outward transport by a specific reserpine-sensitive carrier located within the vesicular membrane [6,92,148]. A disturbed neuronal energy metabolism leads to dissipation of the proton potential and a loss of noradrenaline from the vesicles into the axoplasm [37,135]. Ongoing glycolysis is sufficient to prevent this vesicular catecholamine loss, even in the absence of functional oxidative phosphorylation [115].

In the presence of oxygen, axoplasmic noradrenaline is readily degraded by monoamine oxidase and the inactive metabolite dihydroxyphenylethyleneglycol (DOPEG) is formed [41]. Thus, both exhaustion of glycolysis and hypoxia are required to cause major noradrenaline accumulation in the axoplasm of the sympathetic neuron.

Increased axoplasmic noradrenaline concentrations are not sufficient to induce outward transport of noradrenaline across the plasma membrane of the neuron, because under normal conditions catecholamine transport via the uptake carrier is directed from extracellular space to cytoplasm (Fig. 4). Energetics and direction of this cotransport of catecholamines with sodium ions depend on the transmembrane sodium gradient [44,46,133,136]. Only an increase of intracellular sodium, which is assumed to occur during early ischemia [4,39], enables a carrier-mediated efflux of noradrenaline into the extracellular space [44,46,133,136]. Blockade of the carrier by tricyclic antidepressants such as desipramine inhibits both inward and outward

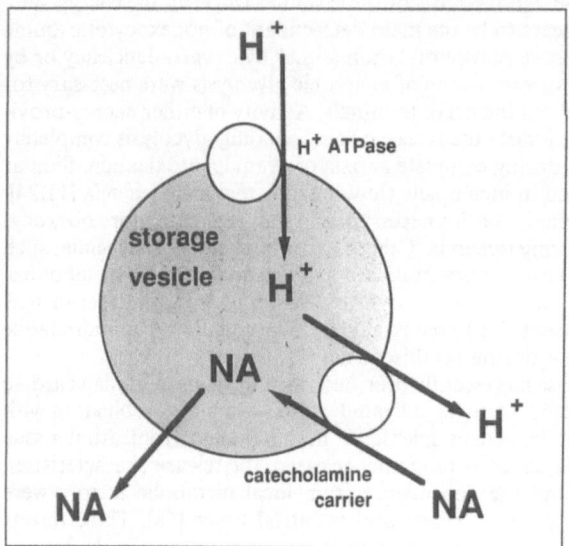

Fig. 3. Scheme of noradrenaline storage in neuronal vesicles.

The driving force of catecholamine storage is an electrochemical proton potential of nearly 200 mV across the vesicle membrane which consists of a proton gradient (inside pH 5.5) and the membrane potential (inside positive). The potential is established by the proton pump activity of vesicular H^+ATPase which in several features resembles the mitochondrial H^+ATPase. The catecholamine inward transport which is coupled with proton extrusion is provided by a specific reserpine-sensitive carrier. The low pH inside the storage vesicles and the presence of a polyanionic matrix result in a trapping of the protonized amine and a catecholamine accumulation to nearly millimolar concentrations inside the vesicle. Blockade of H^+ATPase by trimethyltin or energy deficiency of the nerve terminal result in a collapse of the proton potential and catecholamine inward transport. In this situation outward leak diffusion is no longer compensated and free noradrenaline accumulates in the axoplasm. (For details see [6,92,148].)

transport of catecholamines, and thus effectively suppresses nonexocytotic noradrenaline release.

In myocardial ischemia the rise of intracellular sodium is caused by both failure of Na^+,K^+-ATPase activity due to progressive energy depletion and increased sodium influx across the plasma membrane. Thus, inhibition of sodium pump activity by digitalis glycosides prior to energy deficiency accelerates and increases non-exocytotic noradrenaline release [67,116].

Sodium-proton-exchange, on the other hand, has been identified as the predominant pathway of sodium entry into the sympathetic nerve endings in ischemia. This carrier-mediated transport system plays a critical role in the regulation of intracellular pH and is maximally activated by intracellular acidosis [3,69]. The extrusion of protons is coupled with sodium entry which leads to intracellular sodium accumulation, especially when Na^+,K^+-ATPase activity is suppressed. Inhibition of Na^+,H^+-exchange by amiloride and, more specifically, by ethyliso-propylamiloride [143] markedly reduced ischemia-induced noradrenaline release [116].

Time-course and local distribution of adrenergic overactivity in cardiac ischemia

Based on the data presented three subsequent phases of adrenergic overactivity may be distinguished during the course of myocardial ischemia [113].

Phase 1 (ischemia up to 10 min): During this phase the release of catecholamines occurs by exocytosis and depends on the activity of the efferent cardiac sympathetic nerves. The extracellular accumulation of noradrenaline is limited by the activity of the neuronal reuptake process and the effect of centrally originating sympathetic neural activity is reduced by presynaptic inhibitory effects of adenosine and noradrenaline itself.

Release and effects of noradrenaline

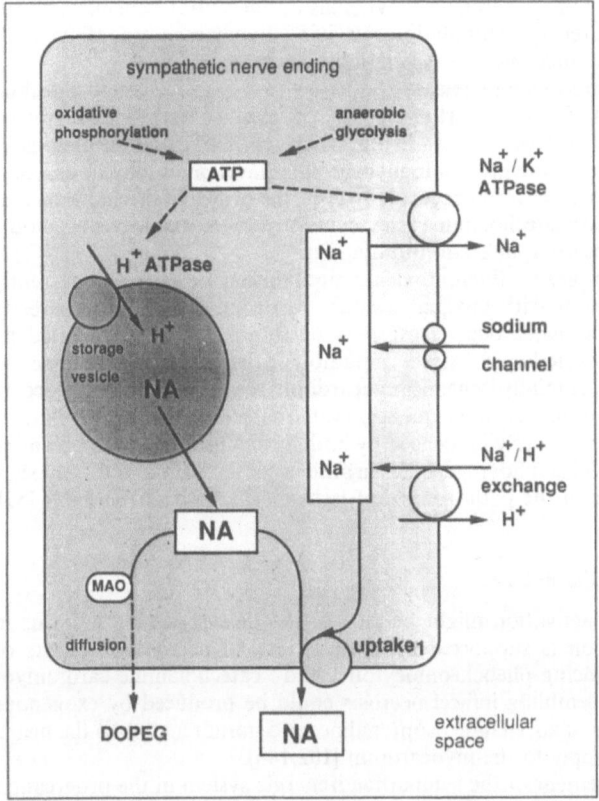

Fig. 4. Scheme of nonexocytotic, calcium-independent noradrenaline (NA) release (left part) and its relation to neuronal sodium homoeostasis (right part).

Energy depletion of the nerve terminal disturbs vesicular storage function and results in a loss of noradrenaline from storage vesicles into the axoplasm. Enhanced axoplasmic noradrenaline concentrations can be detected by increased dihydroxyphenylethyleneglycol (DOPEG) release. Enhanced axoplasmic noradrenaline concentrations do not cause relevant noradrenaline release from the nerve ending as long as neuronal sodium homoeostasis is undisturbed. The combined increase of noradrenaline and sodium within the cytoplasma results in major noradrenaline release via uptake$_1$ which reverses its normal transport direction under these conditions. Blockade of uptake$_1$ by desipramine-like agents inhibits nonexocytotic noradrenaline release, but not DOPEG overflow.

Several mechanisms may interfere with neuronal sodium homoeostasis (right part) and thereby modify nonexocytotic noradrenaline release. Inhibition of Na^+,K^+-ATPase by digitalis glycosides and/or increased sodium influx into the neuron cause noradrenaline release, if high cytoplasmic noradrenaline concentrations are present. A disturbed energy state, induced by ischemia or anoxia of the sympathetic nerve terminal, interferes with both vesicular function and sodium homoeostasis and therefore results in nonexocytotic noradrenaline release. Na^+,H^+-exchange has been identified to be a major route of neuronal sodium influx in early ischemia. (For details see [115,116].)

Phase 2 (10–40 min of ischemia): A massive accumulation of noradrenaline occurs in the extracellular space of the ischemic myocardium. The release demonstrates the characteristics of a carrier-mediated efflux and is determined by local energy exhaustion rather than by centrally originating factors. In combination with a temporarily enhanced supersensitivity to catecholamines of the myocyte the excessive local noradrenaline concentrations are considered to accelerate the propagation of the wavefront of irreversible cell damage.

Phase 3: When ischemia lasts longer than 40 min, the structurally damaged sympathetic neurons are progressively depleted of noradrenaline. Whereas in the central ischemic area the tissue has become insensitive to adrenergic stimulation, surviving myocytes may continue to be affected by diffusion of catecholamines into the ischemic border zones.

This time-course of cardiac noradrenaline release and myocyte responsiveness to catecholamines is derived from studies in the isolated heart of various species and from studies in human cardiac tissue. It cannot be extrapolated directly to "in vivo" conditions. In situations of variable and incomplete ischemia that are present in myocardial infarction, it may be assumed that a temporally dispersed time-course would occur. However, the principal mechanisms that have been postulated to account for catecholamine release and myocyte responsiveness would be expected to also apply to myocardial infarction in humans.

The situation is even more complex as different release mechanisms are active concurrently in closely related myocardial regions with unequal residual perfusion flow [23]. In severely ischemic areas nonexocytotic local metabolic release may be found, which is inhibited by desipramine-like drugs and is unaffected by classical sympatholytic drugs such as clonidine. At the same time, in the non-ischemic or mildly ischemic myocardium, release preferentially occurs via exocytosis and agents like desipramine result in an aggravated catecholamine accumulation. The border zone in between is unpredictably influenced by both mechanisms via catecholamine diffusion. Such chronological and local inhomogeneities in adrenergic activation of the myocardium may play an important role in the pathogenesis of ischemia-induced arrhythmias [59].

Role of catecholamines in myocardial damage

The hypothesis that sympathetic activation might be part of the deleterious mechanisms in myocardial ischemia and infarction is supported by observations of necroses in hearts of patients with catecholamine-producing pheochromocytoma and "catecholamine cardiomyopathy" [102,138]. Cell damage resembling infarct-necrosis could be produced by exogenous application of high concentrations of adrenaline, isoprenaline, and noradrenaline to the heart, without interrupting the blood supply to the myocardium [102,144].

In order to confirm the involvement of the sympathoadrenergic system in the progression of myocardial ischemia different surgical and pharmacological interventions have been studied. In 1931, Leriche and Fontaine demonstrated beneficial effects of stellate gangliectomy during experimental myocardial ischemia and in patients with angina pectoris [71]. These early findings were strengthened by several other authors who found a significant reduction in infarct size when coronary ligation was combined with cardiac sympathectomy or cardiac denervation. In these studies chronic denervation was much more effective than acute denervation [5,19,62,151]. After acute sympathectomy the size of infarction was reduced by about 25% vs 90% after chronic denervation [62]. Moreover, pharmacological catecholamine depletion by reserpine prior to ischemia improved cardiac cell metabolism and prevented the drop of myocardial ATP [55]. However, there are conflicting results concerning cardiac denervation in myocardial ischemia, and recent studies do not show significant protective effects by denervation when myocardial infarction was induced in conscious dogs (Shen et al., this volume).

The publications concerning pharmacological interventions in experimental ischemia revealed similar controversy. Most authors reported that β-adrenergic blockade decreased the degree of ischemic damage and reduced the necrotic area [1,8,96,123,126]. However, protective effects of propranolol were not documented in other studies [32,61,91]. These contradictory results might be due to different methods to describe the degree of ischemic injury after coronary ligation. Furthermore, it is likely that β-adrenergic blockade delays the progression of ischemic injury more effectively than it reduces the final extension of myocardial infarction. Therefore, the protective effect of β-receptor blockade may depend on the time interval following coronary occlusion.

The efficacy of β-adrenoceptor blockade in human myocardial infarction is less controversial. Cumulative data from randomized clinical trials indicate a 14% reduction of mortality within the first week following myocardial infarction [57,81]. Reduction of mortality by β-blockers corresponds with a decrease of infarct size as assessed from cumulative enzyme release [81]. Recent analyses of the mechanisms for the early mortality reduction by β-blockade in the ISIS-I study elucidate another feature of β-adrenergic blockers in myocardial infarction, which was not apparent in previous experimental studies. A significant reduction of myocardial rupture was reported in patients allocated to atenolol [58]. β-adrenoceptor blockers have also been investigated for secondary prevention after acute myocardial infarction. Numerous studies show that these drugs improve survival after myocardial infarction and reduce the incidence of nonfatal reinfarction [overview: 20; 152].

Given the protective effects of β-adrenoceptor blockers during and after myocardial infarction, evidence exists that adrenergic overactivity affects myocardial damage in ischemia. The dosage of the β-blockers used and the negative results with isomeres which do not interact with the β-adrenoceptor support the conclusion that these agents reduce myocardial ischemic injury through β-adrenergic blockade rather than through unspecific actions [20,96,152].

While the contribution of β-adrenergic stimulation to the ischemic injury is accepted by most authors, there is considerable controversy about possible mechanisms involved in this effect. In part, the injury of marginal, still contracting areas of ischemia may be explained by the augmented mechanical work caused by catecholamine release and β-adrenergic overstimulation which result in enhanced heart rate and contractility. β-adrenergic stimulation may also increase the wall stress on ischemic and necrotic myocardium and enhance the risk of cardiac rupture [58]. In addition, there is evidence that adrenergic stimulation causes impeded blood flow to the ischemic border zone. The vasoconstrictor effects have been shown to be mediated through vascular α_2-receptors [51]. This issue is discussed in detail by various authors in this volume.

Changes in mechanical work or blood supply cannot provide a reasonable explanation for the protection in the central zone of ischemia because heart muscle in this area is not contracting and blood flow is virtually nonexistent. Based on experimental findings, it may be assumed that adrenergic stimulation further reduces high-energy phosphates, even in the noncontracting part of the ischemic zone. Adrenergic stimulation is still capable of activating sodium-potassium-ATPase and inducing formation of cyclic AMP under these conditions. Formation of cyclic AMP is an energy-consuming process itself, and it activates various intracellular processes which result in further degradation of high-energy phosphates. Among the energy consuming processes in non-contracting cells an activation of the sodium-potassium-pump is of major relevance. During the early phase of myocardial ischemia, this activation is associated with a secondary decrease of extracellular potassium concentration [145] and may be due to catecholamine excess. It remains unclear, however, whether the effect of catecholamines on sodium-potassium-ATPase is a direct one or is mediated via intracellular messengers such as cyclic AMP [33].

A further harmful effect of catecholamines in ischemia may be mediated by cellular calcium overload, which has been identified as playing a crucial role in the pathogenesis of myocardial necrosis [102]. In this context the finding of an α-receptor-dependent accumulation of calcium in reperfused myocardium [122] is in accordance with the involvement of α-receptor stimulation in ischemic cell damage.

Adrenergic influences on arrhythmogenesis in myocardial ischemia

The relationship between catecholamines and ischemia-induced arrhythmias represents a fascinating and confusing issue in clinical and experimental cardiology. It has enormous practical consequences for the antiarrhythmic treatment and prophylaxis in patients suffering from coronary heart disease. More than half of the patients who do not survive an acute myocardial infarction die from primary ventricular fibrillation during the first 6 h following the first symptoms, and two thirds of these die within the first hour (Fig. 5) [88]. In various studies conventional antiarrhythmic drugs were utilized for prophylaxis of sudden cardiac death in high-risk patients during the postinfarct period [12]. Beneficial effects, however, could not be demonstrated for these agents, and several antiarrhythmic drugs even revealed adverse effects

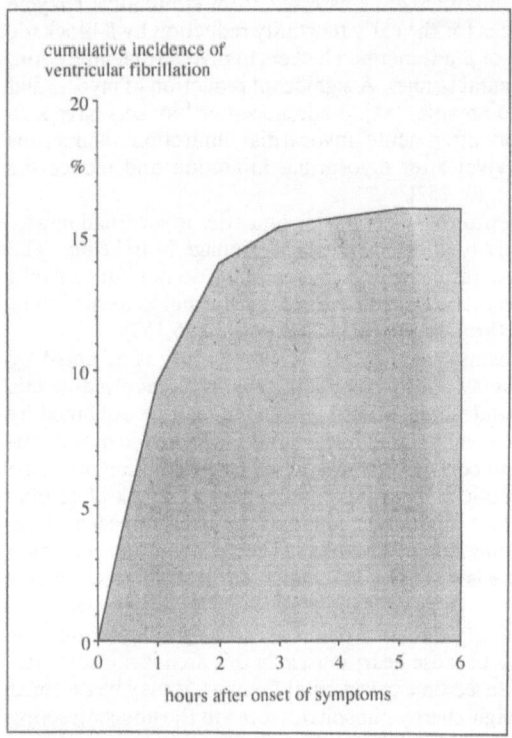

Fig. 5. Cumulative incidence of primary ventricular fibrillation during the first 6 h of acute myocardial infarction in 294 patients. (Data from [88].)

and increased the sudden death rate more than twofold [12]. So far, only β-adrenoceptor blocking agents showed beneficial effects in the postinfarct period in randomized studies. In 14 studies involving 15,819 patients the rate of sudden cardiac death was reduced by 31.6% [20,152]. This long-term protection by β-blockade against primary ventricular fibrillation in coronary heart disease clearly emphasizes the role of adrenergic mechanisms in the pathogenesis of ischemia-induced arrhythmias.

First evidence for a causal relationship between adrenergic mechanisms and ischemia-associated arrhythmias was derived from experiments utilizing both surgical and pharmacological interventions to reduce sympathetic activity in myocardial infarction. Chronic cardiac sympathetic denervation suppressed the incidence of ventricular fibrillation, whereas acute sympathetic ablation failed to reduce ventricular fibrillation associated with ischemia (Fig. 6) [31,106]. This apparent paradox might be explained by the finding that chronic denervation resulted in nearly total depletion of myocardial catecholamines, whereas acute denervation did not [31]. Thus, acute sympathetic denervation prevents catecholamine release evoked by central sympathetic activity, but not release induced by local metabolic processes. After chronic denervation, however, the heart is protected from any kind of local catecholamine release.

In accordance with the results obtained after surgical denervation, pharmacological depletion of myocardial catecholamines with 6-hydroxydopamine reduced the incidence of ventricular fibrillation following complete obstruction of the left anterior descending coronary artery in dogs [120]. Comparable results were obtained in isolated heart preparations from rats pretreated with 6-hydroxydopamine [1,27].

Another approach to limit noradrenaline release during myocardial ischemia is the use of

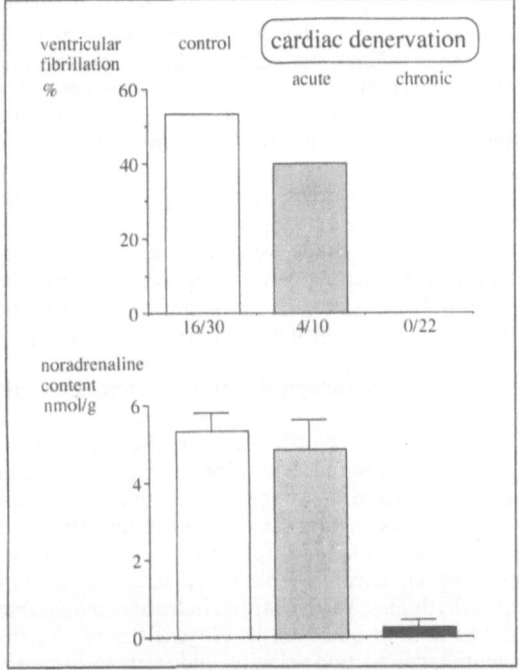

Fig. 6. Incidence of ventricular fibrillation (upper panel) and cardiac noradrenaline content (lower panel) in experimental myocardial infarction in canines. (Data from [31].)

Acute sympathectomy failed to reduce ventricular fibrillation caused by myocardial ischemia, whereas chronic sympathetic denervation suppressed the incidence of ventricular fibrillation. This protective effect of chronic sympathectomy was associated with a nearly complete catecholamine depletion of the heart. After acute denervation, however, cardiac catecholamine stores were maintained.

uptake$_1$-inhibitors that suppress local metabolic noradrenaline release in ischemia (see above). As demonstrated for catecholamine depletion, uptake$_1$-inhibitors, such as desipramine, nisoxetine, and amitriptyline, significantly reduced the incidence of ventricular fibrillation following coronary artery ligation in perfused rat hearts [27] and in dogs [146]. The same nanomolar concentrations of desipramine which revealed potent antifibrillatory effects during myocardial ischemia significantly increased the incidence of ventricular fibrillation in rat hearts if arrhythmias were caused by perfusion with exogenous noradrenaline. In this case local catecholamine concentrations were increased by the drug. These data are consistent with the hypothesis that local metabolic noradrenaline release in ischemia is critically involved in the pathogenesis of arrhythmias.

This concept was further strengthened by the pronounced antifibrillatory effects of sodium/proton exchange inhibitors. Amiloride and its derivatives ethylisopropylamiloride and dimethylamiloride significantly reduced ischemia-induced ventricular fibrillation in the rat heart in the same concentration range that suppressed local metabolic noradrenaline release in this experimental model [111]. Furthermore, antiarrhythmic effects of amiloride have recently been demonstrated during the postinfarct phase in canines [29] and in a small group of patients suffering from ventricular tachycardias resistant to conventional antiarrhythmic therapy [30].

The situation becomes more complicated when looking at the effects of unilateral and bilateral gangliectomy on ischemia-induced arrhythmias. Schwartz et al. demonstrated that blockade of the left stellate ganglion provided an antiarrhythmic effect during ischemia, while blockade of the right stellate ganglion resulted in arrhythmogenic effects [118], and acute bilateral gangliectomy failed to influence arrhythmias [31]. It becomes apparent from these findings that the asymmetry in cardiac sympathetic innervation contributes to arrhythmogenesis during ischemia. The dominant effect of left stellate output is a shortening of ventricular refractory period, a phenomenon that is likely to enhance the propagation of ventricular arrhythmias due to reentry mechanisms [119].

Several findings suggest that adrenergic effects on arrhythmogenesis in ischemia are mediated by the activation of β-adrenergic receptors of the cardiomyocyte: Adrenaline-induced vulnerability to ventricular fibrillation was accompanied by an increase in tissue cAMP [93,119], and β-adrenergic blockade with propranolol, atenolol, and metoprolol attenuated ventricular arrhythmias in experimental myocardial infarction [38,80,90,120].

Since these antiarrhythmic effects of β-adrenergic blockade were variable, whereas procedures depleting catecholamines or inhibiting their release appeared to be more effective, the role of α-adrenergic stimulation was evaluated. Indeed, α-receptor blockade by phentolamine and prazosin significantly reduced ventricular arrhythmias during coronary occlusion, especially during reperfusion in various species [17,124,128,147].

So far, it has remained speculative which electrophysiological actions of noradrenaline contribute to the pathogenesis of ventricular arrhythmias during myocardial ischemia [15,18,59,77]. On the cellular level, catecholamines interfere via stimulation of β-receptors with various components of the action potential, such as amplitude, duration, refractoriness, and conduction velocity depending on the type of cardiac tissue. α-Adrenergic mechanisms increase slow inward current and, therefore, may initiate delayed after-depolarization and triggered activity [59,70]. Experiments utilizing exogenous catecholamines for coronary perfusion demonstrate that under normoxic conditions in an electrically stable heart, micromolar concentrations are required to induce malignant arrhythmias. In myocardial ischemia comparable local catecholamine concentrations are reached and, in addition, heterogeneities of electrophysiological parameters such as resting potential, conduction velocity, and refractoriness are present, which are important preconditions for the occurrence of reentry. These inhomogeneities correspond to metabolic and ionic gradients in cellular content of high-energy phosphates and glycogen, in extracellular potassium concentration, and in intra- and extracellular pH between closely related areas with different ischemic injury [13,60]. Superimposed are local and temporary inhomogeneities in noradrenaline release and catecholamine responsiveness of the myocytes (see above) that may be considered to promote ischemia-induced arrhythmias by aggravating these variabilities of conduction and refractoriness.

Apart from these direct electrophysiological effects, excess sympathetic stimulation facilitates arrhythmias by indirect actions [107]. Enlargement of the ischemic area caused by sympathetic overstimulation may facilitate arrhythmias, and increased heart rate has been shown to unmask heterogeneities of conduction velocity in the ischemic zone [59,141]. Thus, by direct and indirect interference with electrophysiological processes in the ischemic heart, excessive adrenergic stimulation promotes malignant arrhythmias and considerably determines the outcome in acute myocardial infarction.

References

1. Abrahamsson T, Almgren O, Svensson L (1981) Local noradrenaline release in acute myocardial ischemia: Influence of catecholamine synthesis inhibition and β-adrenoceptor blockade on ischemic injury. J Cardiovasc Pharmacol 3: 807–817
2. Abrahamsson T, Almgren O, Carlsson L (1983) Ischemia-induced noradrenaline release in the isolated rat heart: Influence of perfusion, substrate, and duration of ischemia. J Mol Cell Cardiol 15: 821–830
3. Aronson PS (1985) Kinetic properties of the plasma membrane Na^+-H^+ exchanger. Ann Rev Physiol 47: 545–560
4. Balschi JA, Frazer JC, Fetters JK, Clarke K, Springer CS, Smith TW, Ingwall JS (1985) Shift reagent and Na-23 nuclear magnetic resonance discriminates between extra- and intracellular sodium pools in ischemic heart. Circulation 72 (suppl III): 355 (abstr.)
5. Barber MJ, Thomas JX, Stephen JR, Jones B, Randall WC (1982) Effect of sympathetic nerve stimulation and cardiac denervation on MBF during LAD occlusion. Am J Physiol 243: H556–H574

6. Beers MF, Carty SE, Johnson RG, Scarpa A (1982) H^+-ATPase and catecholamine transport in chromaffin granules. Ann NY Acad Sci 402: 116–133
7. Benedict CR, Graham-Smith DG (1979) Plasma adrenaline and noradrenaline concentrations and dopamine-β-hydroxylase activity in myocardial infarction with and without cardiogenic shock. Br Heart J 42: 214–220
8. Bernauer W (1985) The effect of β-adrenoceptor blocking agents on evolving myocardial necrosis in coronary ligated rats with and without reperfusion. Naunyn-Schmiedeberg's Arch Pharmacol 328: 288–294
9. Bertel O, Bühler FR, Baitsch G, Ritz R, Burkart F (1982) Plasma adrenaline and noradrenaline in patients with acute myocardial infarction. Relationship to ventricular arrhythmias of varying severity. Chest 82: 64–68
10. Carlsson L, Abrahamsson T, Almgren O (1985) Local release of myocardial norepinephrine during acute ischemia: An experimental study in the isolated perfused rat heart. J Cardiovasc Pharmacol 7: 791–798
11. Carlsson L (1988) A crucial role of ongoing anaerobic glycolysis in attenuating acute ischemia-induced release of myocardial noradrenaline. J Mol Cell Cardiol 20: 247–253
12. The Cardiac Arrhythmia Suppression Trial (CAST) Investigators (1989) Preliminary report: Effect of encainide and flecainide on mortality in a randomized trial of arrhythmia suppression after myocardial infarction. N Engl J Med 321: 406–412
13. Coronel R, Fiolet JWT, Wilms-Schopman FJG, Schaapherder AFM, Johnson TA, Gettes LS, Janse MJ (1988) Distribution of extracellular potassium and its relation to electrophysiologic changes during acute myocardial ischemia in isolated perfused porcine heart. Circulation 77: 1125–1138
14. Corr PB, Crafford WA (1981) Enhanced alpha-adrenergic responsiveness in the myocardium: Role of alpha adrenergic blockade. Am Heart J 102: 605–614
15. Corr PB, Gillis RA (1978) Autonomic neural influences on the dysrhythmias resulting from myocardial infarction. Circ Res 43: 1–9
16. Corr PB, Shayman JA, Kramer JB, Kipnis RJ (1981) Increased α-adrenergic receptors in ischemic cat myocardium. J Clin Invest 67: 1232–1236
17. Corr PB, Witkowski FX, Sobel BE (1978) Mechanisms contributing to malignant dysrhythmias induced by ischemia in the cat. J Clin Invest 61: 109–119
18. Corr PB, Yamada KA, Witkowski FX (1986) Mechanisms controlling cardiac autonomic function and their relation to arrhythmogenesis. In: Fozzard HA, Jennings RB, Haber E, Katz AM, Morgan HE (Eds), Heart and Cardiovascular System. Raven Press, New York, pp 1343–1403
19. Cox WV, Robertson HF (1936) The effect of stellate ganglionectomy on the cardiac function of intact dogs and its effect on the extent of myocardial infarction and on cardiac function following coronary artery occlusion. Am Heart J 12: 285–300
20. Cruickshank JM, Prichand BN (1987) Beta-blockers in clinical practice. Churchill Livingstone, Edinburgh, London, Melbourne and New York.
21. Culling W, Penny WJ, Lewis MJ, Middleton K, Sheridan DJ (1984) Effects of myocardial catecholamine depletion on cellular electrophysiology and arrhythmias during ischaemia and reperfusion. Cardiovasc Res 18: 675–682
22. Dart AM (1988) Influence of myocardial ischaemia on exocytotic noradrenaline release. In: Brachmann J, Schömig A (Eds) Adrenergic System and Ventricular Arrhythmias in Myocardial Infarction. Springer Verlag, New York-Berlin-Heidelberg, pp 34–43
23. Dart AM, Riemersma RA (1988) Origins of endogenous noradrenaline overflow during reperfusion of the ischaemic rat heart. Clin Sci 74: 269–274
24. Dart AM, Riemersma RA, Schömig A, Ungar A (1987) Metabolic requirements for release of endogenous noradrenaline during myocardial ischaemia and anoxia. Br J Pharmacol 90: 43–50
25. Dart AM, Schömig A, Dietz R, Mayer E, Kübler W (1984) Release of endogenous catecholamines in the ischemic myocardium of the rat. Part B: Effect of sympathetic nerve stimulation. Circ Res 55: 702–706
26. Daugherty A, Frayn KN, Redfern WS, Woodward B (1986) The role of catecholamines in the production of ischaemia-induced ventricular arrhythmias in the rat in vivo and in vitro. Br J Pharmacol 87: 265–277
27. Dietz R, Offner B, Dart AM, Schömig A (1989) Ischaemia-induced noradrenaline release mediates ventricular arrhythmias. In: Brachmann J, Schömig A (Eds) Adrenergic System and Ventricular Arrhythmias in Myocardial Infarction. Springer Verlag, Berlin-Heidelberg-New York, pp 313–321
28. Du XJ, Riemersma RA (1990) Reduced neuronal noradrenaline overflow in the innervated ischaemic rat heart: Importance of the severity of coronary flow reduction. Basic Res Cardiol (in press)

29. Duff HJ, Lester WM, Rahmberg M (1988) Amiloride. Antiarrhythmic and electrophysiological activity in the dog. Circulation 78: 1469–1477
30. Duff HJ, Mitchell LB, Kavanagh KM, Manyari DE, Gillis AM, Wyse DG (1989) Amiloride. Antiarrhythmic and electrophysiologic actions in patients with inducible sustained ventricular tachycardia. Circulation 79: 1257–1263
31. Ebert PA, Vanderbeek RB, Allgood RJ, Sabiston Jr DC (1970) Effect of chronic cardiac denervation on arrhythmias after coronary artery ligation. Cardiovasc Res 4: 141–147
32. Edoute Y, Sanan D, Lochner A, Graney D, Kotze JCN (1981) Effects of propranolol on myocardial ultrastructure, mitochondrial function and high energy phosphates of isolated working rat hearts with coronary artery ligation. J Mol Cell Cardiol 13: 619–639
33. Ellingsen O, Sejersted OM, Leraand S, Ilebekk A (1987) Catecholamine-induced myocardial potassium uptake mediated by β_1-adrenoceptors and adenylate cyclase activation in the pig. Circ Res 60: 540–550
34. Ellis SG, Henschke Cl, Sandor T, Wynne J, Braunwald E, Kloner RA (1983) Time course of functional and biochemical recovery of myocardium salvaged by reperfusion. J Am Coll Cardiol 1: 1047–1055
35. Emanuelsson H, Mannheimer C, Waagstein F (1990) Changes in arterial levels and myocardial metabolism of catecholamines during pacing-induced angina pectoris. Clin Cardiol (in press)
36. Esler M, Jennings G, Korner P, Blombery P, Sacharias N, Leonard P (1984) Measurement of total and organ-specific norepinephrine kinetics in humans. Am J Physiol 247: E21–E28
37. Euler von US, Lishajko F (1963) Effect of adenine nucleotides on catecholamine release and uptake in isolated adrenergic nerve granules. Acta Physiol Scand 59: 454–461
38. Fearon RE (1967) Propranolol in the prevention of ventricular fibrillation due to experimental coronary artery occlusion. Am J Cardiol 20: 222–228
39. Fiolet JWT, Baartscheer A, Schumacher CA, Coronel R, ter Welle HF (1984) The change of the free energy of ATP hydrolysis during global ischemia and anoxia in the rat heart. Its possible role in the regulation of transsarcolemmal sodium and potassium gradients. J Mol Cell Cardiol 16: 1023–1036
40. Forfar JC, Riemersma RA, Oliver MF (1983) Alpha-adrenoceptor control of norepinephrine release from acutely ischemic myocardium: Effects of blood flow, arrhythmias, and regional conduction delay. J Cardiovasc Pharmacol 5: 752–759
41. Fowler CJ, Oreland L (1980) The nature of the substrate-selective interaction between rat liver mitochondrial monoamine oxidase and oxygen. Biochem Pharmacol 29: 2225–2233
42. Franco-Cereceda A, Saria A, Lundberg JM (1989) Differential release of calcitonin gene-related peptide and neuropeptide Y from the isolated heart by capsaicin, ischaemia, nicotine, bradykinin and ouabain. Acta Physiol Scand 135: 173–187
43. Gazes PC, Richardson JA, Woods EF (1959) Plasma catecholamine concentrations in myocardial infarction and angina pectoris. Circulation 19: 657–661
44. Graefe KH (1989) On the mechanism of non-exocytotic release of noradrenaline from noradrenergic neurons. In: Brachmann J, Schömig A (Eds) Adrenergic System and Ventricular Arrhythmias in Myocardial Infarction. Springer Verlag, New York-Berlin-Heidelberg, pp 44–52
45. Graefe KH, Fuchs G (1979) On the mechanism of neuronal efflux of axoplasmatic ^{3}H-(-)noradrenaline. In: Usdin E, Kopin IJ, Barchas J (Eds) Catecholamines: Basic and Clinical Frontiers. Vol 1, Pergamon Press, New York, Oxford, Toronto, Sydney, Frankfurt, Paris, pp 268–270
46. Graefe KH, Zeitner CJ, Fuchs G, Keller B (1984) Role played by sodium in the membrane transport of ^{3}H-noradrenaline across the axonal membrane of noradrenergic neurones. In: Fleming WW (Ed) Neuronal and Extraneuronal Events in Autonomic Pharmacology. Raven Press, New York, pp 51–62
47. Griffiths J, Leung F (1971) The sequential estimation of plasma catecholamines and whole blood histamine in myocardial infarction. Am Heart J 82: 171–179
48. Haass M, Cheng B, Richardt G, Lang RE, Schömig A (1989) Characterization and presynaptic modulation of stimulation-evoked exocytotic co-release of noradrenaline and neuropeptide Y in guinea pig heart. Naunyn-Schmiedeberg's Arch Pharmacol 339: 71–78
49. Haass M, Hock M, Richardt G, Schömig A (1989) Neuropeptide Y differentiates between exocytotic and nonexocytotic noradrenaline release in guinea-pig heart. Naunyn-Schmiedeberg's Arch Pharmacol 340: 509–515
50. Haass M, Richardt G, Lang RE, Schömig A (1990) Common features of NPY and noradrenaline release in guinea pig heart. Ann N Y Acad Sci USA (in press)
51. Heusch G, Deussen A (1983) The effects of cardiac sympathetic nerve stimulation on perfusion of stenotic coronary arteries in the dog. Circ Res 53: 8–15
52. Hirche HJ, Franz C, Bös L, Bissig R, Lang R, Schramm M (1980) Myocardial extracellular K^+ and H^+ increase and noradrenaline release as possible cause of early arrhythmias following acute coronary artery occlusion in pigs. J Mol Cell Cardiol 12: 579–593
53. Holmgren S, Abrahamsson T, Almgren O (1985) Adrenergic innervation of coronary arteries and

ventricular myocardium in the pig: Fluorescence microscopic appearance in the normal state and after ischemia. Basic Res Cardiol 80: 18–26

54. Holmgren S, Abrahamsson T, Almgren O, Eriksson BM (1981) Effect of ischaemia on the adrenergic neurons of the rat heart: A fluorescence histochemical and biochemical study. Cardiovasc Res 15: 680–689

55. Humprey SM, Gavin JB, Herdson PB (1982) Catecholamine-depletion and the no-reflow phenomenon in anoxic and ischaemic rat hearts. J Mol Cell Cardiol 14: 151–161

56. Insel PA, Maisel AS (1989) Alpha$_1$- and beta-adrenergic receptors in myocardial ischemia and injury. In: Brachmann J, Schömig A (Eds) Adrenergic System and Ventricular Arrhythmias in Myocardial Infarction. Springer Verlag, New York-Berlin-Heidelberg, pp 81–90

57. ISIS-I Collaborative Group (First International Study of Infarct Survival) (1986) Randomized trial of intravenous atenolol among 16027 cases of suspected acute myocardial infarction: ISIS-I. Lancet 2: 57–66.

58. ISIS-1 (First International Study of Infarct Survival) Collaborative Group (1988) Mechanisms for the early mortality reduction produced by beta-blockade started early in acute myocardial infarction: ISIS-1. Lancet 1: 921–923

59. Janse MJ (1989) Why is increased adrenergic activity arrhythmogenic? In: Brachmann J, Schömig A (Eds) Ventricular Arrhythmias in Myocardial Infarction. Springer Verlag, New York-Berlin-Heidelberg, pp 353–363

60. Janse MJ, Cinca J, Morena H, Fiolet JWT, Kleber AG, de Vries GP, Beckert AE, Durrer D (1979) The border zone in myocardial ischemia. An electrophysiological, metabolic and histological correlation in the pig heart. Circ Res 44: 576–588

61. Jesmok GJ, Warltier DC, Gross GJ, Harman HF (1978) Effect of propranolol on enzymatic and histochemical estimates of infarct size in experimental myocardial infarction. Basic Res Cardiol 73: 559–570

62. Jones CE, Beck LY, DuPont E, Barnes GE (1978) Effect of coronary ligation on the chronically sympathectomized dog ventricle. Am J Physiol 235: H429–H434

63. Karlsberg RP, Cryer PE, Roberts R (1981) Serial plasma catecholamine response early in the course of clinical acute myocardial infarction: Relationship to infarct extent and mortality. Am Heart J 102: 24–29

64. Karlsberg RP, Penkoske PA, Cryer PE, Corr PB, Roberts R (1979) Rapid activation of the sympathetic nervous system following coronary artery occlusion: Relationship to infarct size, site, and haemodynamic impact. Cardiovasc Res 13: 523–531

65. Kleber AG (1983) Extracellular potassium accumulation in acute myocardial ischemia. J Mol Cell Cardiol 16: 389–394

66. Kopin IJ, Zukowska-Grojec Z, Bayorh MA, Goldstein DS (1984) Estimation of intrasynaptic norepinephrine concentrations at vascular neuroeffector junctions in vivo. Naunyn-Schmiedeberg's Arch Pharmacol 325: 298–305

67. Kranzhöfer R, Haass M, Kurz T, Richardt G, Schömig A (1990) Effect of digitalis glycosides on noradrenaline release in the heart: Dual mechanism of action. (submitted for publication)

68. Kurz Th, Said W, Saggau W, Richardt G, Schömig A (1990) Energy deficiency induces nonexocytotic release of endogenous noradrenaline in human atrial tissue. Naunyn-Schmiedeberg's Arch Pharmacol (suppl) 341: R84 (abstr.)

69. Lazdunski M, Frelin C, Vigne P (1985) The sodium/hydrogen exchange system in cardiac cells: Its biochemical and pharmacological properties and its role in regulating internal concentrations of sodium and internal pH. J Mol Cell Cardiol 17: 1029–1042

70. Lazzara R, Marchi S (1989) Electrophysiological mechanisms for the generation of arrhythmias with adrenergic stimulation. In: Brachmann J, Schömig A (Eds) Adrenergic System and Ventricular Arrhythmias in Myocardial Infarction. Springer Verlag, New York-Berlin-Heidelberg, pp 231–238

71. Leriche R, Fontaine R (1931) Les résultats actuels du traitement chirurgical de l'angine de pointrine. J Chir 38: 785–815

72. Lindmar R, Löffelholz K (1974) Neuronal and extraneuronal uptake and efflux of catecholamines in the isolated rabbit heart. Naunyn-Schmiedeberg's Arch Pharmacol 284: 63–92

73. Lorenz RR, Vanhoutte PM (1975) Inhibition of adrenergic neurotransmission in isolated veins of the dog by potassium ions. J Physiol 246: 479–500

74. Louis JC, Magael E, Yavin E (1988) Proteinkinase C alterations in fetal rat brain after global ischemia. J Biol Chem 263: 19282–19285

75. Maisel AS, Motulsky NJ, Insel PA (1985) Externalization of beta-adrenergic receptors promoted by myocardial ischemia. Science 230: 183–186.

76. Malliani A, Schwartz PJ, Zanchetti A (1969) A sympathetic reflex elicited by experimental coronary occlusion. Am J Physiol 217: 703–709

77. Malliani A, Schwartz PJ, Zanchetti A (1980) Neural mechanisms in life-threatening arrhythmias. Am Heart J 100: 705–715
78. Martins JB, Kerber RE, Marcus ML, Laughlin DL, Levy DM (1980) Inhibition of adrenergic neurotransmission in ischaemic regions of the canine left ventricle. Cardiovasc Res 14: 116–124
79. McGrath BP, Lim SP, Leversha L, Shanahan A (1981) Myocardial and peripheral catecholamine responses to acute coronary artery constriction before and after propranolol treatment in the anaesthetised dog. Cardiovasc Res 15: 28–34
80. Menken U, Wiegand V, Bucher P, Meesmann W (1979) Prophylaxis of ventricular fibrillation after acute experimental coronary occlusion by chronic beta-adrenoceptor blockade with atenolol. Cardiovasc Res 13: 588–594
81. MIAMI Trial Research Group (1985) Metoprolol in acute myocardial infarction. (MIAMI) A randomized placebo-controlled international trial. Eur Heart J 6: 199–226
82. Miyazaki T, Zipes DP (1990) Presynaptic modulation of efferent sympathetic and vagal neurotransmission in the canine heart by hypoxia, high K^+, low pH, and adenosine. Possible relevance to ischemia-induced denervation. Circ Res 66: 289–301
83. Mukherjee A, McCoy KE, Duke RJ, Hogan M, Hagler HK, Buja LM, Willerson JT (1982) Relationship between beta adrenergic receptor numbers and physiological responses during experimental canine myocardial ischemia. Circ Res 50: 735–741
84. Muntz KH, Hagler HK, Boulas HJ, Buja LM (1984) Redistribution of catecholamines in the ischemic zone of the dog heart. Am J Pathol 114: 64–78
85. Nadeau RA, de Champlain J (1979) Plasma catecholamines in acute myocardial infarction. Am Heart J 98: 548–554
86. Nayler WG, Sturrock WJ (1984) An inhibitory effect of verapamil and diltiazem on the release of noradrenaline from ischaemic and reperfused hearts. J Mol Cell Cardiol 16: 331–344
87. Nayler WG, Sturrock WJ (1985) Inhibitory effect of calcium antagonists on the depletion of cardiac norepinephrine during postischemic reperfusion. J Cardiovasc Pharmacol 7: 581–587
88. Pantridge JF, Webb SW, Adgey AAJ (1981) Arrhythmias in the first hours of acute myocardial infarction. Prog Cardiovasc Dis 23: 265–278
89. Paton DM (1973) Mechanism of efflux of noradrenaline from adrenergic nerves in rabbit atria. Br J Pharmacol 49: 614–627
90. Pentecost BL, Austen WG (1966) Beta-adrenergic blockade in experimental myocardial infarction. Am Heart J 72: 790–796
91. Peter T, Heng MK, Singh BH, Ambler P, Nisbet H, Elliot R, Norris RM (1978) Failure of high doses of propranolol to reduce experimental myocardial damage. Circulation 57: 534–540
92. Phillips JH (1982) Dynamic aspects of chromaffin granule structure. Neurosci 7: 1595–1609
93. Podzuweit T, Darby AJ, Cherry GW, Opie LH (1978) Cyclic AMP levels in ischemic and non-ischemic myocardium following coronary artery ligation: Relation to ventricular fibrillation. J Mol Cell Cardiol 10: 81–94
94. Puig M, Kirpekar SM (1971) Inhibitory effects of low pH on norepinephrine release. J Pharmacol Exp Ther 176: 134–138
95. Raiteri M, del Carmine R, Bertollini A, Levi G (1977) Effect of desmethylimipramine on the release of ^{3}H-norepinephrine induced by various agents in hypothalamic synaptosomes. Mol Cell Pharmacol 13: 746–758
96. Reimer KA, Rasmussen MM, Jennings RB (1976) On the nature of protection by propranolol against myocardial necrosis after temporary coronary occlusion in dogs. Am J Cardiol 37: 520–527
97. Richardt G, Haass M, Schömig A (1990) Calcium antagonists and cardiac noradrenaline release in ischemia. (submitted for publication)
98. Richardt G, Lumpp U, Haass M, Schömig A (1990) Propranolol inhibits nonexocytotic noradrenaline release in myocardial ischemia. Naunyn-Schmiedeberg's Arch Pharmacol 341: 50–55
99. Richardt G, Waas W, Kranzhöfer R, Cheng B, Lohse MJ, Schömig A (1989) Interaction between the release of adenosine and noradrenaline during sympathetic stimulation: A feed-back mechanism in rat heart. J Mol Cell Cardiol 21: 269–277
100. Richardt G, Waas W, Kranzhöfer R, Mayer E, Schömig A (1987) Adenosine inhibits exocytotic release of endogenous noradrenaline in the rat heart: A protective mechanism in early myocardial ischemia. Circ Res 61: 117–123
101. Rochette L, Didier J-P, Moreau D, Brallet J (1980) Effect of substrate on release of myocardial norepinephrine and ventricular arrhythmias following reperfusion of the ischemic isolated working rat heart. J Cardiovasc Pharmacol 2: 267–279
102. Rona G (1985) Catecholamine cardiotoxicity. J Mol Cell Cardiol 17: 291–306
103. Ross SB, Kelder D (1979) Release of ^{3}H-noradrenaline from the rat vas deferens under various in vitro conditions. Acta Physiol Scand 105: 338–349

104. Saffitz JE, Corr PB (1989) Mechanisms of altered adrenergic responsiveness contributing to arrhythmogenesis during myocardial ischemia. In: Brachmann J, Schömig A (Eds) Adrenergic System and Ventricular Arrhythmias in Myocardial Infarction. Springer Verlag, New York-Berlin-Heidelberg, pp 112–122

105. Sammet S, Graefe KH (1979) Kinetic analysis of the interaction between noradrenaline and Na$^+$ in neuronal uptake: Kinetic evidence for co-transport. Naunyn-Schmiedeberg's Arch Pharmacol 309: 99–107

106. Schaal SF, Wallace AG, Sealy WC (1969) Protective influence of cardiac denervation against arrhythmias of myocardial infarction. Cardiovasc Res 3: 241–244

107. Scherlag BJ, El-Sherif N, Hope RR, Lazzara R (1974) Characterization and localization of ventricular arrhythmias resulting from myocardial ischemia and infarction. Circ Res 35: 372–383

108. Schömig A (1988) Adrenergic mechanisms in myocardial infarction: Cardiac and systemic catecholamine release. J Cardiovasc Pharmacol 12 (suppl 1): 1–7

109. Schömig A (1989) Increase of cardiac and systemic catecholamines in myocardial ischemia. In: Brachmann J, Schömig A (Eds) Adrenergic System and Ventricular Arrhythmias in Myocardial Infarction. Springer Verlag, New York-Berlin-Heidelberg, pp 61–67

110. Schömig A (1990) Catecholamines in myocardial ischemia: Systemic and cardiac release. Circulation 82 suppl II: II-13-II-22

111. Schömig A, Beyer Th, Rehmert G (1989) Amiloride and analogues suppress noradrenaline release and malignant arrhythmias in the ischemic rat heart. Circulation 80 (suppl II): 202 (abstr.)

112. Schömig A, Dart AM, Dietz R, Kübler W, Mayer E (1985) Paradoxical role of neuronal uptake for the locally mediated release of endogenous noradrenaline in the ischemic myocardium. J Cardiovasc Pharmacol 7 (suppl 5): S40–S44

113. Schömig A, Dart AM, Dietz R, Mayer E, Kübler W (1984) Release of endogenous catecholamines in the ischemic myocardium of the rat. Part A: Locally mediated release. Circ Res 55: 689–701

114. Schömig A, Dietz R, Strasser R, Dart AM, Kübler W (1982) Noradrenaline release and inactivation in myocardial ischemia. In: Caldarera CM, Harris P (Eds) Advances in Studies on Heart Metabolism. Bologna, CLUEB, pp 239–244

115. Schömig A, Fischer S, Kurz Th, Richardt G, Schömig E (1987) Nonexocytotic release of endogenous noradrenaline in the ischemic and anoxic rat heart: Mechanism and metabolic requirements. Circ Res 60: 194–295

116. Schömig A, Kurz Th, Richardt G, Schömig E (1988) Neuronal sodium homoeostasis and axoplasmic amine concentration determine calcium-independent noradrenaline release in normoxic and ischemic rat heart. Circ Res 63: 214–226

117. Schömig A, Ness G, Mayer E, Katus H, Dietz R (1984) Sympathetic activity in patients with acute myocardial infarction before and after intracoronary thrombolytic therapy. Eur Heart J 5 (suppl 1): 39 (abstr.)

118. Schwartz PJ, Snebold NG, Brown AM (1976) Effects of unilateral cardiac sympathetic denervation on the ventricular fibrillation threshold. Am J Cardiol 37: 1034–1040

119. Schwartz PJ, Verrier RL, Lown B (1977) Effect of stellectomy and vagotomy on ventricular refractoriness in dogs. Circ Res 40 (suppl 6): 536–540

120. Sethi V, Haider B, Ahmed SS, Oldewurtel HA, Regan TJ (1973) Influence of beta blockade and chemical sympathectomy on myocardial function and arrhythmias in acute ischaemia. Cardiovasc Res 7: 740–747

121. Shahab L, Wollenberger A, Haase M, Schiller U (1969) Noradrenalinabgabe aus dem Hundeherzen nach vorübergehender Okklusion einer Koronararterie. Acta Biol Med Germ 22: 135–143

122. Sharma AD, Saffitz JE, Lee BI, Sobel BE, Corr PB (1983) Alpha adrenergic-mediated accumulation of calcium in reperfused myocardium. J Clin Invest 72: 802–818

123. Shatney CH, MacCarter DL, Lillehei RC (1976) Effects of allopurinol, propranolol and methylprednisolone on infarct size in experimental myocardial infarction. Am J Cardiol 37: 572–580

124. Sheridan DJ, Penkoske PA, Sobel BE, Corr PB (1980) Alpha adrenergic contributions to dysrhythmia during myocardial ischemia and reperfusion in cats. J Clin Invest 65: 161–171

125. Siggers DC, Salter C, Fluck DC (1971) Serial plasma adrenaline and noradrenaline levels in myocardial infarction using a new double isotope technique. Br Heart J 33: 878–883

126. Sommers HM, Jennings R (1972) Ventricular fibrillation and myocardial necrosis after transient ischemia. Effect of treatment with oxygen, procainamide, reserpine, and propranolol. Arch Intern Med 129: 780–789

127. Starke K (1977) Regulation of noradrenaline release by presynaptic receptor systems. Rev Physiol Biochem Pharmacol 77: 1–124

128. Stewart JR, Burmeister WE, Burmeister J, Lucchesi BR (1980) Electrophysiologic and antiarrhythmic effects of phentolamine in experimental coronary artery occlusion and reperfusion in the dog. J Cardiovasc Pharmacol 2: 77–91

129. Strange RC, Rowe MJ, Oliver MF (1978) Lack of relation between venous plasma total catecholamine concentration and ventricular arrhythmias after acute myocardial infarction. Br Med J 2: 921–922

130. Strasser RH, Dullaeus B, Marquetant R (1990) β-Receptor independent sensitization of the adenylylcyclase system in acute myocardial ischemia, Br J Pharmacol (in press)

131. Strasser RH, Krimmer J, Marquetant R (1988) Regulation of β-adrenergic receptors: Impaired desensitization in myocardial ischemia. J Cardiovasc Pharmacol 12 (suppl 1): S15–S24

132. Strasser RH, Marquetant R, Kübler W (1989) Sensitization of the adrenergic system in early myocardial ischemia: Independent regulation of β-adrenergic receptors and adenylate cyclase. In: Brachmann J, Schömig A (Eds) Adrenergic System and Ventricular Arrhythmias in Myocardial Infarction, Springer Verlag, New York-Berlin-Heidelberg, pp 98–111

133. Stute N, Trendelenburg U (1984) The outward transport of axoplasmic noradrenaline induced by a rise of the sodium concentration in the adrenergic nerve endings of the rat vas deferens. Naunyn-Schmiedeberg's Arch Pharmacol 327: 124–132

134. Thames MD, Klopfenstein HS, Abboud FM, Mark AL, Walker JL (1978) Preferential distribution of inhibitory cardiac receptors with vagal afferents to the inferoposterior wall of the left ventricle activated during coronary occlusion in the dog. Circ Res 43: 512–519

135. Toll L, Howard BD (1978) Role of Mg^{2+}-ATPase and a pH gradient in the storage of catecholamines in synaptic vesicles. Biochemistry 17: 2517–2523

136. Trendelenburg U (1989) The dynamics of adrenergic nerve endings. In: Brachmann J, Schömig A (Eds) Adrenergic System and Ventricular Arrhythmias in Myocardial Infarction, Springer Verlag, New York-Berlin-Heidelberg, pp 53–60

137. Uchida Y, Murao S (1974) Excitation of afferent cardiac sympathetic nerve fibers during coronary occlusion. Am J Physiol 226: 1094–1099

138. van Vliet PD, Burchell HB, Titus JF (1966) Focal myocarditis associated with pheochromocytoma. N Engl J Med 274: 1102–1108

139. Vatner DE (1989) Uncoupling of the β-adrenergic receptor by myocardial ischemia. In: Brachmann J, Schömig A (Eds) Adrenergic System and Ventricular Arrhythmias in Myocardial Infarction. Springer Verlag, New York-Berlin-Heidelberg, pp 91–97

140. Vatner DE, Knight D, Shen YT, Thomas JX, Homcy CJ, Vatner SF (1988) One hour of myocardial ischemia in conscious dogs increases beta-adrenergic receptors, but decreases adenylate cyclase activity. J Mol Cell Cardiol 20: 75–82

141. Verrier RL, Thompson PL, Lown B (1974) Ventricular vulnerability during sympathetic stimulation: role of heart rate and blood pressure. Cardiovasc Res 8: 602–610

142. Videbaek J, Christensen NJ, Sterndorff B (1972) Serial determination of plasma catecholamines in myocardial infarction. Circulation 46: 846–855

143. Vigne P, Frelin C, Cragoe Jr EJ, Lazdunski M (1983) Ethylisopropylamiloride: A new and highly potent derivative of amiloride for the inhibition of the Na^+/H^+ exchange system in various cell types. Biochem Biophys Res Commun 116: 86–90

144. Waldenström AP, Hjalmarson AC, Thornell L (1978) A possible role of noradrenaline in the development of myocardial infarction. Am Heart J 95: 43–51

145. Wilde AAM, Peters RJG, Janse MJ (1988) Catecholamine release and potassium accumulation in the isolated globally ischemic rabbit heart. J Mol Cell Cardiol 20: 887–896

146. Wilkerson RD, Sanders PW (1978) The antiarrhythmic action of amitriptyline on arrhythmias associated with myocardial infarction in dogs. Eur J Pharmacol 51: 193–198

147. Williams LT, Guerrero JL, Leinbach RC, Gold HK (1982) Prevention of reperfusion dysrhythmias by selective coronary alpha adrenergic blockade. Am J Cardiol 49: 1046 (abstr.)

148. Winkler H, Apps DK, Fischer-Colbrie R (1986) The molecular function of adrenal chromaffin granules: Established facts and unresolved topics. Neurosci 18: 261–290

149. Wollenberger A, Shahab L (1965) Anoxia-induced release of noradrenaline from the isolated perfused heart. Nature 207: 88–89

150. Yates JC, Beamish RE, Dhalla NS (1981) Ventricular dysfunction and necrosis produced by adrenochrome metabolite of epinephrine: Relation to pathogenesis of catecholamine cardiomyopathy. Am Heart J 102: 210–221

151. Yodice A (1941) Sympathectomy and experimental occlusion of a coronary artery. Am Heart J 22: 545–548

152. Yusuf S, Peto R, Lewis J, Collins R, Sleight P (1985) Betablockade during and after myocardial infarction: An overview of the randomized trials. Prog Cardiovasc Dis 27: 335–371

Authors' address:

Prof. Dr. A. Schömig, Department of Cardiology, Bergheimer Straße 58, 6900 Heidelberg, Germany

Modulation of α-Adrenergic Receptors and their Intracellular Coupling in the Ischemic Heart*

P. B. Corr, K. A. Yamada and S. D. DaTorre

Cardiovascular Division, Department of Internal Medicine and Department of Pharmacology, Washington University School of Medicine, St. Louis, Missouri, USA

Summary

The α_1-adrenergic receptor exists as at least two distinct subtypes, α_{1a} and α_{1b}. Based on hydrophobic exclusion studies and limited proteolysis of the cloned receptor, it appears to possess characteristics analogous to other membrane-bound receptors including seven membrane spanning domains, three extracellular, and three intracellular loops, with extensive glycosylation near the extracellular amino terminus. Although the receptor is coupled to phospholipase C in cardiac myocytes, with activation resulting in the production of inositol trisphosphate (IP_3) and diacylglycerol, recent findings suggest that the receptor may also be linked to phospholipase A_2, phospholipase D, and cyclic nucleotide phosphodiesterase. The α_1-adrenergic receptor has been shown to increase in response to myocardial ischemia in a number of different species and to mediate not only positive inotropic effects, but also to contribute substantially to arrhythmogenesis. The increase in α_1-adrenergic receptors can also occur in isolated adult ventricular myocytes in response to hypoxia, a mechanism which appears to be secondary to the sarcolemmal accumulation of long-chain acylcarnitines. This increase in α_1-adrenergic receptors in hypoxic myocytes is also linked to an enhanced increase in IP_3 in response to receptor stimulation. These and other findings obtained in vivo during ischemia suggest that α_1-adrenergic mechanisms can become prominent in myocardium under pathophysiologic conditions in which a depressed contractile state exists and may therefore serve as a secondary inotropic system. However, the arrhythmogenic effects of stimulation of the α_1-adrenergic receptor in the ischemic heart in man may contribute substantially to arrhythmogenesis and, thereby, to the incidence of sudden cardiac death.

Introduction

In 1948, Ahlquist demonstrated that adrenergic receptors could be classified, based on pharmacologic criteria, into α- and β-adrenergic subtypes [1]. This seminal contribution has subsequently led to a plethora of studies dealing with the development of specific agents which block these receptors, and to more recent findings indicating that there are additional subclasses for both the α- and β-adrenergic receptor [51, 79]. More recently, α- and β-adrenergic receptor subtypes have been cloned and their structures within membranes have been proposed based on hydrophobic exclusion studies [14, 40, 41, 61]. Results over the last decade have also indicated that these adrenergic receptor subtypes can exist in a variety of different states and can be cycled intracellularly, leading to both sensitization and desensitization of the receptor [71]. Findings have also revealed that adrenergic receptors are coupled to intracellular events

*Research from the authors' laboratory was supported in part by National Institutes of Health grant HL-17646, SCOR in Ischemic Heart Disease and NIH grants HL-28995 and HL-36773.

through a variety of different G-proteins which are critical to the mediation of intracellular responses [23, 54]. This brief review will focus exclusively on the α-adrenergic receptor, its subtypes and their structure as well as pathways for intracellular coupling. A major emphasis of the review will be the modulation of the density and intracellular coupling of the α_1-adrenergic receptor in the ischemic heart. Although there is substantial experimental evidence to suggest that the α-adrenergic receptor, particularly the α_1-adrenergic receptor, contributes to arrhythmogenesis in the ischemic heart, that particular topic is discussed elsewhere in this volume, as well as previous reviews [12, 26], and will be discussed only briefly here.

Subtypes of the α-adrenergic receptor and their three-dimensional membrane structure

Current evidence would suggest that α-adrenergic receptors can be divided into at least four different subtypes [51]. In the case of the α_1-adrenergic receptor, both α_{1a} and α_{1b} subtypes are present in a variety of different tissues. The α_{1a}-adrenergic receptor has a high affinity for WB-4101 and is not inactivated by alkylation using chlorethylclonidine up to a concentration of $10\,\mu M$ [51]. In contrast, the α_{1b}-adrenergic receptor has a much lower affinity for WB-4101 and is inactivated by chlorethylclonidine at concentrations of only $1\,\mu M$ [51]. Likewise, the α_2-adrenergic receptor also appears to exist as two subtypes, the α_{2a}- and the α_{2b}-adrenergic receptor. The α_{2a}-adrenergic receptor has a higher affinity for oxymetazoline, whereas the α_{2b}-adrenergic receptor has a higher affinity for prazosin and ARC-239 [51]. Both subtypes of the α_2-adrenergic receptor have recently been cloned [40, 61]. The α_1-adrenergic receptor has also been cloned and is probably the α_{1b}-adrenergic receptor subtype [14]. Thus, α-adrenergic receptor subtypes, which exist in at least four distinct forms, have an even greater diversity than β-adrenergic receptors. However, much less is known regarding the coupling of α_1-adrenergic receptors to intracellular events. For example, although α_2-adrenergic receptors appear to couple through G_i and, thereby, produce an inhibitory effect on adenylate cyclase, the G-protein which couples intracellular events to α_1-adrenergic receptors has not been characterized fully [8].

The α_1-adrenergic receptor is a complex glycoprotein of a molecular weight of about 85 KDa [14]. The overall structure of the α_1-adrenergic receptor has recently been obtained after cloning of the receptor protein, which indicates that the structure is very similar to other surface membrane receptors, including the β_1- and β_2-adrenergic receptors, the muscarinic receptor subtypes, and the serotonin receptor subtypes, among others [45, 46]. Thus, the α_1-adrenergic receptor protein is composed of 515 amino acids and has seven membrane-spanning domains consisting of the seven hydrophobic portions of the receptor protein which are enbedded in the phospholipid bilayer of the surface plasma membrane [14]. In addition, the receptor has three extracellular loops and three intracellular loops (Fig. 1). It should be noted that the structural representation of the α_1-adrenergic receptor shown has been inferred, based on several indirect approaches, including electron diffraction studies. The structure of the receptor is also consistent with experiments using limited proteolysis for the β_2-adrenergic receptor which exhibits similar sequence homology [18]. The extracellular amino terminus of the protein contains sites for glycosylation [45, 46]. The extracellular carbohydrate moiety is critical for optimal insertion of the receptor into the membrane. The intracellular carboxyl terminus, as well as the third intracellular loop are potential sites for phosphorylation of the receptor which may, under certain conditions, be responsible for alterations in the three-dimensional structure of the protein and/or coupling to intracellular events [45, 46]. The intracellular events for the α_1-adrenergic receptor are mediated by coupling to a G-protein which appears to differ between the α_1- and α_2-adrenergic receptor and may, in fact, differ between the α_{1a}- and α_{1b}-adrenergic receptor [8, 19, 75].

Studies using chimeric receptors composed of partial sequences from structurally similar clones are now being performed to delineate not only the binding sites for agonists and antagonists, but also the sites for binding to G-proteins [45, 46]. These types of studies should provide the detailed information required to elucidate the relationships between the structure of the receptors and its biologic response. This type of data will also be essential for determining

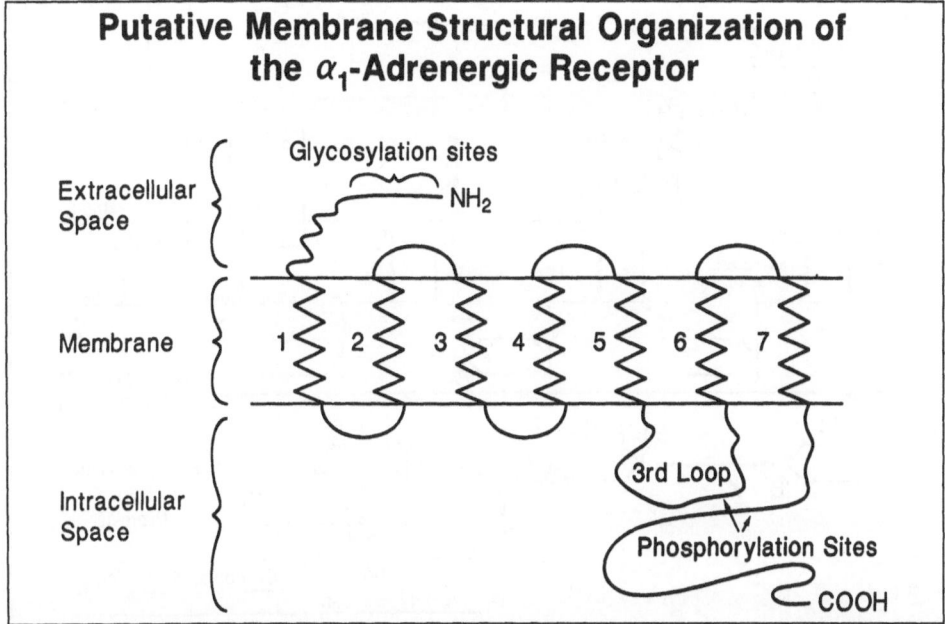

Fig. 1. Diagrammatic representation of the membrane structural organization of the α_1-adrenergic receptor. The membrane-spanning domains of the protein are depicted and labeled as 1 through 7 based on data reported in [14, 45].

how pathophysiologic changes lead to alterations in the structure and coupling of these membrane bound receptors.

Coupling of the α_1-adrenergic receptor to intracellular events

There are substantial data in the literature to indicate the α_1-adrenergic receptors are coupled to phospholipase C, resulting in cleavage of phosphatidylinositol bisphosphate (PIP_2) to form inositol trisphosphate (IP_3) and diacylglycerol (Fig. 2). IP_3 has been shown to mobilize intracellular calcium from sarcoplasmic reticulum through a receptor located on the sarcoplasmic reticulum [6]. The coupling of the α-adrenergic receptor to an increase in contractility appears to be related to an increase in IP_3 in rat [64] and human atria [42], and in some studies appears not to be mediated through a pertussis toxin-sensitive G-protein [7, 63]. This result suggests that if a G-protein mediates the response, it is pertussis toxin-insensitive. However, findings by Rosen and colleagues indicate that in the neonatal heart stimulation of the α_1-adrenergic receptor results in an increase in chronotropy that apparently is not sensitive to ADP-ribosylation by pertussis toxin, whereas in the adult heart a 41 KDa protein appears, dependent on innervation and the presence of nerve growth factor, which then results in a decrease in chronotropy in response to α_1-adrenergic stimulation secondary to stimulation of Na^+/K^+ ATPase activity [50, 62, 73, 85]. The ADP-ribosylation with pertussis toxin of this 41 KDa protein in adult tissue results in inhibition of the negative chronotropic response to α_1-adrenergic stimulation and the appearance of a positive chronotropic response [50, 62]. Therefore, it appears that cardiac tissue may possess both a pertussis toxin-sensitive and -insensitive pathway for mediation of α_1-adrenergic responses. It remains to be determined whether these responses may be mediated by two different subtypes of the α_1-adrenergic receptor, α_{1a} and α_{1b}. Diacylglycerol can stimulate protein kinase C which can phosphorylate a variety of intracellular proteins. One protein which can be phosphorylated by protein kinase C is a transsarcolemmal calcium channel, thereby, producing an increase in intracellular calcium from the

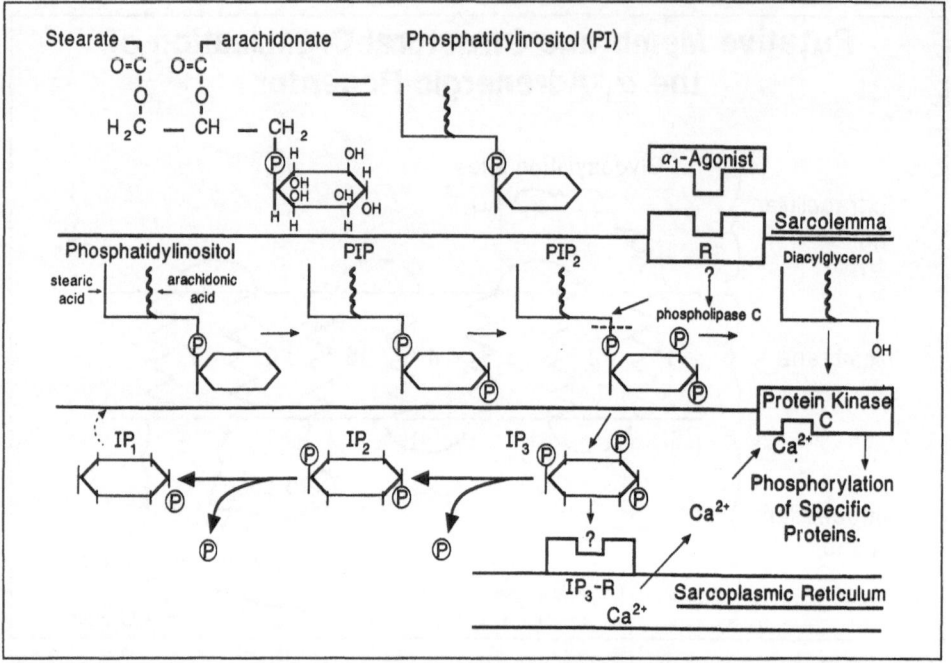

Fig. 2. Pathways for the formation of inositol, 1,4,5-trisphosphate (IP$_3$) and diacylglycerol (DG) as second messengers mediating the effects of α_1-adrenergic stimulation in cardiac tissue. Phosphatidylinositol 4,5-bisphosphate (PIP$_2$) is formed by sequential phosphorylation of phosphatidylinositol (PI) and phosphatidylinositol 4-phosphate (PIP). After binding of the α_1-agonist to the receptor (R), membrane-bound phospholipase C is activated and produces the second messengers IP$_3$ and diacylglycerol from PIP$_2$. Subsequently, IP$_3$ binds to a receptor on the sarcoplasmic reticulum and enhances Ca^{2+} release. The action of IP$_3$ is curtailed by sequential dephosphorylations that remove phosphate groups to form inositol 1,4-bisphosphate (IP$_2$) and inositol 1-phosphate (IP$_1$). Protein kinase C is activated by diacylglycerol and an increased Ca^{2+} concentration, resulting in increased phosphorylation of specific proteins. (Reproduced from [26] with the permission of Futura Publishing Co.)

extracellular space [68]. However, the positive inotropic effect of α-adrenergic stimulation in rabbit heart is not associated with an increase in the voltage-dependent, slow inward current carried by Ca^{2+} [32].

More recent evidence suggests that the α_1-adrenergic receptor may be coupled to intramembranous proteins other than phospholipase C. For example, the α_1-adrenergic receptor may also be coupled to phospholipase A$_2$ (PLA$_2$) which cleaves phosphatidylcholine (PC) to produce lysophosphatidylcholine (LPC) and a free fatty acid, in particular, arachidonic acid [8, 72]. In heart tissue, at least 50% of the phosphatidylcholine is plasmalogen PC (possessing a vinyl ether linkage to the sn-1 carbon of glycerol) which contains the majority of arachidonic acid in myocardium. This coupling of the α_1-adrenergic receptor to phospholipase A$_2$ may be critical, not only in the release of arachidonic acid, but also the accumulation of LPC, an event known to occur in ischemic myocardium [10]. Alpha$_1$-adrenergic receptors have also been shown to be coupled to phospholipase D, which results in cleavage of PC to phosphatidic acid, and which may also increase intracellular calcium due to enhanced transsarcolemmal flux [25]. Finally, α_1-adrenergic receptors have also been shown to be linked to cyclic nucleotide phosphodiesterase, which would be expected to decrease the levels of cyclic AMP, thereby also down-modulating the β-adrenergic receptor pathway [9]. Based on several lines of evidence, one could hypothesize that the α_{1a}-adrenergic receptor is predominately responsible for entry of calcium across the sarcolemma in cardiac tissue and that the α_{1b}-adrenergic receptor is responsible for increasing 1,4,5-IP$_3$, resulting in the release of intracellular calcium from stores within the sarcoplasmic reticulum [25]. Thus, the α_1-adrenergic receptor not only exists in at least two different subtypes, but the same receptor, either α_{1a} or α_{1b}, may be involved in a variety of intracellular and transsarcolemmal events by virtue of variations in intracellular coupling, all of which could contribute to its physiologic and pathophysiologic effects.

Regulation of α-receptors in the ischemic heart

There is a significant amount of experimental evidence to suggest that β-adrenergic receptors can be up- and down-regulated in response to a variety of perturbations. This occurs secondary to phosphorylation of the surface receptor by a specific kinase, β-adrenergic receptor kinase (BARK), followed by internalization and recycling of the receptors, prior to their return to the surface membrane [46]. In contrast, the α_1-adrenergic receptor appears to be modulated differently, particularly in cardiac tissue. Although the receptor can be phosphorylated, as pointed out above [44], and its density at surface membrane sites can be altered in response to agonists and in response to pathophysiologic conditions, it does not appear to cycle through a pathway analogous to the β-adrenergic receptor (see below).

In 1981, we reported that the number of α_1-adrenergic receptors increased two-fold in ischemic myocardium [13]. The time-course of increase in α_1-adrenergic receptors correlated with an increase α_1-adrenergic responsivity assessed electrophysiologically [69]. During subsequent reperfusion of the ischemic myocardium, there was a reversal of the increase in α_1-adrenergic receptors to control values with a time-course which was nearly identical to the decrease in α_1-adrenergic responsivity in vivo [13, 69]. Subsequent studies by several groups also indicated that α_1-adrenergic receptors increased with ischemia in the cat [13, 17], dog [52], and guinea pig [49], but in the rat heart both an increase [3] and no change [17] have been reported. Maisel and colleagues reported that ischemia in the guinea pig heart was associated with an increase in both β- and α-adrenergic receptors in the sarcolemma [49]. The increase in β-adrenergic receptors in the sarcolemma was associated with a corresponding decrease in the density of β-receptors in the light vesicle or intracellular fraction, suggesting a translocation of the receptor to the surface plasma membrane [49]. These results suggested that the β-adrenergic receptor was transported from an intracellular light vesicle fraction to the surface sarcolemma in response to ischemia and was coupled to an increase in intracellular cyclic AMP in response to β-adrenergic stimulation [49]. In the case of the β-adrenergic receptor, the increase in the sarcolemma with ischemia was reversible if reperfusion was performed within 60 min of ischemia. In contrast, under control conditions, the sarcolemma appeared to contain only approximately 20% of the measured total α_1-adrenergic receptors [49]. Although the α_1-adrenergic receptors increased in the sarcolemmal fraction during ischemia, there was no change in the density of α_1-adrenergic receptors in the intracellular light vesicle fraction [49]. These findings indicate that the pathway for α_1-adrenergic receptor trafficking in response to ischemia was markedly different than that for the β-adrenergic receptor. This conclusion is also supported by findings in other cell systems where the trafficking for the α- and β-adrenergic receptors differed [33, 77]. In concert, these findings suggested that in response to ischemia, the α_1-adrenergic receptor was originating from a different intracellular site than the light vesicle fraction or, alternatively, that the α_1-adrenergic receptor was present in the sarcolemma, even under control normoxic conditions, but that the binding site to the α_1-adrenergic receptor was being exposed through some mechanism occurring in the ischemic heart.

Mechanisms underlying the changes in α₁-adrenergic receptors in ischemic myocardium

The cellular mechanisms responsible for the changes in α_1-adrenergic receptors and their increased responsivity in the ischemic heart has not been delineated with certainty. However, recent insights suggest that the cellular mechanisms may depend not only on changes in the density of α_1-adrenergic receptors, but also on their coupling to intracellular events in the ischemic heart. An understanding of the cellular events responsible for the change in α-adrener-

Table 1. Studies indicating antiarrhythmic effects of α_1-adrenergic blockade

Study	Species	Agent	Effectiveness
1. Sheridan et al. J Clin Invest 65:161, 1980 [69]	cat	phentolamine prazosin	VT,VF
2. Davey J Cardiovasc Pharm 2:287, 1980 [16]	cat	prazosin	VT, VF
3. Stewart et al. J Cardiovasc Pharm 2:77, 1980 [74]	dog	phentolamine	VT
4. Williams et al. Am J Cardiol 49:1046, 1982 [82]	dog	phentolamine	VT, VF
5. Thandroyen et al. J Am Coll Cardiol 1:1056, 1983 [76]	rat	prazosin	VT, VF
6. Benfey et al. Br J Pharmacol 82:717, 1984 [5]	dog, pig	prazosin	VT, VF
7. Kane et al. Br J Pharmacol 82:349, 1984 [36]	rat	prazosin	VT, VF
8. Wilber et al. J Cardiovasc Pharm 10:96, 1987 [81]	conscious dog	prazosin	VF
9. Schwartz et al. Am Heart J 109:937, 1985 [65]	cat	prazosin	PVCs, VT
10. Culling et al. J Mol Cell Cardiol 19:251, 1987 [15]	guinea pig	phentolamine	VT, VF'

gic responsivity in the ischemic heart is also likely to provide an understanding of the mechanisms underlying the antiarrhythmic efficacy of α_1-adrenergic blockade in the ischemic heart (see Table 1 for summary, and [5, 15, 16, 36, 65, 69, 74, 76, 81, 82]). Although the number of α_1-adrenergic receptors increases in the ischemic region in the cat, dog, guinea pig, and in one study in the rat, all of the reports to date have evaluated alterations in adrenergic receptors after homogenization of tissue derived from the ischemic region [3, 13, 17, 49, 52]. A major problem with experiments using these types of membrane preparations is that the membrane fractions would be derived from both myocytes and nonmyocytic cells. Furthermore, it is likely that during preparation of the membrane fractions, changes that occurred in intact tissue may partially or completely revert. This could explain the lack of a measured increase in α_1-adrenergic receptor density until 30 min of ischemia, despite the fact that enhanced α_1-adrenergic responsivity can occur more rapidly during ischemia in vivo [13, 69]. To elucidate the mechanisms responsible for the changes in α_1-adrenergic receptors during ischemia, we developed an isolated adult canine myocyte preparation in which hypoxia could be induced that was sufficient to inhibit β-oxidation of fatty acids analogous to that seen during the first few minutes of ischemia in vivo. In this preparation, hypoxia of 10 min duration at 37°C resulted in nearly a three-fold increase in surface α_1-adrenergic receptors with no change in receptor affinity. This increase was completely reversible with reoxygenation and occurred independent of any evidence of irreversible cell damage. In fact, if irreversible cell damage occurred, as was evidenced by release of intracellular enzymes including creatine kinase and lactate dehydrogenase, there was no net increase in surface α_1-adrenergic receptors [29]. Therefore, a transient yet marked exposure of the α_1-adrenergic receptor occurred under hypoxic conditions, analogous to that seen during ischemia in vivo. This model system has been used to probe the underlying mechanisms contributing to the changes in α_1-adrenergic receptors in response to ischemia.

Prior to considering the mechanisms underlying the changes in α_1-adrenergic receptors in the ischemic heart, it is important to first review briefly the metabolism of fatty acids in the myocardium and how this is altered in response to ischemia. Fatty acids constitute the major metabolic fuel of the myocardium. Oxidation of fatty acids accounts for 60% to 80% of the total energy requirements of the myocardium. Fatty acid metabolism in myocytes begins with the uptake of free fatty acids by the cell. Extracellularly in blood, fatty acids are bound primarily to albumin, the bound form existing in equilibrium with a small amount of free fatty acids. This free, unbound fatty acid is capable of entering the myocyte by passive diffusion and, possibly, by an active carrier mechanism. Once within the cytoplasm, the fatty acid is bound to a fatty acid binding protein [22]. As a result of binding to this cytosolic protein, free fatty acid concentration in the cytosol does not rise significantly and an inward concentration gradient is maintained to permit further uptake of free fatty acids from the extracellular space.

Free fatty acids in the cell are converted in the sarcoplasmic reticulum and outer mitochondrial membrane to acyl-CoA thioesters by acyl-CoA synthetase [55]. The inner mitochondrial membrane is impermeable to acyl-CoA and, as a result, the acyl-CoA thioesters are unable to enter directly the mitochondrial matrix where the enzymes responsible for β-oxidation are located. To overcome this diffusion barrier, a specific transport system has evolved in which fatty acids are transported to the mitochondrial matrix in the form of O-acylcarnitine esters, or long-chain acylcarnitines. The first step in the translocation process occurs at the outer portion of the inner mitochondrial membrane and involves the conversion of acyl-CoA to long-chain acylcarnitines and free CoA catalyzed by carnitine acyltransferase I (CAT-I). The long-chain acylcarnitine is then translocated across the inner mitochondrial membrane to the mitochondrial matrix by a specific transport mechanism in exchange for free (unesterified) carnitine. This reaction is mediated by carnitine:acylcarnitine translocase. Long-chain acylcarnitines are then transesterified back to acyl-CoA and free carnitine on the matrix side of the inner mitochondrial membrane by carnitine acyltransferase II (CAT-II), thereby completing the process of translocation. The acyl-CoA product within the mitochondrial matrix is then available for β-oxidation, which results in sequential cleavage of two carbon acetyl-CoA units and production of the reduced forms of nicotinamide and flavin adenine dinucleotides (NADH and $FADH_2$, respectively). Both NADH and $FADH_2$ must then be oxidized through the electron transport chain within the mitochondria. The terminal enzyme of electron transport, cytochrome oxidase, has an absolute requirement for molecular oxygen. Under conditions of ischemia or hypoxia, when oxygen delivery to the mitochondria is reduced severely, electron transport is inhibited, resulting in a marked increase in $[FADH_2]$ and [NADH]. Flux through the β-oxidation pathway is inhibited by the accumulation of $FADH_2$ and NADH through a negative feedback mechanism resulting in a marked increase in the acyl-CoA and long-chain acylcarnitine content of the myocardium [11, 34, 47, 70, 80]. The increase in acyl-CoA is confined primarily to the mitochondrial matrix during early ischemia and would, therefore, not have direct access to the sarcolemma [34]. In contrast, long-chain acylcarnitine accumulates primarily in the cytosol and may, therefore, have access to the sarcolemma [34].

We have demonstrated that hypoxia in isolated myocytes results in a marked yet reversible increase in the amphiphile long-chain acylcarnitine [39] analogous to the increase in this amphiphile that occurs during ischemia in vivo. Using electron microscopic autoradiography of cells prelabeled with ^{3}H-carnitine, we have shown that endogenous long-chain acylcarnitines increase up to 70-fold in the sarcolemma of isolated myocytes in response to hypoxia (Fig. 3). Since long-chain acylcarnitines are amphiphiles that can perturb a variety of membrane systems, we hypothesized that long-chain acylcarnitines might alter the movement or orientation of the α_1-adrenergic receptor protein within the sarcolemma and, thereby, the receptor density, measured with an extracellular surface ligand.

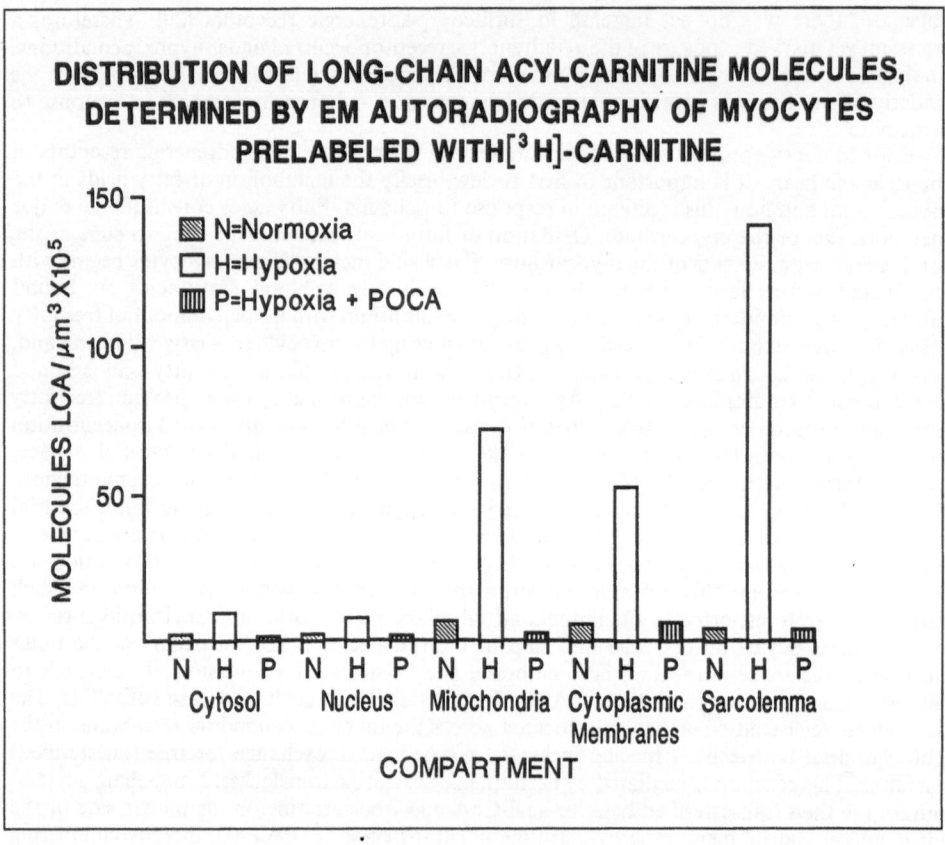

Fig. 3. The number of long-chain acylcarnitine (LCA) molecules per μm^3 calculated from grain densities of electron microscopic autoradiographs after tissue processing of [^{3}H]-carnitine prelabeled myocytes and subsequent removal of acid-soluble carnitine. The myocytes were exposed to normoxia (N), hypoxia (H) or hypoxia plus POCA (P). Hypoxia resulted in an increased number of LCA molecules/μm^3 in the cytosol, nucleus, mitochondria, cytoplasmic membranes, and particularly in the sarcolemma. The presence of the carnitine acyltransferase I inhibitor, POCA (10 μM) completely inhibited the increase in LCA. (Data used for this figure were derived from [39] with permission of the American Heart Association.)

Inhibition of the accumulation of long-chain acylcarnitines by pretreatment of myocytes with POCA, a specific inhibitor of carnitine acyltransferase I, not only prevented the endogenous accumulation of long-chain acylcarnitines, but also the increase in α_1-adrenergic receptor density (Fig. 4). To assess whether the effect of POCA was due specifically to inhibition of long-chain acylcarnitine production, normoxic myocytes were exposed to low levels of exogenous long-chain acylcarnitine [29]. Palmitoyl carnitine evoked an increase in α_1-adrenergic receptors, even in normoxic myocytes, in the presence or absence of POCA [29]. Similar results have been reported in the intact rat heart exposed to global ischemia wherein pretreatment with POCA or another inhibitor of CAT-I, 2-tetradecylglycidic acid (TGDA) also prevented the increase in α_1-adrenergic receptors [3]. These results would suggest that the accumulation of long-chain acylcarnitines within the sarcolemma may be one major mechanism responsible for the measured increase in α_1-adrenergic receptors in response to ischemia. Since the incorporation of long-chain acylcarnitines into the sarcolemma can alter membrane fluidity [21], this may contribute to the increase in accessability of the α_1-adrenergic receptor to the surface of the cell membrane and, thereby, to the ligand.

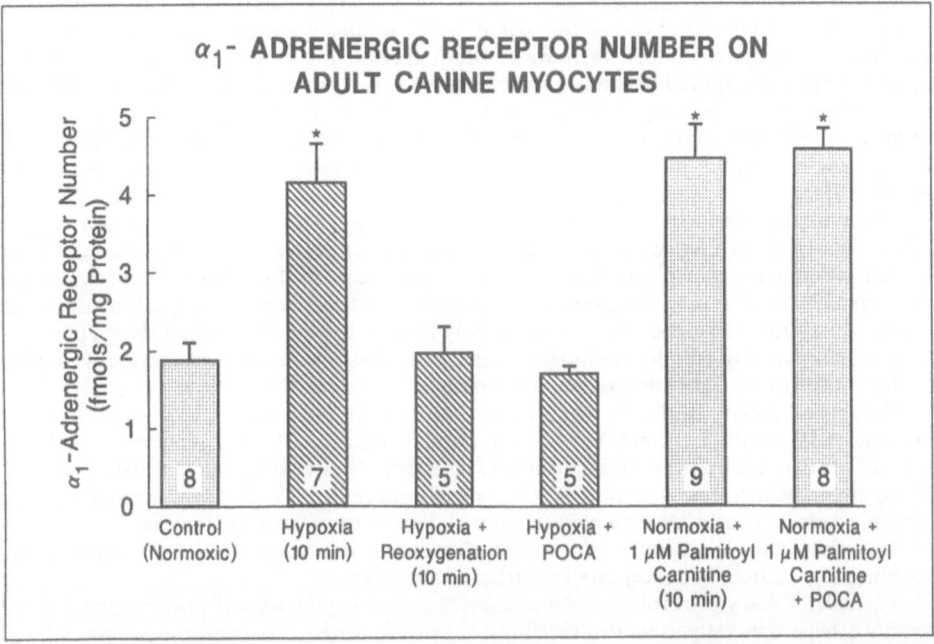

Fig. 4. Effect of hypoxia (10 min at 37°C), hypoxia followed by reoxygenation (10 min) or incubation with exogenous palmitoyl carnitine (1 μM) on α_1-adrenergic receptor number, expressed as fmols/mg protein, on isolated adult canine myocytes in the presence or absence of POCA. Values are given for control (normoxic) myocytes (left), myocytes exposed to hypoxia (10 min), hypoxia plus reoxygenation (10 min) and hypoxia (10 min) in the presence of POCA. Data for normoxic myocytes incubated with palmitoyl carnitine (1 μM) or normoxic myocytes with palmitoyl carnitine (1 μM) in the presence of POCA are also given in the last two bars. (Data used for this figure were derived from [29] with permission of the American Heart Association.)

Alterations in coupling of the α_1-adrenergic receptor under hypoxic conditions

Although these results demonstrate that hypoxia results in a marked increase in surface α_1-adrenergic receptors on isolated adult myocytes, the question of whether the exposed receptors are coupled to intracellular events and their overall relation to arrhythmogenesis in the ischemic heart remain. As indicated above, one of the major mechanisms for intracellular coupling for the α_1-adrenergic receptor in many tissue types involves the hydrolysis of membrane PIP$_2$ to produce 1,4,5-IP$_3$ and diacylglycerol (Fig. 2). IP$_3$ can release calcium directly from the sarcoplasmic reticulum and diacylglycerol can stimulate protein kinase C which can phosphorylate a variety of intracellular enzymes critical in increasing the transsarcolemmal movement of calcium (Fig. 2). The question which we have addressed recently was whether hypoxia altered the response of these isolated myocytes to the production of 1,4,5-IP$_3$.

Previous studies using ^{3}H-inositol prelabeling of isolated cardiac myocytes have suggested that α_1-adrenergic receptors are indeed coupled to an increase in IP$_3$. However, several methodologic problems associated with the lack of uniform prelabeling of PIP$_2$ with ^{3}H-inositol do not permit precise characterization of the absolute time course of the second messenger response. To circumvent these methodologic problems, we developed a procedure to measure directly the actual mass of individual inositol phosphates (IP$_1$, IP$_2$, IP$_3$ and IP$_4$) after short intervals of norepinephrine stimulation of isolated myocytes under normoxic and hypoxic conditions [30]. Using these procedures, we demonstrated that IP$_3$ increased four-fold within 30 s of stimulation with norepinephrine, returning to baseline levels by 60 s with subsequent sequential increases in IP$_2$ and IP$_1$ [31]. In addition, stimulation of these cells with norepinephrine also results in a marked but transient increase in IP_4 [27]. When these isolated adult canine myocytes were exposed to hypoxia for 10 min, which results in a two- to three-fold increase in α_1-adrenergic receptor number, there was a 100-fold reduction in the threshold concentration of norepinephrine required to elicit an increase in IP_3 [31]. Likewise, the effective concentration of norepinephrine to produce a 50% increase in IP_3 also decreased eight-fold in the presence of hypoxia [31]. Importantly, both in normoxic and hypoxic myocytes, the initial increase in IP_3 was exclusively the physiologically active isomer, $1,4,5$-IP_3, as determined using HPLC methods [31].

These findings indicate that the α_1-adrenergic receptor on myocytes is in fact coupled intracellularly to an increase in IP_3 and that hypoxia results in a marked reduction in the concentration of norepinephrine required to elicit an increase in IP_3. The data also suggest that the increase in α_1-adrenergic responsivity in the ischemic heart is not simply a function of the increase in α_1-adrenergic receptors, which is only two- to three-fold, but that there appears to be an enhancement of coupling to intracellular events, possibly mediated by increased coupling to the G-protein for activation of phospholipase C.

Recent preliminary studies, in which left stellate nerve stimulation in vivo leads to a selective increase in IP_3 in the ischemic, but not the normal region of the left ventricular wall, also support the concept that this mechanism may be operative in vivo as well [28]. The increase in IP_3 in the ischemic region in vivo was completely prevented by the α_1-adrenergic receptor blocking agent prazosin, or by pretreatment with POCA to prevent the increase in long-chain acylcarnitines [28]. These preliminary results provide the first biochemical data demonstrating coupling of α_1-adrenergic receptors in cardiac tissue in vivo.

The role of diacylglycerol as a second messenger for stimulation of protein kinase C will require extensive investigation. Protein kinase C exists in cardiac tissue in at least four different subforms [84] and has been shown to phosphorylate a variety of intracellular proteins [68]. Findings by Lindemann and colleagues have demonstrated that α_1-adrenergic stimulation of myocardial tissue leads to phosphorylation of a 15 KDa sarcolemmal protein, likely mediated through protein kinase C, and appears to result in an increase in intracellular calcium [48, 59]. Preliminary evidence suggests that norepinephrine may stimulate protein kinase C in myocytes only if the cell is depolarized [35], as would occur in ischemic tissue in vivo. It is possible that during myocardial ischemia modulation of the activity of protein kinase C may be involved in the increase in α_1-adrenergic receptors. Kushida and colleagues [43] have demonstrated in rabbit papillary muscle that stimulation of protein kinase C with phorbol esters actually decreased the positive inotropic response to α_1-adrenergic stimulation and decreased the extent of binding of ^{3}H-prazosin in membrane fractions isolated from this tissue. This finding is consistent with the known phosphorylation of the α_1-adrenergic receptor by protein kinase C [44] which may thereby result in sequestration of the receptor to an intramembranous or intracellular site. Since LPC has been shown to inhibit protein kinase C at higher concentrations [2] and accumulates within minutes in ischemic myocardium in vivo [11], this mechanism may, in part, mediate the increase or lack of physiologic decrease in α_1-adrenergic receptors in response to agonist stimulation in the ischemic heart in vivo.

Potential arrhythmogenic effects of α_1-adrenergic stimulation

Under normal physiologic conditions, α_1-adrenergic stimulation results in electrophysiologic effects which are primarily antiarrhythmic as opposed to arrhythmogenic. These include a decrease in normal automaticity, an increase in the fibrillation threshold, a lengthening in repolarization time and effective refractory period and, under some conditions, hyperpolarization of the resting membrane potential. The increase in repolarization time or action potential duration in response to α_1-adrenergic stimulation has been shown to be mediated by a decrease in the amplitude of the outward K^+ currents (I_{out}) without effect on the voltage gated Ca^{2+} current (I_{si}) [4, 20]. This effect on I_{out} appears to be mediated by activation of protein kinase C since it also occurs in response to treatment with phorbol esters [4]. Although this effect of α_1-adrenergic stimulation to increase the repolarization time may be considered antiarrhythmic, if the effect is heterogenous due to differences in sympathetic innervation it could lead to an increase in the heterogeneity of repolarization across the ventricle, thereby contributing to arrhythmogenesis. Alternatively, a decrease in outward repolarizing currents in response to α_1-adrenergic stimulation could permit the development of early afterdepolarizations and triggered rhythms. In addition, the increase in action potential duration could permit intracellular calcium levels to rise sufficiently to induce the transient inward current (I_{Ti}) carried by either electrogenic Na^+/Ca^{2+} exchange or the nonspecific inward Na^+ current, and, thereby, lead to the development of delayed afterdepolarizations and triggered rhythms [83]. We have recently demonstrated that during ischemia and, particularly, during subsequent reperfusion in vivo, nonreentrant or focal mechanisms contribute importantly to the development of lethal arrhythmias, including ventricular tachycardia and fibrillation [56, 57, 58]. Although the precise mechanisms underlying this nonreentrant activity are unknown, their occurrence is consistent with a triggered rhythm.

At least two additional mechanisms may also contribute to the arrhythmogenic effects of α_1-adrenergic stimulation, particularly under ischemic conditions. First, the increase in IP_3 which leads to release of intracellular Ca^{2+} from sarcoplasmic reticulum may be insufficient under normal physiologic conditions to activate I_{Ti} and, thereby, induce triggered rhythms. However, under ischemic conditions, when intracellular Ca^{2+} may rise or the sarcoplasmic reticulum may have a reduced capacity to sequester the released Ca^{2+}, this could lead to a greater and more prolonged increase in intracellular Ca^{2+} sufficient to activate I_{Ti} and, thereby, produce either early or delayed afterdepolarizations and triggered rhythms. This conclusion is supported by recent preliminary data from our laboratory which indicate that under hypoxic conditions in vitro in isolated adult canine myocytes in which intracellular transmembrane action potentials are recorded simultaneously, that norepinephrine in the presence of the β-adrenergic blocking agent nadolol can elicit both delayed afterdepolarizations and triggered rhythms during reoxygenation [60]. These results indicate an enhanced α_1-adrenergic responsivity during hypoxia since this arrhythmogenic effect never occurred in normoxic cells in response to α_1-adrenergic stimulation. Analogous findings have recently been shown in isolated cardiac tissue exposed to a simulated ischemic environment consisting of hypoxia, acidosis, and hyperkalemia [24], as well as in Purkinje fibers isolated from areas underlying a previous infarct [38]. These results are also consistent with previous findings from our laboratory and others indicating that α_1-adrenergic blockade can attenuate or prevent the increase in intracellular calcium occurring during reperfusion of reversibly injured tissue [53, 67]. Future investigations will be required to ascertain the precise changes in intracellular Ca^{2+} occurring during ischemia or hypoxia, the subcellular biochemical events and the precise ionic channels being activated in response to α_1-adrenergic stimulation as well as the specific subtypes of the receptors involved. Since protein kinase C can directly modulate K^+ and Ca^{2+} ion channel activity in a number of tissues including heart [68], the relative effects of α_1-adrenergic stimulation mediated through IP_3, as opposed to protein kinase C, will also require clarification.

Rationale for the presence of α_1-adrenergic receptors in cardiac myocytes

Although β-adrenergic receptors are the primary mediator of the effects of catecholamines in cardiac myocytes under physiologic conditions, one could hypothesize that the α_1-adrenergic receptor system exists as a secondary or reserve inotropic system in the heart. Under pathophysiologic conditions, when the inotropic state of the myocardium is reduced, the density and intracellular coupling of α_1-adrenergic receptors may increase. Although the α_1-adrenergic receptor density appears to increase in response to myocardial ischemia and hypoxia, the increase has also been shown to occur in experimental congestive heart failure [37], cardiomyopathy in man [78], and by some investigators in response to hypothyroidism [66]. The increase in receptor number and coupling could augment the inotropic state of the myocyte by increasing intracellular calcium via intracellular pathways independent of β-adrenergic receptor coupling to adenylate cyclase. Although an increase in contractile state could be beneficial, the arrhythmogenic effects of increasing intracellular calcium may be devastating. Further studies will be required to assess the role of protein kinase C, as well as the potential coupling of the α_1-adrenergic receptor to other phospholipases including phospholipase A_2 and phospholipase D. Studies will also be required to define precisely the intracellular or intramembranous trafficking of the α_1-adrenergic receptor, particularly under pathophysiologic conditions. One approach could be the use of electron microscopic autoradiography of the α_1-adrenergic receptor to assess its movement under hypoxic and ischemic conditions and more precisely define the mechanisms contributing to the changes in receptor number. Elucidation of the individual G-proteins responsible for coupling of the various α-adrenergic receptor subtypes will be critical in designing new modalities to selectively modify the response to stimulation. Finally, these results would suggest that more extensive evaluation of the efficacy of α_1-adrenergic antagonists for the treatment and prevention of sudden cardiac death in patients with ischemic heart disease and chronic heart failure may be warranted.

References

1. Ahlquist RP (1948) A study of adrenotropic receptors. Am J Physiol 153:586–600
2. Aishi K, Raynor RL, Charp PA, Kuo JF (1988) Regulation of protein kinase C by lysophospholipids. Potential role in signal transduction. J Biol Chem 263:6865–6871
3. Allely MC, Brown CM (1988) The effects of POCA and TGDA on the ischaemia-induced increase in α_1-adrenoceptor density in the rat left ventricle. Br J Pharm 95:705P (abstr.)
4. Apkon M, Nerbonne JM (1988) α_1-Adrenergic agonists selectively suppress voltage-dependent K^+ currents in rat ventricular myocytes. Proc Natl Acad Sci USA 85:8756–8760
5. Benfey BG, Elfellah MS, Ogilvie RI, Varma DR (1984) Anti-arrhythmic effects of prazosin and propranolol during coronary artery occlusion and reperfusion in dogs and pigs. Br J Pharmacol 82:717–725
6. Berridge MJ (1987) Inositol trisphosphate and diacylglycerol: two interacting second messengers. Ann Rev Biochem 56:159–193
7. Böhm M, Schmitz W, Scholz H (1987) Evidence against a role of a pertussis toxin-sensitive guanine nucleotide-binding protein in the alpha$_1$-adrenoceptor-mediated positive inotropic effect in the heart. Naunyn Schmiedeberg's Arch Pharmacol 335:476–479
8. Burch RM, Luini A, Axelrod J (1986) Phospholipase A_2 and phospholipase C are activated by distinct GTP-binding proteins in response to alpha$_1$-adrenergic stimulation in FRTL5 thyroid cells. Proc Natl Acad Sci USA 83:7201–7205
9. Buxton ILO, Brunton LL (1985) Action of the cardiac alpha$_1$-adrenergic receptor. Activation of cyclic AMP degradation. J Biol Chem 260:6733–6737
10. Corr PB, Gross RW, Sobel BE (1984) Amphipathic metabolites and membrane dysfunction in ischemic myocardium. Circ Res 55:135–154
11. Corr PB, Creer MH, Yamada KA, Saffitz JE, Sobel BE (1989) Prophylaxis of early ventricular fibrillation by inhibition of acylcarnitine accumulation. J Clin Invest 83:927–936
12. Corr PB, Yamada KA, Witkowski FX (1986) Mechanisms controlling cardiac autonomic function and their relation to arrhythmogenesis. In: Fozzard HA, Haber E, Jennings RB, Katz AM, Morgan HE (eds) The Heart and Cardiovascular System: Scientific Foundations; Raven Press, New York, pp 1343–1403
13. Corr PB, Shayman JA, Kramer JB, Kipnis RJ (1981) Increased α-adrenergic receptors in ischemic cat myocardium: A potential mediator of electrophysiological derangements. J Clin Invest 67:1232–1236
14. Cotecchia S, Schwinn DA, Randall RR, Lefkowitz RJ, Caron MG, Kobilka BK (1988) Molecular cloning and expression of the cDNA for the hamster alpha$_1$-adrenergic receptor. Proc Natl Acad Sci USA 85:7159–7163
15. Culling W, Penny WJ, Cunliffe G, Flores NA, Sheridan DJ (1987) Arrhythmogenic and electrophysiological effects of alpha adrenoceptor stimulation during myocardial ischaemia and reperfusion. J Mol Cell Cardiol 19:251–258
16. Davey MJ (1980) Relevant features of the pharmacology of prazosin. J Cardiovasc Pharmacol 2 (suppl 3):S287–S301
17. Dillon JS, Gu XH, Nayler WG (1988) Alpha$_1$ adrenoceptors in the ischaemic and reperfused myocardium. J Mol Cell Cardiol 20:725–735
18. Dohlman HG, Bouvier M, Benovic JL, Caron MG, Lefkowitz RJ (1987) The multiple membrane spanning topography of the β_2-adrenergic receptor: localization of the sites of binding, glycosylation and regulatory phosphorylation by limited proteolysis. J Biol Chem 262:14282–14288

19. Exton JH (1985) Mechanisms involved in alpha-adrenergic phenomena. Am J Physiol 248:E633–E647
20. Fedida D, Shimoni Y, Giles WR (1989) A novel effect of norepinephrine on cardiac cells is mediated by α_1-adrenoceptors. Am J Physiol 256:H1500–H1504
21. Fink KL, Gross RW (1984) Modulation of canine myocardial sarcolemmal membrane fluidity by amphiphilic compounds. Circ Res 55:585–594
22. Fournier NC, Zuker M, Williams RE, Smith ICP (1983) Self-association of the cardiac fatty acid binding protein. Influence on membrane-bound, fatty acid-dependent enzymes. Biochemistry 22:1863–1872
23. Gilman AG (1987) G proteins: transducers of receptor-generated signals. Ann Rev Biochem 56:615–649
24. Hamra M, Rosen MR (1988) α-Adrenergic receptor stimulation during simulated ischemia and reperfusion in canine cardiac Purkinje fibers. Circulation 78:1495–1502
25. Han G, Abel PW (1987) Alpha$_1$-adrenoceptor subtypes linked to different mechanisms for increasing Ca^{++} in smooth muscle. Nature 325:333–335
26. Heathers GP, Yamada KA, Pogwizd SM, Corr PB (1988) The contribution of α- and β-adrenergic mechanisms in the genesis of arrhythmias during myocardial ischemia and reperfusion. In: Kulbertus HE, Franck G (eds) Neurocardiology; Futura Publishing, Mount Kisco, New York, pp 143–178
27. Heathers GP, Corr PB, Rubin LJ (1988) Transient accumulation of inositol (1,3,4,5)-tetrakisphosphate in response to α_1-adrenergic stimulation in adult cardiac myocytes. Biochem Biophys Res Commun 156:485–492
28. Heathers GP, Evers AS, Corr PB (1988) Sympathetic stimulation selectively increases inositol trisphosphate levels in ischemic myocardium in vivo. Circulation 78 (suppl II):II-460 (abstr.)
29. Heathers GP, Yamada KA, Kanter EM, Corr PB (1987) Long-chain acylcarnitines mediate the hypoxia-induced increase in α_1-adrenergic receptors on adult canine myocytes. Circ Res 61:735–746
30. Heathers GP, Juehne T, Rubin LJ, Corr PB, Evers AS (1989) Anion exchange chromatographic separation of inositol phosphates and their quantification by gas chromatography. Anal Biochem 176:109–116
31. Heathers GP, Evers AS, Corr PB (1989) Enhanced inositol trisphosphate response to α_1-adrenergic stimulation in hypoxic cardiac myocytes. J Clin Invest 83:1409–1413
32. Hescheler J, Nawrath H, Tang M, Trautwein W (1988) Adrenoceptor-mediated changes of excitation and contraction in ventricular heart muscle from guinea-pigs and rabbits. J Physiol 397:657–670
33. Hughes RJ, Insel PA (1986) Agonist-mediated regulation of alpha$_1$- and beta$_2$-adrenergic receptor metabolism in a muscle cell line, BC3H1. Mol Pharmacol 29:521–530
34. Idell-Wenger JA, Grotyohann LW, Neely JR (1978) Coenzyme A and carnitine distribution in normal and ischemic hearts. J Biol Chem 253:4310–4318
35. Kaku T, Lakatta E, Filburn CR (1986) Effect of α_1-adrenergic stimulation on phosphoinositide metabolism and protein kinase C (PK-C) in rat cardiomyocytes. Fed Proc 45:209 (abstr.)
36. Kane KA, Parratt JR, Williams FM (1984) An investigation into the characteristics of reperfusion-induced arrhythmias in the anesthetized rat and their susceptibility to antiarrhythmic agents. Br J Pharmacol 82:349–357
37. Karliner JS, Barnes P, Brown M, Dollery C (1980) Chronic heart failure in the guinea pig increases cardiac α$_1$- and β-adrenoceptors. Eur J Pharmacol 67:115–118
38. Kimura S, Bassett AL, Kohya T, Kozlovskis PL, Myerburg RJ (1987) Automaticity, triggered activity, and responses to adrenergic stimulation in cat subendocardial Purkinje fibers after healing of myocardial infarction. Circulation 75:651–660
39. Knabb MT, Saffitz JE, Corr PB, Sobel BE (1986) The dependence of electrophysiological derangements on accumulation of endogenous long-chain acylcarnitine in hypoxic neonatal rat myocytes. Circ Res 58:230–240
40. Kobilka BK, Matsui H, Kobilka TS, Yang-Feng TL, Francke U, Caron MG, Lefkowitz RJ, Regan JW (1987) Cloning, sequencing, and expression of the gene coding for the human platelet alpha$_2$-adrenergic receptor. Science 238:650–656
41. Kobilka BK, Kobilka TS, Daniel K, Regan JW, Caron MG, Lefkowitz RJ (1988) Chimeric α_2-β_2-adrenergic receptors: delineation of domains involved in effector coupling and ligand binding specificity. Science 240:1310–1316
42. Kohl C, Schmitz W, Scholz H, Scholz J, Toch M, Doring V, Kalmar P (1989) Evidence for α_1-adrenoceptor-mediated increase of inositol trisphosphate in the human heart. J Cardiovasc Pharm 13:324–327
43. Kushida H, Hiramoto T, Satoh H, Endoh M (1988) Phorbol ester does not mimic, but antagonizes the alpha-adrenoceptor-mediated positive inotropic effect in the rabbit papillary muscle. Naunyn-Schmiedeberg's Arch Pharmacol 337:169–176

44. Leeb-Lundberg LMF, Cotecchia S, Lomasney JW, DeBernardis JF, Lefkowitz RJ, Caron MG (1985) Phorbol esters promote α_1-adrenergic receptor phosphorylation and receptor uncoupling from inositol phospholipid metabolism. Proc Natl Acad Sci USA 82:5651–5655

45. Lefkowitz RJ, Kobilka BK, Caron MG (1989) The new biology of drug receptors. Biochem Pharm 38:2941–2948

46. Lefkowitz RJ, Kobilka BK, Benovic JL, Bouvier M, Cotecchia S, Hausdorff WP, Dohlman HG, Regan JW, Caron MG (1988) Molecular biology of adrenergic receptors. Cold Spring Harbor Symposia on Quantitative Biology 53:507–514

47. Liedtke AJ, Nellis S, Neely JR (1978) Effects of excess free fatty acids on mechanical and metabolic function in normal and ischemic myocardium in swine. Circ Res 43:652–661

48. Lindemann JP (1986) α-Adrenergic stimulation of sarcolemmal protein phosphorylation and slow responses in intact myocardium. J Biol Chem 261:4860–4867

49. Maisel AS, Motulsky HJ, Ziegler MG, Insel PA (1987) Ischemia- and agonist-induced changes in α- and β-adrenergic receptor traffic in guinea pig hearts. Am J Physiol 253:H1159–H1167

50. Malfatto G, Rosen TS, Steinberg SF, Ursell PC, Sun LS, Daniel S, Danilo P Jr, Rosen MR (1990) Sympathetic neural modulation of cardiac impulse initiation and repolarization in the newborn rat. Circ Res 66:427–437

51. Minneman KP (1988) Alpha$_1$-adrenergic receptor subtypes, inositol phosphates, and sources of cell Ca^{++}. Pharmacol Rev 40:87–119

52. Mukherjee A, Hogan M, McCoy K, Buja LM, Willerson JT (1980) Influence of experimental myocardial ischemia on alpha$_1$-adrenergic receptors. Circulation 62 (Suppl III):149 (abstr.)

53. Nayler WG, Gordon M, Stephens DJ, Sturrock WJ (1985) The protective effect of prazosin on the ischaemic and reperfused myocardium. J Mol Cell Cardiol 17:685–699

54. Neer EJ, Clapham DE (1988) Role of G protein subunits in transmembrane signalling. Nature 333:129–133

55. Oram JF, Wenger JI, Neely JR (1975) Regulation of long chain fatty acid activation in heart muscle. J Biol Chem 250:73–78

56. Pogwizd SM, Corr PB (1987) Reentrant and nonreentrant mechanisms contribute to arrhythmogenesis during early myocardial ischemia: Results using three-dimensional mapping. Circ Res 61:352–371

57. Pogwizd SM, Corr PB (1987) Electrophysiologic mechanisms underlying arrhythmias due to reperfusion of ischemic myocardium. Circulation 76:404–426

58. Pogwizd SM, Corr PB (1990) Mechanisms underlying the development of ventricular fibrillation during early myocardial ischemia. Circ Res 66:672–695

59. Presti CF, Scott BT, Jones LR (1985) Identification of an endogenous protein kinase C activity and its intrinsic 15-kilodalton substrate in purified canine cardiac sarcolemmal vesicles. J Biol Chem 260:13879–13889

60. Priori SG, Yamada KA, Corr PB (1989) α-Adrenergic induced delayed afterdepolarizations in hypoxic isolated ventricular canine myocytes. Circulation 80(Suppl II):II-202 (abstr.)

61. Regan JW, Kobilka TS, Yang-Feng TL, Caron MG, Lefkowitz RJ, Kobilka BK (1988) Cloning and expression of a human kidney cDNA for an alpha$_2$-adrenergic receptor subtype. Proc Natl Acad Sci USA 85:6301–6305

62. Rosen MR, Steinberg SF, Chow Y-K, Bilezikian JP, Danilo P Jr (1988) Role of a pertussis toxin-sensitive protein in the modulation of canine Purkinje fiber automaticity. Circ Res 62:315–323

63. Schmitz W, Scholz H, Scholz J, Steinfath M, Lohse M, Puurunen J, Schwabe U (1987) Pertussis toxin does not inhibit the α_1-adrenoceptor-mediated effect on inositol phosphate production in the heart. Eur J Pharm 134:377–378

64. Scholz J, Schaefer B, Schmitz W, Scholz H, Steinfath M, Lohse M, Schwabe U, Puurunen J (1988) Alpha-1 adrenoceptor-mediated positive inotropic effect and inositol trisphosphate increase in mammalian heart. J Pharm Exp Ther 245:327–335

65. Schwartz PJ, Vanoli E, Zaza A, Zuanetti G (1985) The effect of antiarrhythmic drugs on life-threatening arrhythmias induced by the interaction between acute myocardial ischemia and sympathetic hyperactivity. Am Heart J 109:937–948

66. Sharma VK, Banerjee SP (1978) α-Adrenergic receptor in rat heart. Effects of thyroidectomy. J Biol Chem 253:5277–5279

67. Sharma AD, Saffitz JE, Lee BI, Sobel BE, Corr PB (1983) Alpha-adrenergic mediated accumulation of calcium in reperfused myocardium. J Clin Invest 72:802–818

68. Shearman MS, Sekiguchi K, Nishizuka Y (1989) Modulation of ion channel activity: A key function of the protein kinase C enzyme family. Pharm Rev 41:211–237

69. Sheridan DJ, Penkoske PA, Sobel BE, Corr PB (1980) α-Adrenergic contributions to dysrhythmia during myocardial ischemia and reperfusion in cats. J Clin Invest 65:161–171

70. Shug AL, Thomsen JH, Folts JD, Bittar N, Klein MI, Koke JR, Huth PJ (1978) Changes in tissue levels of carnitine and other metabolites during myocardial ischemia and anoxia. Arch Biochem Biophys 187:25–33
71. Sibley DR, Benovic JL, Caron MG, Lefkowitz RJ (1987) Regulation of transmembrane signaling by receptor phosphorylation. Cell 48:913–922
72. Slivka SR, Insel PA (1987) Alpha$_1$-adrenergic receptor-mediated phosphoinositide hydrolysis and prostaglandin E_2 formation in Madin-Darby canine kidney cells. Possible parallel activation of phospholipase C and phospholipase A_2. J Biol Chem 262:4200–4207
73. Steinberg SF, Drugge ED, Bilezikian JP, Robinson RB (1985) Acquisition by innervated cardiac myocytes of a pertussis toxin-specific regulatory protein linked to the alpha$_1$ receptor. Science 230:186–188
74. Stewart JR, Burmeister WE, Burmeister J, Lucchesi BR (1980) Electrophysiologic and antiarrhythmic effects of phentolamine in experimental coronary artery occlusion and reperfusion in the dog. J Cardiovasc Pharmacol 2:77–91
75. Terman BI, Slivka SR, Hughes RJ, Insel PA (1987) α_1-Adrenergic receptor-linked guanine nucleotide-binding protein in muscle and kidney epithelial cells. Mol Pharmacol 31:12–20
76. Thandroyen FT, Worthington MG, Higginson LM, Opie LH (1983) The effect of alpha- and beta-adrenoceptor antagonist agents on reperfusion ventricular fibrillation and metabolic status in the isolated perfused rat heart. J Am Coll Cardiol 1:1056–1066
77. Toews ML (1987) Comparison of agonist-induced changes in beta- and alpha$_1$-adrenergic receptors of DDT$_1$ MF$_2$ cells. Mol Pharmacol 31:58–68
78. Vago T, Bevilacqua M, Norbiato G, Baldi G, Chebat E, Bertora P, Baroldi G, Accinni R (1989) Identification of α_1-adrenergic receptors on sarcolemma from normal subjects and patients with idiopathic dilated cardiomyopathy: Characteristics and linkage to GTP-binding protein. Circ Res 64:474–481
79. Weiner N (1985) Norepinephrine, epinephrine, and the sympathomimetic amines. In: Gilman AS, Goodman LS, Rall TW, Murad F (eds); Goodman and Gilman's The Pharmacological Basis of Therapeutics; MacMillan Publishing, New York, pp 145–180
80. Whitmer JT, Idell-Wenger JA, Rovetto MJ, Neely JR (1978) Control of fatty acid metabolism in ischemic and hypoxic hearts. J Biol Chem 253:4305–4309
81. Wilber DJ, Lynch JL, Montgomery DG, Lucchesi BR (1987) α-Adrenergic influences in canine ischemic sudden death. Effects of α_1-adrenoceptor blockade with prazosin. J Cardiovasc Pharmacol 10:96–106
82. Williams LT, Guerrero JL, Leinbach RC (1982) Prevention of reperfusion dysrhythmias by selective coronary alpha-adrenergic blockade. Am J Cardiol 49:1046 (abstr.)
83. Wit AL, Rosen MR (1986) Afterdepolarizations and triggered activity. In: Fozzard HA, Haber E, Jennings RB, Katz AM, Morgan HE (eds); The Heart and Cardiovascular System, Raven Press, New York, pp 1449–1490
84. Wolf RA (1989) Cardiolipin-sensitive phospholipase C in subcellular fractions of rabbit myocardium. Am J Physiol 257:C926–C935
85. Zaza A, Kline RP, Rosen MR (1990) Effects of α-adrenergic stimulation on intracellular sodium activity and automaticity in canine Purkinje fibers. Circ Res 66:416–426

Authors' address

Dr. Peter B. Corr, Washington University School of Medicine, Cardiovascular Division – Box 8086, 660 South Euclid Avenue, St. Louis, Missouri 63110, USA

β-Adrenergic Receptors and the G_s Protein in Myocardial Ischemia and Injury[*]

A. S. Maisel, L. A. Ransnäs and P. A. Insel

Division of Cardiology, Department of Medicine and Department of Pharmacology, University of California, San Diego, and Veterans Administration Medical Center, San Diego, California, USA, and University of Goteborg, Sweden

Summary

The purpose of this study was to explore alterations in the life cycle of adrenergic receptors and the G_s protein in the heart of ischemic animals. In initial experiments left anterior descending coronary artery occlusion was performed in guinea pigs. Sarcolemmal (SL) and light vesicle (LV) (presumably intracellular) fractions were prepared. Both fractions contained a substantial number of β-adrenergic receptors and α_1-adrenergic receptors: the relative proportion of β-adrenergic receptors in LV/SL was greater than for α_1-adrenergic receptors. Myocardial ischemia produced a rapid externalization of β-adrenergic receptors from LV to SL. α_1-adrenergic receptors also increased in SL but without an apparent externalization from LV. Pretreatment of animals with either the non-selective β-antagonist propranolol or the β_1-selective antagonist atenolol increased the number of SL β-receptors and blunted the ischemia-induced increase in SL β-adrenergic receptors. Treatment with the partial agonist pindolol did not cause up-regulation of β-receptors, and did not block ischemia-induced externalization.

In the second part of this study, we have begun to examine post-receptor events in a rat model of myocardial ischemia. Ligation of the distal left main coronary artery in the rat led to an increase in SL β-receptors. As G proteins play a pivotal role in transducing receptor occupancy to activation of effector molecules, we measured levels of G_s which stimulates adenylate cyclase activity, using an ELISA technique. In rat SL the amount of α_s markedly decreased within 15 min of coronary occlusion. There was no transfer of G_s activity to the light vesicle fraction.

These studies indicate the dynamic nature of adrenergic receptors and the α_s protein in the sarcolemma in myocardial ischemia. Changes in adrenergic receptor number and in G protein expression may contribute to the altered pathophysiology of the ischemic heart.

Introduction

The response of the myocardium to neurotransmitters and hormones is critically dependent on expression of sarcolemmal receptors which initiate cellular responses to neurohormonal factors. The steady-state level of receptor expression in target cells, such as myocardial cells, represents the interplay of cellular processes involved in receptor synthesis and storage, transport and insertion into the plasma membrane (externalization), internalization and degradation (the receptor life cycle) [9].

Myocardial ischemia appears to be a setting where an altered cellular environment may produce changes in receptor expression as a consequence of alterations in the receptor life cycle.

Work from the authors' laboratories is supported by grants from NIH and the Veterans Administration.

Ischemia produces dramatic effects on the myocardium, including increased sensitivity to the arrhythmogenic and other effects of catecholamines [3, 4, 12, 15].

The heart contains α_1- and β-adrenergic receptors that regulate both rate and force of cardiac contraction, and changes in the number and affinity of these receptors have been found in various physiological, pharmacological, and disease settings. This review will focus mainly on our observations concerning the cardiac β_1-receptor [8, 10, 11]. Our working hypothesis is that alterations in the life cycle of adrenergic receptors in myocardial cells might provide insight into the altered responsiveness of the heart to catecholamines in ischemia. Additionally, we believe that a possible mechanism for the beneficial effects of β-antagonists in ischemic heart disease might also be explained by alterations in the receptor life cycle. In other studies, we have begun to examine changes in levels of the signal transducing G_s protein in myocardial ischemia.

Methods

Male Hartley guinea pigs (300–500 g, 4–6 weeks old) were used for some of the experiments. In other experiments, Sprague Dawley rats were used. Animals were anesthetized with sodium pentobarbital (30 mg/kg i.p.), intubated, and ventilated with supplemental oxygen on a Harvard respirator. A left thoracotomy was performed, the pericardium was opened, and a 6-0 prolene suture was used to ligate the left anterior descending coronary artery (guinea pig) or left main coronary artery (rat). This ligation produced cyanosis of the anterior left ventricle and evidence of myocardial ischemia and injury, as manifested by ST-segment elevation on an epicardial ECG lead sutured over the area. Ischemia was maintained for up to 90 min; in some experiments a reversible ligature was placed and then removed at various time points to allow reperfusion of the ischemic zone. Sham-operated animals were used as controls in all studies. For biochemical studies, hearts were rapidly removed, the left ventricle was separated, and attempts were made to remove the non-ischemic areas. The heart was washed in iced 50 mM Tris-HCl (pH 7.4) and then either used immediately or else frozen in liquid nitrogen (typically < 4 weeks before thawing and use in subsequent studies.)

The principal membrane fractions that were prepared were a purified sarcolemmal fraction and a light vesicle, presumably intracellular fraction. Details of the preparation are presented elsewhere [8, 11]. Briefly, minced left ventricular tissue was extracted with 750 mM NaCl, 10 mM histidine (pH 7.5) to remove contractile proteins, and was then centrifuged twice at 17 000 g with an intermediate wash with 10 mM Na HCO$_3$, 10 mM histidine (pH 7.5). The pellet was homogenized in 250 mM sucrose, 10 mM histidine, and was then centrifuged at 45 000 g for 30 min. The pellet was washed, centrifuged at 17 000 g, and the supernatant centrifuged at 137 000 g; the rim above the pellet was used as the purified sarcolemmal membranes. The supernatant from the 45 000 g spin was centrifuged at 137 000 g for 90 min and the pellet was used as the light vesicle fraction.

Radioligand binding assays were conducted using membrane fractions and [^{3}H]prazosin, an α_1-adrenergic antagonist, to identify α_1-receptors and [^{125}I]iodocyanopindolol, a β-adrenergic antagonist, to identify β-adrenergic receptors. Phentolamine (10 μM) and (−)propranolol (1 μM) were used to define nonspecific binding to α_1- and β-adrenergic receptors, respectively. Binding reactions were conducted at 25 °C for 1 h and were terminated by dilution, filtration, and washing of samples over Whatman GF/B filters that had been presoaked in 2% polyethyleneimine. Filters were counted in a gamma counter for [^{125}I]iodocyanopindolol and in a liquid scintillation counter for [^{3}H] prazosin. Experiments to define receptor number and equilibrium dissociation constants were conducted using multiple concentrations of radioligand with Scatchard analysis of saturation binding isotherms. Equilibrium dissociation constants were not different in experiments with membranes from ischemic, drug-treated, or control animals.

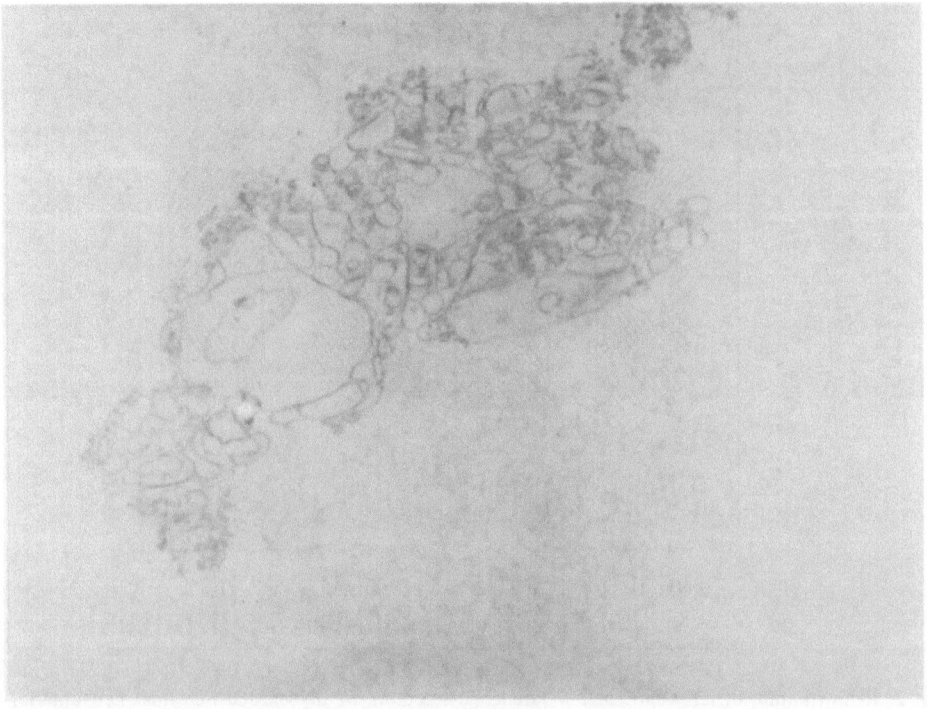

Fig. 1. Electron microscopic photograph of light vesicle membranes (magnification × 46 000).

Results

Characterization of membrane fractions

In initial experiments we sought to characterize the sarcolemmal and light vesicle fractions. An electron microscopic photograph of the light vesicle fraction is presented in Fig. 1 and shows scattered vesicles without clear attachments to the plasma membrane. The plasma membrane marker Na^+/K^+-ATPase was enriched over five-fold in the sarcolemma as compared with the light vesicle fraction. The light vesicle fraction showed no enrichment in endoplasmic reticulum, as assessed by glucose-6-phosphatase activity, and had no detectable presence of lysosomes, as determined by acid phosphatase measurements. Moreover, the absence of guanine nucleotide-mediated regulation of agonist binding to β-adrenergic receptors in the light vesicle fraction suggests that this fraction also lacks function of the stimulatory guanine nucleotide regulatory protein G_s [5, 10].

Changes in β-adrenergic receptors with myocardial ischemia: guinea pig model

It has previously been shown that myocardial ischemia is associated with an increased number (up-regulation) of β-adrenergic receptors in crude cardiac membrane preparations [12]. We

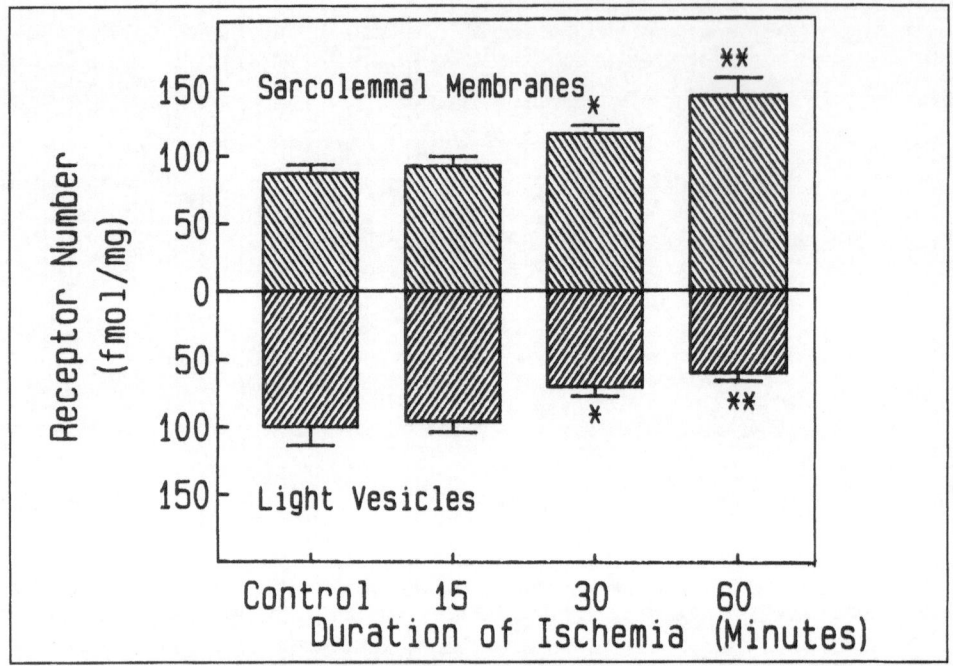

Fig. 2. Redistribution of left ventricular β-adrenergic receptors by myocardial ischemia. Left anterior descending coronary artery ligation was performed in guinea pigs and at the indicated times; sarcolemmal and light vesicle membrane fractions were prepared as described in Methods. Binding of [^{125}I]iodocyano-pindolol to β-adrenergic receptors was assessed in membrane fractions at each time point using 8–10 concentrations of radioligands (50–600 pM.) Maximal binding from Scatchard analysis is shown. The data shown are mean ± SEM of at least three experiments, by unpaired two-tailed t-test and correction by the Bonferroni method: *p < 0.09; **p < 0.03. (Adapted from [8].)

asked whether this up-regulation of β-receptors in ischemia might result from an "externalization" of β-receptors from an intracellular compartment to the cell surface.

Ligation of the left anterior descending coronary artery produced a rapid onset of ischemia, characterized by obvious cyanosis of the heart and prominent ST-segment changes. Within 30 min of the onset of ischemia, there was an increase in sarcolemmal β-adrenergic receptors (Fig. 2). Associated with this increase was a concomitant decrease in the number of light vesicle β-adrenergic receptors. We have interpreted these results as compatible with the hypothesis that there is an apparent translocation of receptors from the light vesicle (presumably intracellular site) to the surface sarcolemma, i.e., an externalization of β-adrenergic receptors. Preliminary studies have revealed that this externalization is potentially reversible, but only if coronary occlusion is reversed within 60 min [7].

In the guinea pig heart, the life cycle of the α_1-adrenergic receptor appears to be different than that of the β-adrenergic receptor. Approximately an equal specific activity of β-adrenergic receptors was found in the light vesicle and sarcolemmal fractions (85 ± 8 fmol/mg vs 93 ± 8 fmol/mg), while α_1-receptors had substantially higher values in sarcolemmal vs light vesicle fractions (86 ± 5 fmol/mg vs 27 ± 4 fmol/g) [11]. Furthermore the increase in sarcolemmal α_1-receptors observed with ischemia (Table 1) did not appear to be mediated by a translocation of α_1-adrenergic receptors from a light vesicle pool.

These results not only defined a possible explanation for enhanced sensitivity of ischemic β-Adrenergic receptors and the G_s protein in myocardial ischemia and injury

Table 1. Percent change in receptors following ischemia

	Crude	Sarcolemma	Light vesicle
Alpha receptors	+60%	+50%	−3%
Beta receptors	+10%	+70%	−52%

Sarcolemma and light vesicle membranes were prepared as described in Methods. The pellet from the 45 000 g centrifugation was resuspended, washed, and centrifuged at 17 000 g. The pellet from this spin was resuspended in 50 mM Tris-HCl, 8 mM $MgCl_2$, and 0.5 mM EGTA, was then washed, and then centrifuged twice at 500 g. The supernatant was used to obtain the crude membrane fraction.

myocardium to the effects of catecholamines, they also provided an experimental model to explore mechanisms that might regulate receptor translocation in ischemia. The β-adrenergic blocking agents may modulate the magnitude of cardiac injury, as well as reduce the likelihood of ventricular arrhythmias in the ischemic setting [2, 14, 16, 20]. Previous results demonstrated that treatment with propranolol leads to an up-regulation of sarcolemmal β-receptors, perhaps by blockade of the down-regulation of β-receptors that occurs in response to ambient sympathetic tone [1, 6]. We thus sought to determine: 1) whether treatment of guinea pigs with β-blockers leads to an externalization of β-receptors; 2) whether treated animals would have an altered response to myocardial ischemia.

Guinea pigs had osmotic minipumps implanted that were designed to deliver one of three drugs for 7 days: the non-selective β-blocker propranolol, the $β_1$-selective antagonist atenolol, and a non-selective partial agonist that possesses intrinsic sympathomimetic activity, pindolol. As can be seen in Fig. 3, both propranolol and atenolol treatment were associated with externalization, that is, an increase in the number of sarcolemmal β-receptors with an accompanying decrease in the number of light vesicle β-receptors. These results imply that cardiac $β_1$-adrenergic receptors are the β-receptors susceptible to externalization by propranolol. This externalization was associated with an increase in isoproterenol-stimulated adenylate cyclase activity without a change in guanine nucleotide- or forskolin-stimulated adenylate cyclase activity (data not shown and [10]). In contrast to treatment with propranolol and atenolol, treatment with the partial agonist pindolol failed to produce up-regulation of sarcolemmal β-adrenergic receptors (Fig. 3).

Another group of animals was treated with propranolol and then subjected to ischemia [10]. This pretreatment appeared to eliminate much of the ischemia-induced increase in sarcolemmal β-receptors while blunting the accompanying decrease in light vesicle β-receptors (data not shown). We also tested whether pre-treatment with either a $β_1$-adrenergic receptor blocking agent or a partial agonist would prevent ischemia-induced externalization. As shown in Fig. 4, responses to atenolol were similar to that of propranolol, that is, there was blunting of ischemia-induced externalization. Treatment with pindolol failed to block ischemia-induced externalization. Thus we believe that propranolol treatment and ischemia, both access and mobilize the same pool of intracellular β-receptors. It is conceivable that this action of propranolol (and atenolol) may contribute to the protective role of β-blockers in preventing arrhythmias or sudden death associated with myocardial ischemia [16, 20].

Changes in β-adrenergic receptors with myocardial ischemia: rat model

In order to begin to look closely at the effects of ischemia on post-receptor events, we have recently switched to a rat model of myocardial ischemia. The reasons are as follows: 1) In the guinea pig the response of adenylate cyclase activity to β-adrenergic agonists is poor, generally less than three-fold. This relatively poor response makes it difficult to do precise quantitative studies when comparing different groups of animals; 2) While guinea pigs have fairly reproducible areas of ischemia, these areas are small due to the presence of collaterals, and often it is

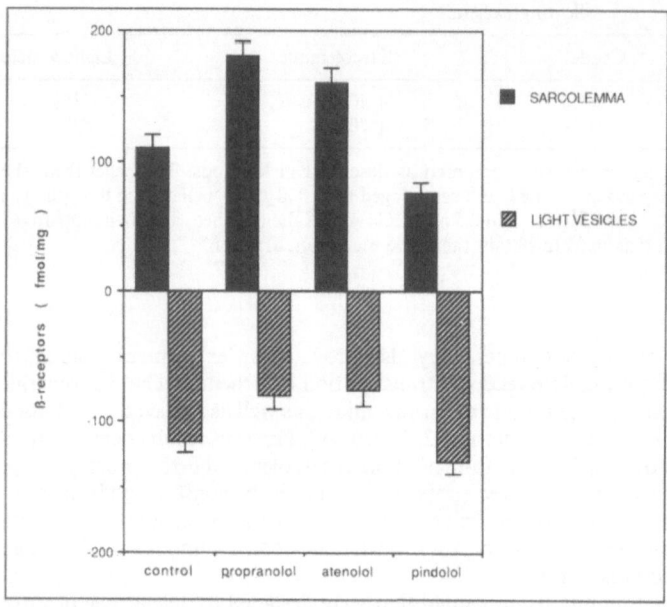

Fig. 3. Effects of propranolol, atenolol, and pindolol treatment on β-adrenergic receptor distribution. Guinea pigs were administered propranolol (0.15 mg kg^{-1} h^{-1}), atenolol (0.06 mg kg^{-1} h^{-1}), pindolol (0.01 mg kg^{-1} h^{-1}) or vehicle (10 mM HCl) by osmotic minipumps implanted subcutaneously for 7 days. Sarcolemmal and light vesicle membrane fractions were prepared and [^{125}I]iodocyanopindolol binding was assessed.

difficult to separate ischemic from normal tissue. Our rat model of ischemia has been very successful. The ligation of the distal left main coronary artery gives a reproducible, global ischemic area of the left ventricular wall, documented by post-mortem staining. Additionally, we can readily separate rat heart into sarcolemmal and light vesicle fractions (Fig. 5). In preliminary experiments we have shown that the sarcolemmal fraction contains an increased density of β-adrenergic receptors after 1 h of ischemia (Fig. 6).

Changes in the G_s protein in myocardial ischemia: rat model

While ischemia can lead to the rapid appearance of an increased number of β-receptors on the cell surface, the ability of these "new" receptors to couple to post-receptor components responsible for transmembrane signalling and second messenger generation is not well defined. Guanine nucleotide binding (G) proteins play a pivotal role in transducing receptor activation into modulation of effector molecules in a variety of systems, but whether such G proteins are altered in myocardial ischemia has not been extensively studied. Two homologous guanine nucleotide binding proteins, G_s and G_i, mediate the stimulatory and inhibitory actions, respectively, of various hormones and neurotransmitters on adenylate cyclase [5, 17, 18]. G_s is a multimeric protein composed of a 45 000–52 000 M_r alpha subunit, a 35 000 M_r beta subunit, and a 5 000–10 000 M_r gamma-subunit, while G_i is a multimeric protein composed of one of at least three 41 000 M_r alpha subunits, and beta and gamma subunits that may be identical to those of G_s. Activation of adenylate cyclase by agonists acting at receptors, such as β-adrenergic receptors, is thought to result from the dissociation of the inactive heterotrimer into an active α_s subunit, which, with GTP bound to it, is able to activate C, the catalytic unit of adenylate cyclase. Inhibition of adenylate cyclase by G_i is thought to result primarily from reformation of the inactive heterotrimer.

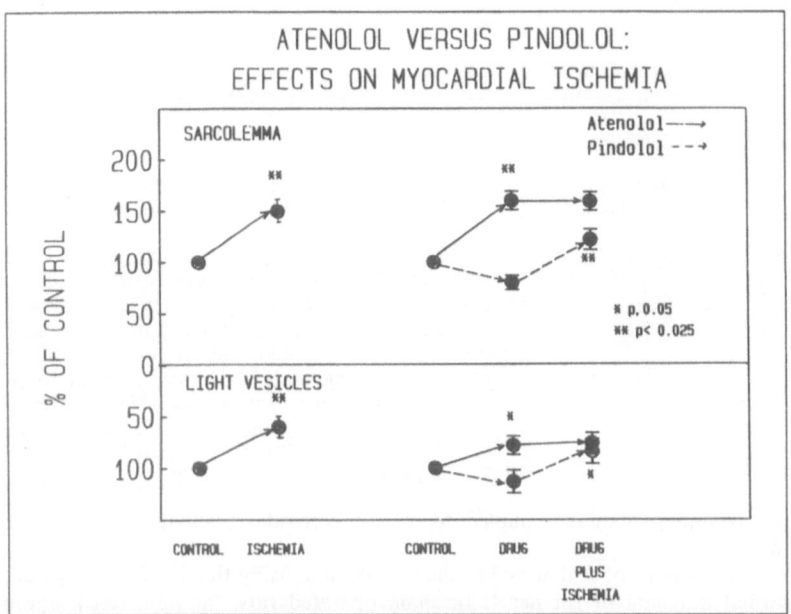

Fig. 4. Effects of atenolol (solid lines) and pindolol (broken lines) treatment on β-adrenergic receptor distribution in the absence and presence of myocardial ischemia. Guinea pigs were administered atenolol (0.06 mg kg^{-1} h^{-1}), pindolol (0.01 mg kg^{-1} h^{-1}) or vehicle (10 mM HCl) by osmotic minipumps implanted subcutaneously for 7 days. Some animals were then subjected to myocardial ischemia for 1 h. Sarcolemmal and light vesicle membrane fractions were prepared and [^{125}I]iodocyanopindolol binding was assessed.

In order to quantitate G$_s$, antipeptide antibodies directed against α$_s$ have been developed in rabbits immunized with a peptide that corresponds to amino acids 28–42 in α$_s$ [13]. The antibody response and membrane levels of α$_s$ were followed by an ELISA technique using a peptide-coated microtiter plate. Membranes solubilized with 1% sodium cholate were mixed with antibodies, applied to a microtiter plate coated with peptide, and incubated overnight. Then, after a thorough washing procedure, an anti-rabbit IgG antibody was added. This second

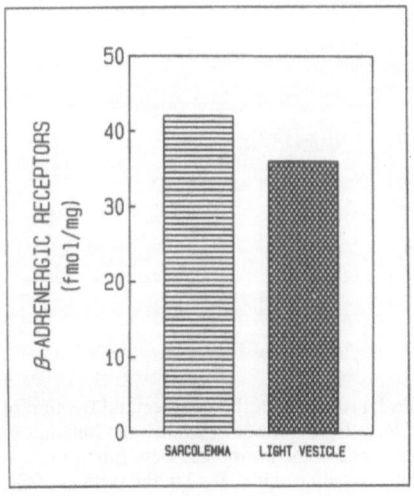

Fig. 5. β-adrenergic receptors in sarcolemmal and light vesicle membranes from rat heart. Membranes were prepared and binding performed as discussed in the legend of Fig. 2.

53

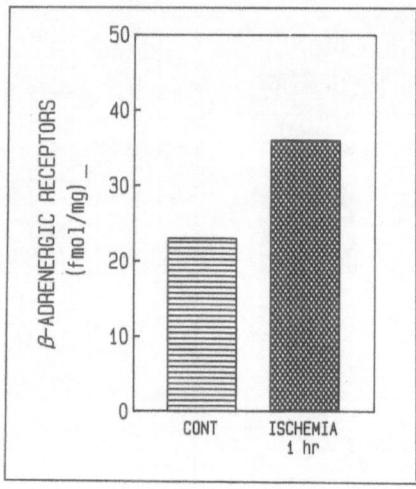

Fig. 6. β-adrenergic receptors in cardiac sarcolemmal membranes from rats undergoing myocardial ischemia. Left main coronary artery ligation was performed and at the indicated times, sarcolemmal and light vesicle membrane fractions were prepared as described in Methods. Binding of [^{125}I]iodocyanopindolol to β-adrenergic receptors was assessed in membrane fractions at each time point using 8–10 concentrations of radioligands (50–600 pM). Maximal binding from Scatchard analysis is shown (N = 3).

antibody has the enzyme peroxidase coupled to it and peroxidase activity is utilized to quantitate bound antibody.

Figure 7 shows the amount of α_s detected in the sarcolemma using this ELISA assay after 15 min of myocardial ischemia of rat heart. In sham operated rats, (at time zero), about 125 pmol/mg G_s was detected in the sarcolemmal membrane preparation. In this sarcolemmal

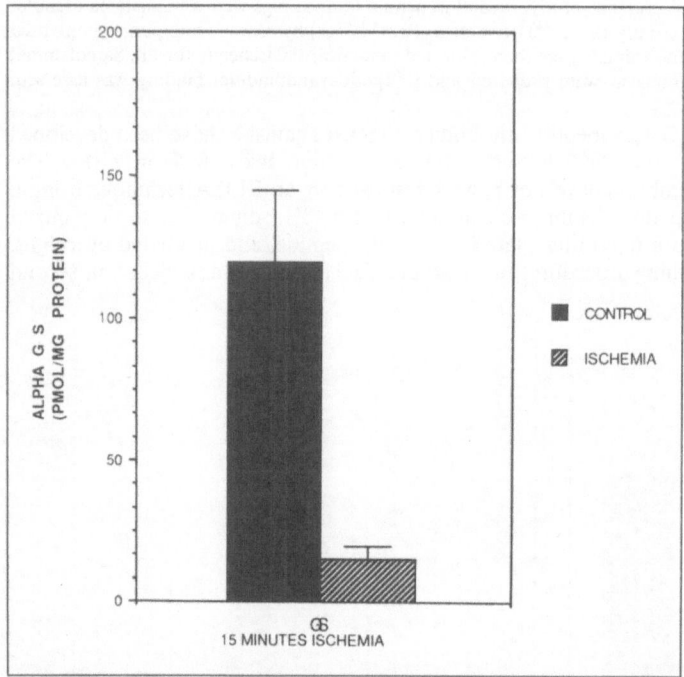

Fig. 7. Quantitation of the amount of α_s at 15 min of myocardial ischemia in the sarcolemmal fraction of rat heart (N = 6 animals with quadruplicate determinations). Polyclonal antibodies against the stimulatory G-protein (G_s) were utilized for quantitation in cardiac membranes. The antibodies were raised in New Zealand white rabbits against a synthetic peptide corresponding to amino acids 28–42 in the subunit of G_s.

fraction α_s markedly decreases within 15 min of occlusion of the left coronary artery. Comparable changes in G_s levels were also observed in crude cardiac membranes that included sarcolemmal and other membrane fractions. We found no transfer of G_s activity to the light vesicle fraction, although this fraction does possess some detectable α_s (data not shown).

Discussion and conclusion

We believe that the foregoing studies offer potentially important insights into the understanding of myocardial ischemia. A key conclusion is that the number of sarcolemmal receptors for catecholamines, as well as the amount of the signal transducing G_s-protein changes dramatically with myocardial ischemia and injury. The external expression of sarcolemmal α_1- and β-receptors within minutes after the onset of myocardial ischemia might contribute to enhanced responsiveness to neurally released and circulating catecholamines if post-receptor components are still fully functional in the ischemic animals. An important future direction will be to define the underlying molecular mechanisms responsible for externalization of β-receptors and to understand why the regulation of cardiac α_1-adrenergic receptors appears to be different from that of cardiac β-adrenergic receptors.

The ability of some β-antagonists (propranolol and atenolol), but not others (pindolol) to blunt ischemia-induced externalization, suggests that the mechanisms involved in the receptor cycling process may be shared by both the antagonists and the ischemic process itself. Presumably, both access a "pool" of β_1-adrenergic receptors that can "move" between the sarcolemma and light vesicle fractions. Since the partial agonist pindolol fails to block ischemia-induced externalization, receptor occupancy alone is not sufficient to blunt the externalization events occurring in ischemia.

It appears that events distal to the cardiac β-adrenergic receptor are important in cellular signalling during myocardial ischemia. Thus, merely having an increase in receptor number will not necessarily lead to enhanced responsivity. Our finding that a rapid decrease in G_s-levels can occur in the setting of myocardial ischemia has important ramifications with respect to subsequent ventricular function, catecholamine-induced lethal arrhythmias, and further ischemia. Conceivably, the reported decrease in β-adrenergic-stimulated adenylate cyclase activity after ischemia may result from a loss in G_s protein [19]. Alternatively, other G_s-regulated pathways (e.g., the voltage-dependent Ca^{++}-channel) might also be altered in the setting of myocardial ischemia. Changes in the amount of effector molecules (adenylate cyclase, ion channels, etc.) might also occur in the setting of myocardial ischemia. Thus, the physiological endpoint in terms of response will represent the interplay of what may be non-coordinate changes produced by ischemia in multiple components of signal transduction.

References

1. Brodde O-E (1989) The β-adrenoceptors. In: Williams M, Glennon RA, Timmermans PBMWM (eds) Receptor Pharmacology and Function. Dekker, New York, pp 207–256
2. Bush LR, Haack DW, Shlafer M, Lucchesi BR (1980) Protective effects of β-adrenergic blockade in isolated ischemic hearts. Eur J Pharmacol 67:209–217
3. Corr PB, Shayman JA, Kramer JB, Kipins RJ (1981) Increased alpha 1-adrenergic receptors in ischemic cat myocardium: a potential mediator of electrophysiological derangements. J Clin Invest 67:1232–1236
4. Curtis MJ, Macleod BA, Walker MJA (1987) Models for the study of arrhythmias in myocardial ischemia and infarction: the use of the rat. J Mol Cell Cardiol 19:399–419
5. Gilman AG (1987) G proteins: transducers of receptor generated signals. Ann Rev Biochem 56:615–649
6. Insel PA (1990) Beta adrenergic receptors in pathophysiological states and in clinical medicine. In Perkins JP (ed) The β-adrenergic receptors. Human Press, Clifton NJ (in press)
7. Insel PA, Maisel AS (1989) α 1 and -β-adrenergic receptors in myocardial ischemia and injury. In: Brachmann J, Schömig A (eds.) Adrenergic System and Ventricular Arrhythmias in Myocardial Ischemia. Springer-Verlag, Berlin pp 81–90

8. Maisel AS, Motulsky HJ, Insel PA (1985) Externalization of β-adrenergic receptors promoted by myocardial ischemia. Science 230:183–186
9. Maisel AS, Motulsky HJ, Insel PA (1987) Life cycles of cardiac α 1- and β-adrenergic receptors. Biochem Pharmacol 36:1–6
10. Maisel AS, Motulsky HJ, Insel PA (1987) Propranolol treatment externalizes β-adrenergic receptors in guinea pig myocardium and prevents further externalization by ischemia. Circ Res. 60:108–112
11. Maisel AS, Motulsky HJ, Ziegler MG, Insel PA (1987) Ischemia and agonist induced changes in α- and β-adrenergic receptor traffic in guinea pig hearts. Am J Physiol 253:H1159–H1167
12. Mukherjee A, Bush LR, McCoy KE, Duke RJ, Hagler H, Buja M, Willerson JT (1982) Relationship between β-adrenergic receptor number and physiologic responses during experimental canine myocardial ischemia. Circ Res 50:735–742
13. Ransnas L, Insel PA (1988) Quantitation of the guanine nucleotide binding regulatory protein G_s in S49 cell membranes using antipeptide antibodies to alpha s. J Biol Chem 263:9482–9485
14. Reimer KA, Rasmussen MM, Jennings RB (1973) Reduction by propranolol of myocardial necrosis following temporary coronary occlusion in dogs. Am J Cardiol 33:353–363.
15. Schwartz PJ, Zuanetti G (1988) Role of the autonomic nervous system in reperfusion arrhythmias. J Mol Cell Cardiol 20(Suppl II):113–118
16. Sleight PC (1986) Use of beta-adrenoceptor blockade during and after acute myocardial infarction. Ann Rev Med 36:415–425
17. Spiegel AM (1987) Signal transduction by guanine nucleotide binding proteins. Mol Cell Endocrinol 49:1–16
18. Stryer L, Bourne HR (1986) G proteins: A family of signal transducers. Ann Rev Cell Biol 2:391–419
19. Vatner DE, Young MA, Knight DR, Vatner SF (1990) β-receptors and adenylate cyclase: comparison of non-ischemic, ischemic, and post-mortem tissue. Am J Physiol 258:H140–H144
20. Vedin JA, Wilhelmsson CE (1983) Beta receptor blocking drugs in the secondary prevention of coronary heart disease. Ann Rev Pharmacol Toxicol 23:29–44

Author's address:

Alan Maisel, M.D., Dept. of Pharmacology, M-036, School of Medicine, University of California, San Diego, La Jolla, CA 92093, U.S.A.

β- and α-Adrenoceptor-Agonists and -Antagonists in Chronic Heart Failure

O.-E. Brodde

Zentrum für Innere Medizin, Medizinische Klinik und Poliklinik, Abteilung für Nieren- und Hochdruckkrankheiten, Universitätsklinikum, Essen, FRG

Summary

In patients with chronic heart failure, cardiac β-adrenoceptor function is decreased, and this decrease is related to the degree of heart failure. Under these conditions, treatment with β-adrenoceptor agonists seems to be of limited value as it might further down-regulate cardiac β-adrenoceptors, resulting, finally, in a loss of therapeutic efficacy. However, β-adrenoceptor antagonists might have beneficial effects, because they can protect the myocardium from the deleterious effects of elevated endogenous catecholamines and can, simultaneously, restore the previously down-regulated β-adrenoceptor function. Stimulation of cardiac α-adrenoceptors, however, seems not to be of any therapeutic value in patients with chronic heart failure, because a) the number of α-adrenoceptors in the human heart is very low and its function is not completely understood, and b) no α-adrenoceptor agonist is presently available that selectively stimulates cardiac α-adrenoceptors without concomitantly activating vascular α-adrenoceptors. In acute myocardial ischemia, cardiac β-adrenoceptors increase; this increase is – at least in early acute myocardial ischemia – accompanied by an increased β-adrenoceptor functional responsiveness; thus, under these conditions, β-adrenoceptor agonists again might not be of clinical value, while β-adrenoceptor antagonists may exert beneficial effects, because they can block (over)activation of the sensitized β-adrenoceptors by elevated endogenous catecholamines.

Introduction

Adrenoceptors were originally subdivided into the subtypes α and β based on the findings that catecholamines evoked their effects on different organs with different orders of potency [3]. α-Adrenoceptors mediate catecholamine-induced vasoconstriction; at these receptors noradrenaline and adrenaline are much more potent than isoprenaline. β-Adrenoceptors mediate the myocardial effects of the catecholamines; at these receptors isoprenaline is more potent than noradrenaline and adrenaline. It later became clear that both α- and β-adrenoceptors can be subdivided into at least two major subtypes: α_1- and α_2- and β_1- and β_2-adrenoceptors (for reviews see [10, 27, 61, 64]).

Cardiac β-adrenoceptors

Beta-adrenoceptors were originally subclassified [40] into cardiac β_1-(noradrenaline and adrenaline equipotent) and vascular and bronchial smooth muscle β_2-adrenoceptors (adrenaline about 10–30 times more potent than noradrenaline). However, it rapidly became apparent that this organ-specific subclassification was an oversimplification; it is now generally accepted that, in a variety of tissues including the heart of various species, β-adrenoceptors are not a homogeneous population but that β_1- and β_2-adrenoceptors coexist [10, 11, 49, 62]. This holds true also for the human heart, where several groups have convincingly demonstrated the coexistence of β_1- and β_2-adrenoceptors, first by radioligand binding studies, and subsequently, by functional experiments [10, 11, 37, 62]. Both β_1- and β_2-adrenoceptors are coupled to adenylate cyclase [6, 20, 38] and can mediate positive inotropic effects of β-adrenoceptor agonists. Among the classical catecholamines isoprenaline and adrenaline cause their positive inotropic effects on the human heart via stimulation of β_1- and β_2-adrenoceptors, while noradrenaline, the main transmitter of the sympathetic nervous system, evokes its positive inotropic effect predominantly, if not exclusively, via β_1-adrenoceptor stimulation [33, 38, 50, 73].

In the last few years evidence has accumulated that in patients with chronic heart failure cardiac β-adrenoceptor density and functional responsiveness are markedly reduced, and the amount of this reduction is related to the degree of heart failure (as judged clinically by NYHA functional class) [5, 7, 9, 13, 19]. Such a decrease might be very likely caused by "endogenous" down-regulation through elevated catecholamines, since it is well known that in chronic heart failure plasma noradrenaline levels are elevated in response to the depressed cardiac function [31, 32]. Interestingly, the etiology of heart failure and/or some other (presently unknown) factors seem to differentially regulate cardiac β_1- and β_2-adrenoceptors in heart failure; while in all kinds of heart failure β_1-adrenoceptors are decreased in number and function, β_2-adrenoceptors are not altered in end-stage idiopathic dilated cardiomyopathy [5, 9, 19] and in patients with aortic valve disease [46], but are down-regulated to a very similar extent as β_1-adrenoceptors in end-stage ischemic cardiomyopathy [13], mitral valve disease [14], and tetralogy of Fallot [13].

We were interested in studying whether such a down-regulation of β_1- and β_2-adrenoceptors may also occur during long-term treatment with β-adrenoceptor agonists. Therefore, we have determined the effects of chronic treatment of healthy volunteers with the selective β_1-adrenoceptor agonist xamoterol [25, 55] and the selective β_2-adrenoceptor agonist procaterol [71, 72] on the number of β-adrenoceptors in circulating lymphocytes (containing exclusively β_2-adrenoceptors, see [18]) and on β_1- and β_2-adrenoceptor-mediated in vivo physiological effects [12]. As β_1-adrenoceptor-mediated in vivo physiological effects, we determined isoprenaline infusion-induced increases in systolic blood pressure and exercise-induced tachycardia; as β_2-adrenoceptor-mediated effect isoprenaline infusion-induced decreases in diastolic blood pressure; in addition, isoprenaline infusion-induced tachycardia was assessed as a mixed (cardiac) β_1- and β_2-adrenoceptor mediated effect [17, 44, 50, 70].

After 14 days oral treatment of healthy volunteers with the selective β_1-adrenoceptor agonist xamoterol, the lymphocyte β_2-adrenoceptor density was not changed (Fig. 1); however, the isoprenaline-induced increase in systolic blood pressure and the exercise-induced tachycardia were significantly decreased (Figs. 2 and 3), while the isoprenaline-induced decrease in diastolic blood pressure was not affected (Fig. 2), indicating that under these conditions β_1-adrenoceptor-mediated effects are attenuated while β_2-adrenoceptor-mediated effects are not. The isoprenaline-induced increase in heart rate (the mixed β_1- and β_2-adrenoceptor mediated effect) was also decreased following the xamoterol treatment (Fig. 3) but to a lesser extent than the pure β_1-adrenoceptor-mediated effects [12].

In contrast to the xamoterol treatment, after 9 days treatment of healthy volunteers with the selective β_2-adrenoceptor agonist procaterol, lymphocyte β_2-adrenoceptor density was significantly reduced by about 35% (Fig. 1). Concomitantly, the isoprenaline-induced decrease in diastolic blood pressure (the β_2-adrenoceptor-mediated effect) was significantly attenuated (Fig. 2) while the isoprenaline-induced increase in systolic blood pressure and the exercise-induced tachycardia (the β_1-adrenoceptor-mediated effects) were not at all affected (Figs. 2 and 3). Isoprenaline-induced tachycardia (the mixed β_1- and β_2-adrenoceptor-mediated effect) was also attenuated (Fig. 3), but to a lesser extent than the pure β_2-adrenoceptor mediated decrease in diastolic blood pressure [12]. These results demonstrate that, obviously, in general long-term treatment of patients with β-adrenoceptor agonists (for example β_2-adrenergic bronchodilators in the therapy of asthma or β_1-adrenergic positive inotropic drugs in the therapy of chronic

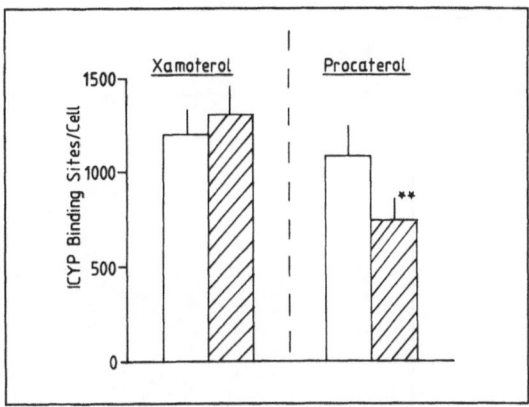

Fig. 1. Lymphocyte β_2-adrenoceptor density in male healthy volunteers before ($\square$) and after a 14 days treatment with xamoterol ($\blacksquare$, 2×200 mg/day, left), and before ($\square$) and after a 9 days treatment with procaterol ($\blacksquare$, $2 \times 50\,\mu$g/day, right), respectively. Ordinate: Lymphocyte β_2-adrenoceptor density, determined by Scatchard-analysis [56] of $(-)$-$[^{125}$I]-iodocyanopindolol (ICYP) binding in *intact cells*, in ICYP binding sites/cell. Given are means $\pm$ SEM of 10 experiments each. **p < 0.01 vs the corresponding pre-drug level. From [12] with modifications.

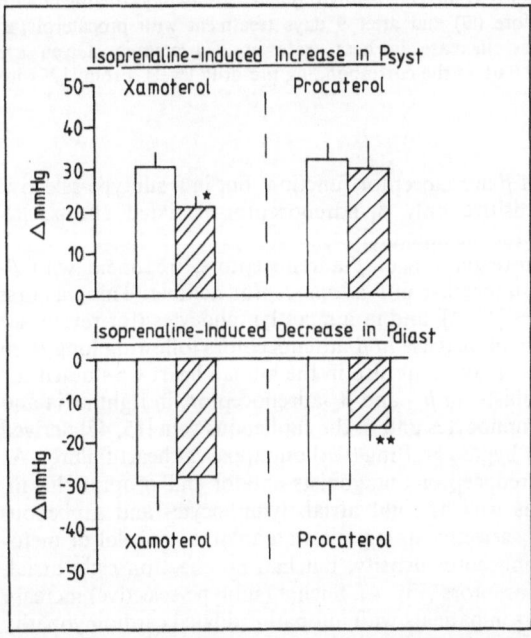

Fig. 2. Isoprenaline infusion-induced maximum increase in systolic blood pressure (P_{syst}, upper panel) and maximum decrease in diastolic blood pressure (P_{diast}, lower panel) in male healthy volunteers before ($\square$) and after a 14 days treatment with xamoterol ($\blacksquare$, 2×200 mg/day, left) and before ($\square$) and after a 9 days treatment with procaterol ($\blacksquare$, $2 \times 50\,\mu$g/day, right), respectively. Ordinate, upper panel: increase in systolic blood pressure in Δ mm Hg; lower panel: decrease in diastolic blood pressure in Δ mm Hg. Given are means $\pm$ SEM of 10 experiments each. **p < 0.01, *p < 0.05 vs the corresponding pre-drug levels. From [12] with modifications.

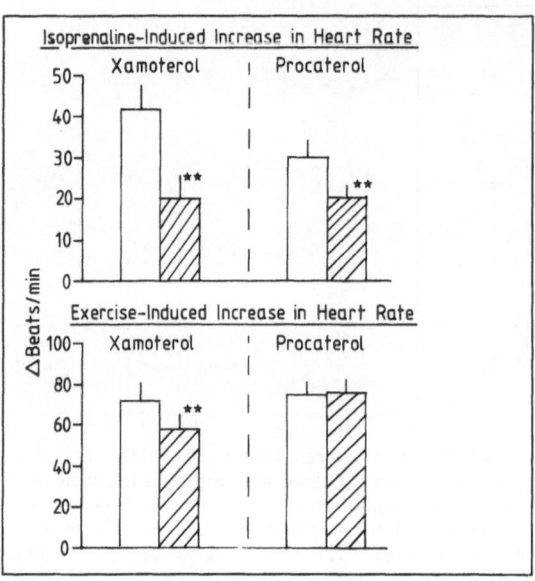

Fig. 3. Isoprenaline infusion-induced increase in heart rate (upper panel) and bicycle exercise-induced increase in heart rate (lower panel) in male healthy volunteers before (□) and after 14 days treatment with xamoterol (■, 2 × 200 mg/day, left) and before (□) and after 9 days treatment with procaterol (■, 2 × 50 μg/day, right), respectively. Ordinates: increase in heart rate in Δ beats/min. Given are means ± SEM of 10 experiments each. **p < 0.01 vs the corresponding pre-drug levels. From [12] with modifications.

heart failure) leads to a desensitization of β-adrenoceptor function, but in a subtype-selective manner: β_1-adrenoceptor agonists desensitize only β_1-adrenoceptor-mediated effects, β_2-adrenoceptor agonists only β_2-adrenoceptor-mediated effects.

In contrast to agonist-induced down-regulation of β-adrenoceptors, treatment with β-adrenoceptor antagonists often leads to an increase in β-adrenoceptor number. This was first shown in rat heart, lung, and lymphocytes [1, 34], and in human lymphocytes (for review see [10, 11, 18, 62]). To find out whether such an increase in β-adrenoceptors following long-term treatment with β-adrenoceptor antagonists may occur also in the human heart we studied the effects of different β-adrenoceptor antagonists on β_1- and β_2-adrenoceptors in right atria and on β_2-adrenoceptors in the circulating lymphocytes and in the saphenous vein [15, 47] derived from patients undergoing coronary artery bypass grafting (without apparent heart failure). As shown in Fig. 4, the non-selective β-adrenoceptor antagonists sotalol and propranolol increased concomitantly right atrial β_1- as well as right atrial, lymphocyte- and saphenous vein-β_2-adrenoceptor density while the β_1-selective antagonists atenolol, bisoprolol or metoprolol increased only right atrial β_1-adrenoceptor density, but had no effect on right atrial, lymphocyte and saphenous vein β_2-adrenoceptors (Fig. 4). Such a (subtype-selective) increase in cardiac β-adrenoceptors that also occurs in patients with idiopathic dilated cardiomyopathy [35, 67] may be one reason for the beneficial effects of low-dose metoprolol treatment in end-stage congestive heart failure [35, 67, 68]. Thus, the present results show that, obviously, in general, in man, long-term application of β-adrenoceptor antagonists (*without ISA*) causes an up-regulation of β-adrenoceptors, but in a β-adrenoceptor subtype-selective manner: only non-selective β-adrenoceptor antagonists (e.g., propranolol or sotalol) increase both β_1- *and* β_2-adrenoceptors, while β_1- or β_2-adrenoceptor selective antagonists affect only β_1- *or* β_2-adrenoceptors, respectively. Such a subtype-selective regulation of β_1- and β_2-adrenoceptors

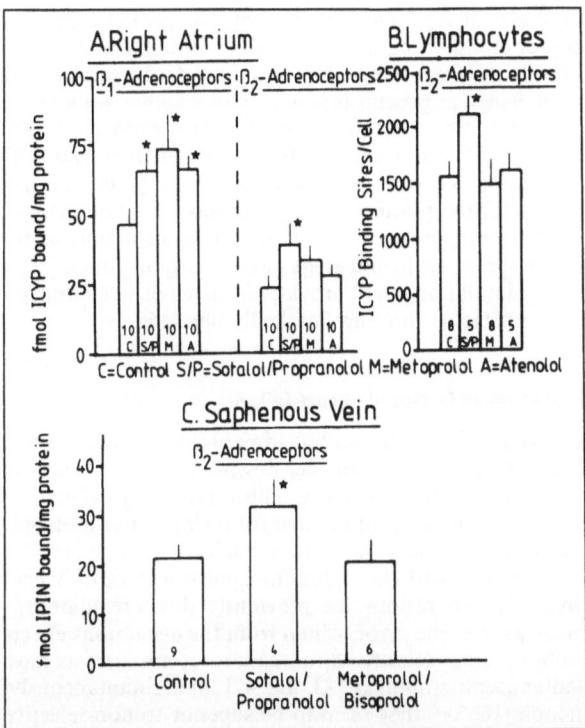

Fig. 4. Effects of chronic β-adrenoceptor antagonist treatment on β-adrenoceptor density in right atria (A), lymphocytes (B), and saphenous veins (C) of patients undergoing coronary artery bypass grafting. Ordinate: upper panel, left: β_1- and β_2-adrenoceptor density in right atrial membranes, assessed from ICI 118,551 competition curves with $(-)$-$[^{125}I]$-iodocyanopindolol (ICYP) binding using the iterative curve fitting program LIGAND [45], in fmol ICYP specifically bound/mg protein; right: β_2-adrenoceptor density in intact lymphocytes in ICYP binding sites/cell; lower panel: β_2-adrenoceptor density in saphenous vein membranes in fmol $(-)$-$[^{125}I]$-iodopindolol (IPIN) specifically bound/mg protein. Given are means ± SEM; Number of experiments at the bottom of the columns. *p < 0.05 vs the corresponding controls (i.e., patients *not* treated with β-adrenoceptor antagonists). Data recalculated from [15] and [47].

should be taken into consideration when treating patients chronically with β-adrenoceptor agonists and/or antagonists in order to achieve an optimal dose regimen.

Cardiac α-adrenoceptors

In the last 20 years it has been convincingly shown that besides β-adrenoceptors, also myocardial α-adrenoceptors can mediate positive inotropic effects in various species [22, 30, 69], including man [2, 23]. In contrast to the β-adrenoceptor-mediated effect, the positive inotropic effect mediated by stimulation of cardiac α-adrenoceptors (that belong exclusively to the α_1-subtype) is not associated with an increase in the intracellular level of cyclic AMP [21, 22, 30, 69]. Recent evidence suggests that cardiac α_1-adrenoceptors might be coupled via a (pertussis toxin-insensitive) G-protein to the phospholipase C/inositol-triphosphate/diacylglycerol pathway [8, 39, 58].

In the human heart the number of α_1-adrenoceptors has been found to be very low (about 10 fmol/mg protein) and seems not to be altered in heart failure [4, 8]. However, whether the functional response to cardiac α_1-adrenoceptor stimulation is attenuated in severe heart failure is still a matter of controversy. While Böhm et al. [4] did not observe any changes in the positive inotropic effect evoked by α-adrenoceptor stimulation, Schmitz et al. [57] observed that with increasing severity of heart failure not only the β-adrenergic but also the α-adrenergic positive inotropic effect was markedly depressed. Thus, at present it is uncertain whether stimulation of cardiac α_1-adrenoceptors in patients with chronic heart failure might provide any benefit for the patient. In this context, it is interesting to note that, in rats, prolonged treatment with isoprenaline resulted in the expected decreases in cardiac β-adrenoceptor density and contractile response; however, α_1-adrenoceptor mediated positive inotropic effects were not changed as was the α_1-adrenoceptor number [24]. Thus, under these circumstances of a shift in the α/β-adrenoceptor ratio towards α-adrenoceptors (as in end-stage dilated cardiomyopathy, see [4, 8]) selective (cardiac) α_1-adrenoceptor stimulation might provide effective positive inotropic support when β-adrenergic increases in contractile force are markedly blunted.

β- and α-Adrenoceptor-agonists and -antagonists in chronic heart failure

Long-term treatment of patients with chronic heart failure with β-adrenoceptor agonists seems to be of limited beneficial effect, because it will lead to a further down-regulation of cardiac β-adrenoceptors, resulting finally in a loss of therapeutic efficacy. This is especially evident for partial agonists, since they are more dependent on receptor number for their positive inotropic effects than are full agonists, and since the human heart has only a small receptor reserve for β-adrenoceptor agonists [51]. On the other hand, β-adrenoceptor antagonists may be of benefit in dilated cardiomyopathy; they obviously can restore the previously down-regulated β-adrenoceptor density and simultaneously protect the myocardium from the deleterious effects of the elevated endogenous catecholamines. Since selective β_1-adrenoceptor antagonists have been shown to increase cardiac β_1-adrenoceptor number [13, 16, 47] and, simultaneously, sensitize cardiac β_2-adrenoceptor function [16, 38], they seem to be superior to non-selective β-adrenoceptor antagonists. Finally, due to their very low number, stimulation of cardiac α_1-adrenoceptors seems not to be of therapeutic use in the treatment of chronic heart failure, especially since, at present, no α-adrenergic drug is available that selectively stimulates cardiac α_1-adrenoceptors without activating vascular α_1-adrenoceptors. However, α_1-adrenoceptors can be subclassified into the subtypes α_{1A}- and α_{1B}-adrenoceptors [48]. Future studies have to show whether in man cardiac and vascular α_1-adrenoceptors belong to different α_1-adrenoceptor subtypes that would offer the possibility to synthesize α_1-adrenoceptor agonists and/or antagonists specific for cardiac α_1-adrenoceptors.

β- and α-Adrenoceptor-agonists and -antagonists in myocardial ischemia

In acute myocardial ischemia large amounts of catecholamines are released from myocardial sympathetic nerve terminals [59]; as discussed above, one should, therefore, expect that cardiac β-adrenergic (and possibly also α-adrenergic) receptors are desensitized. However, the opposite occurs: it has consistently been found that in acute myocardial ischemia β-adrenoceptor number in dog, guinea-pig, and rat heart is increased [28, 42, 43, 52, 54, 63, 65]. The mechanism underlying this increase in β-adrenoceptor number in the face of elevated endogenous catecholamines is not completely understood: it might be due to an enhanced externalization [42, 43] or an impaired internalization [63] of the receptor. This increase in β-adrenergic receptor number seems to be accompanied in early myocardial ischemia by an increased activity of the adenylate cyclase [43], while during prolonged ischemia adenylate cyclase activity decreases [28, 52, 65]. Whether similar changes occur in man is not known at present. However, the well-known clinical observations of persistent malignant arrhythmias in acute myocardial infarction, which are at least partially sensitive to β-blockade, support the idea of an overreactive cardiac β-adrenergic system also in man [36, 60, 66]. Interestingly, Maisel et al. [41] have shown that in guinea-pig heart propranolol and atenolol can prevent the ischemia-induced increase in β-adrenergic receptor number. Thus, β-blockers may have beneficial effects in human myocardial ischemia for two reasons: they can block the (over)activation of β-adrenergic receptors by the elevated endogenous catecholamines and may, in addition, prevent the increase in receptor number.

Similar to β-adrenoceptors, cardiac α-adrenoceptor number has been found to increase in cat [26], dog [53], and guinea-pig heart [42], but not in rat heart [29] during myocardial ischemia. As mentioned earlier, in the human heart cardiac α-adrenoceptor number is very low and its functional role is not completely understood. Thus, it is uncertain whether changes in cardiac α-adrenoceptors might play any pathophysiological role in the ischemic human myocardium.

Acknowledgements: The author's work was supported by the SANDOZ-Stiftung für Therapeutische Forschung (Nürnberg, FRG) and the Gesellschaft zur Erforschung und Bekämpfung des hohen Blutdrucks (Essen, FRG).

References

1. Aarons RD, Molinoff PB (1982) Changes in the density of beta adrenergic receptors in rat lymphocytes, heart and lung after chronic treatment with propranolol. J Pharmacol Exp Ther 221:439–443
2. Aass H, Skomedal T, Osnes J-B, Fjeld NB, Klingen G, Langslet A, Svennegig J, Semb G (1986) Noradrenaline evokes an α-adrenoceptor-mediated inotropic effect in human ventricular myocardium. Acta Pharmacol et Toxicol 58:88–90
3. Ahlquist RP (1948) A study of the adrenotropic receptors. Am J Physiol 153:586–600
4. Böhm M, Diet F, Feiler G, Kemkes B, Erdmann E (1988) α-Adrenoceptors and α-adrenoceptor-mediated positive inotropic effects in failing human myocardium. J Cardiovasc Pharmacol 12:357–364
5. Böhm M, Pieske B, Schnabel P, Schwinger R, Kemkes B, Klövekorn W-P, Erdmann E (1989) Reduced effects of dopexamine one force of contraction in the failing human heart despite preserved β_2-adrenoceptor subpopulation. J Cardiovasc Pharmacol 14:549–559
6. Bristow MR, Hershberger RE, Port JD, Minobe W, Rasmussen R (1989) β_1- and β_2-adrenergic receptor mediated adenylate cyclase stimulation in nonfailing and failing human ventricular myocardium. Mol Pharmacol 35:295–303
7. Bristow MR, Port JD, Hershberger RE, Gilbert EM, Feldman AM (1989) β-adrenergic receptor-adenylate cyclase complex as a target for therapeutic intervention in heart failure. Eur Heart J 10 (Suppl. B):45–54
8. Bristow MR, Minobe W, Rasmussen R, Hershberger RE, Hoffman BB (1988) Alpha-1 adrenergic receptors in the nonfailing and failing human heart. J Pharmacol Exp Ther 247:1039–1045
9. Bristow MR, Ginsburg R, Umans V, Fowler M, Minobe W, Rasmussen R, Zera P, Menlove R, Shah P, Jamieson S, Stinson EB (1986) β_1- and β_2-adrenergic-receptor subpopulations in nonfailing and failing human ventricular myocardium: coupling of both receptor subtypes to muscle contraction and selective β_1-receptor down-regulation in heart failure. Circ Res 59:297–309
10. Brodde O-E (1989) β-Adrenoceptors. In: Williams M, Glennon RA, Timmermans PBMWM (Eds) Receptor pharmacology and function. Marcel Dekker, New York, pp 207–255
11. Brodde O-E (1987) Cardiac beta-adrenergic receptors. ISI Atlas Sci: Pharmacol 1:107–112
12. Brodde O-E, Daul A, Michel-Reher M, Boomsma F, Man in't Veld AJ, Schlieper P, Michel MC (1990) Agonist-induced desensitization of β-adrenoceptor function in man. Subtype-selective reduction in β_1- or β_2-adrenoceptor-mediated physiological effects by xamoterol or procaterol. Circulation 81:914–921
13. Brodde O-E, Zerkowski H-R, Borst HG, Maier W, Michel MC (1989) Drug- and disease-induced changes of human cardiac β_1- and β_2-adrenoceptors. Eur Heart J 10 (Suppl B):38–44
14. Brodde O-E, Zerkowski H-R, Doetsch N, Motomura S, Khamssi M, Michel MC (1989) Myocardial beta-adrenoceptor changes in heart failure: Concomitant reduction in beta$_1$- and beta$_2$-adrenoceptor function related to the degree of heart failure in patients with mitral valve disease. J Am Coll Cardiol 14:323–331
15. Brodde O-E, Zerkowski H-R, Doetsch N, Khamssi M (1989) Subtype-selective up-regulation of human saphenous vein β_2-adrenoceptors by chronic β-adrenoceptor antagonist treatment. Naunyn-Schmiedeberg's Arch Pharmacol 339:479–482
16. Brodde O-E, Motomura S (1989) Effects of chronic β-adrenoceptor antagonist treatment on β-adrenergic and muscarinic receptor function in the human heart. J Mol Cell Cardiol 21 (Suppl. III): S3 (abstr.)
17. Brodde O-E, Daul A, Wellstein A, Palm D, Michel MC, Beckeringh JJ (1988) Differentiation of β_1- and β_2-adrenoceptor-mediated effects in humans. Am J Physiol 254:H199–H206

18. Brodde O-E, Beckeringh JJ, Michel MC (1987) Human heart β-adrenoceptors: a fair comparison with lymphocyte β-adrenoceptors? Trends Pharmacol Sci 8:403–407
19. Brodde O-E, Schüler S, Kretsch R, Brinkmann M, Borst HG, Hetzer R, Reidemeister JChr, Warnecke H, Zerkowski H-R (1986) Regional distribution of β-adrenoceptors in the human heart: coexistence of functional β_1- and β_2-adrenoceptors in both atria and ventricles in severe congestive cardiomyopathy. J Cardiovasc Pharmacol 8:1235–1242
20. Brodde O-E, O'Hara N, Zerkowski H-R, Rohm N (1984) Human cardiac β-adrenoceptors: both β_1- and β_2-adrenoceptors are functionally coupled to the adenylate cyclase in right atrium. J Cardiovasc Pharmacol 6:1184–1191
21. Brodde O-E, Motomura S, Endoh M, Schümann HJ (1978) Lack of correlation between the positive inotropic effect evoked by α-adrenoceptor stimulation and the levels of cyclic AMP and/or cyclic GMP in the isolated ventricle strip of the rabbit. J Mol Cell Cardiol 10:207–219
22. Brückner R, Mügge A, Scholz H (1985) Existence and functional role of alpha$_1$-adrenoceptors in the mammalian heart. J Mol Cell Cardiol 17:639–645
23. Brückner R, Meyer W, Mügge A, Schmitz W, Scholz H (1984) α-Adrenoceptor-mediated positive inotropic effect of phenylephrine in isolated human ventricular myocardium. Eur J Pharmacol 99:345–347
24. Chang HY, Klein RM, Kunos G (1982) Selective desensitization of cardiac beta receptors by prolonged in vivo infusion of catecholamines in rats. J Pharmacol Exp Ther 221:784–789
25. Cook N, Richardson A, Barnett DB (1984) Comparison of the β_1-selective affinity of prenalterol and corwin demonstrated by radioligand binding. Eur J Pharmacol 98:407–412
26. Corr PB, Shayman JA, Kramer JB, Kipnis RJ (1981) Increased α_1-adrenergic receptors in ischemic cat myocardium. J Clin Invest 67:1232–1236
27. Daly MJ, Levy GP (1979) The subclassification of β-adrenoceptors: evidence in support of the dual β-adrenoceptor hypothesis. In: Kalsner S (Ed) Trends in autonomic pharmacology Vol. I, Urban & Schwarzenberg, Baltimore Munich, pp 347–382
28. Devos C, Robberecht P, Nokin P, Waelbroeck M, Clinet M, Camus JC, Beaufort P, Schoenfeld P, Christophe J (1985) Uncoupling between beta-adrenoceptors and adenylate cyclase in dog ischemic myocardium. Naunyn-Schmiedeberg's Arch Pharmacol 331:71–75
29. Dillon JS, Gu XH, Nayler WG (1988) Alpha-1 adrenoceptors in ischaemic and reperfused myocardium. J Mol Cell Cardiol 20:725–735
30. Endoh M (1982) Adrenoceptors and the myocardial inotropic response: Do alpha and beta receptor sites functionally coexist? In: Kalsner S (Ed) Trends in autonomic pharmacology, Vol 2. Urban & Schwarzenberg, Baltimore Munich, pp 304–322
31. Francis GS, Cohn JN (1986) The autonomic nervous system in congestive heart failure. Ann Rev Med 37:235–247
32. Francis GS (1985) Neurohumoral mechanisms involved in congestive heart failure. Am J Cardiol 55:15A–21A
33. Gille E, Lemoine H, Ehle B, Kaumann AJ (1985) The affinity of (−)-propranolol for β_1- and β_2-adrenoceptors of human heart. Differential antagonism of the positive inotropic effects and adenylate cyclase stimulation by (−)-noradrenaline and (−)-adrenaline. Naunyn-Schmiedeberg's Arch Pharmacol 331:60–70
34. Glaubiger G, Lefkowitz RJ (1977) Elevated beta-adrenergic receptor number after chronic propranolol treatment. Biochem Biophys Res Commun 78:720–725
35. Heilbrunn SM, Shah P, Bristow MR, Valantine HA, Ginsburg R, Fowler MB (1989) Increased β-receptor density and improved hemodynamic response to catecholamine stimulation during long-term metoprolol therapy in heart failure from dilated cardiomyopathy. Circulation 79:483–490
36. Hjalmarson A, Elmford D, Herlitz J, Holmberg S, Malek I, Nyberg G, Ryden L, Svedberg K, Vedin A, Waagstein F, Waldenstroem A, Waldenstroem J, Wedel H, Wilhelmsson L, Wilhelmsson C (1981) Effect on mortality of metoprolol in acute myocardial infarction. Lancet 2:823–827
37. Jones CR, Molenaar P, Summers RJ (1989) New views of human cardiac β-adrenoceptors. J Mol Cell Cardiol 21:519–535
38. Kaumann AJ, Hall JA, Murray KJ, Wells FC, Brown MJ (1989) A comparison of the effects of adrenaline and noradrenaline on human heart: the role of β_1- and β_2-adrenoceptors in the stimulation of adenylate cyclase and contractile force. Eur Heart J 10 (Suppl B):29–37
39. Kohl C, Schmitz W, Scholz H, Scholz J, Toth M, Döring V, Kalmar P (1989) Evidence for alpha$_1$-adrenoceptor-mediated increase of inositol trisphosphate in the human heart. J Cardiovasc Pharmacol 13:324–327
40. Lands AM, Arnold A, McAuliff JP, Luduena FP, Brown TG (1967) Differentiation of receptor system activated by sympathomimetic amines. Nature 214:597–598

41. Maisel AS, Motulsky HJ, Insel PA (1987) Propranolol treatment externalizes β-adrenergic receptors in guinea pig myocardium and prevents further externalization by ischemia. Circ Res 60:108–112

42. Maisel AS, Motulsky HJ, Ziegler MG, Insel PA (1987) Ischemia- and agonist-induced changes in α- and β-adrenergic receptor traffic in guinea pig hearts. Am J Physiol 253:H1159–H1167

43. Maisel AS, Motulsky HJ, Insel PA (1985) Externalization of β-adrenergic receptors promoted by myocardial ischemia. Science 230:183–186

44. McDevitt DG (1989) In vivo studies on the function of cardiac β-adrenoceptors in man. Eur Heart J 10 (Suppl B):22–28

45. McPherson GA (1985) Analysis of radioligand binding experiments: a collection of computer programs for the IBM PC. J Pharmacol Methods 14:213–228

46. Michel MC, Maisel AS, Brodde O-E (1990) Mitigation of β_1- and/or β_2-adrenoceptor function in human heart failure. Br J Clin Pharmacol (Suppl), in press

47. Michel MC, Pingsmann A, Beckeringh JJ, Zerkowski H-R, Doetsch N, Brodde O-E (1988) Selective regulation of β_1- and β_2-adrenoceptors in the human heart by chronic β-adrenoceptor antagonist treatment. Br J Pharmacol 94:685–692

48. Minneman KP (1988) α_1-Adrenergic receptor subtypes, inositol phosphates, and sources of cell Ca^{2+}. Pharmacol Rev 40:87–119

49. Minneman KP, Pittman RN, Molinoff PB (1981) β-Adrenergic receptor subtypes: properties, distribution and regulation. Ann Rev Neurosci 4:419–461

50. Motomura S, Zerkowski H-R, Daul A, Brodde O-E (1990) On the physiologic role of beta$_2$-adrenoceptors in the human heart: in vitro and in vivo studies. Am Heart J 119:608–619

51. Motomura S, Khamssi M, Zerkowski H-R, Brodde O-E (1989) Is there a receptor reserve for isoprenaline in the human heart? Eur Heart J 10 (Suppl):427 (abstr.)

52. Mukherjee A, Bush LR, McCoy KE, Duke RJ, Hagler H, Buja LM, Willerson JT (1982) Relationship between β-adrenergic receptor numbers and physiological responses during experimental canine myocardial ischemia. Circ Res 50:735–741

53. Mukherjee A, Hogan M, McCoy K, Buja LM, Willerson JT (1980) Influence of experimental myocardial ischemia on alpha$_1$-adrenergic receptors. Circulation 64 (Suppl III):III-149 (abstr.)

54. Mukherjee A, Wong TM, Buja LM, Lefkowitz RJ, Willerson JT (1979) Beta-adrenergic and muscarinic cholinergic receptors in canine myocardium. J Clin Invest 64:1423–1428

55. Nuttall A, Snow HM (1982) The cardiovascular effects of ICI 118,587: a β_1-adrenoceptor partial agonist. Br J Pharmacol 77:381–388

56. Scatchard G (1949) The attraction of proteins for small molecules and ions. Ann N Y Acad Sci 51:660–672

57. Schmitz W, Kohl C, Neumann J, Scholz H, Scholz J (1989) On the mechanism of positive inotropic effects of alpha-adrenoceptor agonists. Basic Res Cardiol 84 (Suppl. 1):23–33

58. Schmitz W, Scholz H, Scholz J, Steinfarth M, Lohse M, Puurunen J, Schwabe U (1987) Pertussis toxin does not inhibit the alpha$_1$-adrenoceptor-mediated effect on inositol phosphate production in the heart. Eur J Pharmacol 134:377–378

59. Schoemig A, Dart AM, Dietz R, Mayer E, Kübler W (1984) Release of endogenous catecholamines in the ischemic myocardium of the rat. Part A: locally mediated release. Circ Res 55:689–701

60. Sleight PC (1986) Use of beta-adrenoceptor blockade during and after acute myocardial infarction. Ann Rev Med 36:415–425

61. Starke K (1981) α-Adrenoceptor subclassification. Rev Physiol Biochem Pharmacol 88:199–236

62. Stiles GL, Caron MG, Lefkowitz RJ (1984) β-Adrenergic receptors: biochemical mechanisms of physiological regulation. Physiol Rev 64:661–743

63. Strasser RH, Krimmer J, Marquetant R (1988) Regulation of β-adrenergic receptors: impaired desensitization in myocardial ischemia. J Cardiovasc Pharmacol 12 (Suppl 1):S15–S24

64. Timmermans PBMWM (1989) α-Adrenoceptors. In: Williams M, Glennon RA, Timmermans PBMWM (Eds) Receptor pharmacology and function. Marcel Dekker, New York, pp 173–205

65. Vatner DE, Knight DR, Shen YT, Thomas JX, Homcy CJ, Vatner SF (1988) One hour of myocardial ischemia in conscious dogs increases β-adrenergic receptors, but decreases adenylate cyclase activity. J Mol Cell Cardiol 20:75–82

66. Vedin JA, Wilhelmsson CE (1983) Beta-receptor blocking drugs in the secondary prevention of coronary heart disease. Ann Rev Pharmacol Toxicol 23:29–44

67. Waagstein F, Caidahl K, Wallentin I, Bergh C-H, Hjalmarson A (1989) Long-term β-blockade in dilated cardiomyopathy. Effects of short- and long-term metoprolol treatment followed by withdrawal and readministration of metoprolol. Circulation 80:551–563

68. Waagstein F, Hjalmarson A, Swedberg K, Wallentin I (1983) Beta-blockers in dilated cardiomyopathy: they work. Eur Heart J 4:173–178

69. Wagner J, Brodde O-E (1979) On the presence and distribution of α-adrenoceptors in the heart of various mammalian species. Naunyn-Schmiedeberg's Arch Pharmacol 302:239–254
70. Wellstein A, Belz GG, Palm D (1988) Beta adrenoceptor subtype binding activity in plasma and beta blockade by propranolol and beta-1 selective bisoprolol in humans. Evaluation with Schild-plots. J Pharmacol Exp Ther 246:328–337
71. Yabuuchi Y (1977) The β-adrenoceptor stimulant properties of OPC 2009 on guinea-pig isolated tracheal, right atrial and left atrial preparations. Br J Pharmacol 61:513–521
72. Yoshizaki S, Tanimura K, Tamada S, Yabuuchi Y, Nagakawa N (1976) Sympathomimetic amines having a carbostyril nucleus. Med Chem 19:1138–1142
73. Zerkowski H-R, Ikezono K, Rohm N, Reidemeister JChr, Brodde O-E (1986) Human myocardial β-adrenoceptors: demonstration of both β_1- and β_2-adrenoceptors mediating contractile responses to β-agonists on the isolated right atrium. Naunyn-Schmiedeberg's Arch Pharmacol 332:142–147

Author's address:

Prof. Dr. Otto-Erich Brodde, Zentrum für Innere Medizin, Medizinische Klinik und Poliklinik, Abteilung für Nieren- und Hochdruckkrankheiten, Universitätsklinikum, Hufelandstr. 55, D-4300 Essen 1, FRG.

Supersensitivity of the Adenylyl Cyclase System in Acute Myocardial Ischemia: Evaluation of Three Independent Mechanisms

R. H. Strasser and R. Marquetant

Department of Cardiology, Medical Center, University of Heidelberg, Heidelberg, FRG

Summary

Malignant arrhythmias and the spreading of the infarcted zone in acute myocardial ischemia may be influenced by the sympathetic system. It has been known for quite some time that acute ischemia leads to an increased release of endogenous catecholamines. Adaptive mechanisms at the postsynaptic level such as receptor desensitization, which are operative under normoxic conditions, are abolished in acute myocardial ischemia. On the contrary, three newly characterized, distinct mechanisms lead to a transiently increased activity of the β-adrenergic system in the early phase of acute ischemia: 1) Functionally coupled β-adrenergic receptors are rapidly and persistently increased at the cell surface due to the impairment of β-agonist-promoted uncoupling and internalization. 2) Despite the reversible increase of inhibitory, muscarinic M_2 receptors, the inhibitory pathway of the adenylyl cyclase systems becomes ineffective since the coupling protein, G_i, is rapidly impaired. Both the G_i-linked GTPase-activity and the binding of $[\gamma\text{-}^{35}S]GTP$ are reduced by 25–30% without any loss of the total protein. Stimulatory effects prevail at the G-protein level since in the early period of ischemia the stimulatory G-protein, G_s, remains intact. 3) The adenylyl cyclase is transiently sensitized by about 30%. This increased activity is closely associated with the partially purified enzyme and may be due to a rapidly reversible covalent modification. Prolonged ischemia, in contrast, results in a general decrease of the cyclase activity notwithstanding any changes at the receptor or G-protein level. The individual mechanisms may play distinct and/or complimentary roles in the early sensitization of the adenylyl cyclase system in acute myocardial ischemia.

Introduction

Acute myocardial ischemia may become a life-threatening event when malignant arrhythmias or the development of heart failure turn unmanageable. Both malignant arrhythmias and the spreading of the infarcted zone followed by heart failure are greatly influenced by the sympathetic system [5, 6, 26, 42]. Adaptive regulation of the sympathetic system could help to prevent the unwanted outcome [47]. But in contrast to expectations, the sympathetic system is dramatically altered at all levels during acute myocardial ischemia. These alterations include presynaptic [4, 50–52] and postsynaptic mechanisms [4, 62]. Presynaptically, the adrenergic system is inadequately activated, leading to an increased outflow of catecholamines, predominantly norepinephrine [4, 50]. Central mechanisms, including pain, induce an increased discharge of presynaptic neurotransmitters.

Moreover, specific local mechanisms such as the reverse orientation of the neuronal uptake

Supported by the Deutsche Forschungsgemeinschaft, Bonn, Sonderforschungsbereich 320 enhance this release of norepinephrine in acute ischemia [51–53]. This mechanism also prevents the removal of neurotransmitters from the subsynaptic cleft [34]. Consequently, large quantities of endogenous catecholamines are present at the receptor site. In consequence, presynaptic regulation mechanisms fail to prevent or counterbalance the increase of sympathetic outflow during acute myocardial ischemia. Additional postsynaptic mechanisms, which are operative under normoxic conditions, could enable the cells to protect themselves quite effectively from a sympathetic overflow. Prolonged exposure to catecholamines induces an unresponsiveness at the postsynaptic, i.e., cellular level [21, 36, 57]. This unresponsiveness, also called tachyphylaxis or desensitization, is intimately linked to activation of receptors by catecholamines or receptor-agonists [1, 2, 36, 65]. For the two most important receptor systems of endogenous catecholamines, the β- and the α-adrenergic systems desensitization has been described in many model systems [11, 35]. This desensitization may not only include the receptors with internalization and downregulation, but also the other components of the adenylyl cyclase system. Briefly, the adenylyl cyclase is regulated by a stimulatory and an inhibitory pathway [15, 17–19]. These pathways include stimulatory receptors, such as β-adrenergic receptors, or inhibitory receptors, such as muscarinic M_2 receptors, and distinct stimulatory or inhibitory guanine nucleotide binding proteins, G-proteins, and finally specific effectors, such as the adenylyl cyclase or phospholipase C. The primary structure of all these components has been revealed by their successful cloning [12, 13, 23, 30–32, 46] and indicates a close relationship of these components.

The primary site for desensitization are the receptors. For both α- and β-adrenergic receptors it has been shown in many model systems that desensitization leads to a rapid loss of receptors at the cell surface by internalization, which after prolonged desensitization is followed by a down regulation of receptors [21, 22, 25]. Bearing in mind that in acute myocardial ischemia large quantities of endogenous catecholamines are present at the receptor site, adaptive regulation should lead to a completely desensitized and unresponsive receptor system with a decrease of receptors at the cell surface. However, a large body of data in several model systems for acute myocardial ischemia has accumulated, indicating that, on the contrary, both receptor systems are largely activated [5, 10, 62, 63]. Corr and his group could demonstrate in a feline model that the density of α_1-adrenergic receptors is increased in the ischemic heart [5, 14]. Consequently, blockade of α_1-adrenergic receptors was able to partially prevent α_1-receptor-mediated arrhythmias in acute myocardial ischemia [3, 24, 43, 54, 67]. Data on the α-adrenergic system are reviewed extensively elsewhere [24]. Focus of the present paper is to briefly review the most recent data on the alterations of the β-adrenergic system and the adenylyl cyclase system in acute myocardial ischemia.

The adenylyl cyclase system in acute myocardial ischemia

In the very early phase of myocardial ischemia the β-adrenergic system is inadequately sensitized [60]. This sensitization involves several components of the signal transduction pathway, i.e., the receptors, the G-proteins and, finally, the effector enzyme, the adenylyl cyclase. Quite distinct mechanisms contribute to this sensitization of the individual components.

1. β-Adrenergic receptors in acute myocardial ischemia

Despite high concentrations of endogenous catecholamines present at the receptor site the density of β-adrenergic receptors at the cell surface rapidly increases in acute myocardial ischemia. This increase has been observed both in vivo and in vitro [9, 16, 28, 39, 40, 62]. The earliest time point tested was after 15 min of ischemia [60, 62]. In several model systems, both in vivo and in vitro, the initial increase amounted to about 30–40% after a 15 min period of ischemia, which was followed by a further rise after 30 and 50 min of ischemia. Even after prolonged periods of ischemia (up to 5 days) this increase of β-adrenergic receptors persisted although the functional response to β-adrenergic stimulation had already dramatically decreased in the infarcted zone [61, 62, 70]. As shown in Fig. 1, the rise of β-adrenergic receptors

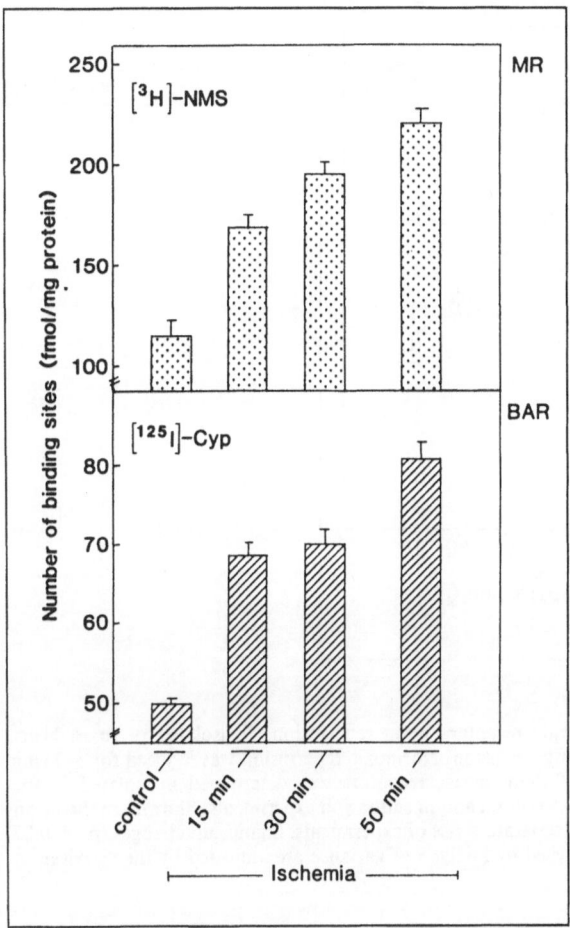

Fig. 1. Increase of muscarinic M_2- and β-adrenergic receptors in acute myocardial ischemia. Isolated rat hearts were perfused according to the Langendorff preparation [33]. Ischemia (15–50 min) was induced by stopping perfusion. The density of muscarinic receptors or β-adrenergic receptors in the plasma membranes of isolated perfused rat hearts was determined by saturation experiments using the radioligand N-methyl-scopolamine ([^{3}H]NMS) or Iodocyanopindolol ([^{125}I]CYP), respectively. Shown are the means ($\pm$ SEM) of maximal binding capacities of six experiments. Binding affinities of the receptors were unchanged.

occurs rapidly and is quite rapidly reversible upon reperfusion (Fig. 2). Similarly, other receptors, such as the muscarinic receptors, increase in the plasma membranes after an ischemic insult (Fig. 1). As shown in Fig. 1, the time-course of this increase is quite similar to that of β-adrenergic receptors. Again, reperfusion leads to a rapid return to control values (Fig. 3). As described by Corr et al. [5, 6] α_1-adrenergic receptors increase in a similar way, although not in all model systems used so far [10, 58]. These data indicate that it may not be a β-receptor-specific effect that leads to an increase of receptors at the cell surface. More likely is a "non-specific" membrane-linked alteration, which may include several receptor-systems independent of their activation. However, an increased responsiveness in the early phase of global ischemia could be demonstrated only for the β-adrenergic system.

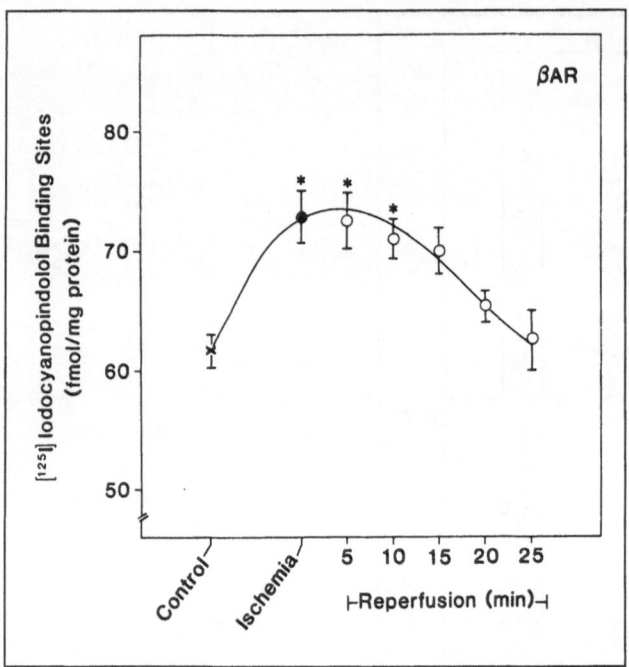

Fig. 2. Rapid normalization of β-adrenergic receptors upon reperfusion. In isolated perfused hearts ischemia was induced for 15 min by stopping perfusion. Normoxic reperfusion was induced for 5–25 min after the ischemic period. The density of β-adrenergic receptors was determined in isolated plasma membranes using [^{125}I]Iodocyanopindolol as radioligand in saturation experiments. Shown are the means (± SEM) of maximal binding sites of three separate series of experiments. Significant changes (p < 0.02) compared to the normoxic control as analyzed by analysis of variance are indicated by the asterisks.

Impairment of receptor internalization

Interestingly, in the ischemic zone the total number of receptors was not altered, but the density of receptors at the cell surface did increase. This increase occurred at the expense of intracellular receptors [37, 38]. This shift of receptors from the intracellular compartment to the plasma membranes was postulated to be due to an increased externalization of β-adrenergic receptors [37, 38]. However, direct determination of externalization of receptors has not been possible so far. Additionally, an activation of externalization, which is known to be an energy-dependent process, appears to be unlikely in acute myocardial ischemia, resulting in the rapid loss of high-energy phosphates [27]. To test the hypothesis, that not an increased externalization, but perhaps an impairment of physiological, agonist-induced internalization may shift the balance towards an increase of receptors at the cell surface, a model for ischemia with energy depletion despite continuous perfusion was developed using isolated perfused rat hearts. In analogy to acute ischemia, perfusion with cyanide in the absence of glucose induces a rapid loss of high-energy phosphates and an increase of β-adrenergic receptors in the plasma membranes [61]. Additional superfusion with β-agonists, which under normoxic conditions leads to a complete desensitization of the β-adrenergic system, failed to provoke an internalization of β-adrenergic receptors [61]. Even at highest concentrations the density of β-adrenergic receptors remained persistently increased in the plasma membranes. These data indicate that, in the energy-depleted, ischemic hearts, not externalization, but the loss of receptor internalization is responsible for the shift of receptors to the cell surface.

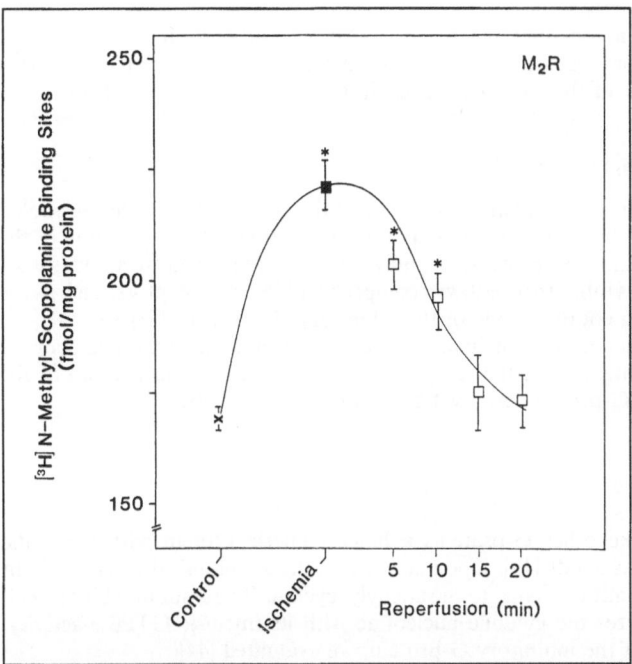

Fig. 3. Effect of reperfusion on muscarinic receptors. After 15 min of ischemia normoxic reperfusion was induced for 5–25 min. In isolated plasma membranes the density of muscarinic receptors was determined using the muscarinic antagonist N-methylscopolamine in saturation experiments. Depicted are the means ($\pm$ SEM) of specific binding sites of three series of experiments. Non-specific binding was determined as the residual binding in the presence of atropin (10^{-5} M). Significant changes (p < 0.01) compared to the controls are indicated by the asterisks. The affinity of the receptors for the ligands did not change with ischemia or reperfusion.

Persistent functional coupling of β-adrenergic receptors

Even the first step of β-agonist-induced desensitization, i.e., the functional uncoupling of the receptors from the stimulatory G-protein, G_s, is impaired in acute myocardial ischemia.

In equilibrium binding studies coupling of the receptors to the G-protein can be determined only indirectly by evaluating the ability of the receptors to form the β-agonist-promoted high affinity state [7, 8, 56]. β-Agonist competition curves in the absence of guanine nucleotides are shallow and biphasic, indicating that in the controls about 50% of the receptors bind the agonist with high affinity, whereas about 50% bind it only with low affinity. In normoxic hearts after perfusion with desensitizing concentrations of catecholamines all the receptors in the so-called high affinity state are transferred to receptors only capable of forming the so-called low affinity state. These data indicate that with desensitization these receptors become functionally uncoupled from the coupling protein G_s [56, 61].

In acute myocardial ischemia, however, even high concentrations of norepinephrine present at the site of the receptor are not able to functionally uncouple the β-adrenergic receptors [61]. These data indicate that in the energy-depleted heart the β-agonist-promoted uncoupling of receptors may be impaired. To test this hypothesis more rigorously, cyanide-perfused hearts were additionally perfused with increasing doses of the β-agonist isoproterenol. Even highest doses (1×10^{-5} M) were unable to induce a loss of receptors in the "high-affinity" state. The agonist-competition curves remained shallow and biphasic, indicating that even the first step of β-agonist-promoted desensitization, i.e., the process of functional uncoupling from the

71

G_s-protein is impaired in acute myocardial ischemia [61]. Consequently, this increased amount of functionally coupled receptors resulting in the persistent sensitization at the receptor level may add to the supersensitivity of the adenylyl cyclase in the ischemic heart (see below).

2. G-Proteins in acute myocardial ischemia

To transmit the neuronal or hormonal signal to the inside of the cell, adenylyl cyclase-coupled receptors initially interact with the so-called coupling or G-proteins [18, 19]. Activation of the adenylyl cyclase via the stimulatory G-protein, G_s, may be counterbalanced by activation of the inhibitory pathway [29]. This inhibitory pathway comprises inhibitory receptors, such as the muscarinic M_2 receptors, which couple to one of the inhibitory G-proteins, G_i [20].

In acute myocardial ischemia both G-proteins are altered, however, at different time-courses, which, in addition to the sensitization at the receptor level, enhance the sensitization of the adenylyl cyclase system at the G-protein level, yet by a different mechanism.

Rapid loss of functional G_i

The inhibitory G-protein is like other G-proteins a heterotrimeric protein with a specific α-subunit and non-specific β/γ-subunits [19]. Upon activation, the α_i-subunit dissociates from the β/γ-subunits to further transmit the signal to the adenylyl cyclase. The α-subunit binds GTP and upon activation it hydrolyzes the guanine nucleotide with its intrinsic GTPase-activity.

To determine the activity of the inhibitory G-protein, we evaluated [44]:
1) its receptor-activated GTPase activity,
2) its ability to bind the non-hydrolyzable $[\gamma\text{-}^{35}S]GTP$, and
3) the total amount of its α-subunit, using Western blot analysis.

Already 15 min after the onset of ischemia the binding of $[\gamma\text{-}^{35}S]GTP$ to the G-protein was decreased by about 30% [44]. $[\gamma\text{-}^{35}S]GTP$ binds to all α-subunits of G-proteins, including the stimulatory G_s-protein. Nonetheless, binding of $[\gamma\text{-}^{35}S]GTP$ in the heart predominantly represents binding to the inhibitory G-protein, since the amount of $G_{i\alpha}$ in the heart outweighs other α-subunits by more than tenfold [15, 19]. Thus, the reduced binding of $[\gamma\text{-}^{35}S]GTP$ indicates that the function of the α-subunit of G_i is rapidly reduced or blunted.

Additionally, this rapid reduction of the $[\gamma\text{-}^{35}S]GTP$ binding is accompanied by a rapid decrease of the carbachol-stimulated GTPase-activity [45]. The slope of the dose-response curve is unaltered after acute myocardial ischemia, however, its total activity is reduced by 25–30%. Interestingly, both basal and maximally stimulated GTPase-activity are decreased. These data indicate that in the very early phase of the ischemic insult the function of the inhibitory G-protein may be impaired. This impairment is completely reversible upon reperfusion [41]. The molecular mechanism of this functional decrease is as yet unidentified. However, using Western blot analysis it could be demonstrated that the total amount of $G_{i\alpha}$ is completely unchanged. These data suggest that a regulatory modulation, possibly a covalent modification of the α-subunit, may be responsible for this functional impairment of the inhibitory G-protein in acute myocardial ischemia. Thus, by losing the tonic inhibition of the adenylyl cyclase, this functional impairment may add a second mechanism to the transient sensitization of the adenylyl cyclase system in the ischemic heart.

Slow decrease of the stimulatory G-protein, G_s

In contrast, the stimulatory G-protein, G_s, loses its function more slowly after the onset of the ischemic insult. As shown in dose-response curves to β-adrenergic stimulation the increased density of functionally coupled receptors is shifted to the left after 15 min of global ischemia [61]. These data demonstrate that the G_s-protein present at this time is capable of transmitting even the increased β-adrenergic stimulation to the cell. After 1 h of ischemia, Susanni et al. [66]

could demonstrate that the amount of the α-subunit of the G_s-protein was decreased. Using Western blot analysis and functional reconstitution they could demonstrate that, at this time point, both the total amount and function of G_s were significantly reduced [66]. With the inhibition maximally reducing the stimulated activity by about 25–30% and predominantly modulating the stimulated but not the basal activity, the impairment of the stimulatory pathway outweighs the preceding reduction of the inhibitory pathway [44].

These data are in good agreement with previously published data showing that after 1 h of global ischemia the response to β-adrenergic stimulation was greatly reduced despite the persistent increase of β-adrenergic receptors at the cell surface [64]. The imbalance of functional G-proteins in the very early period of ischemia – with the rapid loss of G_i preceding the slower loss of the G_s-protein – may further contribute to the genesis of malignant arrhythmias. Such electrophysiological effects may be mediated by direct activation of calcium channels [72], sodium channels [49, 55], or potassium channels [71–73].

3. Regulation of the adenylyl cyclase in acute myocardial ischemia

The third site of regulation of the adenylyl cyclase system in acute myocardial ischemia is the adenylyl cyclase itself. The transient changes of the adenylyl cyclase activity in acute myocardial ischemia call for a detailed temporal analysis. Therefore, we divided the time-course into an early phase of ischemia (0–20 min) and a period after prolonged duration of ischemia (20–60 min).

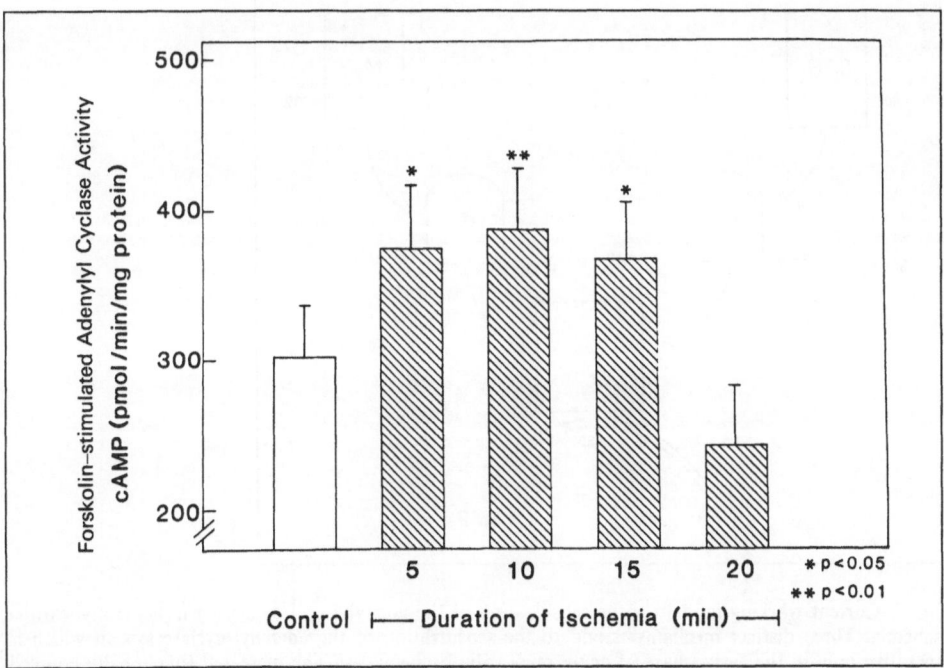

Fig. 4. Transient sensitization of the adenylyl cyclase in early myocardial ischemia. In isolated perfused hearts global ischemia was induced for 5–20 min. In purified plasma membranes adenylyl cyclase in response to forskolin (10^{-4} M) was determined by the method of Salomon et al. [48]. Shown are the means ($\pm$ SEM) of six series of experiments with triplicate determinations.

Transient sensitization of the adenylyl cyclase in early myocardial ischemia

In the very early phase of global ischemia (5–15 min) the activity of the adenylyl cyclase is rapidly increased [61]. This increased activity includes basal adenylyl cyclase activity, β-agonist-, sodium fluoride- or forskolin-stimulated adenylyl cyclase activity [64]. The detailed time-course of the sensitization process is shown in Fig. 4. Since in intact plasma membranes the effect of other components of the adenylyl cyclase system could not be excluded (including the effect of G-proteins, see above), the adenylyl cyclase was isolated from the plasma membranes and was partially purified using solubilization and affinity chromatography. Interestingly, the sensitization of basal adenylyl cyclase was no longer present in the purified preparation, indicating that for this sensitization the loss of membrane interaction (possibly, the loss of inhibition) plays a crucial role. In contrast, stimulated adenylyl cyclase activities remained sensitized to the same extent after partial purification [61]. These data indicate that acute myocardial ischemia leads to an enzyme-linked sensitization of the adenylyl cyclase system, which remains tightly associated to the partially purified enzyme. Up to now, it was not possible to directly identify the molecular mechanism of this unique sensitization process of the adenylyl cyclase. However, it is a process which, upon reperfusion, is very rapidly reversible and is closely associated with the partially purified protein [59]. These data suggest that it may be a covalent modification of the enzyme which, as the third mechanism, contributes to the sensitization of the adenylyl cyclase in the early period of ischemia. The three different mechanisms are briefly summarized in Fig. 5.

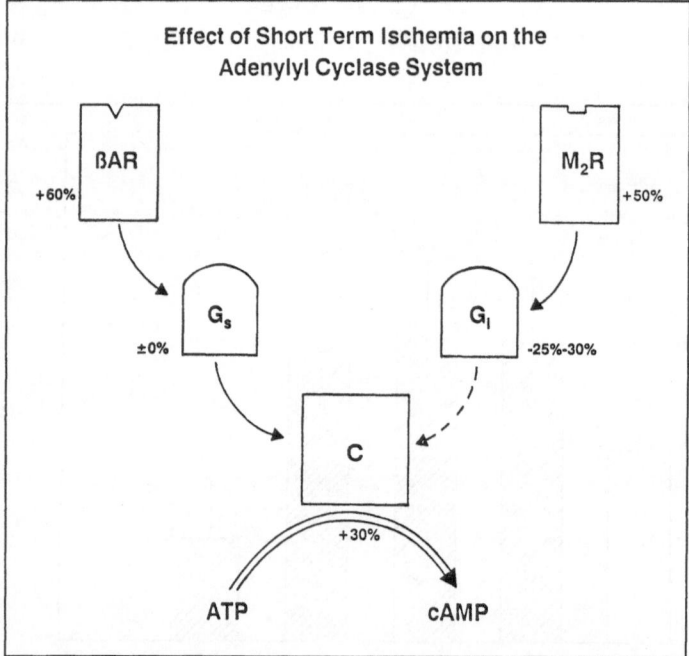

Fig. 5. Current working model for the sensitization of the adenylyl cyclase system in acute myocardial ischemia. Three distinct mechanisms add to the sensitization of the adenylyl cyclase system which is operative only in the early phase of acute myocardial ischemia: a) The increased functionally coupled β-adrenergic receptors (βAR) mediate an augmented responsiveness to β-adrenergic stimulation via the intact G_s-protein. b) Tonic inhibition is lost due to the functional impairment of the inhibitory G-protein, G_i, despite the increase of inhibitory muscarinic M_2 receptors (M_2R). c) Independently, the catalyst adenylyl cyclase (C) transiently has increased activity due to a covalent modification of the enzyme. All three sensitization processes are rapidly reversible upon reperfusion.

Inactivation of adenylyl cyclase after prolonged periods of myocardial ischemia

With prolonged ischemia this supersensitivity of the adenylyl cyclase is rapidly replaced by a decreased activity of the enzyme [61, 64, 68, 69]. In global ischemia this decrease is already obvious after 20–30 min [61]. All stimulated activities are greatly reduced, including basal, β-agonist-stimulated, sodium fluoride-stimulated (i.e., via the G_s-protein) and the forskolin-stimulated adenylyl cyclase activity (Fig. 4). It is noteworthy that the isoproterenol-stimulated adenylyl cyclase activity is greatly reduced although the β-adrenergic receptors remain persistently increased and sensitized (see above). In consequence, the sensitization at the receptor level after prolonged periods of ischemia has no functional significance with regard to the activation of the adenylyl cyclase. At this point it cannot be decided if perhaps other signal transduction pathways, especially channels, may be influenced by these persistently sensitized β-adrenergic receptors.

In model systems with less complete ischemia or local ischemia the time-courses of this inactivation of the adenylyl cyclase were observed to be only slightly different. Thus, Vatner et al. [70] observed that in vivo ligation of the left anterior descending coronary artery in dogs resulted in a decreased adenylyl cyclase activity 1 h after the onset of the ischemic insult. In contrast to the global ischemic model in the isolated perfused rat, they found a loss of β-adrenergic receptors in the high-affinity state. However, they had very low concentrations of functionally coupled receptors in the controls to start with [70]. These slight differences in time-course with the general regulation mechanisms being quite similar may explain that, even within a single heart, areas with a supersensitive and/or unresponsive sympathetic system may be in close proximity. Varying degrees of collateral flow, especially in the borderzone, may add to this local inhomogeneity of the adrenergic system. The differing effects on membrane potential and conductivity may promote malignant arrhythmias in the ischemic heart.

To date, the molecular mechanism that leads to the reduced adenylyl cyclase activity in late ischemia is not known. Preliminary data, however, indicate that this decreased enzyme activity is not rapidly reversible. This may suggest that the enzyme itself is degraded, probably due to proteolytic activity which is released in the severely ischemic myocardium.

References

1. Benovic JL, Bouvier M, Caron, MG Lefkowitz RJ (1988) Regulation of adenylyl cyclase-coupled β-adrenergic receptors. Annu Rev Cell Biol 4:405–428
2. Benovic JL, Strasser RH, Caron MG, Lefkowitz RJ (1986) Beta-adrenergic receptor kinase: Identification of a novel protein kinase which phosphorylates the agonist-occupied form of the receptor. Proc Natl Acad Sci USA 83:2797–2801
3. Bolli R, Fisher DJ, Taylor AA, Young JB, Miller RR (1984) Effect of α-adrenergic blockade on arrhythmias induced by acute myocardial ischemia and reperfusion in the dog. J Mol Cell Cardiol 16: 1101–1117
4. Carlsson L, Abrahamson T, Almgren O (1985) Local release of myocardial norepinephrine during acute ischemia: An experimental study in the isolated rat heart. J Cardiovasc Pharmacol 7:791–798
5. Corr PB, Shayman JA, Kramer JB, Kipnis RJ (1982) Increased α-adrenergic receptors in ischemic cat myocardium. A potential mediator of electrophysiological derangements. J Clin Invest 67:1232–1236
6. Corr PB, Witkowski FX, Sobel BE (1978) Mechanisms contributing to malignant dysrhythmias induced by ischemia in the cat. J Clin Invest 61:109–119
7. De Lean A, Munson PJ, Rodbard D (1978) Simultaneous analysis of families of sigmoidal curves: Application to bioassay and physiological dose-response curves. Am J Physiol 235:E97–E102
8. De Lean A, Stadel JM, Lefkowitz RJ (1980) A ternary complex model explains the agonist-specific binding properties of the adenylate cyclase-coupled beta-adrenergic receptor. J Biol Chem 255:7108–7117
9. Devos C, Robberecht P, Waelbroeck M, Clienet M, Beaufort P, Schoenfield P, Christophe J (1985) Uncoupling between β-adrenoceptors and adenylate cyclase in dog ischemic myocardium. Naunyn-Schmiedeberg's Arch Pharmacol 331:71–75
10. Dillon JS, Gu XH, Nayler WG (1988) Alpha$_1$ adrenoceptors in the ischemic and reperfused myocardium. J Mol Cell Cardiol 20:725–735
11. Dillon-Carter O, Chuang D-M (1989) Homologous desensitization of muscarinic cholinergic, histami-

nergic, adrenergic, and serotonergic receptors coupled to phospholipase C in the cerebellar granule cells. J Neurochem 52:598–603

12. Dixon RAF, Kobilka BK, Strader DJ, Benovic JL, Dohlman HG, Frielle T, Bolanowski MA, Bennett CD, Rands E, Diehl RE, Mumford RA, Slater EE, Sigal IS, Caron MG, Lefkowitz RJ, Strader CD (1986) Cloning of the gene and cDNA for mammalian beta-adrenergic receptor and the homology with rhodopsin. Nature 321:75–79

13. Dixon RAF, Sigal IS, Candelore MR, Register RB, Scattergood W, Rands E, Strader CD (1987) Structural features required for ligand binding to the β-adrenergic receptor. EMBO J 6:3269–3275

14. Dobmeyer DJ, Kekec BK, Sobel BE, Corr PB (1988) Alpha-1 adrenergic mediated accumulation of lysophosphatidylcholine in isolated adult canine myocytes. Circulation 78 4:II-483 (abstr.)

15. Freissmuth M, Casey PJ, Gilman AG (1989) G proteins control diverse pathways of transmembrane signaling. FASEB J 3:2125–2131

16. Freissmuth M, Schütz W, Weindlmayer-Göttel M, Zimpfer M, Spiss CK (1987) Effects of ischemia on the canine myocardial β-adrenoceptor-linked adenylate cyclase. J Cardiovasc Pharmacol 10:568–574

17. Gilman AG (1984) G proteins and dual control of adenylate cyclase. Cell 36:577–579

18. Gilman AG (1987) G proteins: Transducers of receptor-generated signals. Annu Rev Biochem 56:615–649

19. Gilman AG (1989) G proteins and regulation of adenylyl cyclase. JAMA 262:1819–1825

20. Goyal RK (1989) Muscarinic receptor subtypes: Physiology and clinical implications. N Engl J Med 321:1022–1029

21. Harden TK (1983) Agonist-induced desensitization of the beta-adrenergic receptor-linked adenylate cyclase. Pharmacol Rev 35:5–32

22. Harden TK, Cotton CU, Waldo GL, Lutton JK, Perkins JP (1980) Catecholamine-induced alteration in the sedimentation behaviour of membrane-bound beta-adrenergic receptors. Science 210:441–443

23. Harris BA, Robishaw JD, Mumby SM, Gilman AG (1985) Molecular cloning of complementary DNA for the alpha subunit of the G protein that stimulates adenylate cyclase. Science 229:1274–1277

24. Heathers GP, Yamada KA, Pogwizd SM, Corr PB (1988) The contribution of α- and β-adrenergic mechanisms in the genesis of arrhythmias during myocardial ischemia and reperfusion. In: Kulbertus HE (ed) Neurocardiology, Futura Publishing, Mount Kisco, New York, pp 143–178

25. Hertel C, Coulter SJ, Perkins JP (1986) The involvement of cellular ATP in receptor-mediated internalization of epidermal growth factor and hormone-induced internalization of beta-adrenergic receptors. J Biol Chem 261:5974–5980

26. Janse MJ (1989) Why is increased adrenergic activity arrhythmogenic? In: Brachmann J, Schömig A (eds) Adrenergic System and Ventricular Arrhythmias in Myocardial Infarction, Springer Verlag, New York–Berlin–Heidelberg, pp 353–363

27. Jennings RB, Steenbergen C (1985) Nucleotide metabolism and cellular damage in myocardial ischemia. Ann Rev Physiol 47:727–749

28. Karliner JS, Stevens MB, Honbo N, Hoffman JIE (1989) Effects of acute ischemia in the dog on myocardial blood flow, beta receptors, and adenylate cyclase activity with and without chronic beta blockade. J Clin Invest 83:474–481

29. Katada T, Oinuma M, Ui M (1986) Mechanisms for inhibition of the catalytic activity of adenylate cyclase by the guanine nucleotide binding protein serving as the substrate of islet-activating protein, pertussis toxin. J Biol Chem 261:5215–5221

30. Kobilka BK, Dixon RAF, Frielle T, Dohlman HG, Bolanowski MA, Sigal IS, Yang-Feng TL, Francke U, Caron MG, Lefkowitz RJ (1987) cDNA for the human β₂-adrenergic receptor: A protein with multiple membrane-spanning domains and encoded by a gene whose chromosomal location is shared with that of the receptor for platelet-derived growth factor. Proc Natl Acad Sci USA 84:46–50

31. Kobilka BK, Frielle T, Dohlman HG, Bolanowski MA, Dixon RAF, Keller P, Caron MG, Lefkowitz RJ (1987) Delineation of the intronless nature of the genes for the human and hamster β₂-adrenergic receptor and their putative promoter regions. J Biol Chem 262:7321–7327

32. Kobilka BK, Matsui H, Kobilka TS, Yang-Feng TL, Francke U, Caron MG, Lefkowitz RJ, Regan JW (1987) Cloning, sequencing, and expression of the gene coding for the human platelet α₂-adrenergic receptor. Science 238:650–656

33. Langendorff O (1895) Untersuchungen am ueberlebenden Saeugetierherzen. Arch Gen Physiol 230:183–186

34. Langer SZ (1974) Presynaptic regulation of catecholamine release. Biochem Pharmacol 23:1793–1800

35. Lefkowitz RJ, Caron MG, Stiles GL (1984) Mechanism of membrane-receptor regulation. Biochemical, physiological, and clinical insights derived from studies of the adrenergic receptors. N Engl J Med 310:1570–1579

36. Lefkowitz RJ, Hausdorff WP, Caron MG (1990) Role of phosphorylation in desensitization of the β-adrenoceptor. Trends Pharmacol Sci 11:190–194
37. Maisel AS, Motulsky HJ, Insel PA (1985) Externalization of β-adrenergic receptors promoted by myocardial ischemia. Science 230:183–186
38. Maisel AS, Motulsky HJ, Ziegler MG, Insel PA (1987) Ischemia- and agonist-induced changes in α- and β-adrenergic receptor traffic in guinea pig hearts. Am J Physiol 253:H1159–H1166
39. Mukherjee A, Haghani Z, Brady J, Bush L, McBride W, Buja LM, Willerson JT (1983) Differences in myocardial α- and β-adrenergic receptor numbers in different species. Am J Physiol 245:H957–H962
40. Mukherjee A, Hogan M, McKoy K, Buja LM, Willerson JT (1980) Influence of experimental myocardial ischemia on alpha₁ adrenergic receptors. Circulation 62 suppl III: III-149 (abstr.)
41. Niroomand F, Weinbrenner Ch, Schwencke C, Bayer Th, Marquetant R, Strasser RH, Hasselbach W, Rauch B (1990) Reversible inhibition of the GTPase activity of the inhibitory G-protein after short time myocardial ischemia. Circulation (in press)
42. Penny WJ (1984) The deleterious effects of myocardial catecholamines on cellular electrophysiology and arrhythmias during ischemia and reperfusion. Eur Heart J 5:960–973
43. Penny WJ, Culling W, Lewis MJ, Sheridan DJ (1985) Antiarrhythmic and electrophysiological effects of alpha adrenoceptor blockade during myocardial ischemia and reperfusion in isolated guinea-pig heart. J Mol Cell Cardiol 17:399–409
44. Rauch B, Weinbrenner C, Marquetant R, Schwencke C, Beyer T, Kübler W, Hasselbach W, Strasser RH (1990) Reduction of GTPase activity of the inhibitory G-protein in acute myocardial ischemia. J Mol Cell Cardiol 22:PW50 (abstr.)
45. Rauch B, Weinbrenner C, Marquetant R, Schwencke C, Beyer T, Kübler W, Hasselbach W, Strasser RH (1990) The inhibitory G-protein in acute myocardial ischemia: Impairment of the carbachol-stimulated GTPase activity. In preparation
46. Robishaw JD, Russell DW, Harris BA, Smigel MD, Gilman AG (1986) Deduced primary structure of the α subunit of the GTP-binding stimulatory protein of adenylate cyclase. Proc Natl Acad Sci USA 83:1251–1255
47. Rona G (1985) Catecholamine cardiotoxicity. J Mol Cell Cardiol 17: 291–306
48. Salomon Y, Londos C, Rodbell MA (1974) A highly sensitive adenylate cyclase assay. Anal Biochem 58:541–548
49. Schubert B, VanDongen AMJ, Kirsch GE, Brown AM (1989) β-Adrenergic inhibition of cardiac sodium channels by dual G-protein pathways. Science 245:516–519
50. Schömig A, Dart AM, Dietz R, Mayer E, Kübler W (1984) Release of endogenous catecholamines in the ischemic myocardium of the rat. Circ Res 55:689–701
51. Schömig A, Fischer S, Kurz T, Richardt G, Schömig E (1987) Nonexocytotic release of endogenous noradrenaline in the ischemic and anoxic heart: Mechanism and metabolic requirements. Circ Res 60:194–205
52. Schömig A, Kurz T, Richardt G, Schömig E (1988) Neuronal sodium homoeostasis and axoplasmic amine concentration determine calcium-independent noradrenaline release in normoxic and ischemic rat heart. Circ Res 63:1–13
53. Schömig A, Strasser RH, Richardt G (1990) Release and effects of catecholamines in myocardial ischemia. In: Piper HM (ed) Pathophysiology of Severe Ischemic Myocardial Injury, Kluwer Academic Press, Dordrecht, pp 381–412
54. Sheridan DJ, Penkoske AA, Sobel BE, Corr PB (1980) Alpha adrenergic contributions to dysrhythmias during myocardial ischemia and reperfusion in cats. J Clin Invest 65:161–171
55. Sokolovsky M, Cohen-Armon M (1988) Cross-talk between receptors: Muscarinic receptors, sodium channels, and guanine nucleotide-binding protein(s) in rat membrane preparations and synaptoneurosomes. Adv Second Messenger Phosphoprotein Res 21:11–18
56. Stadel JM, De Lean A, Lefkowitz RJ (1980) A high affinity agonist-beta-adrenergic receptor complex is an intermediate for catecholamine stimulation of adenylate cyclase in turkey erythrocyte membranes. J Biol Chem 255:1436–1441
57. Strasser RH (1989) Phosphorylation of the beta-adrenergic receptor: Mechanisms of densensitization. In: Moudgil VK (ed) Receptor Phosphorylation, CRC Press, Boca Raton, Florida, pp 199–226
58. Strasser RH, Braun-Dullaeus R, Marquetant R (1990) Acute myocardial ischemia activates the α₁-adrenergic pathway without any change of α₁-adrenergic receptor density. J Mol Cell Cardiol (in press)
59. Strasser RH, Braun-Dullaeus R, Walendzik H, Marquetant R (1990) Receptor-independent activation of protein kinase C in acute myocardial ischemia: Mechanism of sensitization of the adenylyl cyclase system. Circulation (in press)
60. Strasser RH, Dullaeus RB, Marquetant R, Kübler W (1988) Dual sensitization of the β-adrenergic

system in early myocardial ischemia: Independent regulation of β-receptors and adenylate cyclase. Circulation 78:II-482 (abstr.)

61. Strasser RH, Krimmer J, Dullaeus BR, Marquetant R, Kübler W (1989) Dual sensitization of the adrenergic system in early myocardial ischemia: Independent regulation of the β-adrenergic receptors and the adenylyl cyclase. J Mol Cell Cardiol 100:100–110

62. Strasser RH, Krimmer J, Marquetant R (1988) Regulation of β-adrenergic receptors: Impaired desensitization in myocardial ischemia. J Cardiovasc Pharmacol 12:S15–S24

63. Strasser RH, Krimmer J, Marquetant R (1988) The β-adrenergic system in early myocardial ischemia: Mechanisms of its dual sensitization. Eur Heart J 9:1388 (abstr.)

64. Strasser RH, Marquetant R, Kübler W (1990) Independent sensitization of β-adrenergic receptors and adenylate cyclase in acute myocardial ischemia. Br J Clin Pharmacol 30 suppl 1: 27S–36S

65. Strasser RH, Sibley DR, Lefkowitz RJ (1986) A novel catecholamine-activated cAMP-independent pathway for beta-adrenergic receptor phosphorylation in wild type and mutant S49 lymphoma cells: Mechanism of homologous desensitization of adenylate cyclase. Biochemistry, USA 25:1371–1377

66. Susanni EE, Knight DR, Vatner DE, Vatner SF, Homcy CJ (1989) One hour of myocardial ischemia is associated with a decrease in the stimulatory guanyl nucleotide binding protein, G_s. Circ Res 65:1145–1149

67. Thandroyen FT, Worthington MG, Higginson LM, Opie LH (1983) The effect of alpha- and beta-adrenoceptor antagonist agents on reperfusion ventricular fibrillation and metabolic status in the isolated perfused rat heart. J Am Coll Cardiol 1:1056–1066

68. Vatner DE, Knight DR, Shen YT, Thomas JXJ, Homcy CJ, Vatner SF (1988) One hour of myocardial ischemia in conscious dogs increases beta-adrenergic receptors, but decreases adenylate cyclase activity. J Mol Cell Cardiol 20:75–82

69. Vatner DE, Vatner SF, Fujii AM, Homcy C (1985) Loss of high affinity cardiac beta adrenergic receptors in dogs with heart failure. J Clin Invest 76:2259–2264

70. Vatner DE, Young MA, Knight DR, Vatner SF (1990) β-Receptors and adenylate cyclase: Comparison of nonischemic, ischemic, and postmortem tissue. Am J Physiol 258:H140–H144

71. Yatani A, Brown AM (1989) Rapid β-adrenergic modulation of cardiac calcium channel currents by a fast G protein pathway. Science 245:71–74

72. Yatani A, Codina J, Imoto Y, Reeves JP, Birnbaumer L, Brown AM (1987) A G protein directly regulates mammalian cardiac calcium channels. Science 238:1288–1292

73. Yatani A, Mattera R, Codina J, Graf R, Okabe K, Padrell E, Iyengar R, Brown AM, Birnbaumer L (1988) The G protein-gated atrial K^+ channel is stimulated by three distinct $G_i\alpha$-subunits. Nature 336:680–682

Authors' address:

Priv. Doz. Dr. Ruth H. Strasser, University of Heidelberg, Medical Center, Bergheimerstr. 58, 6900 Heidelberg, FRG

II. Adrenergic Coronary Vasomotion

11. Adrenergic Coronary Vasomotion

α-Adrenoceptor Subtypes in the Coronary Circulation

J. Holtz

Institut für Angewandte Physiologie und Balneologie, Freiburg, FRG

Summary

The pathophysiological role of sympathetic coronary innervation in myocardial ischemia is not clear, probably due to the complexities of adrenergic vascular control. In the canine coronary bed in vivo under β-adrenergic blockade, α_1- as well as α_2-adrenoceptor-mediated constrictions can be elicited with predominance of the former in the epicardial conductance arteries, and of the latter in coronary resistance vessels. However, this distribution of functional responsiveness cannot indicate distribution of receptor density and cannot remain unchanged under differing conditions. First, each of these two classes of α-adrenoceptors consists of a mixture of different, interacting subtypes; second, the smooth muscular responsiveness to these two classes of α-adrenoceptors is differently modulated by contractile preactivation, by β_2-blockade, and by the influence of sympathetic cotransmitters; third, α-adrenoceptors on endothelial cells and on sympathetic nerve endings can substantially modulate sympathetic coronary constriction. Thus, this neurogenic coronary control possesses a great functional plasticity, which is not yet fully evaluated with the presently available pharmacological tools.

Introduction

More than 25 years ago, a grand old man of coronary physiology, the late Donald E. Gregg, asked a panel of clinical and experimental researchers: "Do you want your sympathetic coronary nerves?" The background of the question was a controversy on the primary coronary vasomotor effect of these nerves: good (β-adrenergic dilation) or bad (α-adrenergic constriction). The panel reacted with laughter and uncertainty.

Meanwhile, there is a general agreement on the constrictive effect of sympathetic coronary innervation. However, the uncertainty remained on the pathophysiologic role of these nerves in clinical myocardial ischemia: good (preventing the dangerous steal of myocardial blood flow away from jeopardized subendocardial myocardial areas towards the subepicardial myocardium?), or bad (reducing flow to critically hypoperfused areas by coronary constriction?), or none (adrenergic coronary constriction in hypoperfused myocardium not effective?).

This uncertainty partially results from the complexities of adrenergic vascular control. Two types of vascular α-adrenoceptors on vascular smooth muscle have been identified, both mediating vasoconstriction; α-adrenoceptors exist also on the endothelium, mediating vasodilation; and they exist on sympathetic nerve endings, attenuating sympathetic neurovascular transmission by autoinhibition of transmitter release. Presently, we learn that each of the two classical α-receptors, in fact, consists of a mixture of differing subtypes, probably activating differing pathways of signal transduction.

Therefore, the present state of knowledge on coronary α-adrenoceptor subtypes will be summarized here, with emphasis on the canine coronary circulation. This is the most frequently used model for the analysis of sympathetic coronary control under pathophysiologic conditions. In this extensively studied standard model, a lot of controversial data exists in the literature, concerning the effect of α-adrenoceptor subtype selective antagonists during coronary hypoperfusion, exercise, or combinations thereof. These controversial pathophysiological observations will be commented on elsewhere in this monograph (see contributions of Drs. Feigl, Jones, Seitelberger and Heusch). I will concentrate on the role of α-adrenoceptor subtypes for adrenergic coronary constriction under basal conditions, on the variability in vascular responsiveness to activation of α-adrenoceptor subtypes, and on the growing diversification of α-adrenoceptor subtypes.

The canine heart in situ as a model for α-adrenergic coronary constriction

The classical experimental approach for the analysis of sympathetic coronary vasoconstriction was the application of β-adrenoceptor blockade in animals [26]. With the suppression of the catecholamine-induced myocardial activation and concomitant metabolic vasodilation by this blockade, the α-adrenergic coronary constriction was unmasked [26]. When it became clear a decade ago that vascular α-adrenoceptors consisted of at least two different subtypes [23, 77], this classical approach helped to identify the α-adrenoceptor-subtypes involved in noradrenaline-induced coronary vasoconstriction [41].

In a canine preparation under unselective β-blockade, the intracoronary injection of norepinephrine and of synthetic α-agonists caused dose-dependent, transient declines in coronary flow and in coronary venous O_2-saturation (Fig. 1). The coronary arteriolar constrictions induced by the α_2-selective agonist azepexole (B-HT 933) were resistant to a high dosage of the α_1-blocker prazosin, but were attenuated by the α_2-antagonist rauwolscine. This is the classical functional test to demonstrate the presence of postjunctional α_2-adrenoceptors in situ [59]. Both blockers attenuated the norepinephrine-induced vasoconstriction (Fig. 1), indicating that intracoronary norepinephrine activated both postjunctional α-adrenoceptor subtypes in the canine coronary bed. The constrictions by the two, more or less, selective agonists were inhibited to a comparable degree by the respective antagonists (Fig. 1). However, the norepinephrine response was much more inhibited by the α_2-blocker rauwolscine than by prazosin. This indicates that circulating norepinephrine constricts the coronary resistance vessels prefentially by activating postjunctional α_2-adrenoceptors [41].

However, this kind of preference may vary between different segments of the coronary bed within the same species [40, 83]. When coronary constriction in anesthetized dogs under β-blockade was elicited by bilateral carotid occlusion, the constriction of the coronary resistance vessels was slightly more attenuated by α_2- than by α_1-receptor blockade, while the constriction of the large epicardial coronary arteries is only sensitive to α_1-blockade (Fig. 2).

Similar characterizations of adrenergic vasoconstriction in the canine coronary bed have been performed by several authors; the results are summarized in Fig. 3. The changes in the coronary pressure/flow ratio are regarded as reflection of overall vasoreactions in the coronary arteriolar segment, though it has been shown that heterogenous reactions to humoral and neurogenic activation within that coronary "resistance" segment do occur [14].

This meta-analysis of various studies suffers from the differences in the applied α_2-blockade with regard to antagonists, dosages and modes of application. It might be possible that the scatter in the sensitivity of small vessel responses to α_2-blockade could be reduced by appropriate corrections for these differences in α_2-blockade. However, only one antagonist dosage was applied in most of these studies and different α_2-antagonists have not been compared in the canine coronary bed in the same study. Therefore, extrapolation of the blockade-induced attenuation to another antagonist or to another dosage is hazardous. Extrapolations from available data in other models should be avoided; as will be discussed below, the response of the coronary vasculature in vivo must be the net result of several opposite reactions and might include more than two α-adrenoceptor subtypes. Therefore, the variability in Fig. 3 cannot be reduced without arbitrary, unjustified assumptions.

In spite of this variability, two generalizations can be obtained from this synopsis of in vivo experiments with dogs: 1) large epicardial coronary arteries are constricted by α_1-selective agonists; their constriction by sympathetic activation is sensitive only to α_1-adrenoceptor

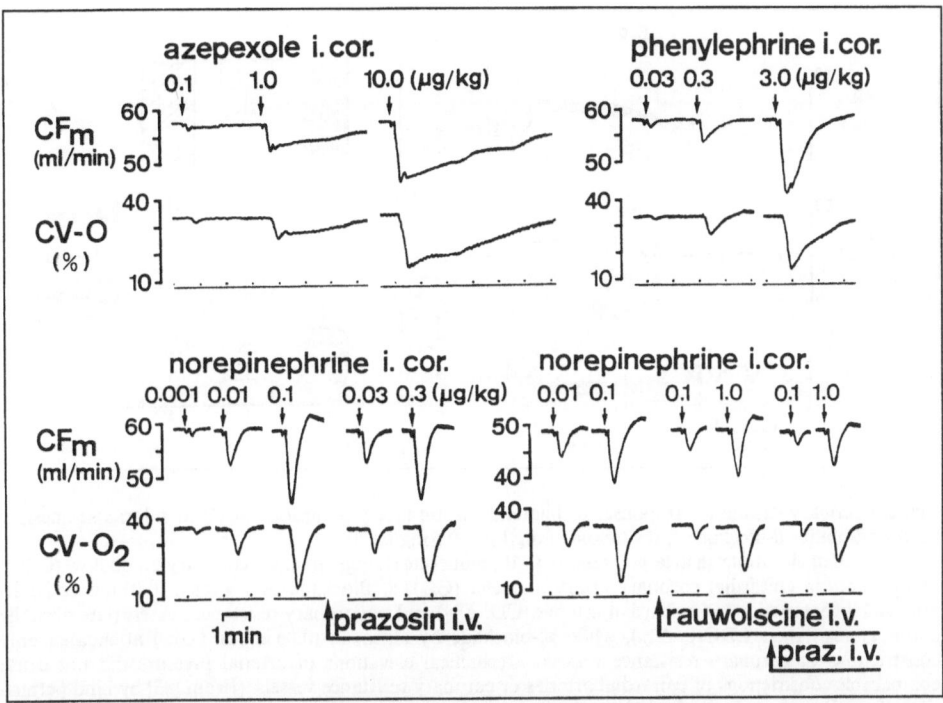

Fig. 1. α-Adrenergic coronary constrictions in pentobarbital-anesthetized, despinalized, vagotomized dogs under β-blockade (nadolol 4 mg/kg i.v.).

Intracoronary bolus injections of the α_2-selective agonist azepexole (BHT 933), of the preferential α_1-agonist phenylephrine, and of norepinephrine caused transient declines in mean coronary flow (CF_m) and in oxygen saturation of coronary venous blood ($CV\text{-}O_2$), while heart rate and arterial pressure remained unchanged.

α_1-Blockade (prazosin 1.2 mg/kg i.v.) caused a shift of the norepinephrine-induced constriction to 2.4-fold higher doses, while the shift induced by α_2-blockade (rauwolscine 0.3 mg/kg i.v.) was 22-fold. The azepexole-induced constrictions were not affected by prazosin, but shifted to 13-fold higher doses by rauwolscine. The constrictions by phenylephrine were shifted by prazosin to 12-fold higher doses and by rauwolscine to four-fold higher doses, the rauwolsine-resistant phenylephrine constriction was shifted 15-fold by prazosin. (From [41] by kind permission of the European Journal of Pharmacology.)

blockade; 2) coronary resistance vessels respond to α_2-selective, as well as to α_1-selective agonists. The adrenergic constriction in this segment of the coronary bed is sensitive to both types of subtype-selective α-blockade with some tendency for preferential α_2-sensitivity; with regard to this α_2-preference, no clear difference is detectable between constrictions induced by exogenous norepinephrine and those induced by sympathetic activation.

A straight forward interpretation of this synopsis would claim: In large epicardial arteries, mainly α_1-adrenoceptors exist on vascular smooth muscle, while in small coronary resistance vessels, both subtypes (α_1- as well as α_2-adrenoceptors) exist with similar localization relative to the sympathetic nerve endings (i.e., with similar "intrasynpatic" and "extrasynaptic" distribution; see below).

At first glance, this conclusion is attractively simple. Unfortunately, however, it is an oversimplification that is not helpful! We must try to integrate the complexities of α-adrenergic vasoconstriction into our view of coronary reactions, such as receptor subtype-specific modulation of functional responses, contributions of non-muscular elements to the reactions of the vascular wall, and the possibility that vascular α-adrenoceptors consist of many more than two receptor subtypes.

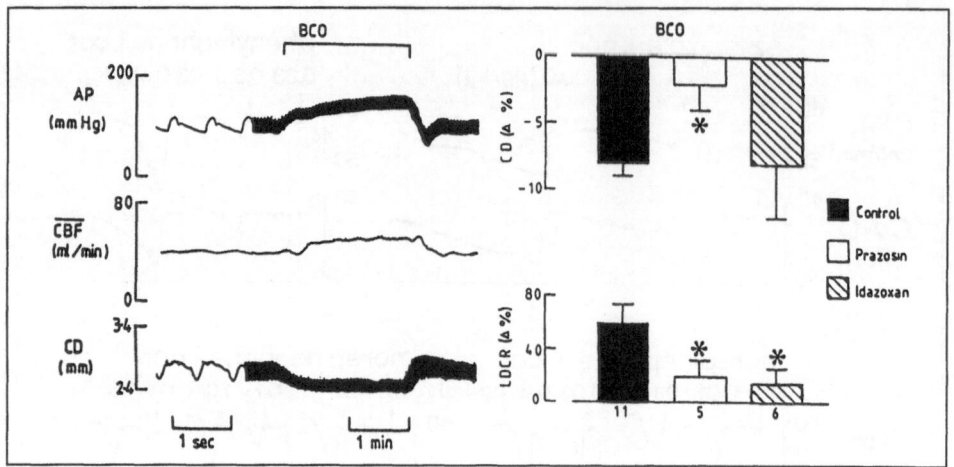

Fig. 2. Coronary vasomotor responses to bilateral carotid artery occlusion (BCO) in chloralose-anesthetized dogs under β-blockade (propranolol 1 mg/kg + 0.5 mg/kg/h).

BCO caused elevations in arterial pressure (AP), moderate changes in mean coronary blood flow (CBF), and declines in epicardial coronary artery diameter (CD). α_1-Blockade by prazosin (0.01 mg/kg i.cor.) attenuated constrictions of epicardial arteries (CD, Δ%) and of coronary resistance vessels (late diastolic coronary resistance, LDCR, Δ%), while α_2-blockade by idazoxan (0.05 mg/kg i.cor.) attenuated only constrictions of coronary resistance vessels. Mechanical elevations of arterial pressure did not cause comparable constrictions of epicardial arteries or coronary resistance vessels. (From [83] by kind permission of the European Journal of Pharmacology.)

Multiple vascular α-adrenoceptors

Subtype heterogeneity or species differences?

The present state of adrenoceptor subclassification is summarized in Fig. 4. The information on the molecular structure of these subtypes was derived from cloned genes encoding these receptors. As can be seen from the scheme, the available information comes from various species. However, species differences for a given receptor would not justify a division of this receptor into different subtypes. There is general agreement that subtype classification requires the parallel existence of both subtypes within the same species and, ideally, within the same tissue.

For all the subtypes listed in Fig. 4, functional studies and/or ligand-binding analyses have documented that they do exist within the same species, and in many tissues mixed receptor populations of subtypes have been documented. However, some cell types normally express only one of the receptor subtypes. Therefore, such cells or tissues became valuable reference tools for characterizing the respective subtype and for its purification. Typical reference tools from different species, such as human platelet (α_{2A}), neonatal rat lung (α_{2B}), human adenocarcinoma cell line (α_{2A}), hamster vas deferens (α_{1A}), rat spleen (α_{1B}), and others (for examples, see [11, 12, 60]) are the reason why the genes for various adrenoceptor subtypes have been isolated and characterized from different species. Nevertheless, the α-adrenoceptor subtype ramification definitely is not a pseudo-problem due to species differences.

The tentative classification presented in Fig. 4 proposes three major classes of adrenoceptors (β, α_1, α_2), each with a variable and still open number of subtypes. The basis for grouping the host of adrenoceptors into three major classes is the assumption that each adrenoceptor class is coupled to a different class of G proteins: G_s in case of the β-adrenoceptors, G_i for the α_2-adrenoceptors, and one or several not yet identified proteins (G_x) coupling α_1-adrenoceptors [10]. This rationale for classification is certainly preliminary and problematic. Several pathways

α-Adrenoceptor subtypes in the coronary circulation

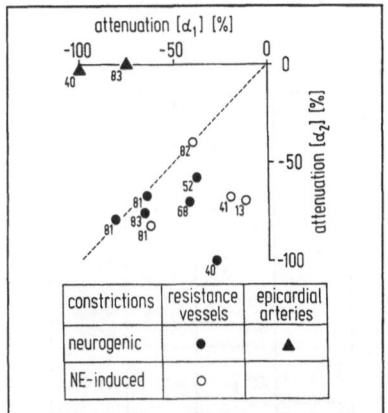

Fig. 3. Effects of α_1- or α_2-receptor blockade on adrenergic vasoconstriction in different segments of the canine coronary bed in vivo.

The position of each symbol relative to the x-axis gives the mean value of vasoconstriction (in % of the mean control response of that study) after α_1-blockade, the position relative to the y-axis indicates the respective response after α_2-blockade. (After combined α_1- and α_2-blockade, the residual responses amounted to 10% or less of the control responses [13, 40, 41, 81].

Closed symbols: neurogenic vasoconstriction by cardiac nerve stimulation [40, 52, 68], by reflex activation [81, 83], or by application of tyramine [81]; Open symbols: vasoconstrictions induced by exogenous norepinephrine. Triangles: responses of epicardial coronary arteries, assessed by sonomicrometry; Circles: responses of coronary resistance vessels, assessed by measuring coronary flow [40, 41, 52, 81–83], coronary venous O_2-saturation [41, 68], or coronary perfusion pressure at constant flow [13].

Experiments were performed under unselective β-blockade with 1–2 mg/kg i.v. propranolol [13, 40, 81–83] or 4 mg/kg i.v. nadolol [41, 52, 68] in anesthetized dogs with vagotomy [13, 40, 41, 52, 68, 81, 83] or in conscious dogs with ganglionic blockade [82]. Only the response to the largest adrenergic stimulus applied in each study is presented in the figure.

For α_1-blockade, 0.4–1.2 mg/kg prazosin i.v. (or 0.01 mg/kg i.cor., [83]) was given; in one study [68], the effect of 1 mg/kg i.v. prazosin was extrapolated from two other prazosin dosages. For α_2-blockade, 0.1–0.3 mg/kg i.v. rauwolscine [40, 41, 68, 82] or idazoxan (0.3–1 mg/kg i.v. [52, 81]; 0.001 mg/kg/min i.cor. [13], or 0.05 mg/kg i.cor. [83]) was given; corrections for differences in drugs, dosages, or routes of application were not considered.

Synthetic, subtype-selective α-adrenoceptor agonists were applied in most of these studies in dosages matching the effect of the analyzed adrenergic stimuli on coronary resistance vessels. For α_1-receptor stimulation, phenylephrine (2–3 μg/kg i.cor. [13, 41, 83], or 1–5 μg/kg/min i.v. [81, 82]) or methoxamine (15 μg/kg i.cor. [40]) was given and for α_2-stimulation the imidazole derivatives B-HT 933 (10 μg/kg i.cor. [13, 41]) or B-HT 920 (2–15 μg/kg i.cor. [40, 83], or 1–2 μg/kg/min i.v. [81, 82]). When different segments of the coronary vasculature were compared in the same study [40, 83], the large epicardial coronary arteries were responsive only to the α_1-agonists, while coronary resistance vessels could be activated by both groups of subtype-selective agonists. The coronary constrictions induced by B-HT 933 or 920 were resistant to α_1-blockade in all studies and were attenuated only by α_2-blockade: phenylephrine-induced responses were attenuated by α_1-blockade, but were also affected to a lesser degree by α_2-blockade.

of signal transduction can be activated by one α-adrenoceptor subtype [9, 19, 30, 55], and not all of them appear to involve the same type of G-protein [9, 30]. Thus, the scheme is given here without certainty about the relevance of the major class assignment. The assignment into the three groups β, α_1, and α_2 may indicate a rational classification of functionally related subtypes, or it may reflect more the casualness of the historical sequence of discoveries.

Subtypes of α_1-adrenoceptors

Though various proposals for α_1-adrenoceptor subclassification have been made in the past, only that given in Fig. 4 has been generally accepted. This subclassification discriminates

	Adrenoceptors G-protein coupled 7 transmembrane regions									
class	α₁		α₂					β		
class-selective antagonists	prazosin		rauwolscine; yohimbine					propranolol		
class-selective agonists	methoxamine circazoline		UK 14304; BHT 920; BHT 933					isoproterenol		
pharmacol.defined subtypes	α₁A	α₁B	α₂A	α₂B		(α₂C)	(α_pre)	β₁	β₂	β₃
cloned subtypes	α₁A	α₁B	α₂C10		α₂C4		(α₂C2)	β₁	β₂	β₃
structure source	466aa bovine library	515aa hamster vas def.	450aa human platelet		461aa human kidney			477aa human library	413aa human library	402aa human library
signal transduction	DHP-sens. channels ↑	PLC ↑	AC ↓	AC ↓	AC ↓		N-type Ca^{2+} channels ↓	AC ↑	AC ↑	AC ↑
subtype selective drugs	WB 4101 (ant.)	chloroethyl-clonidine (ant.)	oxymetazoline (ag.)	ARC 239 (ant.)			not sensitive to SK&F 104078	atenolol (ant.) CGP 207RA (ant.) xamoterol (ag.)	procaterol (ag.) ICI 118551 (ant.)	BRC 37344 (ag.) pindolol (ag.)

Fig. 4. Tentative scheme for subclassification of presently identified adrenoceptors (modified from [10, 60, 79]).

Abbreviations: aa = amino acids; DHP = dihydropyridine; PLC = phospholipase C; AC = adenylate cyclase; ant = antagonist; ag = agonist. Chemical names of the drugs given as code numbers can be found in the above references.

Subtype selective drugs are selective within their classes, but selectivity is not always given if several classes of adrenoceptors or other receptors are considered.

Within the class of α_2-adrenoceptors, the classification is still open, subtypes in brackets are not clearly defined; for the putative α_{2C}-subtype, affinities of a number of ligands do not correlate with the affinities for the α_{2A}- or α_{2B}-subtype [10]; the presynaptic α-adrenoceptor on sympathetic nerve endings (α_{pre}) has low affinity for the α_2-antagonist SK&F 104078 in some (e.g., [22]), but not in all studies [17], and an α_2-like gene (α_2C2) on chromosome 2 has been recognized under low stringency conditions by a probe made from the human platelet α_2-receptor gene [51], but has not yet been analyzed.

subtypes of α_1-adrenoceptors by differences in the signal transduction mechanisms increasing intracellular Ca^{2+}-concentration [38, 60] and by differing affinities for the competitive antagonists WB 41101 and 5-methyl-urapidil, and for the irreversible alkylating agent chloroethyl-clonidine [37, 39, 60]. Enantiomers of the novel 1,4-dihydropyridine Ca^{2+}-antagonist niguldipine can also discriminate between these two α_1-adrenoceptor subtypes [34]. Both subtypes have been cloned and sequenced recently, the α_{1A}-subtype from the bovine genome [79] and the α_{1B}-subtype from a cell line derived from hamster vas deferens [18]. In agreement with the subdivision based on functional studies [38], expression of the cloned α_{1B}-receptor demonstrated its coupling to inositol phospholipid metabolism [18]. This α_{1B}-subtype predominates in rat myocardium [39], suppressing voltage-dependent K^+-currents, probably via activation of protein kinase C [5]. However, myocardial Ca^{2+}-channels are activated by stimulation of protein kinase C [54] and this has been shown also for dihydropyridine-sensitive Ca^{2+}-channels in vascular smooth muscle [27]. It is not yet known whether this effect of protein kinase C on dihydropyridine-sensitive Ca^{2+}-channels can be achieved by activation of α_{1B}-adrenoceptors in myocardium or in smooth muscle.

Activation of the other α_1-adrenoceptor subtype, the α_{1A}-receptor, promoted a nifedipine-sensitive influx of extracellular calcium in rat vas deferens without causing inositol phosphate accumulation [38]. Such an α_1-subtype specific activation of dihydropyridine-sensitive Ca^{2+}-channels is exciting in view of numerous previous observations indicating Ca^{2+}-influx-dependent, as well as independent components in the α_1-adrenoceptor-mediated constriction of vascular smooth muscle (for review see [60]). In very elegant experiments it was shown recently that noradrenaline contracts arteries by activating voltage-dependent, dihydropyridine-sensitive Ca^{2+}-channels through increasing their open-state probability; thus, calcium entry would regulate muscle contraction at physiological membrane potentials [63]. Unfortunately, the α-adrenoceptor subtype mediating this activation of vascular Ca^{2+}-channels was not determined. It could result from direct activation of Ca^{2+}-channels via $α_{1A}$-receptors, from indirect modification of the channels due to activation of protein kinase C via $α_{1B}$-receptors, or from the not yet exactly analyzed enhancement of Ca^{2+}-influx by $α_2$-receptors. Recent attempts to characterize the $α_1$-adrenoceptor subtypes involved in constrictions of several arteries in vitro by functional analyses with antagonists [62] could not clearly identify $α_{1A}$- or $α_{1B}$-subtypes. This is not surprising, since the tested arteries might contain mixed populations of $α_{1A}$- and $α_{1B}$-subtypes, and since α-adrenoceptor subtypes could interact at the postreceptor level (see below).

Subtypes of $α_2$-adrenoceptors

The scheme in Fig. 4 indicates that at least three subtypes of $α_2$-receptors do exist, but that the subclassification is still rather open and less clear than for $α_1$-subtypes. Proposals for the separation of $α_2$-adrenoceptor-subtypes came from three independent lines of research, which only partially appear to converge at present. At first, functional studies indicated that presynaptic (or prejunctional) $α_2$-adrenoceptors on the nerve endings of the peripheral sympathetic nervous system can be separated from postsynaptic (postjunctional) $α_2$-receptors on the vasculature [22]. Secondly, radioligand binding studies allow a clear division of two subtypes ($α_{2A}$ and $α_{2B}$), and, probably, a third subtype ($α_{2C}$) that can be characterized by subtype-selective ligands [11, 85]. Finally, the human platelet $α_2$-receptor was cloned and expressed recently [51]; probes made from this gene recognized three different related genes in Southern blot analysis with low stringency, localized to different chromosomes [51]. Meanwhile, two of these three genes have been cloned, sequenced, expressed, and characterized as $α_2$-receptor subtypes [66].

For the moment it is not yet possible to unify these different subtype classifications into one encompassing scheme. There is general agreement that the human platelet $α_2$-adrenoceptor subtype [51], which is localized on the long arm of chromosome 10 ($α_2$C10), is identical to the $α_{2A}$-subtype characterized by functional studies and ligand binding analyses [10, 12, 66]. However, the $α_2$C4 receptor cloned from a human kidney cDNA library and characterized after expression in a mammalian cell line [66] probably does not correspond to the $α_{2B}$-subtype as identified from pharmacological studies [10, 12]. Furthermore, the subtype encoded by the third $α_2$-adrenoceptor gene identified by Southern blot test and localized on chromosome 2 [51] has not yet been expressed and characterized; it may or may not correspond to the $α_{2C}$-subtype derived from functional studies [10]. Therefore, it appears reasonable to use the nomenclature $α_2$C10, $α_2$C4, and $α_2$C2 for data from cloning techniques and to define functionally characterized subtypes as $α_{2A}$, $α_{2B}$ or $α_{2C}$ [10], with an established identity only for $α_2$C10 = $α_{2A}$.

The identity of the presynaptic $α_2$-adrenceptor subtype localized on the peripheral sympathetic nerve endings is a special problem. Hieble and colleagues suggested that these prejunctional $α_2$-receptors can be differentiated from other postjunctional $α_2$-receptors by means of their low affinity for benzazepine derivatives such as SK&F 104078 [22]. However, this antagonist did not discriminate between expressed $α_2$C10- and $α_2$C4-adrenoceptors [66] and has not been tested in analyses of $α_{2A}$-, $α_{2B}$- or $α_{2C}$-subtypes. Furthermore, the reported selectivity of SK&F 104078 for postjunctional over prejunctional $α_2$-receptors has not been confirmed in recent studies (see [17]), and these discrepancies have been partially explained by differences in the operation of the negative feedback for noradrenaline release between the applied models [17]. Thus, the position of the prejunctional sympathetic $α_2$-receptor within the scheme of α-adrenoceptor subtypes is entirely open.

The distribution of $α_2$-receptor subtypes in vascular smooth muscle of different origin is unknown and the subtype of the endothelial $α_2$-receptor (see "Modulation by non-muscular elements of the vessel wall") has not been clarified. Though $α_2$-mediated vasoconstriction is generally associated with extracellular Ca^{2+}-influx in vivo and in vitro, there are no detailed studies available on the precise mechanism of $α_2$-receptor interaction with vascular dihydropyridine-sensitive Ca^{2+}-channels. Only in one α_2-mediated reaction, in the autoinhibition of sympathetic neurotransmission, has the interaction with Ca^{2+}-influx been analyzed in detail: in this autoinhibition, noradrenaline acts by reducing the activity of N-type Ca^{2+}-channels via changes in rapid gating kinetics [56].

In previous experiments on α-adrenergic coronary constrictions, the applied pharmacological tools could only discriminate between α_1- and α_2-adrenoceptors, but not between subtypes within these two classes. Therefore, some mechanisms modulating the responsiveness of vascular smooth muscle to activation of the α_1- or α_2-class of adrenoceptors will be considered in the next chapter. However, one should keep in mind that the logic of grouping α-adrenoceptors into an α_1- and an α_2-class is debatable (see above).

Variability in vascular responsiveness to α-adrenoceptor subtype activation

Modulation by contractile preactivation

Vascular α_2-adrenoceptors non-functional in vitro? With very few exceptions (e.g., cutaneous arteries and arterioles from humans [28, 64]) vascular preparations from arteries do not demonstrate α_2-adrenoceptor-mediated vasoconstriction in vitro, while this type of constriction can be elicited in many venous preparations. The data on adrenergic vasoconstriction of large epicardial coronary arteries in vivo (Fig. 3) are in line with this general experience in vitro (though, occasionally, the responsiveness of canine epicardial coronary arteries to α_2-agonists in vitro has been regarded as an α_2-mediated functional response [61]). Originally, it was not clear whether this absence of α_2-mediated constrictions was due to the absence of α_2-adrenoceptors in large arteries or to differences in the coupling mechanisms, rendering α_2-activations less responsive under certain conditions. In the past, very few ligand binding studies with membranes from vascular preparations have been performed because of the high unspecific binding and the large amount of tissue, which is necessary.

Recently, however, a procedure of preparing enriched plasma membrane fractions from vascular smooth muscle was used to evaluate the relation between α-adrenoceptor binding parameters and contractile responses in various canine vessels [70]. Specific binding sites for ^{3}H-prazosin and for ^{3}H-rauwolscine were demonstrated in arterial and venous preparations. The density-ratio of rauwolscine/prazosin binding sites varied by more than 10-fold between different vessels and was higher in veins than in arteries [70]. Only vessels with high α_2/α_1 density ratios contracted in response to an α_2-selective agonist, and the maxima of these α_2-agonist-induced constrictions were similar to the maxima of norepinephrine-induced constrictions after α_1-blockade by prazosin [70]. While these data demonstrate some parallelism between α_2-adrenoceptor density and contractile responsiveness to α_2-simulation, they do not explain why (and under which conditions) vessels with an α_2-density not in excess of the α_1-density appear to be unresponsive to α_2-adrenoceptor stimulation.

In this context, it is of interest that the contractile responsiveness of isolated arteries to α_2-adrenoceptor stimulation is augmented in the presence of non-adrenergic agonists such as angiotensin, vasopressin, prostaglandin $F_{2\alpha}$ or Bay K 8644 [25, 31, 74, 76], even if the agonist itself does not affect baseline resting tension [25]. Thus, functional populations of postjunctional α_2-adrenoceptors may be uncovered by the presence of some contractile activation in those vessels, which do not respond to α_2-stimulation without such preactivation.

Interactions of α_1- and α_2-adrenoceptors: Recent observations indicate that a similar functional interaction also exists between α-adrenoceptor subtypes on vascular [20, 21, 84] and nonvascular [69] smooth muscle. In preparations with postjunctional α_1- and α_2-adrenoceptors, α_1-mediated responses appear to be potentiated by "subthreshold" activation of α_2-adrenoceptors and vice versa [20, 21, 69, 84]. The α_2-mediated potentiation of arterial responses to α_1-selective agonists is abolished in Ca^{2+}-free media and by a Ca^{2+}-antagonist [84]. This interaction at a post-receptor site in the contractile response to α-adrenergic activation means that an α_2-mediated response apparently may become sensitive to α_1-selective blockade and vice versa [20, 21], provided that a mixed population of postjunctional α-adrenoceptors is activated.

Such functional interactions of α-adrenoceptor subtypes during adrenergic vasoconstriction, probably at the postreceptor level, can explain a lot of previous paradoxical observations, which apparently indicated atypical α-adrenoceptors, variable receptor affinities, or noncompetitive antagonisms by competitive blockers. These interactions may also be relevant for the concept of response-dependence on the receptor reserve, and they can explain why in many preparations in vitro, functional α_2-adrenoceptors have been falsely excluded because of their indirect sensitivity to α_1-selective blockade. Such subtype-interactions may become even more important in our understanding of adrenergic vasoconstriction in view of the possibility that vascular α-adrenoceptors consist of even more than two subtypes (see above).

β_2-Blockade: Another kind of α-subtype specific interaction with other signals is obviously relevant for the data summarized in Fig. 3: Acute unselective β-blockade as well as β_2-selective blockade augmented α_2-mediated vasoconstrictions in vivo, but not those induced by α_1-adrenoceptor stimulation [80]; β_1-selective blockade did not preferentially enhance α_2-mediated constriction [80]. On the other hand, acute β_2-agonist-induced vasodilation attenuated α_2-mediated vasoconstriction more than α_1-constriction [47]. Finally, β_2-desensitization by chronic β_2-agonist application augmented α_2-mediated vasoconstriction [48], and β_2-blockade intensified the clonidine withdrawal syndrome in rats [46]. A cellular mechanism explaining these preferential antagonisms of β_2- and α_2-mediated effects in vivo has not been identified so far, though speculations on interferences at the level of cAMP-formation are near at hand.

The role of this β_2-α_2-antagonism in the coronary system has not been studied. All the examples shown in Fig. 3, demonstrating α_2-mediated constrictions of coronary resistance vessels, were obtained under unselective β-adrenoceptor blockade with a β_2-blocking component. This β_2-component should have accentuated the α_2-constriction relative to the α_1-constriction to an unknown degree. In the pithed rat [80], β_2-blockade enhanced the systemic α_2-vasoconstriction by a factor of 1.8. Assuming a similar enhancement in the coronary system would imply that the balance of α_1- vs α_2-vasoconstriction in the heart (Fig. 3) without β-blockade or under β_1-selective blockade is shifted substantially towards α_1-constriction.

In one study with anesthetized, despinalized dogs, the effect of α_1- and α_2-blockade on myocardial oxygen balance (i.e., O_2-saturation of coronary venous blood) during sympathetic nerve stimulation was monitored with and without unselective β-blockade [68]. These data did not indicate an accentuation of the α_2-mediated vasoconstriction by the unselective β-blockade [68]. However, this analysis was not specific enough to exclude such a potential accentuation, since presynaptic β_2-mediated effects of the β-blockade were not ruled out.

NPY: The potentiation of adrenergic vasoconstriction by neuropeptide Y (NPY) may also have relevance for the segmental differences in α-subtype mediated coronary responses shown in Fig. 3. The brain peptide NPY can be demonstrated in pericoronary nerves [35], it is a potent coronary constrictor [1, 29] acting preferentially on small coronary resistance vessels in the human and canine coronary bed when applied luminally [15, 57, 58]. Similar as in other vessels, NPY is a strong potentiator of the catecholamine-induced coronary vasoconstriction in vitro and in vivo [36, 57]. This NPY-induced potentiation probably does not depend on endothelium-derived signals [7]. Selective, potentiating interactions between NPY-receptors and α_2-receptors have been observed in neurons of rat nucleus locus coeruleus [45]. In these cells, norepinephrine and an α_2-selective agonist induced hyperpolarization and inhibition of spontaneous firing, and these effects were potentiated by NPY; furthermore, α_2-blockade antagonized the NPY-induced depression of firing [45]. Similar α_2-subtype selective interactions of NPY in arterial smooth muscles have not yet been studied. However, in rat mesenteric arterioles, NPY selectively enhanced the dihydropyridine-sensitive component of constrictions induced by norepinephrine and phenylephrine, which are regarded as α_1-mediated in this tissue [3]. Since NPY also potentiated other dihydropyridine-sensitive contractions in these arterioles (2), a similar potentiation of the Ca^{2+}-influx-dependent α_2-mediated constrictions can be assumed in vessels, which do respond to α_2-stimulation. It is certainly conceivable that NPY-potentiation contributes to the predominance of α_2-mediated functional responses in the coronary resistance vessels in situ (Fig. 3). Information on the role of ATP (the other sympathetic cotransmitter) for vascular tone in situ is not available, due to the lack of potent, reversible antagonists for

P_2-receptor subtypes (for review see [65]). ATP constricts vascular smooth muscle in vitro by activation of P_{2x}-receptors, causing depolarization and opening of Ca^{2+}-selective channels [6]. It is unknown whether this activation modulates the responsiveness to α-adrenoceptor subtypes.

In summary, the sensitivity of adrenergic vasoconstriction against selective α_1- or α_2-blockade can be substantially modulated by many vasoactive signals. Therefore, the ratio of α_1-mediated/α_2-mediated constrictions in a certain segment of the coronary bed must vary under different circumstances and cannot indicate a pattern of α-adrenoceptor subtype densities on coronary vascular smooth muscle. Important examples of these modulations are: the potentiating interactions of α-adrenoceptor subtypes at the postreceptor level in mediating adrenergic vasoconstriction, the augmentation of α_2-mediated responses by β_2-blockade and by nonadrenergic vasoconstrictors at subthreshold concentrations, and the modification of adrenergic vasoconstriction by other potential sympathetic transmitters such as NPY.

Modulation by non-muscular elements of the vessel wall

Endothelial α-adrenoceptors: Stimulation of α_2-adrenoceptors of precontracted epicardial coronary arteries of pigs and dogs in vitro causes a dilation [4, 16]. This dilation requires the presence of a functionally intact endothelium and can be demonstrated in various vessels (for examples see [78]). This reaction is generally regarded as the result of the release of EDRF (endothelium-derived relaxing factor) due to activation of endothelial α_2-adrenoceptors, since it is abolished by pre-incubation with hemoglobin [8]. Therefore, it is categorized within a group of α_2-mediated prosecretory effects on excitation-secretion coupling [55]. Though this is probably correct, it should be stressed for sake of clarity that other possibilities of endothelium-mediated antagonism to vasoconstriction (for recent review see [32]) have not been excluded. However, α_2-mediated EDRF-release is the most likely mechanism. A comparable endothelium-dependent dilation is not induced by α_1-adrenoceptor activation [4].

In vivo, the contractile state of epicardial coronary arteries of humans and dogs is modulated by the flow-dependent, endothelium-dependent dilation [24, 42]. This is the result of shear-induced EDRF-release [67] and demonstrates the responsiveness of epicardial arteries to EDRF in vivo. Thus, it is possible that circulating α_2-agonists activate opposite actions within the vascular wall, which may balance each other accidentally in their effect on epicardial coronary artery diameter in dogs (Fig. 3).

Recent evidence indicates that EDRF can also modulate resistance vessels of the coronary bed [49], and α_2-adrenoceptor activation probably exerts a dual effect on coronary flow in vivo [43, 44, 71]. Such a dual control was shown directly in open-chest dogs under β-blockade and adenosine-induced coronary dilation: low doses of intracoronary α_2-stimulation (clonidine or norepinephrine in the presence of α_1-blockade) elicited a further coronary resistance vessel dilation, while higher dosages caused constrictions. Intracoronary α_1-stimulation caused only constriction over the whole dosage range [44].

While these observations can be easily explained by assuming interference of endothelial and smooth muscular α_2-adrenoceptor activation, other mechanisms might have contributed to these observations. Hori et al. [43, 44] discussed an amplification of the adenosine-induced cAMP-formation in the muscle cells by α_2-stimulation, as has been documented in other cell types. Clonidine might have partially caused dilation by activating coronary H_2-receptors [71]. Thus, the dilatory role of endothelial α_2-receptors in the regulation of coronary flow cannot yet be quantified. However, the α_2-mediated coronary resistance vessel constriction induced by humoral norepinephrine is probably underestimated in Fig. 3 due to this endothelial component.

Sympathetic presynaptic α-adrenoceptors: The α_2-mediated component of the neurogenic constriction must be underestimated in Fig. 3, since α_2-blockade suppresses the physiologic autoinhibition of stimulus-induced norepinephrine release (for reviews see [72, 73]). None of the α_2-blockers applied in the experiments in Fig. 3 distinguishes between blockade of postsynaptic vascular α_2-adrenoceptors, thereby attenuating sympathetic constriction, and blockade of presynpatic α_2-autoreceptors on sympathetic nerve terminals, which mediate the feedback inhibition of norepinephrine release. This presynaptic effect of α_2-blockade enhances the norepinephrine release per action potential, thereby counteracting the postsynaptic attenuation of neurogenic coronary constriction. Since the norepinephrine concentration in the biophase around the presynaptic and postsynaptic receptors during nerve stimulation is unknown in these in vivo experiments (Fig. 3), it is not possible to quantify the degree to which the vascular α_2-blockade is counteracted by the presynaptic effects of the α_2-blockade. Though some reports exist on α_1-blockade augmenting stimulus-induced norepinephrine release, this is probably an unspecific effect of high doses. The available "evidence is not very favourable for presynaptic α_1-receptors" (see [73]).

In the experiments with reflex activation of coronary constriction in Fig. 3, a further complexity should be considered: α_2-blockers, as applied in Fig. 3, enhance the sympathetic firing rate in response to a hypotensive stimulus [75] through central CNS and peripheral ganglionic sites of action. Taken together, presynaptic, ganglionic, and central effects of α_2-blockade are the reason why the synopsis in Fig. 3 underestimates that component of sympathetic coronary constriction, which is due to activation of vascular α_2-adrenoceptors.

In the past, there has been some debate over whether vascular α_1-adrenoceptors are located intrajunctionally (or are intrasynaptic), whereas vascular α_2-adrenoceptors are extrasynaptic. The basis for these speculations were observations that neurogenic vasoconstriction can be more easily blocked by α_1-antagonists than constrictions induced by humoral norepinephrine, while the opposite preference was claimed for α_2-antagonists. The synopsis in Fig. 3 clearly demonstrates that this is not the case in the canine coronary bed. Probably, it is very difficult to conclude from functional data on the location of α-adrenoceptor subtypes relative to the nerve endings. With regard to the disputed preferential innervation of vascular α_1-adrenoceptors, McGrath et al. concluded recently: "Too many contrary observations have been made to sustain such a generalization" [59].

A direct analysis of the autonomic innervation of epicardial coronary arteries in the dog by electron microscopy [53] yielded an additional argument for caution against simplifying conclusions concerning innervated and non-innervated vascular receptors. In that analysis 50% of the nerve terminals (i.e., varicosities uncovered by the Schwann sheet) were at a distance to vascular smooth muscle cells of between 0.5 and 4.5 µm, while the other "synaptic distances" were larger. This rather large gap between nerve endings and smooth muscles fits with the moderate range of sympathetic constrictive control of epicardial coronary arteries in dogs [33, 40, 83]. However, very often the varicosities lose the Schwann cells on the abluminal site of the vessel wall, facing a fibroblast in close contact. The distance between nerve terminals and fibroblasts was between 0.03 and 0.2 µm in 50% of the contacts (i.e., one-tenth of the distance to coronary smooth muscle cells). Thus, the analysis demonstrated a preferential autonomic innervation of pericoronary fibroblasts, the functional role of which is totally unknown. It does not demonstrate a preferential innervation of vascular smooth muscles and gives little support for the possibility of a preferential innervation of a certain receptor subtype on smooth muscles.

In summary, α-adrenoceptors on sympathetic nerve endings and on endothelial cells can modulate the response to neurogenic and humoral adrenergic activation of the coronary vasculature. The tight autonomic innervation of pericoronary fibroblasts reminds us of the incompleteness of our understanding of neurovascular control.

Outlook

From a simplistic point of view, one might consider the tone of coronary vessels as the net result of two major opposite modulatory influences: metabolic vasodilation and sympathetic vasoconstriction. These two "exogenous" regulatory antagonists interfere with the "intrinsic" control of the vascular system by myogenic autoregulation (Bayliss-effect) and by flow-dependent dilation (Schretzenmayr-effect).

The α-adrenergic vasoconstriction, as one major component of the sympathetic coronary control, involves a variety of α-adrenoceptor subtypes, the number of which still cannot be estimated. In this situation of incomplete understanding, it may be helpful to combine the α-adrenoceptors into the two classical subclasses: α_1- and α_2-adrenoceptors. Mainly, practical reasons argue for this subclassification, since the pharmacological tools available for application in vivo can only discriminate between these subclasses. With these tools it can be shown that the role of these two subclasses of vascular α-adrenoceptors varies between different segments of the coronary bed (Fig. 3) and it may vary also between different species. More important, however, is the extreme variability of the vascular responsiveness to subclass-specific receptor activation. More research on the multiple α-adrenoceptor subtypes is necessary to proceed from a descriptive discussion of this variability to a rational understanding of the functional plasticity of adrenergic vasomotion. Such a progress is mandatory for better understanding the role of adrenergic coronary constriction in pathological situations with a strong activation of its regulatory antagonist, the metabolic vasodilation.

Due to the multitude of closely related α-adrenoceptor subtypes, adrenergic coronary constriction may possess a great functional plasticity with many surprising aspects. Probably, the activation of α-adrenoceptor subtypes on coronary smooth muscle involves not only a functional antagonism against the effect of mediators of metabolic vasodilation, but also some kind of synergism. Kitakaze et al. proposed (with a solid experimental basis) that the activation of coronary vascular α_2-adrenoceptors sensitized the coronary smooth muscles to the dilatory effect of adenosine by enhancing the adenosine-induced intracellular accumulation of cAMP [43, 44, 50]. Under certain conditions of myocardial hypoperfusion, this α_2-mediated enhancement of adenosine dilation would override the direct α_2-mediated vasoconstriction [50].

Is activation of coronary α-adrenoceptors good or bad? Dr Gregg's provocative question remains unanswered and as stimulating as ever!

References

1. Aizawa Y, Murata M, Hayashi M, Funazaki T, Ito S, Shibata A (1985) Vasoconstrictor effects of neuropeptide Y (NPY) on canine coronary artery. Jap Circ J 49:584–588
2. Andriantsitohaina R, Stoclet JC (1988) Potentiation by neuropeptide Y of vasoconstriction in rat resistance arteries. Br J Pharmacol 95:219–228
3. Andriantsitohaina R, Stoclet JC (1990) Enhancement by neuropeptide Y (NPY) of the dihydropyridine-sensitive component of the response to α_1-adrenoceptor stimulation in rat isolated mesenteric arterioles. Br J Pharmacol 99:389–395
4. Angus JA, Cocks TM, Satoh K (1986) The α-adrenoceptors on endothelial cells. Fed Proc 45:2355–2359
5. Apkon M, Nerbonne JM (1988) α_1-Adrenergic agonists selectively suppress voltage-dependent K^+ currents in rat ventricular myocytes. Proc Natl Acad Sci USA 85:8756–8760
6. Benham CD, Tsien RW (1987) A novel receptor-operated Ca^{++}-permeable channel activated by ATP in smooth muscle. Nature 328:275–278
7. Budai D, Vu HQ, Duckles SP (1989) Endothelium removal does not affect potentiation by neuropeptide Y in rabbit ear artery. Eur J Pharmacol 168:97–100
8. Bullock GR, Taylor SG, Weston AH (1986) Influence of the vascular endothelium on agonist-induced contractions and relaxations in rat aorta. Br J Pharmacol 89:819–830
9. Burch RM, Luini A, Axelrod J (1986) Phospholipase A_2 and phospholipase C are activated by distinct GTP-binding proteins in response to α_1-adrenergic stimulation in FRTL5 thyroid cells. Proc Natl Acad Sci USA 83:7201–7205
10. Bylund DB (1988) Subtypes of α_2-adrenoceptors: pharmacological and molecular biological evidence converge. Trends Pharmacol Sci 9:356–361
11. Bylund DB, Ray-Prenger C, Murphy TJ (1988) Alpha-2A and alpha-2B adrenergic receptor subtypes: antagonist binding in tissues and cell lines containing only one subtype. J Pharmacol Exp Ther 245:600–607
12. Bylund DB, Ray-Prenger C (1989) Alpha-2A and alpha 2B adrenergic receptor subtypes: attenuation of cyclic AMP production in cell lines containing only one receptor subtype. J Pharmacol Exp Ther 251:640–644
13. Chen DG, Dai XZ, Zimmerman BG, Bache RJ (1988) Postsynaptic α_1- and α_2-adrenergic mechanisms in coronary vasoconstriction. J Cardiovasc Pharmacol 11:61–67
14. Chilian WM, Layne SM, Eastham CL, Marcus ML (1989) Heterogeneous microvascular coronary α-adrenergic vasoconstriction. Circ Res 64:376–399

15. Clarke JG, Kerwin R, Larkin S, Lee Y, Yacoub M, Davies GJ, Hackett D, Dawbarn D, Bloom SR, Maseri A (1987) Coronry artery infusion of neuropeptide Y in patients with angina pectoris. Lancet 1:1057–1059

16. Cocks TM, Angus JA (1983) Endothelium-dependent relaxation of coronary arteries by noradrenaline and serotonin. Nature 305:627–630

17. Connaughton S, Docherty JR (1990) No evidence for differences between pre- and postjunctional α_2-adrenoceptors in the periphery. Br J Pharmacol 99:97–102

18. Cotecchia S, Schwinn DA, Randall RR, Lefkowitz RJ, Caron MG, Kobilka BK (1988) Molecular cloning and expression of the cDNA for the hamster α_1-adrenergic receptor. Proc Natl Acad Sci USA 85:7159–7163

19. Cotecchia S, Kobilka BK, Daniel KW, Nolan RD, Lapetina EY, Caron MC, Lefkowitz RJ, Regan JW (1990) Multiple second messenger pathways of α-adrenergic subtypes expressed in eukaryotic cells. J Biol Chem 265:63–69

20. Daly CJ, Dunn WR, McGrath JC, Wilson VG (1988) An attempt at selective protection from phenoxybenzamine of postjunctional α-adrenoceptor subtypes mediating contractions to noradrenaline in the rabbit isolated saphenous vein. Br J Pharmacol 95:501–511

21. Daly CJ, McGrath JC, Wilson VG (1988) Pharmacological analysis of postjunctional α-adrenoceptors mediating contractions to (−)-noradrenaline in the rabbit isolated saphenous vein can be explained by interacting responses to simultaneous activation of α_1- and α_2-adrenoceptors. Br J Pharmacol 95:485–500

22. Daly RN, Sulpizio AC, Levitt B, DeMarinis RM, Regan JW, Ruffolo RR, Hieble JP (1988) Evidence for heterogeneity between pre- and postjunctional alpha-2 adrenoceptors using 9-substituted 3-benzazepines. J Pharmacol Exp Ther 247:122–128

23. Drew GM, Whiting SB (1979) Evidence for two distinct types of postsynaptic α-adrenoceptor in vascular smooth muscle in vivo. Br J Pharmacol 67:207–215

24. Drexler H, Zeiher AM, Wollschläger H, Meinertz T, Just H, Bonzel T (1989) Flow-dependent coronary artery dilatation in humans. Circulation 80:466–474

25. Dunn WR, McGrath JC, Wilson VG (1989) Expression of functional postjunctional α_2-adrenoceptors in rabbit isolated distal saphenous artery – a permissive role for angiotensin II? Br J Pharmacol 96:259–261

26. Feigl EO (1983) Coronary physiology. Physiol Rev 63:1–205

27. Fish RD, Sperti G, Colucci WS, Clapham DE (1988) Phorbol ester increases the dihydropyridine-sensitive calcium conductance in a vascular smooth muscle cell line. Circ Res 62:1049–1054

28. Flavahan NA, Cooke JP, Shepherd JT, Vanhoutte PM (1987) Human postjunctional alpha-1 and alpha-2 adrenoceptors: Differential distribution in arteries of the limbs. J Pharmacol Exp Ther 241:361–365

29. Franco-Cereceda A, Lundberg JM, Dahlof C (1985) Neuropeptide Y and sympathetic control of heart contractility and coronary vascular tone. Acta Physiol Scand 124:361–369

30. Fraser CM, Arakawa S, McCombie RW, Venter JC (1989) Cloning, sequence analysis, and permanent expression of a human α_2-adrenergic receptor in chinese hamster ovary cells: evidence for independent pathways of receptor coupling to adenylate cyclase attenuation and activation. J Biol Chem 264:11754–11761

31. Furata T (1988) Precontraction-induced contractile response of isolated canine portal vein to alpha-2 adrenoceptor agonists. Naunyn-Schmiedeberg's Arch Pharmacol 337:525–530

32. Furchgott RF, Vanhoutte PM (1989) Endothelium-derived relaxing and contracting factors. FASEB J 3:2007–2018

33. Gerová M, Barta E, Gero J (1979) Sympathetic control of major coronary artery diameter in the dog. Circ Res 44:459–467

34. Graziadei I, Zernig G, Boer R, Glossman H (1989) Stereoselective binding of niguldipine enantiomers to α_{1A}-adrenoceptors labeled with (3H)5-methyl-urapidil. Eur J Pharmacol (Mol Pharmacol Sect) 172:329–337

35. Gu J, Adrian TE, Tatemoto K, Polak JM, Allen JM, Bloom SR (1983) Neuropeptide tyrosine (NPY): a major cardiac neuropeptide. Lancet 1:1008–1010

36. Han C, Abel PW (1987) Neuropeptide Y potentiates contraction and inhibits relaxation of rabbit coronary arteries. J Cardiovasc Pharmacol 9:675–681

37. Han C, Abel PW, Minneman KP (1987) Heterogeneity of α_1-adrenergic receptors revealed by chlorethylclonidine. Mol Pharmacol 32:505–510

38. Han C, Abel PW, Minneman KP (1987) α_1-Adrenoceptor subtypes linked to different mechanisms for increasing intracellular Ca^{2+} in smooth muscle. Nature 329:333–335

39. Hanft G, Gross G (1989) Subclassification of α_1-adrenoceptor recognition sites by urapidil derivatives

and other selective antagonists. Br J Pharmacol 97:691–700

40. Heusch G, Deussen A, Schipke J, Thämer V (1984) α_1- and α_2-adrenoceptor-mediated vasoconstriction of large and small canine coronary arteries in vivo. J Cardiovasc Pharmacol 6:961–968

41. Holtz J, Saeed M, Sommer O, Bassenge E (1982) Norepinephrine constricts the canine coronary bed via postsynaptic α_2-adrenoceptors. Eur J Pharmacol 82:199–202

42. Holtz J, Förstermann U, Pohl U, Giesler M, Bassenge E (1984) Flow-dependent, endothelium-mediated dilation of epicardial coronary arteries in conscious dogs: Effects of cyclooxygenase inhibition. J Cardiovasc Pharmacol 6:1161–1169

43. Hori M, Kitakaze M, Tamai J, Koretsune Y, Iwai K, Iwakura K, Kagiya T, Kitabatake A, Inoue M, Kamada T (1988) α_2-Adrenoceptor activity exerts dual control of coronary blood flow in canine coronary artery. Am J Physiol 255:H250–H260

44. Hori M, Kitakaze M, Tamai J, Iwakura K, Kitabatake A, Inoue M, Kamada T (1989) α_2-Adrenoceptor stimulation can augment coronary vasodilation maximally induced by adenosine in dogs. Am J Physiol 257:H132–H140

45. Illes P, Regenold JT (1990) Interaction between neuropeptide Y and noradrenaline on central catecholamine neurons. Nature 344:62–63

46. Jonkman FAM, Man PW, Breurkes R, Van Zwieten PA (1989) β_2-Adrenoceptor antagonists intensify clonidine withdrawal syndrome in conscious rats. J Cardiovasc Pharmacol 14:886–891

47. Kazanietz MG, Gutkind JS, Enero MA (1986) Interaction between beta-2 and alpha-2-adrenoceptor responses in the vascular system: effect of clenbuterol. Eur J Pharmacol 130:119–124

48. Kazanietz MG, Gutkind JS, Puyo A, Armado I, Enero MA (1989) Further evidence of interaction between vasodilator β_2- and vasoconstrictor α_2-adrenoceptor-mediated responses in maintaining vascular tone in anesthetized rats. J Cardiovasc Pharmacol 14:847–880

49. Kelm M, Schrader J (1988) Nitric oxide release from the isolated guinea pig heart. Eur J Pharmacol 155:317–321

50. Kitakaze M, Hori M, Tamai J, Iwakura K, Koretsune Y, Kagiya T, Iwai K, Kitabatake A, Inoue M, Kamada T (1987) α_1-Adrenoceptor activity regulates release of adenosine from the ischemic myocardium in dogs. Circ Res 60:631–639

51. Kobilka BK, Matsui H, Kobilka TS, Yang-Feng TL, Francke U, Caron MG, Lefkowitz RJ, Regan JW (1987) Cloning, sequencing, and expression of the gene coding for the human platelet α_2-adrenergic receptor. Science 238:650–656

52. Kopia GA, Kopaciewicz LJ, Ruffolo RR (1986) Alpha Adrenoceptor regulation of coronary artery blood flow in normal and stenotic canine coronary arteries. J Pharm Exp Ther 239:641–647

53. Kristek F, Gerová M (1987) Autonomic nerve terminals in relation to contractile and non-contractile structures in the conduit coronary artery of the dog. Acta Anat 129:149–154

54. Lacerda AE, Rampe D, Brown AM (1988) Effects of protein kinase C activators on cardiac Ca^{2+} channels. Nature 335:249–251

55. Limbird LE (1988) Receptors linked to inhibition of adenylate cyclase: additional signaling mechanisms. FASEB J 2:2686–2695

56. Lipscombe D, Kongsamut S, Tsien RW (1989) α-adrenergic inhibition of sympathetic neurotransmitter release mediated by modulation of N-type calcium-channel gating. Nature 340:639–642

57. Macho P, Pérez R, Huidobro-Toro JP, Domenech RJ (1989) Neuropeptide Y (NPY): a coronary vasoconstrictor and potentiator of catecholamine-induced coronary constriction. Eur J Pharmacol 167:67–74

58. Maturi MF, Greene R, Speir E, Burrus C, Dorsey LMA, Markle DR, Maxwell M, Schmidt W, Goldstein SR, Patterson RE (1989) Neuropeptide-Y: a peptide found in human coronary arteries constricts primarily small coronary arteries to produce myocardial ischemia in dogs. J Clin Invest 83:1217–1224

59. McGrath JC, Brown CM, Wilson VG (1989) Alpha-adrenoceptors: A critical review. Med Res Rev 9:407–533

60. Minneman KP, Han C, Abel PW (1988) Comparison of α_1-adrenergic receptor subtypes distinguished by chlorethylclonidine and WB 4101. Mol Pharmacol 33:509–514

61. Müller-Schweinitzer E (1983) Tissue specific susceptibility of alpha-adrenoceptor mediated vasoconstriction to nifedipine. Naunyn-Schmiedeberg's Arch Pharmacol 324:64–69

62. Muramatsu I, Ohmura T, Kigoshi S, Hashimoto S, Oshita M (1990) Pharmacological subclassification of α_1-adrenoceptors in vascular smooth muscle. Br J Pharmacol 99:197–201

63. Nelson MT, Standen NB, Brayden JE, Worley JF (1988) Noradrenaline contracts arteries by activating voltage-dependent calcium channels. Nature 336:382–385

64. Nielsen H, Mortensen FV, Mulvany MJ (1990) Responses to noradrenaline in human subcutaneous resistance arteries are mediated by both α_1- and α_2-adrenoceptors. Br J Pharmacol 99:31–34

65. Pearson JD, Gordon JL (1989) P_2-Purinoceptors in the blood vessel wall. Biochem Pharmacol 38:4157–4163

66. Regan JW, Kobilka TS, Yang-Feng TL, Caron MG, Lefkowitz RJ, Kobilka BK (1988) Cloning and expression of a human kidney cDNA for an α_2-adrenergic receptor subtype. Proc Natl Acad Sci USA 85:6301–6305

67. Rubanyi GM, Romero JC, Vanhoutte PM (1986) Flow-induced release of endothelium-derived relaxing factor. Am J Physiol 250:H1145–H1149

68. Saeed M, Holtz J, Elsner D, Bassenge E (1985) Sympathetic control of myocardial oxygen balance in dogs mediated by activation of coronary vascular α_2-adrenoceptors. J Cardiovasc Pharmacol 7:167–173

69. Shepperson NB (1984) α_2-Adrenoceptor agonists potentiate responses mediated by α_1-adrenoceptors in the cat nictitating membrane. Br J Pharmacol 83:463–471

70. Shi AG, Kwan CY, Daniel EE (1989) Relation between density (maximum binding) of alpha adrenoceptor binding sites and contractile response in four canine vascular tissues. J Pharmacol Exp Ther 250:1119–1124

71. Söhngen W, Winbury MM, Kitzen JM, Ventura A, Lucchesi BR (1988) The mechanism for the clonidine-induced coronary artery dilatation in the canine heart. J Cardiovasc Pharmacol 12:689–700

72. Starke K (1987) Presynaptic α-autoreceptors. Rev Physiol Biochem Pharmacol 107:73–146

73. Starke K, Göthert M, Kilbinger H (1989) Modulation of transmitter release by presynaptic autoreceptors. Physiol Rev 69:864–989

74. Sulpizio A, Hieble JP (1987) Demonstration of α_2-adrenoceptor-mediated contraction in the isolated canine saphenous artery treated with Bay K 8644. Eur J Pharmacol 135:107–110

75. Szabo B, Hedler L, Starke K (1989) Peripheral presynaptic and central effects of clonidine, yohimbine and rauwolscine on the sympathetic nervous system in rabbits. Naunyn-Schmiedeberg's Arch Pharmacol 340:48–657

76. Templeton AGB, MacMillan J, McGrath JC, Storey ND, Wilson VG (1989) Evidence for prazosin-resistant, rauwolscine-sensitive α-adrenoceptors mediating contractions in the isolated vascular bed of the rat tail. Br J Pharmacol 97:563–571

77. Timmermans PBMWM, Kwa HY, Van Zwieten PA (1979) Possible subdivision of postsynaptic α-adrenoceptors mediating pressor responses in the pithed rat. Naunyn-Schmiedeberg's Arch Pharmacol 310:189–193

78. Vanhoutte PM, Miller VM (1989) Alpha-2-adrenoceptors and endothelium-derived relaxing factor. Am J Med 87 (suppl 3C):1S–5S

79. Watson S, Abbott A (1990) TIPS receptor nomenclature supplement. Trends Pharmacol Sci 11 (supplement):1–30

80. Wilffert B, Gouw MAM, Timmermans PBMWM, Van Zwieten PA (1983) Interaction between beta-2-adrenoceptor-mediated vasodilation and alpha-2-adrenoceptor-mediated vasoconstriction in the pithed normotensive rat. J Cardiovasc Pharmacol 5:822–828

81. Woodman OL (1987) The role of α_1- and α_2-adrenoceptors in the coronary vasoconstrictor responses to neuronally released and exogenous noradrenaline in the dog. Naunyn-Schmiedeberg's Arch Pharmacol 336:161–168

82. Woodman OL, Vatner SF (1987) Coronary vasoconstriction mediated by α_1- and α_2-adrenoceptors in conscious dogs. Am J Physiol 253:H388–H393

83. Woodman OL (1988) Adrenoceptor subtypes involved in the baroreceptor reflex constriction of large coronary arteries in the anesthetized dog. Eur J Pharmacol 158:37–42

84. Xiao XH, Rand MJ (1989) α_2-Adrenoceptor agonists enhance vasoconstrictor responses to α_1-adrenoceptor agonists in the rat tail artery by increasing the influx of Ca^{2+}. Br J Pharmacol 98:1032–1038

85. Young P, Berge J, Chapman H, Cawthorne MA (1989) Novel α_2-adrenoceptor antagonists show selectivity for α_{2A}- and α_{2B}-adrenoceptor subtypes. Eur J Pharmacol 168:381–386

Author's address:

Prof. Dr. med. J. Holtz, Institut für Angewandte Physiologie und Balneologie, Universität Freiburg, Hermann – Herder – Str. 7, 7800 Freiburg.

α- and β-Adrenergic Control of Large Coronary Arteries in Conscious Calves

M. A. Young, D. E. Vatner and S. F. Vatner

Departments of Medicine and Pediatrics, Harvard Medical School, Brigham and Women's Hospital, Boston and New England Regional Primate Research Center, Southborough, Massachusetts, USA

Summary

Large and small coronary arteries are subject to control by α- and β-adrenergic mechanisms. However, controversy exists as to the distribution and physiological effects of α- and β-adrenergic receptor subtypes in large coronary arteries. Studies in our laboratory have addressed these questions in conscious calves, chronically instrumented to measure large coronary artery diameter and coronary blood flow. Additionally, adrenergic receptor subtype distribution was determined using ligand binding assays in membrane preparations isolated from large coronary arteries of calves. Physiological results demonstrate, in contrast to the results of most previous studies, that both α_1- and α_2-adrenergic receptors elicit constriction of the large coronary artery. Studies with ganglionic blockade indicate that the constriction was unaltered by autonomic reflexes or presynaptic release of neurotransmitters. Selective β-adrenergic receptor activation demonstrated that both β_1- and β_2-adrenergic receptors elicit dilation of large coronary arteries, and that the vasodilation was direct, i.e., it was not mediated by increases in coronary blood flow. Biochemical characterization of adrenergic subtype density indicated the presence of both α_1- and α_2-, as well as β_1- and β_2-adrenergic receptor subtypes. Thus, both biochemical and physiological data support the concept that large coronary arteries are regulated by both α_1- and α_2-, as well as β_1- and β_2-adrenergic receptor subtypes.

Introduction

The traditional concept that large coronary arteries serve only a conduit function with little contribution to total coronary vascular resistance has been challenged by recent studies demonstrating that a significant fraction of total resting coronary resistance can be attributed to vessels larger than 100 μm [3, 4]. This observation, combined with the recognition that localized coronary artery vasospasm is a major factor in the development of myocardial ischemia, angina pectoris, myocardial infarction, and sudden death, has focused greater attention on the mechanisms controlling large coronary artery caliber. Clinical studies have implicated alterations in the autonomic control of the circulation in the pathogenesis of congestive heart failure, hypertension, ischemic heart disease, Raynaud's syndrome, and other diseases. However, despite extensive study, controversy still exists concerning the autonomic control of large coronary arteries, particularly with regard to the relative effects of α_1- vs α_2-, and β_1- vs β_2-adrenergic receptor activation. To address this controversy, studies performed over several years in our laboratory have examined the physiological responses of large coronary arteries to intracoronary administration of selective α_1- and α_2-, and β_1- and β_2-adrenergic receptor

Supported in part by USPHS Grants # HL 33107, 38070, 38402, and RR 00168.

agonists in conscious, chronically instrumented calves. The physiological responses were compared with the results of biochemical characterization of α- and β-adrenergic receptor subtype distribution performed in vitro [20, 28].

Methods

The methods used in these studies and the results reported below have been published in detail previously [20, 28].

Physiological studies

Female calves, 6–10 weeks old, fully weaned and weighing 60–80 kg, were anesthetized with halothane following pre-anesthesia with sodium thiamylal. Through a left thoracotomy, two miniature ultrasonic dimension transducers were implanted on the left circumflex coronary artery. Distal to the dimension transducers, a Doppler ultrasonic blood flow transducer and an hydraulic occluder were implanted around the vessel. Proximal to the dimension transducers, a silastic catheter was implanted through the wall of the circumflex artery. Tygon catheters were implanted in the descending aorta and left atrium. The wires and catheters were externalized to the intrascapular space. The incision was closed, the chest evacuated, and the animal was allowed to recover. Catheters were flushed daily with heparin, and the animals were given 6 million units of penicillin daily for 10 days.

Measurements were recorded of left circumflex coronary artery diameter, aortic pressure, and left circumflex coronary blood flow. In eight calves, selective α_1-adrenergic receptor stimulation was accomplished with intracoronary (i.c.) bolus injection of phenylephrine (0.1, 1.0, and 5.0 μg/kg) and selective α_2-adrenergic receptor stimulation with B-HT 920 (0.1, 1.0, and 5.0 μg/kg). Combined α_{1+2}-adrenergic receptor stimulation was accomplished with i.c. norepinephrine (0.01, 0.1, and 0.5 μg/kg) following β-blockade with propranolol (1 mg/kg i.v. bolus followed by 1 mg/kg/hr i.v. infusion). Injections were repeated on separate days in the presence of selective α_1-adrenergic receptor blockade with prazosin (0.01 mg/kg/min, i.c.), or selective α_2-adrenergic receptor blockade with rauwolscine (0.005 mg/kg/min, i.c.). In four animals, responses to phenylephrine, B-HT 920, and norepinephrine were also determined following ganglionic blockade with hexamethonium (10 mg/kg, i.v.) to minimize effects of reflex changes on coronary artery diameter, as well as the prejunctional effects of α-adrenergic receptor stimulation or blockade.

Combined β_1- and β_2-adrenergic receptor stimulation was accomplished with intracoronary (i.c.) bolus injection of isoproterenol (0.0025 μg/kg). Selective β_1-adrenergic stimulation was achieved with prenalterol (0.4 μg/kg, i.c.) and selective β_2-adrenergic stimulation with pirbuterol (0.025 μg/kg, i.c.). On separate days, the effects of these agonists were studied following selective β_1-adrenergic receptor blockade with atenolol (0.01 mg/kg, i.c.) or betaxolol (0.001 mg/kg, i.c.), and following selective β_2-adrenergic blockade with ICI 118,551 (50 μg/min, i.c. constant infusion).

Biochemical studies

The methods used for preparation of coronary artery sarcolemmal membranes and subsequent receptor binding studies have been described in detail in earlier publications from this laboratory [20, 28].

α-Adrenergic receptor studies

Binding of [³H]-prazosin and [³H]-rauwolscine to the membrane preparations was performed in triplicate in a total volume of 500 μl by incubating 400 μl membranes (1 mg/ml for [³H]-

Adrenergic control of large coronary arteries

rauwolscine and 0.5 mg/ml for [^{3}H]-prazosin), with 50 μl of the radioligand and 50 μl of either buffer or various concentrations of an unlabeled competing drug at 25 °C for 45 min. Unbound radioligand was separated by vacuum filtration on Whatman G/C filters which were washed four times with 4 ml aliquots of Buffer B. The amount of ligand bound was determined by liquid scintillation techniques. The concentration range of radioligand utilized in saturation binding studies was 0.5 to 30 nM for [^{3}H]-rauwolscine and 0.01 to 3 nM for [^{3}H]-prazosin. In these studies, nonspecific binding was determined in the presence of 1 μM prazosin for the α_1-antagonist, and 1 μM rauwolscine for the α_2-antagonist. Antagonist competition binding curves were performed with 0.7 nM [^{3}H]-prazosin and 5 nM [^{3}H]-rauwolscine, using prazosin, rauwolscine, and phentolamine (10^{-10} to 10^{-4} M). Non-specific binding was 48% for α_1-agonist and antagonist binding, and 56% for α_2-binding. The K_D and B_{max} for each radioligand was calculated from Scatchard analysis and from the non-transformed binding data with the mass action law-based, weighted nonlinear curve fitting program, LIGAND.

β-Adrenergic receptor studies

For saturation binding studies, 25 μg of membrane protein (100 μl) per assay were incubated in a shaking water bath at 37 °C for 30 min in the presence of increasing concentration of ^{125}I-CYP (0.025–1.0 nM) with or without added unlabelled (−)-propranolol (10 μM). Following incubation, the reaction mixture was filtered and the filters counted in a gamma counter for 1 min. Saturation analysis was performed by the method of Scatchard and, in addition, analyzed by the LIGAND program [13]. To determine relative potency of sympathomimetic amines for the β-adrenergic receptors in the large coronary artery, the following sympathomimetic amines were used in the same experiment: (−)-isoproterenol (10^{-9}–10^{-4} M); (−)-epinephrine (10^{-8}–10^{-4} M); and norepinephrine (10^{-7}–10^{-4} M). Using these agonists, binding data from the coronary artery were compared to that from the heart, which contains primarily β_1-adrenergic receptors, and the lung, which contains primarily β_2-adrenergic receptors.

The percent of β_1- and β_1-receptor subtypes in the sarcolemmal membranes was determined by competitive inhibition studies with the nonselective antagonist radioligand ^{125}I-CYP (0.2 nM), the selective β_1-agonist prenalterol, the selective β_2-agonist pirbuterol, the selective β_1-antagonist betaxolol, and the selective β_2-antagonist ICI 118,551. The high- and low-affinity binding constants and the ratio of high- to low-affinity binding sites were determined for each agonist and antagonist in lung, calf heart and large coronary artery smooth muscle by computer modeling as described by Munson and Rodbard [13].

Results

α-Adrenergic receptor studies

Figure 1 shows the effects of α_2-receptor stimulation following intracoronary injection of B-HT 920. Figure 2 shows the effects of α_1-selective stimulation with phenylephrine following β-blockade, α_2-selective stimulation with B-HT 920 following β-blockade, and mixed $\alpha_1 + \alpha_2$-stimulation with norepinephrine in the presence of β-blockade. Figure 2 also shows the effects of either α_1-blockade with prazosin or α_2-blockade with rauwolscine on the constriction with phenylephrine, B-HT 920 and norepinephrine. The dose-dependent constriction with phenylephrine was unchanged following α_2-blockade, but was abolished following α_1-blockade (p < 0.05). Conversely, the constriction in response to B-HT 920 was unchanged in the presence of α_1-blockade with prazosin, but abolished (p < 0.05) following administration of the selective α_2-adrenergic receptor blocker rauwolscine. As expected, the constriction with norepinephrine was attenuated by either α_1- or α_2-blockade, demonstrating the mixed agonist activity of norepinephrine. When α_1- and α_2-blockade were administered in combination, the constriction with norepinephrine was abolished.

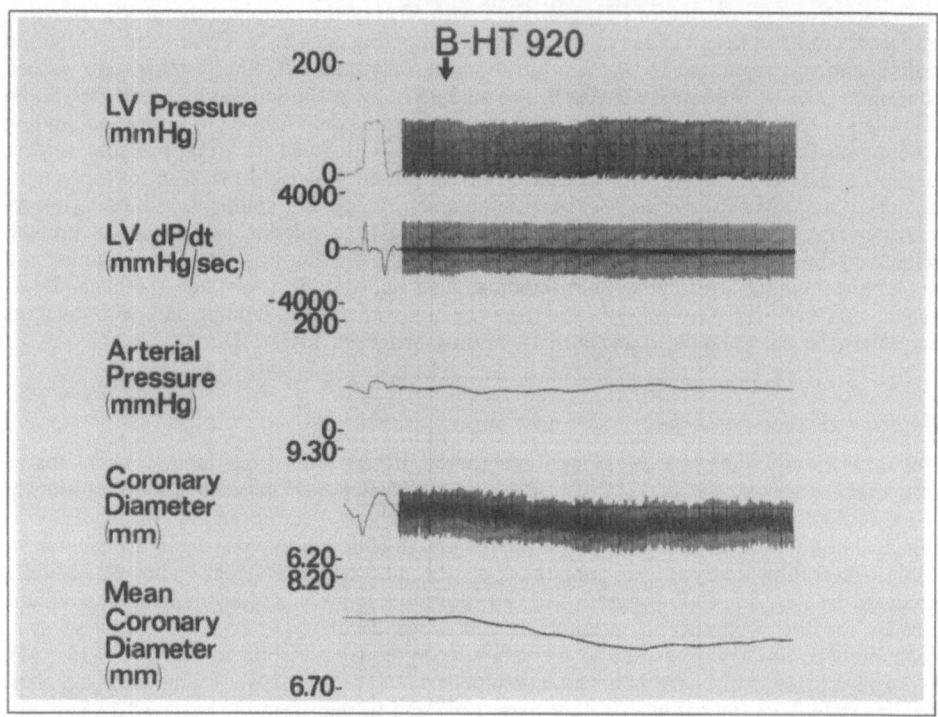

Fig. 1. Illustration of the response of left ventricular pressure, left ventricular dP/dt, mean arterial pressure, and coronary artery diameter to intracoronary injection of B-HT 920 (5.0 µg/kg). α_2-Stimulation constricted the coronary artery despite minor changes in systemic hemodynamics. (Reproduced with permission from [28].)

To assess potential effects of reflex activation or the presynaptic effects of these compounds, we administered norepinephrine, phenylephrine, and B-HT 920 in the presence of β-adrenergic blockade with propranolol and ganglionic blockade with hexamethonium. In four animals, the constrictions before and following ganglionic blockade were similar for phenylephrine ($-7.2 \pm 1.3\%$ vs $-8.3 \pm 2.5\%$), B-HT 920 ($-6.3 \pm 1.1\%$ vs $-7.7 \pm 1.5\%$) or norepinephrine ($-6.6 \pm 0.5\%$ vs $-7.6 \pm 2.3\%$).

Scatchard analysis of binding studies with the calf coronary artery membrane preparation (n = 7) indicated that the α_1-adrenergic receptor density was 15 ± 3.1 fmol/mg protein, and the dissocation constant (K_D) for [^{3}H]-prazosin binding 0.65 ± 0.15 nM. [^{3}H]-rauwolscine binding to the same membrane preparations indicated that the α_2-adrenergic receptor density was 68 ± 5.1 fmol/mg protein, and the K_D was 7.35 ± 1.19 nM (Fig. 3).

The rank order of potency for displacement of [^{3}H]-prazosin with a series of unlabelled antagonists was appropriate for an α_1-adrenergic receptor ligand. For example, with displacement of [^{3}H]-prazosin, the K_i for prazosin was 6 pM, for phentolamine it was 425 pM, and for rauwolscine it was 30,000 pM. With [^{3}H]-rauwolscine, an α_2-adrenergic receptor ligand, the reverse order of potency was seen, i.e., the K_i for rauwolscine was 540 pM, for phentolamine 55,000 pM, and for prazosin 544,000 pM.

β-Adrenergic receptor studies

Figure 4 illustrates the effects of injection of isoproterenol on large coronary artery diameter. In five calves, i.c. injection of isoproterenol increased mean coronary blood flow ($134 \pm 5.9\%$)

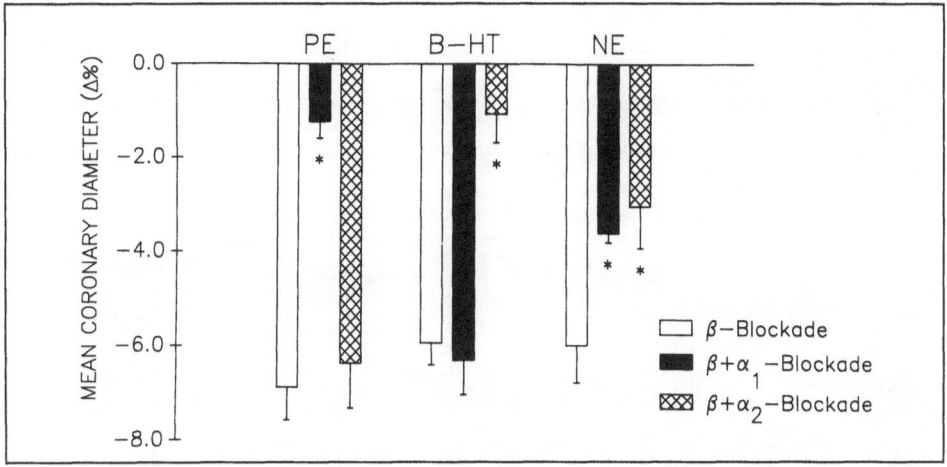

Fig. 2. The effects of selective α_1-receptor activation with phenylephrine (PE) and selective α_2-receptor activation with B-HT 920 and mixed α_{1+2}-receptor activation with norepinephrine (NE) following β-blockade alone, and following $\beta + \alpha_1$-blockade or $\beta + \alpha_2$-blockade. The coronary constriction with PE was abolished by α_1-blockade, but unchanged following α_2-blockade. The coronary constriction with B-HT 920 was abolished by α_2-blockade, but unchanged following α_1-blockade. The constriction with NE was attenuated by either α_1- or α_2-blockade ($p < 0.05$). (Figure redrawn from data in [28].)

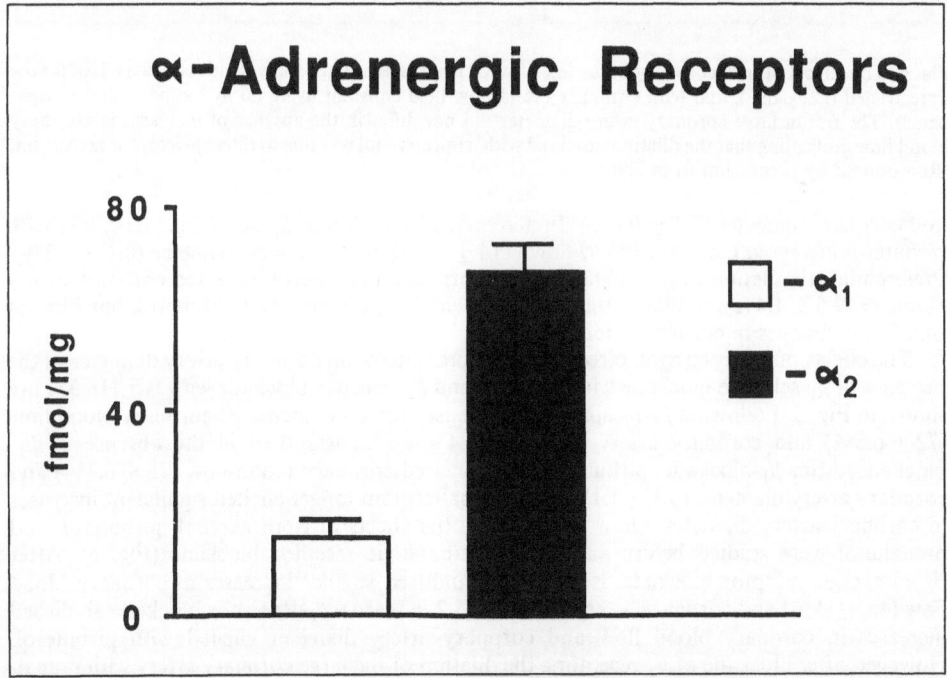

Fig. 3. Comparison of α_1- and α_2-adrenergic receptor density, as determined by binding of [^{3}H]-prazosin and [^{3}H]-rauwolscine. Receptor density is greater for α_2-adrenergic receptor. (Reproduced by permission from [28].)

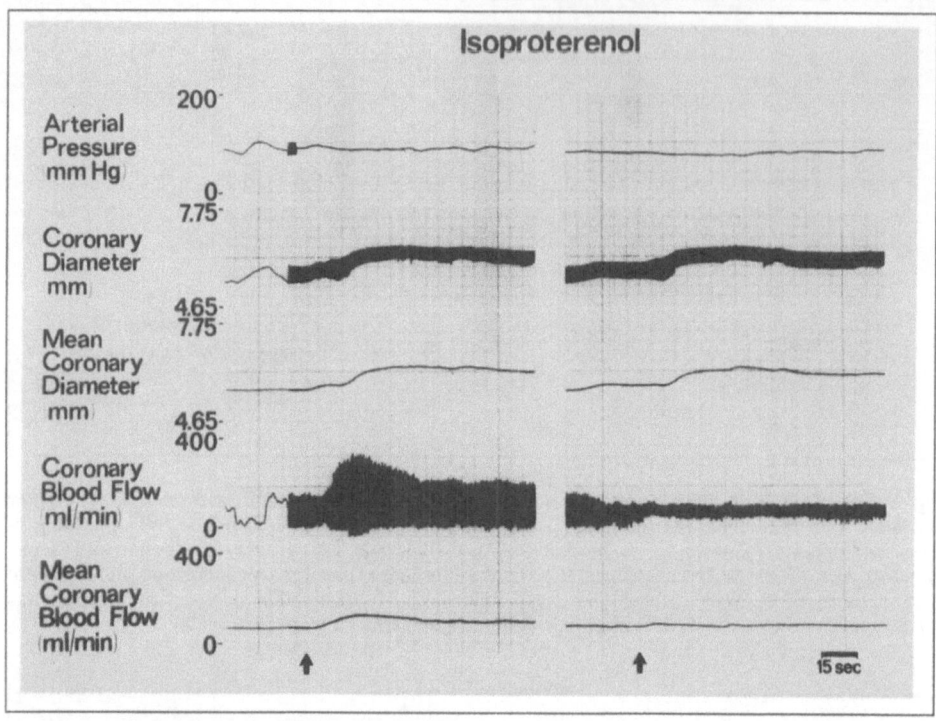

Fig. 4. The effects of intracoronary bolus injection of isoproterenol are shown with coronary blood flow unrestricted (left panel), and with coronary blood flow held constant using an hydraulic occluder (right panel). The rise in large coronary artery diameter did not differ in the absence of increases in coronary blood flow, indicating that the dilation observed with isoproterenol was due to direct β-receptor activation. (Reproduced by permission from [20].)

and coronary diameter (7.1 $\pm$ 0.8%). Preferential stimulation of β_2-adrenergic receptors with pirbuterol increased coronary blood flow (114 $\pm$ 19%) and coronary diameter (8.1 $\pm$ 1.2%). Preferential stimulation of β_1-adrenergic receptors with prenalterol increased coronary artery diameter (9.5 $\pm$ 1.4%), similar to that observed with isoproterenol and pirbuterol, but without significant changes in coronary blood flow.

The effects of isoproterenol, pirbuterol, and prenalterol on coronary artery diameter in the presence of β_1-selective blockade with atenolol, and β_2-selective blockade with ICI 118,551 are shown in Fig. 5. Following β_1-receptor blockade, isoproterenol increased coronary blood flow (72 $\pm$ 6.5%) and coronary artery diameter (4.4 $\pm$ 1.7%) less than in the absence of β_1-blockade. After β_1-blockade, pirbuterol still increased coronary blood flow (75 $\pm$ 8.3%) and coronary artery diameter (5.8 $\pm$ 1.1%), but prenalterol no longer elicited significant increases in coronary artery diameter. On a separate day the effects of isoproterenol, pirbuterol, and prenalterol were studied before and after β_2-adrenergic receptor blockade (Fig. 5). After β_2-adrenergic receptor blockade, isoproterenol induced smaller increases in coronary blood flow (45 $\pm$ 15%) and coronary artery diameter (5.2 $\pm$ 0.4%). β_2-Receptor blockade abolished increases in coronary blood flow and coronary artery diameter elicited with pirbuterol. However, after blockade of β_2-receptors, the dilation of the large coronary artery with prenalterol was still observed.

In the experiments designed to discriminate the effects of increases in coronary blood flow on large artery diameter, isoproterenol, pirbuterol, and prenalterol were injected while holding coronary blood flow constant with an hydraulic occluder. With coronary blood flow restricted,

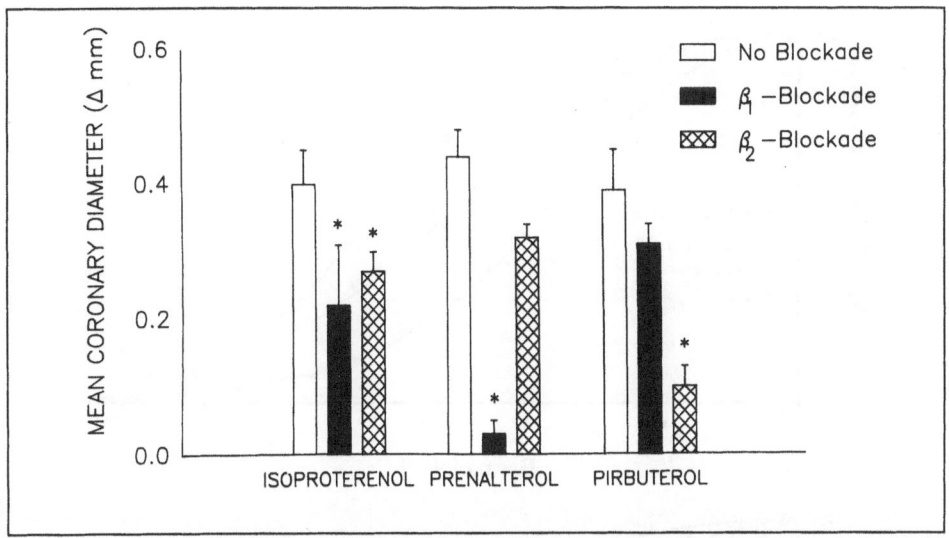

Fig. 5. Comparison of the effects of isoproterenol, pirbuterol, and prenalterol on large coronary artery diameter in the absence of β-blockade (open bars), following β_1-blockade with atenolol (shaded bars), and following β_2-blockade with ICI 118,551 (hatched bars). The dilation with isoproterenol was partially reduced with either β_1- or β_2-blockade. The dilation with pirbuterol was abolished by ICI 118,551, but unaffected by atenolol. Conversely, dilation with prenalterol was unchanged following ICI 118,551 but abolished following atenolol ($p < 0.05$). (Figure redrawn from data in [20].)

the dilation of the large coronary artery did not differ from that observed when blood flow was unrestricted. These experiments support the conclusion that β_1- and β_2-activation directly increase large coronary artery caliber.

Scatchard analyses of saturation binding studies of the large coronary artery membrane preparation indicated that the receptor density was 49 ± 4.1 fmol/mg protein, and the dissociation constant (K_D) for ^{125}I-CYP was 0.17 ± 0.04 nM. To determine the proportion of β_1- to β_2-adrenergic receptors in the large coronary artery, competitive inhibition binding with ^{125}I-CYP and subtype selective antagonists betaxolol (β_1) and ICI 118,551 (β_2) were performed (Fig. 6). Computer modeling of the data shown in Fig. 6 indicated that the β_1-selective antagonist betaxolol competed for ^{125}I-CYP with the highest affinity in heart, followed by lower affinity in calf coronary artery, and with the lowest affinity in the lung. The percent of high affinity receptors for betaxolol was $85 \pm 0.3\%$ in heart, $19 \pm 2.5\%$ in lung, and $64 \pm 6.1\%$ in coronary artery. In contrast, the β_2-selective antagonist ICI 118,551 competed for ^{125}I-CYP with the highest affinity in lung, with lower affinity in the coronary artery, and the lowest affinity in the heart. Computer modeling again indicated the percent of high affinity receptors for ICI 118,551 to be greatest in lung ($78 \pm 3.1\%$), least in heart ($24 \pm 3.5\%$) and intermediate in the coronary artery ($41 \pm 0.8\%$). Similar results were observed in competition binding studies using the catecholamines norepinephrine, isoproterenol, and epinephrine, as well as with the selective agonists pirbuterol (β_2) and prenalterol (β_1). Taken together, these data indicate that both β-adrenergic receptor subtypes are present in calf coronary artery, with a predominance of the β_1-adrenergic subtype.

Discussion

α-Adrenergic receptor studies

The overwhelming majority of studies examining α-adrenergic constriction of large coronary arteries conclude that constriction is exclusively the result of α_1-adrenergic receptor activation, with little or no contribution by α_2-adrenergic receptors [5, 6, 9, 11, 16, 18]. This conclusion is

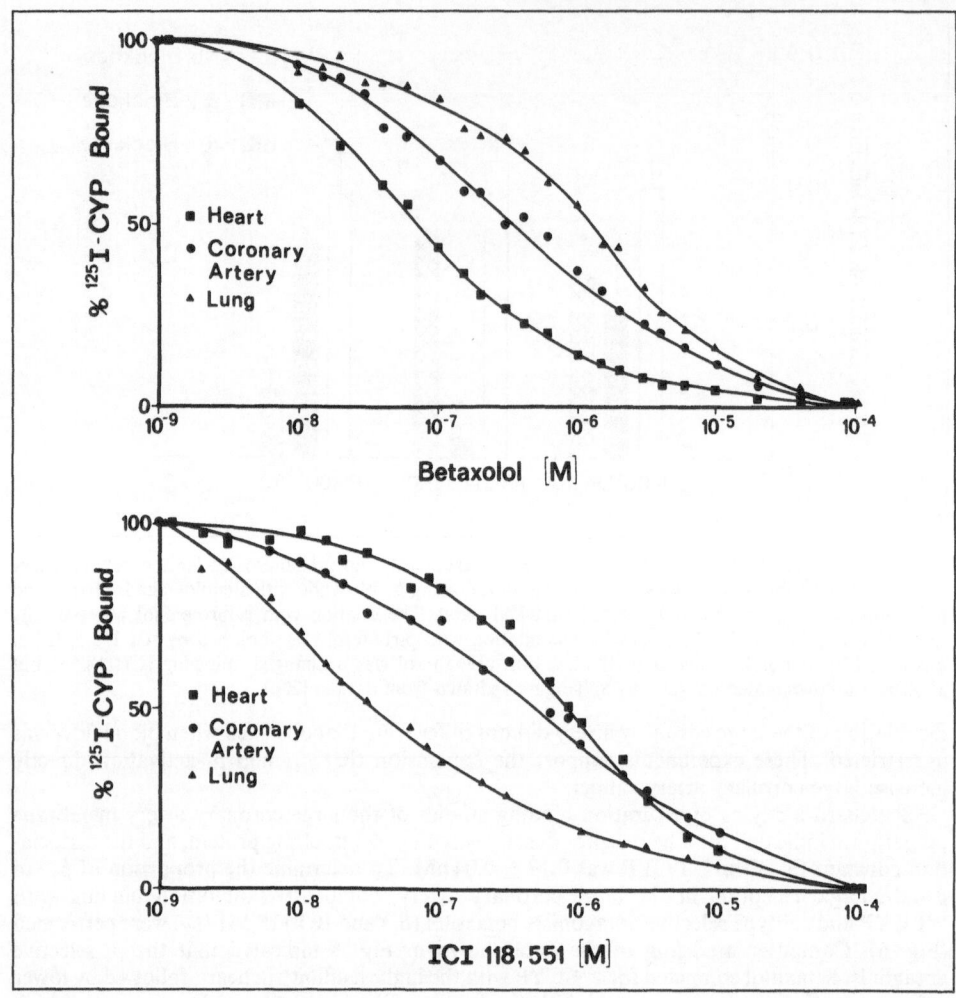

Fig. 6. The effects of increasing concentrations of betaxolol (top panel) and ICI 118,551 (lower panel) are shown on binding of ^{125}I-CYP to membranes prepared from heart (squares), coronary artery (circles), and lung (triangles). These data suggest that receptors in the coronary artery are a mixture of the β_1- and β_2-receptor subtypes. (Reproduced by permission from [20].)

based on functional studies demonstrating the lack of contractile responses to α_2-specific agonists and the relative inability of α_2-specific antagonists to inhibit norepinephrine-induced contractions [1, 5, 6, 9, 18, 19]. In fact, precontracted isolated coronary arteries of dogs and pigs have been reported to respond to α_2-receptor activation with relaxation, due to the presence of α_2-receptors on coronary endothelial cells [1, 5, 6]. However, the observation of α_2-mediated dilation of large coronary arteries has not been substantiated in vivo, and few studies have compared the effects of selective α_1- and α_2-adrenergic activation on large coronary artery caliber.

Biochemical studies of radioligand binding have demonstrated the presence of either α_1- [2, 7, 14] or α_2-adrenergic receptors [2, 24] in membranes from a variety of arterial tissues. Recently, Nishimura et al. [15] demonstrated equal α_1- and α_2-adrenergic receptor density in porcine coronary artery membranes. That study did not examine receptor-mediated changes in smooth muscle tone. In fact, few studies have compared the contractile activity of α_1- and α_2-adrenergic receptor activation with the biochemical characterization of receptor subtypes, and no study has compared adrenergic function with receptor subtype distribution in coronary arteries.

From our recent studies [20, 28], three lines of evidence suggest that both α_1- and α_2-adrenoreceptors are present in bovine large coronary arteries, and that both receptor subtypes are capable of eliciting vasoconstriction: 1) The selective α_1-agonist (phenylephrine) and α_2-agonist (B-HT 920) produce equivalent, dose-dependent reductions in large coronary diameter in the presence and absence of ganglionic blockade; 2) these constrictions with phenylephrine and B-HT 920 are abolished by the selective antagonists prazosin or rauwolscine, while the response to the mixed α_{1+2} agonist norepinephrine is only partially attenuated by either of the selective antagonists; and 3) ligand-binding studies demonstrate the presence of both α_1- and α_2-adrenoreceptors in coronary sarcolemmal membranes [28].

The studies reviewed in this report demonstrate similar dose-response curves for the large coronary artery to the selective α_1-agonist phenylephrine, the selective α_2-agonist B-HT 920, and the mixed α_{1+2} agonist norepinephrine [28]. The conclusion that phenylephrine produces large coronary artery constriction via the specific α_1-receptor subtype is further substantiated by the demonstration that the response to phenylephrine was selectively antagonized with prazosin, but was unaltered in the presence of α_2-blockade with rauwolscine (Fig. 2). Similar specificity of B-HT 920 was demonstrated by the near complete blockade of constriction following rauwolscine, while the response to B-HT 920 was unaffected by α_1-blockade with prazosin (Fig. 2). The partial attenuation of the response to norepinephrine by either α_1- or α_2-adrenergic receptor blockade alone, and the abolition following combined α_{1+2}-blockade provides further support for the concept that both α_1- and α_2-adrenergic receptors, which can elicit large coronary vasoconstriction, are present postjunctionally. The fact that constriction of the large artery in response to any of the three agonists did not differ in the presence of hexamethonium suggests that autonomic reflexes do not mediate or modulate the observed effects.

Although the functional data of coronary constriction suggest equivalent reduction in coronary diameter with α_1- and α_2-adrenergic stimulation, the ligand-binding data demonstrate a four-fold greater number of α_2-adrenergic receptors. The results from these two techniques are not necessarily in conflict. Prejunctional or spare α_2-receptors may comprise a large portion of those receptors assayed biochemically, yet would be ineffective physiologically. Secondly, agonist affinity, receptor occupation, and the stoichiometry relating receptor occupation to activation of contractile mechanisms may be different for the α_1- and α_2-agonists. It must also be recognized that definitive dose-response curves comparing molar concentrations and maximal responsiveness of several agents are impractical in conscious animals due to systemic effects and pharmacokinetic differences between compounds. For example, the fact that intra-coronary phenylephrine and norepinephrine produced increases in arterial pressure which were abolished by intracoronary prazosin, suggests that the coronary effects of agonists and antagonists are complicated by systemic effects.

Nevertheless, the data from our earlier study [28] clearly demonstrate the ability of both α_1- and α_2-agonists to act postjunctionally eliciting coronary constriction. A recent study of ligand binding to porcine coronary artery membranes [15] has reported equal numbers of α_1- and α_2-adrenergic receptors. Both α_1- and α_2-adrenergic receptor densities in that study were 12 fmol/mg protein, which agrees with our data for α_1-adrenergic receptors, but is roughly 25% that for α_2-adrenergic receptors. While that study did not examine contractile responses, a recent study by Cohen et al [6] did examine the contractile response of porcine coronary arteries in vitro to α_1- and α_2-adrenergic receptor stimulation, and found that norepinephrine elicited contraction due to both receptor subtypes. However, that study [6] was able to demonstrate α_2-mediated vasoconstriction only in coronary rings denuded of endothelium, suggesting that α_2-adrenergic constriction in coronary arteries is normally masked by endothelium-mediated vasodilation. Other studies have demonstrated that stimulation of α_2-adrenergic receptors on endothelial cells relaxes precontracted coronary arteries [1, 5, 6] or limits the contraction of rat aorta [8, 12]. However, at no dose of B-HT 920 did we observe any vasodilation of the large coronary artery. Similarly, only vasoconstriction was observed with norepinephrine administration in the presence of prazosin and propranolol, conditions which would result in stimulation of only α_2-adrenergic receptors. It is unlikely that the endothelium was damaged in our studies, since only minimal dissection was performed during surgery 2–4 weeks prior to the study. Secondly, our studies with hexamethonium rule out the possibility that autonomic reflexes or presynaptic postganglionic neurons influenced the observed effects on large coronary constriction. The difference between our results in vivo and those which suggest α_2-mediated vasodilation in vitro remains to be elucidated.

β-Adrenergic receptor studies

Similar controversy exists concerning the β-adrenergic receptor subtype distribution and the effects of receptor activation in large coronary arteries. Studies examining β-adrenergic receptor stimulation in vivo have concluded that coronary vasodilation assessed by increases in coronary blood flow is controlled by the β_2-adrenergic receptor subtype [27]. Conversely, in vitro studies have generally concluded that β_1-adrenergic receptors exert primary control over large coronary arteries [26, 27]. Potentially, these differences can be ascribed to differences in the study of epicardial vs resistance vessels, and few studies have examined the physiological effects of selective β_1- and β_2-agonists and antagonists on large coronary arteries in vivo. Moreover, while prior in vitro studies have examined β_1- and β_2-receptor subtype distribution in large coronary arteries, no studies have compared the biochemical characterization of receptor subtype distribution with the physiological responses to receptor activation.

The physiological studies reviewed in this report indicate that the large coronary arteries of the calf contain both β_1- and β_2-adrenergic receptors [20]. This conclusion is based on finding similar increases in large coronary artery diameter in response to isoproterenol, which stimulates both β_1- and β_2-adrenergic receptors; to prenalterol, which primarily stimulates β_1-adrenergic receptors; and to pirbuterol, which primarily stimulates β_2-adrenergic receptors. Furthermore, the preferential β_1- (atenolol and betaxolol) and β_2- (ICI 118,551) receptor antagonists abolished the responses to their respective agonists, but did not abolish responses to the mixed agonist isoproterenol. Prior studies from this laboratory in conscious dogs have suggested that both β_1- and β_2-adrenergic receptor stimulation and blockade can alter large coronary artery caliber [21, 22]. Those studies were not conclusive since there were concomitant changes in coronary blood flow, as well as systemic changes in hemodynamics, since the agonists and antagonists were administered systemically rather than directly into the coronary artery. Furthermore, those studies also demonstrated that a significant portion of the effect of β-stimulation on large coronary arteries resulted from changes in myocardial metabolism. The subsequent studies from our laboratory discussed in this review avoided these concerns by employing intracoronary injection to minimize cardiac and peripheral hemodynamic changes. Furthermore, since increases in blood flow can dilate large coronary arteries [26], the experiments were also conducted with blood flow held constant. These experiments with restricted blood flow demonstrated no reduction in the large coronary arterial vasodilation in response to any of the agonists tested. Thus, large coronary artery dilation was due directly to stimulation of the β_1- and β_2-adrenergic receptors in the large coronary arteries, and not to a secondary, indirect mechanism involving increases in blood flow.

While the physiological studies clearly identified both β_1- and β_2-mediated vasodilation in the coronary arteries, it was not possible from those studies to conclude which mechanism predominated. To reconcile this problem, ligand-binding studies were performed to quantitate β_1- and β_2-adrenergic subtype populations [20]. The results observed in the calf coronary artery preparations were compared with the data from a relative homogeneous β_1-preparation (calf heart) and a relatively homogeneous β_2-preparation (calf lung). Using isoproterenol, epinephrine, and norepinephrine competition curves, the classical pattern for binding affinity to β_2-receptors, isoproterenol > epinephrine > norepinephrine, was demonstrated in the calf

Adrenergic control of large coronary arteries

lung. Conversely, the classical pattern for β_1-receptors, isoproterenol > epinephrine = norepinephrine, was observed in the calf heart. Membranes from the calf coronary artery preparation demonstrated an intermediate pattern, which was more similar to that of the heart, indicating both β_1- and β_2-receptors, with a greater percentage of β_1-receptors. This was further substantiated by examining binding competition curves with the subtype selective agonists prenalterol and pirbuterol, and the selective antagonists betaxolol and ICI 118,551. Computer modeling consistently demonstrated that the β_1-subtype was present in excess (1.5–2.0:1.0) in the calf coronary artery. Thus, the combined results from both the physiological and biochemical examination of β-adrenergic subtype distribution indicate that both β_1- and β_2-adrenergic receptors contribute to regulation of large coronary artery caliber in conscious calves. However, it is important to emphasize that these studies were conducted with the exogenous agonists administered intracoronarily. The responses of large coronary arteries to the effects of cardiac sympathetic nerve stimulation have not been examined.

Control of coronary resistance vessels

The studies conducted in conscious calves were designed to examine adrenergic control mechanisms in large coronary arteries [20, 28]. However, as reviewed previously [27], earlier studies exploring the receptor subtypes regulating coronary resistance vessels have concluded that coronary blood flow is controlled primarily by β_2-mediated dilation and by α_2-mediated adrenergic constriction. The conclusion that the β_2-receptor subtype is primarily responsible for coronary vasodilation is derived from studies demonstrating substantial increases in coronary blood flow with isoproterenol, despite the presence of β_1-adrenergic receptor blockade [27]. Similar conclusions were drawn by Vatner et al. [21] who studied the changes in coronary vascular resistance with isoproterenol, and the selective agonists pirbuterol (β_2-) and prenalterol (β_1-) before and after β_1-blockade with atenolol. However, the design of those studies using β_1-adrenergic blockade to eliminate the effects of metabolic increases in coronary blood flow is constrained to demonstrating only β_2-adrenergic control of resistance vessels. In contrast, the studies in conscious calves employed the selective antagonists atenolol (β_1-) and ICI 118,551 (β_2-) to distinguish the effects of β-adrenergic subtypes on changes in coronary blood flow. In those studies, isoproterenol still elicited significant increases in coronary blood flow in the presence of either atenolol or ICI 118,551, indicating that both β_1- and β_2-receptors increased coronary blood flow. Furthermore, since the agonists were delivered intracoronarily, those changes in coronary blood flow were associated with minor changes in LV dP/dt and heart rate, suggesting that cardiac metabolic effects did not influence the increase in coronary blood flow. More recent studies have shown that β_1-receptors elicit vasodilation in the peripheral circulation of conscious dogs [23]. Thus, while earlier studies have concluded that β_2-adrenergic receptors are primarily responsible for control of coronary resistance vessels, studies from this laboratory have recently suggested that both β_1- and β_2-receptor subtypes can be activated by pharmacological stimulation to elicit coronary vasodilation. Whether similar effects are observed with neural stimulation remains to be elucidated.

It is well recognized that pharmacologic or neurally induced α-adrenergic receptor stimulation increases coronary vascular resistance, and in some cases, actually reduces coronary blood flow [27]. Yet, as with β-adrenergic receptors, there is controversy regarding the α-receptor subtype responsible for vasoconstriction of coronary resistance vessels. Holtz et al. [10], and Heusch et al. [9] reported that constriction of coronary resistance vessels was predominantly mediated by α_2-adrenergic receptors. Recent studies by Woodman and Vatner [25] from this laboratory have examined the effects of selective stimulation of α_1- and α_2-receptors on coronary resistance vessels. Those studies were conducted in the presence of ganglionic blockade, as well as cholinergic and β-adrenergic receptor blockade, and demonstrated that the selective α_1-agonist phenylephrine and the selective α_2-agonist B-HT 920 both elicit coronary vasoconstriction. Moreover, those studies suggest that both receptor subtypes contribute to the coronary vasoconstriction observed with norepinephrine.

Conclusion

It is important to recognize that conclusions derived from studies designed to elucidate adrenergic receptor subtypes are limited to varying degrees by the models employed (in vivo vs in vitro, awake vs anesthetized, species differences) and the selectivity of the adrenergic agonists and antagonists. The present studies employed: 1) selective α- and β-adrenergic agonists and antagonists to minimize the complicating influences of non-selective receptor activation; and 2) intracoronary injection to minimize the systemic hemodynamic effects typically observed with intravenous administration. Additionally, conscious animals were used to avoid the untoward effects of anesthesia on vascular reactivity. Finally, the unique feature of these studies was the characterization of adrenergic receptor subtypes using both biochemical and physiological approaches. The results demonstrate that β_1- and β_2-adrenergic receptors are present in the large coronary artery, with a predominance of the β_1-receptor subtype. Moreover, large coronary artery dilation was elicited with stimulation of either the β_1- or β_2-receptor subtype. These effects were observed in the absence of increases in coronary blood flow, indicating that vasodilation was a direct effect on large vessel caliber. Ligand-binding studies also demonstrated the presence of both α_1- and α_2-receptors in large coronary arteries, which was supported by the observation of vasoconstriction using both α_1- and α_2-selective agonists. Collectively, these results differ from previous studies and indicate that significant receptor heterogeneity exists in large coronary arteries. Additionally, these studies suggest that each subtype of adrenergic receptor is important for regulation of large coronary arteries.

References

1. Angus JA, Cocks TM, Satoh K (1986) The α-adrenoceptors on endothelial cells. Fed Proc 45:2355–2359
2. Bobik A (1982) Identification of α-adrenoceptor subtypes in dog arteries by (^{3}H) yohimbine and (^{3}H)prazosin. Life Sci 30:219–228
3. Chilian WM, Eastham CL, Marcus ML (1986) Microvascular distribution of coronary vascular resistance in the beating left ventricle. Am J Physiol 251:H779–H789
4. Chilian WM, Layne SM, Eastham CL, Marcus ML (1989) Heterogeneous microvascular coronary α-adrenergic vasoconstriction. Circ Res 64:376–388
5. Cocks TM, Angus JA (1983) Endothelium-dependent relaxation of coronary arteries by noradrenaline and serotonin. Nature 305:627–630
6. Cohen RA, Zitnay KM, Weisbrod RM, Tesfamariam B (1987) Influence of the endothelium on tone and the response of isolated pig coronary artery to norepinephrine. J Pharmacol Exp Ther 244:550–555
7. Colucci WS (1986) Adenosine 3′,5′-cyclic-monophosphate-dependent regulation of α_1-adrenergic receptor number in rabbit aortic smooth muscle cells. Circ Res 58:292–297
8. Godfraind T, Egleme C, Osachie IA (1985) Role of endothelium in the contractile response of rat aorta to α-adrenoceptor agonists. Clin Sci 68(Suppl. 10):65s–71s
9. Heusch G, Deussen A, Schipke J, Thaemer V (1984) α_1- and α_2-adrenoceptor-mediated vasoconstriction of large and small canine coronary arteries in vivo. J Cardiovasc Pharmacol 6:961–968
10. Holtz J, Saeed M, Sommer O, Bassenge E (1982) Norepinephrine constricts the canine coronary bed via postsynaptic α_2-adrenoceptors. Eur J Pharmacol 82:199–202
11. Kaumann AJ (1983) Yohimbine and rauwolscine inhibit 5-hydroxytryptamine-induced contraction of large coronary arteries of calf through blockade of 5-HT$_2$ receptors. Naunyn-Schmiedeberg's Arch Pharmacol 323:149–154
12. Miller RC, Mony M, Schini V, Schoeffter P, Stoclet JC (1984) Endothelial mediated inhibition of contraction and increase in cyclic GMP levels evoked by the α-adrenoceptor agonist B-HT 920 in rat isolated aorta. Br J Pharmacol 83:903–908
13. Munson PJ, Rodbard D (1980) Ligand: A versatile computerized approach for characterization of ligand-binding systems. Anal Biochem 107:220–239
14. Muntz KH, Garcia C, Hagler HK (1985) α_1-receptor localization in rat heart and kidney using autoradiography. Am J Physiol 249:H512–H519
15. Nishimura J, Kanaide H, Nakamura M (1987) Characteristics of adrenoceptors and [3H] nitrendipine receptors of porcine vascular smooth muscle: differences between coronary artery and aorta. Circ Res 60:837–844

16. Rimele TJ, Rooke TW, Aarhus LL, Vanhoutte PM (1983) α_1-Adrenoceptors and calcium in isolated canine coronary arteries. J Pharmacol Exp Ther 226:668–672
17. Toda N (1986) α-Adrenoceptor subtypes and diltiazem actions in isolated human coronary arteries. Am J Physiol 250:H718–H724
18. Toda NT, Okamura T, Nakajima M, Miyazaki M (1984) Modification by yohimbine and prazosin of the mechanical response of isolated dog mesenteric, renal, and coronary arteries to transmural stimulation and norepinephrine. Eur J Pharmacol 98:69–75
19. Toda N (1983) Alpha adrenergic receptor subtypes in human, monkey and dog cerebral arteries. J Pharmacol Exp Ther 226:861–868
20. Vatner DE, Knight DR, Homcy CJ, Vatner SF, Young MA (1986) Subtypes of β-adrenergic receptors in bovine coronary arteries. Circ Res 59:463–473
21. Vatner SF, Hintze TH, Macho P (1982) Regulation of large coronary arteries by β-adrenergic mechanisms in the conscious dog. Circ Res 51:56–66
22. Vatner SF, Hintze TH (1983) Mechanism of constriction of large coronary arteries by β-adrenergic receptor blockade. Circ Res 53:389–400
23. Vatner SF, Knight DR, Hintze TH (1985) Norepinephrine-induced β_1-adrenergic peripheral vasodilation in conscious dogs. Am J Physiol 249:H49–H56
24. Weiss RJ, Webb RC, Smith CB (1983) α_2-Adrenoceptors on arterial smooth muscle: selective labeling by (^{3}H) Clonidine. J Pharmacol Exp Ther 225:599–605
25. Woodman OL, Vatner SF (1987) Coronary vasoconstriction mediated by α_1- and α_2-adrenoceptors in conscious dogs. Am J Physiol 253:H388–H393
26. Young MA, Vatner SF (1986) Brief review: Regulation of large coronary arteries. Circ Res 59:579–595
27. Young MA, Knight DR, Vatner SF (1987) Autonomic control of large coronary arteries and resistance vessels. Prog Cardiovasc Dis 30:211–234
28. Young MA, Vatner DE, Knight DR, Graham RM, Homcy CJ, Vatner SF (1988) α-Adrenergic vasoconstriction and receptor subtypes in large coronary arteries of calves. Am J Physiol 255:H1452–H1459

Authors' address:

Mark A. Young, Ph.D., GENSIA Pharmaceuticals, 11025 Roselle St., San Diego, CA 92121, USA.

Adrenergic Vasomotion in the Coronary Microcirculation

W. M. Chilian*

Department of Medical Physiology, Microcirculation Research Institute, Texas A&M University, College Station, Texas, USA

Summary

The goal of this study was to determine the α-adrenergic receptor subtype(s) responsible for constriction at different microvascular levels in the coronary circulation. To accomplish these goals, the epicardial coronary microcirculation of intact beating hearts was viewed through an intravital microscope using stroboscopic epi-illumination. An initial study was designed to establish sites of α-adrenergic constriction to norepinephrine in preparations with intact vasomotor tone. For the primary experimental goal, coronary microvascular responses to selective α_1-adrenergic (phenylephrine) or α_2-adrenergic (BHT-933) agonists were evaluated, when coronary autoregulatory escape mechanisms were blunted during hypoperfusion. Infusion of norepinephrine decreased diameter of arterial vessels greater than $100\,\mu m$ in diameter, but downstream coronary arterioles dilated significantly, representing autoregulatory escape from adrenergic vasoconstriction. In studies designed to examine the adrenergic receptor subtype (during hypoperfusion), phenylephrine produced modest constriction of vessels throughout the microcirculation (6–9% decrease in diameter), whereas BHT-933 produced marked constriction of small coronary microvessels, those less than $100\,\mu m$ in diameter (24% decrease in diameter). From these results we conclude: 1) norepinephrine infusion causes disparate responses in the coronary microvasculature: constriction occurs in vessels greater than $100\,\mu m$ in diameter, but dilation, via autoregulatory escape, predominates in vessels less than $100\,\mu m$ in diameter; 2) α_1-adrenergic receptors are located in coronary arterioles and arteries; and 3) α_2-adrenergic receptors are preferentially located in small coronary arterioles. Thus, α_1- and α_2-adrenergic activation can produce dissimilar constrictor effects in the coronary microcirculation during hypoperfusion.

Introduction

The microcirculation of the heart is composed of several anatomically and functionally distinct vessel segments arranged in the series. Although these microvascular segments are characterized by different functional and anatomical features, i.e., varying layers of smooth muscle, a common feature of the coronary microcirculation is that all segments contain sympathetic innervation [12]. Most of our previous knowledge concerning sympathetic neural control of the different vascular segments is based on studies of pressure-flow relationships during a variety of physiological and pharmacological interventions. Within this context, α-adrenergic coronary vasoconstriction has been demonstrated during infusion of catecholamines [1, 18, 21, 26, 27] and physiological interventions including exercise, hemorrhage, and carotid sinus hypotension [2, 4, 12, 13, 20, 25]. Despite the plethora of information regarding α-adrenergic modulation of coronary resistance and pressure-flow relationships, there has been a relative paucity of

Supported by the following grants from the National Heart, Lung, and Blood Institute of the U.S. Public Health Service: HL32788 and HL01570. Dr. Chilian is a recipient of a Research Career Development Award from the National Heart, Lung, and Blood Institute.

information regarding coronary microvascular locations or the specific α-adrenergic receptor subtype involved in the constriction. Within the last several years, there have been reports suggesting that coronary vasoconstriction may be mediated exclusively by α_1-adrenergic receptors, α_2-adrenergic receptors, or a combination of the two receptor subtypes [14–17, 23, 28]. Only very recently have microvascular locations of α_1- and α_2-adrenergic receptors in the microcirculation of skeletal muscle been elucidated [11] and to date there have been no studies of the different α-adrenergic receptor subtypes within the coronary microcirculation.

The primary goal of this study was to establish microvascular locations of α_1- and α_2-adrenergic constriction within the coronary microcirculation. As an initial study, the coronary microvascular responses to α-adrenergic receptor activation were documented in the intact, autoregulating coronary vasculature during norepinephrine infusion. The experimental approach was to measure coronary microvascular diameters during different experimental manoeuvres to precisely document microvascular sites that are receptive to α-adrenergic stimuli, in the presence and absence of autoregulatory adjustments.

Methods

Experimental model

Studies were completed in anesthetized cats or dogs. After initial sedation of the animal, anesthesia was induced with sodium pentobarbital, 30–35 mg/kg. The procedures for the general preparation of the animals have been described previously [5–8]. In brief, systemic hemodynamics and blood gases are measured (aortic pressure, left ventricular pressure, left ventricular dP/dt, heart rate, pO_2, pCO_2, pH) in anesthetized, open-chest animals. Jet ventilation is synchronized with the cardiac cycle to minimize the effects of motion of the lungs imposed on cardiac movement.

Microvascular preparation

Measurements of microvascular diameters are accomplished in the beating heart using intravital video-microscopic techniques in conjunction with stroboscopic epi-illumination synchronized to the heart beat [5, 7]. Synchronization of the strobe creates an illusion that the heart is motionless, because the strobe is flashed once per cardiac cycle for only a brief moment (15–20 μs) at the same point (late diastole) during successive cardiac cycles. Using different sized objective lenses ($4\times$, $10\times$, $20\times$), a variety of different sized arteriolar microvessels can be viewed and accurately measured (resolution with the highest power objective lens is approximately 1 μm).

All measurements of microvascular diameter were made using fluorescence techniques, because the lumen, i.e., internal diameter, of the vessel is illuminated by the fluorochrome. This enabled precise and accurate visualization of diameters of coronary arteries and arterioles ranging in diameter from 300 to 20 μm in diameter. Figure 1 illustrates fluorescent images of coronary arterioles with the microvascular lumen well-defined by the fluorochrome. Fluorescence microscopy is also critical for distinguishing between small arterioles and venules. Following injection of the labelled dextran, arterioles and venules illuminate sequentially, i.e., arterioles illuminating initially and, because of the transit time for plasma flow, venules illuminating a few seconds later. Fluorescein isothiocyanate dextran (molecular weight 500 000–2 000 000) was activated with the Leitz H2 filter.

Experimental protocols

Measurements of microvascular diameters in arterioles were completed, in cats, during combined α_1- and α_2-adrenergic activation (norepinephrine infusion, iv, 1.0–2.0 μg/kg·min) at normal perfusion pressures with coronary tone intact. In the second series of experiments, in

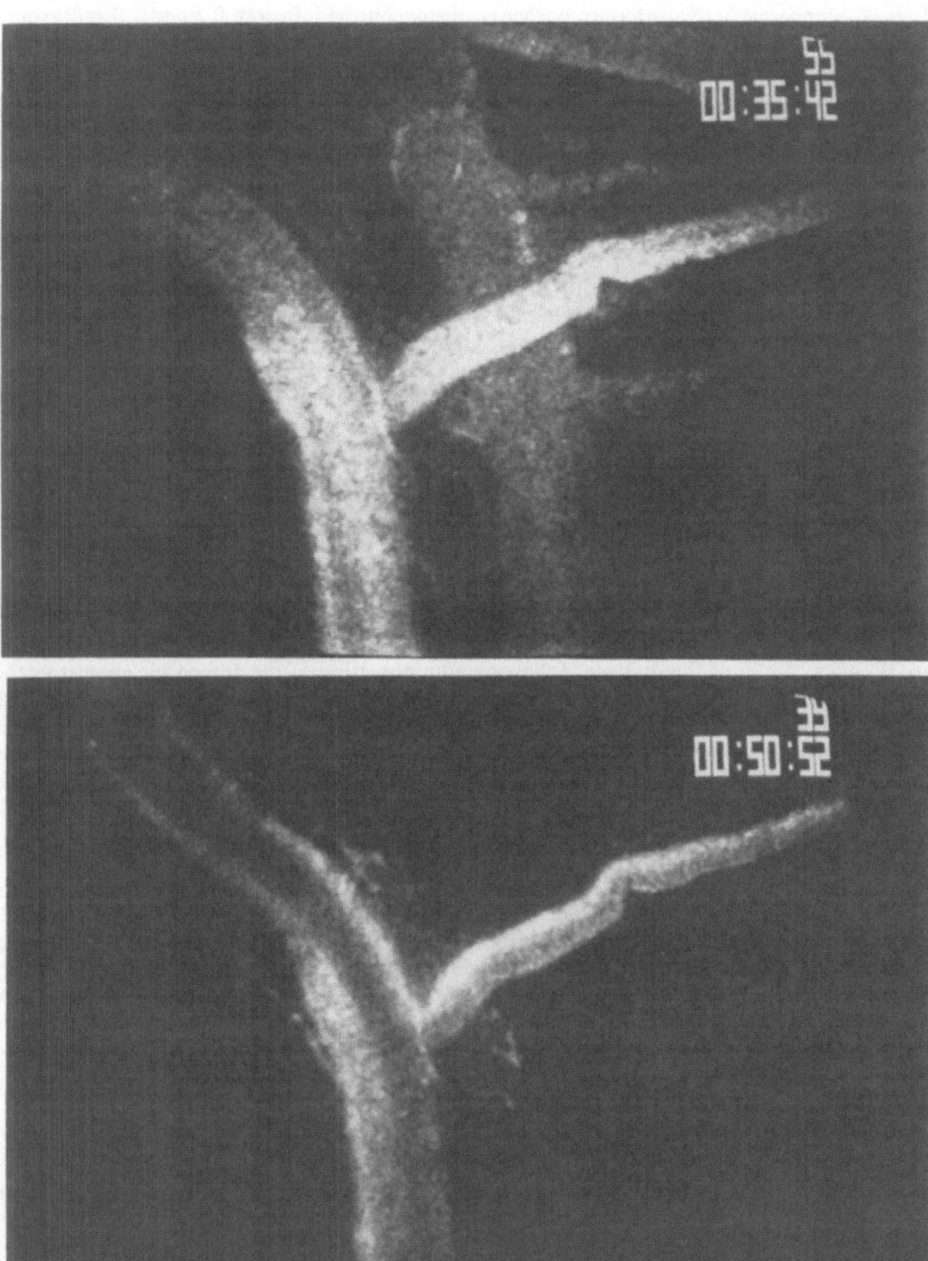

Fig. 1. Illustrations of coronary arteriolar responses to the α_2-adrenergic agonist, BHT-933, 1.0 μg/kg·min. The top panel shows the control response, and the bottom panel shows the response following the intracoronary administration of BHT-933. Note the constriction of the side branch during administration of the adrenergic agonist. Under control conditions, the diameter of the side branch was 72 μm, and during infusion of the adrenergic agonist, the diameter decreased to 60 μm.

113

dogs, α_1-adrenergic activation (intracoronary phenylephrine infusion, 0.2 and 1.0 μg/kg·min) or α_2-adrenergic activation (BHT-933, 0.2 and 1.0 μg/kg·min) was performed during coronary hypoperfusion. During systemic norepinephrine infusion, β-adrenergic blockade was performed with propranolol (1.0 mg/kg) to eliminate the increased myocardial metabolic demands associated with β-adrenergic activation induced by norepinephrine. Arterial pressure increased initially during norepinephrine infusion, but was returned to control by tightening a snare situated around the inferior vena cava.

In studies using selective α_1- and α_2-adrenergic agonists, autoregulatory escape was eliminated by administration of the drugs during coronary hypoperfusion (coronary perfusion pressure = 40 mm Hg). Hypoperfusion at this pressure greatly attenuates vasodilator reserve of coronary arteries and arterioles [5]. During *intracoronary* administration of the selective α-adrenergic agonists, systemic hemodynamics (during hypoperfusion) did not change; thus, it was not necessary to administer β-adrenergic antagonists or instrument the animal with snares to control arterial pressure. Intracoronary administration of the agonists was possible, because these studies were conducted in dogs which have much larger epicardial coronary arteries than cats, allowing coronary cannulation.

Statistical analysis

Data during norepinephrine-induced α-adrenergic activation were analyzed by expressing the change of microvascular diameter during adrenergic stimulation/activation as a percentage of control. These data are shown as a plot of percent change during adrenergic activation vs initial diameter. Changes in microvascular diameter during selective α_1- or α_2-adrenergic activation are shown as a plot of the diameter following administration of the adrenergic agonist vs the initial (pre-administration) diameter. Constrictor or dilator responses are shown as points below or above the line of identity, respectively. Vessels were divided into two groups, greater or less than 100 μm in diameter, and percent changes in the diameter of the different groups were analyzed by Scheffe's test following analysis of variance. All hemodynamic data were analyzed by analysis of variance.

Results

In the initial series of experiments during baseline conditions in the presence of β-adrenergic blockade, heart rate and mean arterial pressure were 140 $\pm$ 7 beats·min^{-1} and 98 $\pm$ 4 mm Hg, respectively. During norepinephrine infusion, heart rate was 156 $\pm$ 7 and arterial pressure was 92 $\pm$ 5 mm Hg. During intracoronary infusion of α_1- and α_2-adrenergic agonists, neither mean arterial pressure nor heart rate were changed from the control values of 82 $\pm$ 6 mm Hg or 165 $\pm$ 7 beats·min^{-1}.

Figure 2 shows the effects of norepinephrine infusion on coronary microvascular diameters. Note that vessels primarily larger than 100 μm in diameter constricted, whereas the smaller arteriolar vessels dilated during α-adrenergic activation. On average, coronary arterioles ($< 100\,\mu$m diameter) dilated by 14 $\pm$ 3% during norepinephrine infusion. This contrasts to upstream coronary arterial vessels, in which vessels constricted by 5 $\pm$ 1% during norepinephrine infusion.

Selective α_1- or α_2-adrenergic activation during coronary hypoperfusion produced disparate microvascular reactions. α_1-Adrenergic activation with phenylephrine resulted in comparable constriction in coronary arterial and arteriolar vessels, -5 ± 2% and -9 ± 4% change in diameter, respectively (Figs. 3 and 4) ($p < 0.05$). In contrast α_2-adrenergic activation produced substantial constriction of coronary arterioles (-24 ± 5% change in diameter, Fig. 3) and less intense constriction of coronary arteries (-4 ± 2% decrease in diameter, Fig. 4). It is noteworthy to emphasize that the magnitude of arteriolar constriction (-24 ± 5%) produced by BHT-933 was greater than that observed for phenylephrine (-9 ± 4%) ($p < 0.05$).

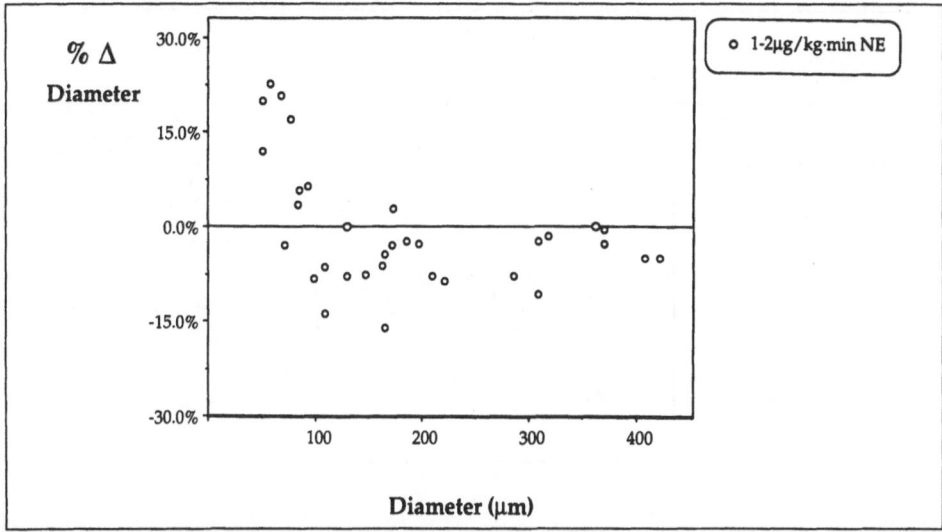

Fig. 2. Coronary microvascular responses to norepinephrine infusion. Data are expressed as a percent change in diameter from the initial diameter.

Discussion

There are several key observations in the present study. *First*, when autoregulatory mechanisms are intact, α-adrenergic activation produces heterogeneous vascular reactions in the coronary microcirculation. Specifically, constriction predominates in coronary arteries and large arterioles, whereas dilation occurs in small coronary arterioles less than 100 μm in diameter. *Second*, the functional distribution of α_1- and α_2-adrenergic receptors appears to be dissimilar in the coronary microcirculation. In the absence of autoregulatory escape, activation of α_1-adrenergic receptors produces relatively uniform constriction in arterioles and small arteries, whereas activation of α_2-adrenergic receptors causes preferential constriction in small coronary arterioles less than 100 μm in diameter. Moreover, the magnitude of arteriolar α_2-adrenergic constriction is almost three-fold that of α_1-adrenergic constriction. Thus, small coronary arterioles possess α_1- and α_2-adrenergic receptors and direct vascular constriction of these blood vessels can be demonstrated when intrinsic autoregulatory mechanisms are blunted. Critical to these conclusions and interpretations are: critique of the experimental approach, considerations of the literature, and α-adrenergic constriction in the coronary circulation.

Critique of the experimental preparation

The accuracy of the experimental measurements and the viability of the preparation are critical to the conclusions. It is possible to resolve pixel sizes of approximately 5 μm with the 4× objective, 2 μm with the 10× objective, and 1 μm with the 20× objective; thus, small changes in microvascular caliber can be detected. Our measurements are calibrated both in vitro with a micrometer grid and in situ with microspheres of various, known diameters (2, 5, 10, and 25 μm) placed on the surface of the heart.

All experimental preparations were characterized by systemic hemodynamics and arterial blood gases within the physiological range. Although the experimental design entailed precise documentation of microvascular segments responsive to α-adrenergic activation within the coronary circulation, a limitation was that pressure measurements were not obtained during the various interventions. Therefore, a decrease in microvascular diameter could, in part, be related to a decrease in intraluminal pressure, i.e., a passive decrease in diameter. Such passive changes

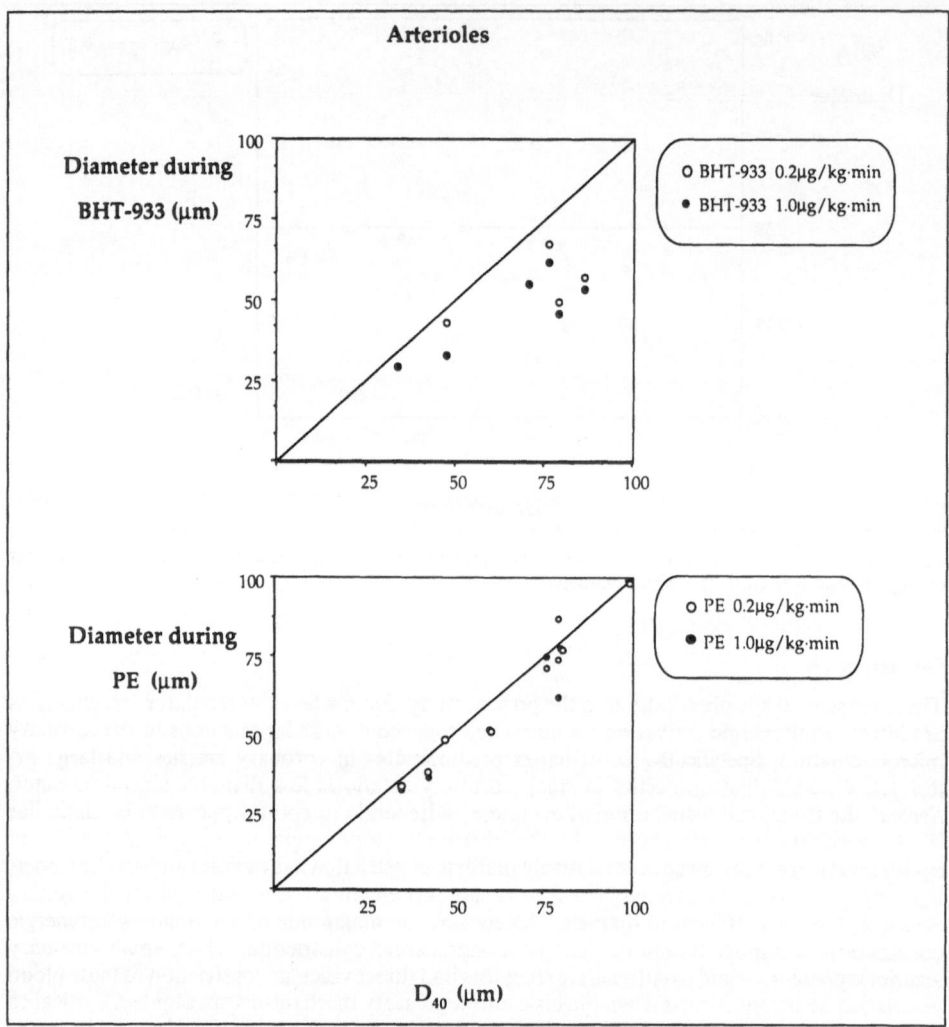

Fig. 3. Responses of coronary small arterioles (vessels less than 100 μm in diameter) to BHT-933 and phenylephrine. Diameters at a coronary perfusion pressure of 40 mm Hg are shown on the X-axis (D_{40}) and diameters during agonist administration are shown on the Y-axis. Note that BHT-933 caused substantial constriction of coronary arterioles, and phenylephrine produced lesser constriction of these small vessels.

in vascular caliber would not have caused the increase in microvascular diameters during norepinephrine-induced α-adrenergic activation, because downstream vessels dilated during constriction of upstream vessels. This must represent an active vasodilatory mechanism. During selective $α_1$- and $α_2$-adrenergic activation, passive responses most likely would not have played any important role in the microvascular reactions. For instance, during selective $α_2$-adrenergic activation, small coronary arterioles constricted without any significant changes in the upstream larger arterioles or small arteries; thus, passive responses of the small vessels would not have occurred. Also, during $α_1$-adrenergic activation, small coronary arteries and large arterioles constricted to a similar extent. A passive response in small vessels would have been less than the active "upstream" response, but the decrease in diameter of the smaller vessels was equivalent to the larger vessels. Furthermore, constriction of the large upstream

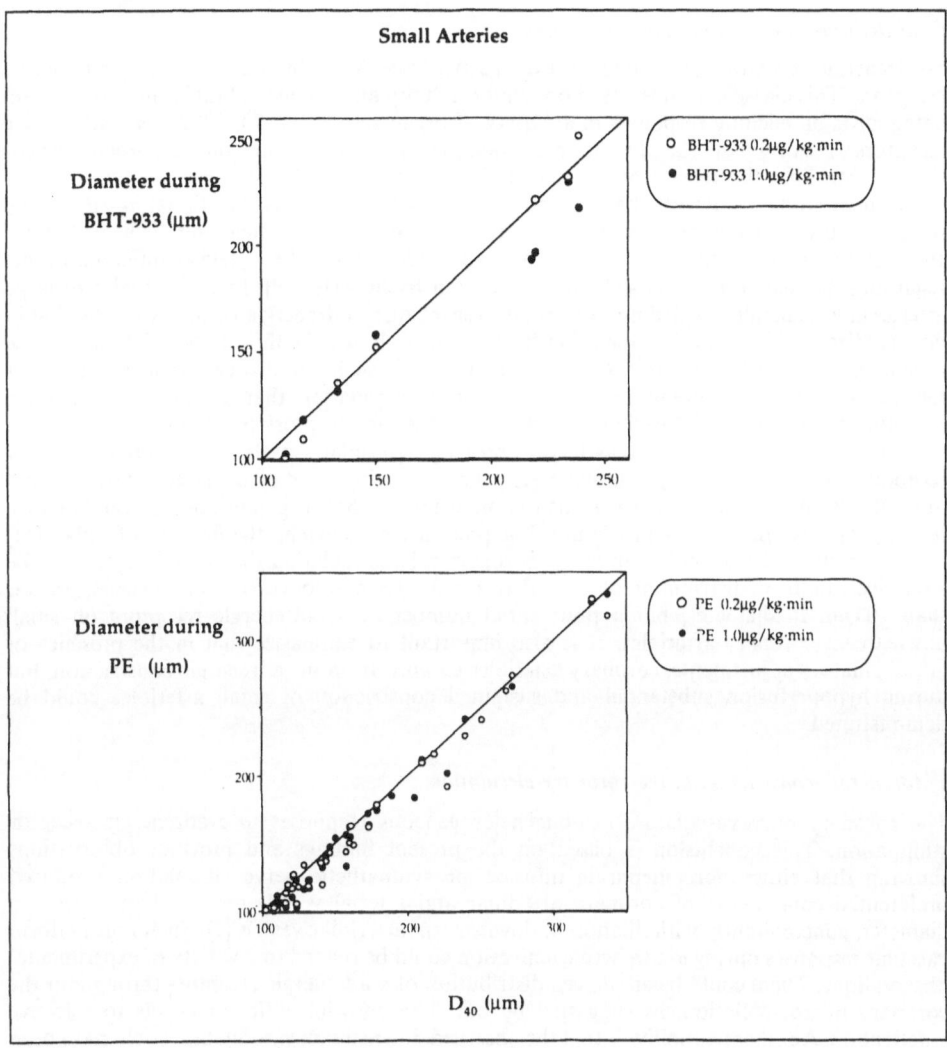

Fig. 4. Responses of large arterioles and small arteries (vessels greater than 100 μm in diameter) to BHT-933 and phenylephrine. Note the variable responses with BHT-933 and uniform constriction to phenylephrine; this is evident by comparing the points to the line to identity; points below the line represent constriction and those above the line represent dilation.

vessels in response to phenylephrine must have been related to an active vasoconstrictor response because coronary perfusion pressure remained constant. It should be emphasized that coronary perfusion pressure remained constant or was maintained constant during measurements of diameter. Collectively, passive alterations in microvascular caliber are not consistent with the observed experimental findings; thus, the microvascular responses to humoral α-adrenergic activation by norepinephrine, and selective α_1- and α_2-adrenergic stimulation are most likely caused by active adrenergic vasoconstriction.

Considerations of α-adrenergic constriction

α-Adrenergic receptors are known to exist as two receptor subtypes, α_1- and α_2-adrenergic receptors. This classification is based on relative affinity and potency of different agonists and antagonists on eliciting responses in a variety of organ systems [9, 10, 19, 24]. α_2-Adrenergic receptors were originally classified as "pre-junctional" receptors located on sympathetic varicosities, controlling the release of norepinephrine during sympathetic nerve stimulation [19]. Subsequent studies, however, have elucidated the existence of post-junctional α_2-adrenergic receptors on vascular smooth muscle in a variety of organ systems, including the heart [10, 15, 16, 19, 28]. Interestingly, in vitro and in vivo studies have often yielded different results regarding the magnitude of α_2-adrenergic vasoconstriction [19, 24]. In vitro studies of large arteries have generally yielded negative results regarding α_2-adrenergic vasoconstriction, but in vivo studies of the coronary circulation have recently shown significant constriction to these adrenergic agonists [14, 15, 16, 19, 24, 28]. To reconcile such paradoxical findings, Faber has recently shown in the cremasteric skeletal muscle circulation that α_1- and α_2-adrenergic receptors are reciprocally distributed; with α_2-adrenergic receptors preferentially located on small coronary arterioles, whereas larger coronary arteriolar vessels possess predominantly receptors of the α_1-adrenergic subtype [11]. Thus, in vitro studies on large arteries would observe selective α_1-adrenergic vasoconstriction, whereas whole organ responses would involve both α_1- and α_2-adrenergic constriction. The present results extend the findings of Faber [11] to the coronary microcirculation: in the coronary microcirculation there would appear to be a relatively uniform distribution of α_1- and α_2-adrenergic receptors on coronary vessels greater than $100\,\mu m$ in diameter, but a preferential number of α_2-adrenergic receptors on small downstream coronary arterioles. It is also important to emphasize that in the presence of autoregulatory adjustments, coronary arterioles escaped from the adrenergic constriction, but during hypoperfusion substantial and sustained constriction of small arterioles could be demonstrated.

α-Adrenergic constriction in the coronary circulation

The coronary microvasculature exhibits heterogeneous responses to α-adrenergic receptor stimulation. This conclusion is based on the present findings and previous observations showing that either norepinephrine infusion or sympathetic nerve stimulation produced preferential constriction of coronary arteriolar and arterial vessels greater than $100\,\mu m$ in diameter, concomitantly with dilation in downstream arteriolar vessels [7]. Such non-uniform vascular responses during α-adrenergic activation could be related to a variety of experimental observations. There could be an uneven distribution of α-adrenergic receptors throughout the coronary microcirculation, as suggested by our data showing different effects to selective α-activation. Another possibility is that the observed downstream vasodilation could have been related to autoregulatory adjustments in vasomotor tone during sustained upstream α-adrenergic constriction. It is possible that coronary arteries and arterioles could have constricted initially during α-adrenergic activation, but "escaped" from the constriction. Escape from sympathetic constriction in other vascular beds has been extensively documented [22]. Our data support this contention, because when autoregulatory mechanisms were blunted during hypoperfusion, sustained α-adrenergic constriction of arterioles was observed; thus, these small vessels are responsive to α-adrenergic agonists under certain conditions.

The magnitude of constriction of arterioles to the α_2-adrenergic agonist, BHT-933, was substantial. On the average, arterioles constricted by $17 \pm 3\,\mu m$ from their initial diameters of $72 \pm 8\,\mu m$. This represents a 24% constriction, which translates to a 76% increase in vascular resistance in this arteriolar segment (assuming radius is related to resistance by a power of four). α_1-Adrenergic vasoconstriction caused a 9% decrease in diameter of small arterioles, and a 6% reduction in diameter in the upstream vessels, which equate to 29% and 19% increases of resistances in arterioles and arteries, respectively. It is noteworthy to emphasize that activation of either α_1- or α_2-adrenergic receptors have the potential to cause direct coronary vasoconstriction.

118

Coronary arteriolar vasoconstriction

In conclusion, α_1- and α_2-adrenergic receptors are distributed differently throughout the coronary microcirculation. α_1-Adrenergic receptors are located throughout the coronary circulation, but α_2-adrenergic receptors are located predominantly in small coronary arterioles, vessels less than $100\,\mu m$ in diameter. Activation of either of these receptor subtypes, in the absence of autoregulatory adjustments, can produce direct coronary vasoconstriction of arterioles, but the magnitude of constriction produced by α_2-adrenergic activation is approximately three-fold that of α_1-adrenergic activation.

Acknowledgements: The author gratefully acknowledges Susan Gard for preparation of this manuscript, and Smith, Kline, and French Laboratories for their generous gift of BHT-933.

References

1. Berne RM (1958) Effect of epinephrine and norepinephrine on coronary circulation. Circ Res 6:644–655
2. Chilian WM, Ackell PH (1988) Transmural differences in sympathetic coronary constriction during exercise in the presence of coronary stenosis. Circ Res 62:216–225
3. Chilian WM, Eastham CL, Marcus ML (1986) Microvascular distribution of coronary vascular resistance in beating left ventricle. Am J Physiol 25:H779–H788
4. Chilian WM, Harrison DG, Haws CW, Snyder WD, Marcus ML (1986) Adrenergic coronary tone during submaximal exercise in the dog is produced by circulating catecholamines: evidence for adrenergic denervation supersensitivity in the myocardium but not in coronary vessels. Circ Res 58:68–82
5. Chilian WM, Layne SM (1990) Coronary microvascular reponses to reductions in perfusion pressure. Evidence for persistent arteriolar tone during hypoperfusion. Circ Res 66:1227–1238
6. Chilian WM, Layne SM, Eastham CL, Marcus ML (1987) Effects of epinephrine on coronary microvascular diameters. Circ Res 61 (Suppl II):II-47–II-53
7. Chilian WM, Layne SM, Eastham CL, Marcus ML (1989) Heterogeneous microvascular coronary α-adrenergic vasoconstriction. Circ Res 64:376–388
8. Chilian WM, Layne SM, Klausner EC, Eastham CL, Marcus ML (1989) Redistribution of coronary microvascular resistance produced by dipyridamole. Am J Physiol 256:H383–H390
9. Curro FA, Greenburg S (1983) Characterization of post-synaptic α_1- and α_2-adrenergic receptors in canine vascular smooth muscle. Can J Physiol Pharmacol 61:893–904
10. Drew GM, Whiting SB (1979) Evidence for two distinct types of post-synaptic α-adrenoceptors in vascular smooth muscle in vivo. Brit J Pharmacol 67:207–215
11. Faber JE (1988) In situ analysis of α-adrenoceptors on arteriolar and venular smooth muscle in rat skeletal muscle microcirculation. Circ Res 62:37–50
12. Feigl EO (1983) Coronary physiology. Physiol Rev 63:1–205
13. Gwirtz PA, Stone HL (1981) Coronary blood flow and myocardial oxygen consumption after alpha-adrenergic blockade during submaximal exercise. J Pharmacol Exp Ther 217:92–98
14. Heusch G, Deussen A (1983). The effects of cardiac sympathetic nerve stimulation on perfusion of stenotic coronary arteries in the dog. Circ Res 53:8–15
15. Heusch G, Deussen A, Schipke J, Thämer V (1984) α_1- and α_2-Adrenoceptor-mediated vasoconstriction of large and small canine coronary arteries in vivo. J Cardiovasc Pharmacol 6:961–968
16. Holtz J, Saeed M, Sommer O, Bassenge E (1982) Norepinephrine constricts the canine coronary bed via post-synaptic α_2-adrenoceptors. Eur J Pharmacol 82:199–202
17. Jones CE, Liang IYS, Gwirtz PA (1987) Effects of α-adrenergic blockade on coronary autoregulation in dogs. Am J Physiol 253:H365–H372
18. Kelley KO, Feigl EO (1978) Segmental α-receptor-mediated vasoconstriction in the canine coronary circulation. Circ Res 43:908–917
19. Langer SZ, Hicks PE (1984) Alpha-adrenoceptor subtypes in blood vessels: physiology and pharmacology. J Cardiovasc Pharmacol 6:S547–S558
20. Murray PA, Vatner SF (1979) Alpha-adrenoceptor attenuation of the coronary vascular response to severe exercise in the conscious dog. Circ Res 45:654–660
21. Pitt B, Elliot EC, Gregg DE (1967) Adrenergic receptor activity in the coronary arteries of the unanesthetized dog. Circ Res 21:75–84
22. Renkin EM (1984) Control of microcirculation and blood-tissue exchange. In: Handbook of Physiology, Section 2: The Cardiovascular System, Volume IV, Part 2. Bethesda, MD, American Physiological Society, pp. 627–687

23. Seitelberger R, Guth BD, Heusch G, Lee J-D, Katayama K, Ross J Jr (1988) Intracoronary α_2-adrenergic receptor blockade attenuates ischemia in conscious dogs during exercise. Circ Res 62:436–442
24. Timmermans PBMWM, van Zwieten PA (1981) The post-synaptic α_2-adrenoceptor. J Auton Pharmacol 1:171–183
25. Vatner SF (1974) Effects of hemorrhage on regional blood flow distribution in dogs and primates. J Clin Invest 54:225–235
26. Vatner SF, Higgins CB, Braunwald E (1974) Effects of norepinephrine on coronary circulation and left ventricular dynamics in the conscious dog. Circ Res 34:812–823
27. Vatner SF, Pagani M, Manders WT, Pasipoularides AD (1980) Alpha-adrenergic vasoconstriction and nitroglycerin vasodilation of large coronary arteries in the conscious dog. J Clin Invest 65:5–14
28. Woodman OL, Vatner SF (1987) Coronary vasoconstriction mediated by α_1- and α_2-adrenoceptors in conscious dogs. Am J Physiol 253:H388–H393

Author's address:

William M. Chilian, Ph.D., Department of Medical Physiology, Texas A&M University College of Medicine, College Station, Texas 77843-1114, USA.

Neurohumoral Regulation of Coronary Collateral Vasomotor Tone

D. G. Harrison*, F. W. Sellke and J. E. Quillen

Cardiovascular Center and Veterans Administration Medical Center, Department of Internal Medicine, University of Iowa College of Medicine, Iowa City, Iowa, USA

Summary

As a result of gradual coronary occlusion, coronary collaterals are stimulated to develop. This maturation process involves not only dilatation of the vessel, but the development of new vascular smooth muscle. Experiments have been performed to examine vasomotor characteristics of mature coronary collaterals from dogs 3 to 6 months following ameroid constrictor placement. Studies in Langendorff blood-perfused hearts have shown that transcollateral resistance does not change during either the administration of α_1- or α_2-adrenergic agonists. Isolated collateral vessels studied as rings in organ chambers do not constrict to either α_1- or α_2-adrenergic agonists. These studies show that mature collateral vessels are not likely to possess functioning α-adrenergic receptors. Subsequent experiments using a cover slip autoradiographic ligand-binding approach have demonstrated a population of β-adrenergic receptors on mature coronary collaterals. Studies of isolated collaterals have demonstrated β-adrenoceptor-mediated relaxation that appears due to a population of mixed β_1- and β_2-adrenergic receptors.

Subsequent studies have demonstrated that mature collateral vessels are hyperresponsive to the vasoconstrictor effects of vasopressin and that concentrations of vasopressin which may be encountered in pathophysiologic conditions can markedly attenuate coronary collateral perfusion. Finally, the microcirculation of the collateral-dependent myocardium develops endothelial cell dysfunction. This results in impaired endothelium-dependent relaxations to adenosine diphosphate and acetylcholine and enhanced vasoconstriction to vasopressin. These alterations of the coronary circulation may have important implications regarding neurohumoral regulation of myocardial perfusion in collateral-dependent myocardium.

In several species including man [4], dog, and pig [16, 17] intermittent or gradual coronary occlusion results in the growth of coronary collateral vessels. Two features of this process are remarkable. First, the lumen of the collateral vessels increases dramatically and rapidly. Six weeks following the onset of gradual coronary occlusion using the ameroid technique, collateral resistance has dropped to near minimal levels [16]. A second major feature is the development of new vascular smooth muscle. By six months following the onset of coronary occlusion in the dog, a well developed tunica media is present (Fig. 1). It is now clear that this tunica media is capable of responding to several neurohumoral substances and that collateral tone may influence perfusion to the collateral-dependent myocardium. In addition, collateral perfusion must traverse not only the collateral vessels but also the proximal (non-occluded) conduit

*Dr. Harrison is an established investigator of the American Heart Association. Supported by NIH grants HL32717, HL39006, Ischemic SCOR HL32295, and a Merit Review Grant from the Veterans Administration.

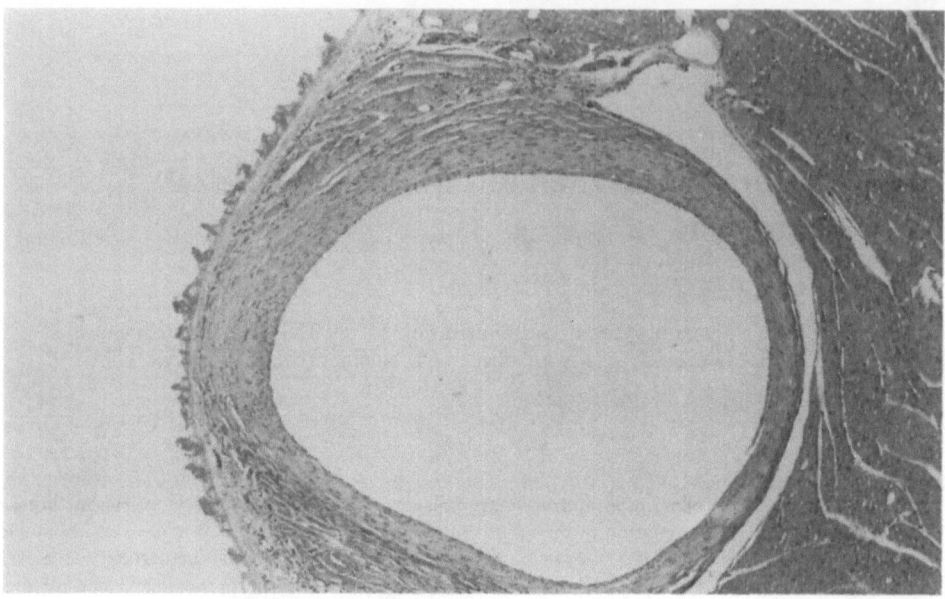

Fig. 1. Histologic section of a mature canine coronary collateral 6 months following ameroid constrictor placement. The vessel has an internal lumen diameter of approximately 1 mm and a well developed tunica media.

coronary vessels and, finally, the resistance circulation distal to collateral vessels. Thus, collateral perfusion is subject to changes in vasomotor tone of these proximal and more distal resistances. During this review the vasomotor characteristics of mature collaterals will be considered along with new information regarding the influence of resistance proximal and distal to the collateral vasculature.

Effect of α-adrenergic stimulation on collateral vasomotor tone

Activation of the sympathetic nervous system may potentially lead to collateral constriction if mature coronary collaterals possess α-adrenergic receptors. Several groups have attempted to address this issue in studies of animals with mature or partially developed collaterals. Schaper and coworkers found that peripheral coronary pressure (the pressure distal to the site of coronary occlusion) in dogs with well developed collaterals did not change during the administration of methoxamine, a selective α_1-adrenergic agonist [18]. We have studied the role of α-adrenergic receptors in dogs with mature collaterals six months after ameroid constrictor placement [7]. Transcollateral resistance was measured in isolated blood-perfused heart preparations during adenosine vasodilatation. Neither the α_1-adrenergic agonist methoxamine nor the α_2-adrenergic agonist clonidine caused collateral constriction, although both produced constriction of the coronary resistance vessels. In separate studies, isolated segments of coronary collaterals studied in organ chambers were found to not respond to either clonidine or phenylephrine, while the left anterior descending coronary segments demonstrated concentration-dependent constrictions to both. Hautamaa and coworkers [8] measured transcollateral resistance in open-chest dogs with immature collaterals and confirmed this finding. In contrast, Maruoka and coworkers studied open-chest, anesthetized dogs in whom collaterals had been stimulated to develop by 2–3-months ameroid constrictor placement [10]. These

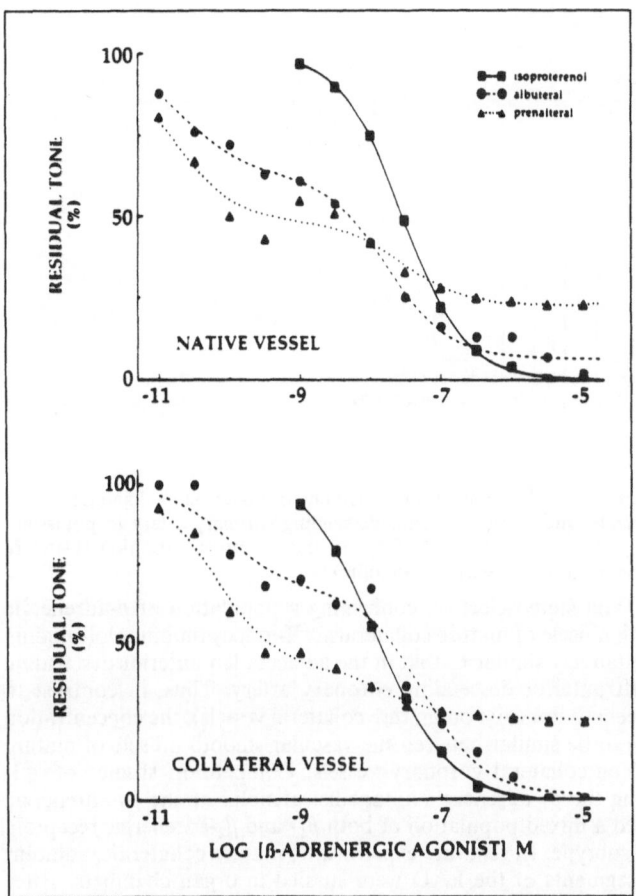

Fig. 2. β-Adrenergic relaxations of segments of the left anterior descending coronary artery (upper panel) and of mature collaterals (lower panel). Vessels were studied as ring segments in isolated organ chambers. Relaxations were obtained following preconstriction with prostaglandin $F_{2\alpha}$. (Data are from [5], used with permission of the American Physiological Society.)

investigators found that norepinephrine (combined α_1- and α_2-adrenergic stimulation) did not alter transcollateral resistance, while a selective α_2-adrenergic agonist (BHT 920) caused an increase in transcollateral resistance. The precise explanation for why a combined α_1- and α_2-adrenergic agonist did not have the same effect as a selective α_2-adrenergic agonist was not offered. Considered together, these studies suggest that circulating or neuronally released norepinephrine is not likely to produce constriction of collateral vessels in vivo. Thus, it is unlikely that α-adrenergic mechanisms modulate collateral perfusion.

The effect of β-adrenergic receptors on mature coronary collaterals

Recently, Feldman et al. [5] examined β-adrenergic receptors in mature canine coronary collaterals (6 months after ameroid constrictor placement). A coverslip ligand-binding technique, similar to that initially described by Muntz et al. was employed [12]. Histologic sections of the left anterior descending coronary segments, coronary collaterals, and adjacent, similar-sized non-collateral vessels were examined. ^{125}I-iodocyanopindolol-binding to coronary col-

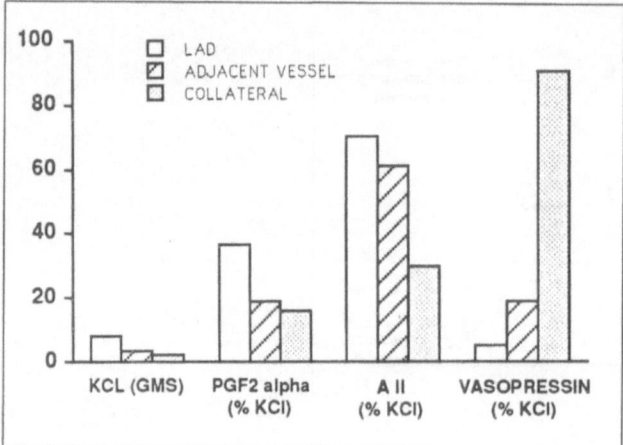

Fig. 3. Constrictions of mature collateral vessels, segments of the left anterior descending diagonal branch immediately adjacent to the collaterals, and the left anterior descending coronary artery to potassium chloride (100 mM), prostaglandin $F_{2\alpha}$ (3 μM), angiotensin II (0.3 μM), and vasopressin (0.2 μM). (Data are from [7], used with permission of the American Heart Association.)

laterals was saturable, specific, and stereoselective, confirming a population of β-adrenergic receptors on the vascular smooth muscle of mature collaterals. ^{125}I-iodocyanopindolol-binding in mature collaterals was quantitatively similar to that in the adjacent left anterior descending and diagonal branch of the left anterior descending coronary artery. Thus, in contrast to α-adrenergic receptors (which seem to not exist on mature collateral vessels), the concentration of β-adrenergic receptors seems to be similar between the vascular smooth muscle of mature collaterals and that of other non-collateral coronary vessels. Competition studies of ^{125}I-iodocyanopindolol-binding using the β_1-adrenergic antagonist atenolol or the β_2-adrenergic antagonist ICI-118,551 suggested a mixed population of both β_1- and β_2-adrenergic receptors with a predominance of the β_1-subtype. In separate experiments, mature collaterals, adjacent LAD diagonal segments, and segments of the LAD were studied in organ chambers. After preconstriction with prostaglandin $F_{2\alpha}$, isoproterenol caused concentration-dependent relaxations of mature collateral segments (Fig. 2). These responses were equally opposed by atenolol and ICI-118,551, again suggesting a mixed population of β-adrenergic receptors. Relaxations of mature collaterals were also observed in response to the selective β_1- and β_2-adrenergic agonists prenalterol and albuterol. β-Adrenergic relaxations of coronary collaterals were virtually identical to those of the left anterior descending or diagonal branches (Fig. 2).

These combined ligand-binding and isolated vessel studies demonstrated a functional population of β-adrenergic receptors in mature collaterals which are quite similar to that present in other coronary vessels. Because mature collaterals do not constrict in response to norepinephrine [7, 10] the effect of sympathohumoral stimulation would seem to favor vasodilatation. Teleologically, this would improve collateral perfusion to potentially ischemic myocardium during periods of physiologic stress.

The effect of other neurohumoral and pharmacologic stimuli on modulation of mature collateral tone: role of vasopressin

In our initial studies, mature coronary collaterals were found to constrict to a variety of substances including KCl, $PGF_{2\alpha}$, and angiotensin II. These responses were generally less than that observed in adjacent segments of diagonal branches or the left anterior descending coronary artery. In contrast, the responses to vasopressin were strikingly different than that observed in non-collateral vessels. Vasopressin produced no response in non-preconstricted, non-collateral vessels while causing marked constrictions on isolated segments of collateral vessels (Fig. 3).

As a result of these findings in isolated vessels, subsequent studies were conducted to determine if collateral perfusion might be affected by levels of vasopressin encountered during physiologic stress [13]. Dogs with mature coronary collaterals were studied 3-6 months following ameroid constrictor placement. An open-chest preparation was employed to allow direct assessment of transcollateral resistance. Myocardial perfusion was measured with radioactive microspheres under basal conditions and during the infusion of increasing concentrations of vasopressin. Infusion rates of vasopressin were employed to increase the intracoronary blood levels to concentrations ranging from 10 to over $1000 \mu U/ml$. The results of this study were striking. Levels of vasopressin which might be encountered during periods of physiologic stress such as hemorrhage, cardiopulmonary bypass, and severe dehydration had no effect on perfusion to non-collateral dependent myocardium while markedly increasing resistance to collateral-dependent myocardium (Fig. 4). This was associated with a decline in the subendocardial to subepicardial ratio of perfusion within collateral-dependent myocardium.

These findings likely have important implications regarding the regulation of perfusion to collateral-dependent myocardium when circulating levels of vasopressin are increased. They may further explain the several reports which have described myocardial ischemia, infarction or arrhythmias occurring following vasopressin administration [2, 9, 11, 15, 23].

The influence of resistances proximal to the collateral vasculature on coronary collateral perfusion

Under conditions of basal coronary blood flow, pressure at the origin of the collateral circulation is similar to aortic pressure [6]. Thus, there is not a substantial portion of the collateral circulation which arises distal to major sites of resistance in the coronary circulation. This does not imply, however, that during increases in coronary flow, minor resistances proximal to the origin of coronary collaterals may not be important. Such resistances may effectively reduce the pressure at the origin of the collateral circulation during the administration of vasodilators. This phenomenon would explain the collateral steal effect (decreases in collateral flow in the setting of increasing flow to normal myocardium) and contribute to worsening myocardial ischemia during exercise or pharmacologically induced coronary vasodilatation. The existence of a collateral steal phenomenon has been suggested in intact animals [3], but the confounding effects of the pharmacologic substances which may induce the steal phenomenon on aortic pressure and systemic hemodynamics have been difficult to control. Recent studies by Simonetti in isolated hearts have shown that during adenosine vasodilatation, collateral perfusion bears a predictable inverse relationship to perfusion to normal myocardium [21]. Collateral flow was found to decline by one-half when normal zone flow was increased from 200 to over 1000 ml/ min per 100 g. This phenomenon could only be explained by the presence of a resistance proximal to the origin of the collateral circulation which, although insignificant under conditions of resting flow, becomes important during intense adenosine vasodilatation. In separate studies, similar degrees of coronary vasodilatation with the calcium channel antagonist diltiazem did not produce this effect, suggesting that this proximal resistor may be regulated differently by different vasodilators [22].

Alteration of vasomotor properties of the coronary resistance circulation produced by chronic perfusion through mature collaterals

During experiments in intact animals with mature collaterals [13] it was noted that vasopressin not only produced collateral constriction, but also seemed to have produced marked constriction of the resistance vessels within the collateral-dependent myocardium. Because vasopressin has a direct effect on smooth muscle and also stimulates the release of endothelium-derived relaxing factor, we hypothesized that chronic perfusion through mature collaterals may alter

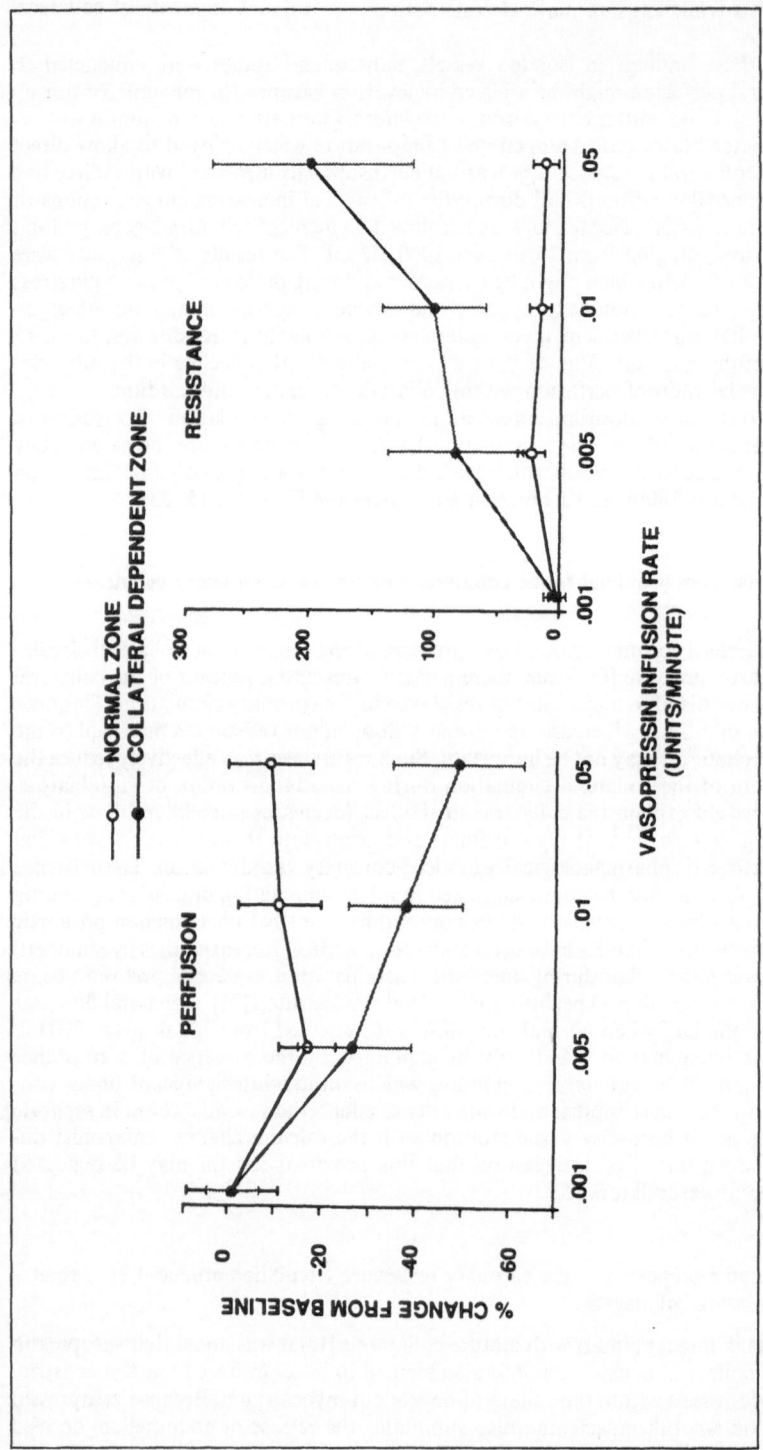

Fig. 4. Changes in normal zone and collateral-dependent zone perfusion (left panel) and resistance (right panel) during vasopressin administration. Studies were performed in open-chest, anesthetized dogs 3–6 months following the onset of ameroid constrictor occlusion of the left circumflex coronary artery. Myocardial perfusion was measured with radioactive microspheres, and vasopressin was administered intracoronarily. The concentrations of vasopressin obtained by intracoronary infusions ranged from 10 to greater than 1000 μU/ml. These concentrations of vasopressin had virtually no effect on non-collateral-dependent myocardium, while markedly increasing resistance to perfusion in the collateral-dependent myocardium. (From [13], used with permission of the American Heart Association.)

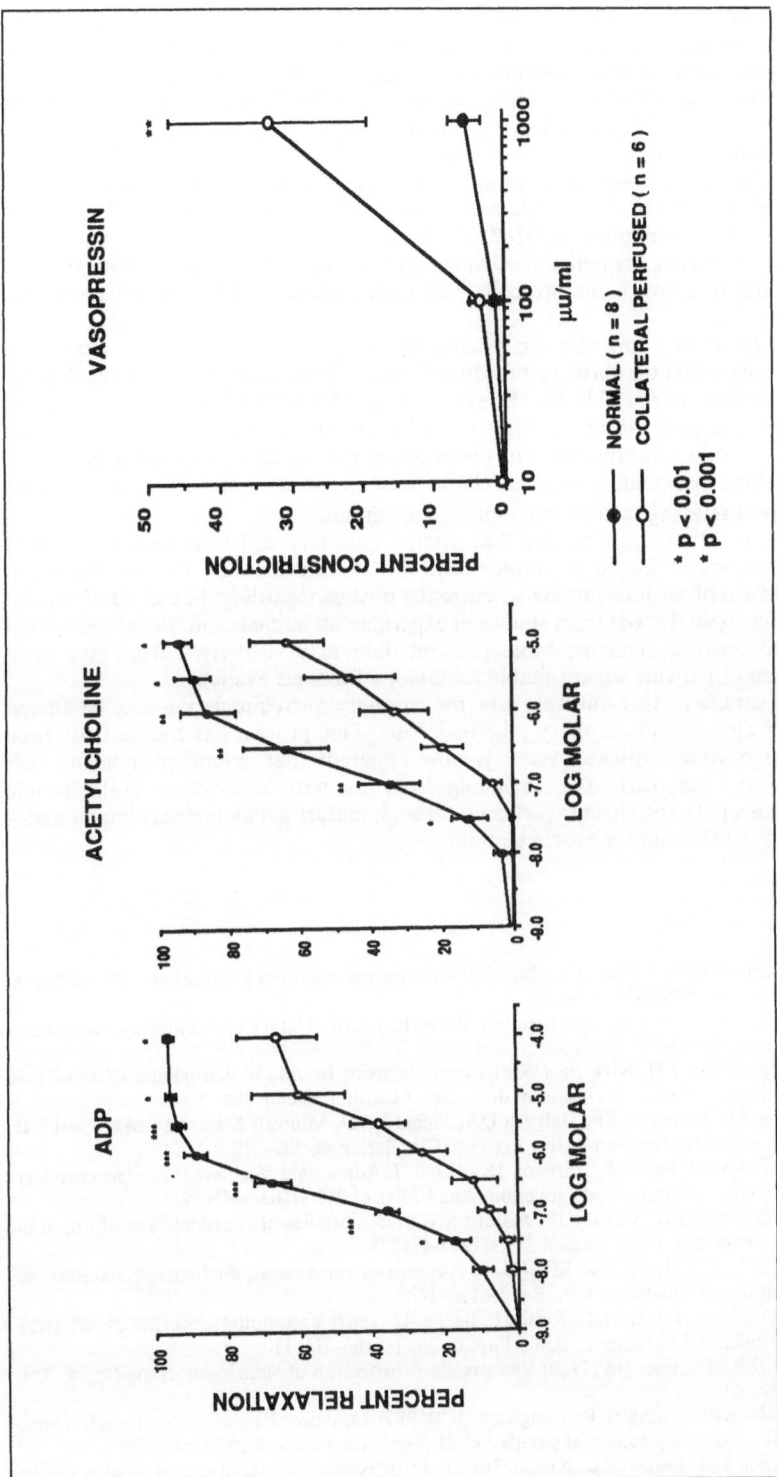

Fig. 5. Relaxations of coronary microvessels from normal myocardium and myocardium perfused by mature coronary collaterals to the endothelium-dependent vasodilators ADP and acetylcholine (left two panels) and constrictions to vasopressin (right panel). Coronary microvessels 100–200 microns in diameter were studied using a Halpern microvessel imaging apparatus. To examine relaxations to acetylcholine and ADP the vessels were preconstricted with the thromboxane-mimetic U46619 (0.01–0.1 μM). Relaxations to adenosine diphosphate and acetylcholine are depressed in vessels removed from collateral-dependent myocardium, while constrictions to vasopressin are substantially increased. (From [20], used with permission of the American Heart Association.)

endothelial function of the recipient resistance circulation. Using an in vitro microvessel imaging apparatus, endothelium-dependent vascular relaxations of normal coronary microvessels and coronary microvessels from collateral-dependent myocardium (6 months after ameroid constrictor placement) were studied. Relaxations to the endothelium-dependent vasodilating agents acetylcholine and ADP were markedly impaired in coronary microvessels from collateral-dependent myocardium (Fig. 5), while relaxations to the calcium ionophore A23187 were unchanged, and relaxations to nitroglycerin were enhanced. Further constrictions to vasopressin were enhanced in arterioles from collateral-dependent myocardium (Fig. 5). The normal relaxations to the calcium ionophore A23187 (which releases the endothelium-derived relaxing factor by directly stimulating the influx of calcium and bypassing membrane receptors) suggest that chronic perfusion through mature collateral vessels alters endothelial cell membrane receptors.

The precise explanation for why this phenomenon occurs is not presently clear. It is interesting to speculate that the chronically reduced perfusion pressure may down-regulate the coupling of the enzyme responsible for the synthesis of EDRF (nitric oxide synthetase) to membrane-bound receptors. Alternatively, growth factors liberated from the upstream collateral vessel may produce proliferation of downstream resistance vessel endothelium. Proliferating endothelium loses many functional characteristics, including the ability to release the endothelium-derived relaxing factor with appropriate stimuli.

In conclusion, these studies illustrate that mature coronary collateral vessels and their recipient vasculature are subject to vasomotor control that is substantially different from that of other vessels. Much of the information we currently possess regarding the collateral circulation of the heart has been derived from studies of experimental animals and, therefore, cannot be directly extrapolated to humans. More relevant data may be derived from studies of collateral perfusion in patients when suitable technology becomes available.

Finally, these studies further illustrate how the coronary microcirculation may be altered by ischemic heart disease and coronary atherosclerosis. This process has traditionally been considered a "large vessel" disease, yet it is now apparent that several phenomena that accompany coronary atherosclerosis, including ischemia with reperfusion [14], chronic hypercholesterolemia [19], and chronic perfusion through mature collaterals may impair endothelial function in the coronary microcirculation.

References

1. Bache RJ, Schwartz JS (1983) Myocardial blood flow during exercise after gradual coronary occlusion in the dog. Am J Physiol 245:H131–H138
2. Beller BM, Trevino A, Uban E (1971) Pitressin-induced myocardial injury and depression in a young woman. Am J Med 51:675–679
3. Cohen MV, Sonnenblick EH, Kirk ES (1976) Coronary steal: Its role in detrimental effect of isoproterenol after acute coronary occlusion in dogs. Am J Cardiol 38:880–888
4. Eng C, Patterson RE, Horowitz SF, Halgash DA, Pichard AD, Midwall J, Herman MV, Gorlin R (1982) Coronary collateral function during exercise. Circulation 66:309–316
5. Feldman RD, Christy JP, Paul ST, Harrison DG (1989) β-Adrenergic receptors on canine coronary collateral vessels: characterization and function. Am J Physiol 257:H1634–H1639
6. Harrison DG, Chapman MP, Christy JP, Marcus ML (1986) Studies of functional site of origin of native coronary collaterals. Am J Physiol 241:H1217–H1224
7. Harrison DG, Chilian WM, Marcus ML (1986) Absence of functioning α-adrenergic receptors in mature canine coronary collaterals. Circ Res 59:133–142
8. Hautamaa PV, Dai X-Z, Homans DC, Robb JF, Bache RJ (1987) Vasomotor properties of immature canine coronary collateral circulation. Am J Physiol 252:H1105–H1111
9. Kelly KJ, Stang JM, Mekhjian HS (1980) Vasopressin provocation of ventricular dysrhythmia. Ann Intern Med 92:205–206
10. Maruoka Y, McKirnan D, Engler R, Longhurst JC (1987) Functional significance of α-adrenergic receptors in mature coronary collateral circulation of dogs. Am J Physiol 253:H582–590
11. Mills MD, Burchell HB, Parker RL, Kirklin BR (1949) Myocardial infarction and sudden deaths

following the administration of pitressin; additional electrocardiographic study of 100 patients given pitressin for cholecystography. Staff Meetings of the Mayo Clinic, Rochester, Minnesota, pp 254–258

12. Muntz KH, Olson EG, Lariviere GR, D'Souza S, Mukherjee A, Willerson JT, Buja LM (1984) Autoradiographic characterization of beta adrenergic receptors in coronary blood vessels and myocytes in normal and ischemic myocardium of the canine heart. J Clin Invest 73:349–357
13. Peters KG, Marcus ML, Harrison DG (1989) Vasopressin and the mature coronary collateral circulation. Circulation 79:1324–1331
14. Quillen JE, Sellke FW, Brooks LA, Harrison DG (1990) Ischemia-reperfusion impairs endothelium-dependent relaxation of coronary microvessels while not affecting large arteries. Circulation 82:586–594
15. Ruskin A (1947) Pitressin test of coronary insufficiency. Am Heart J 30:569–579
16. Schaper W (1971) The Collateral Circulation of the Heart. Amsterdam, North Holland Publishing Co
17. Schaper W, Vandesteen R (1967) The rate of growth of interarterial anastomoses in chronic coronary artery occlusion. Life Sci 6:1673–1680
18. Schaper W (Ed) (1979) The Pathophysiology of Myocardial Perfusion. Elsevier/North-Holland Biomedical Press
19. Sellke FW, Armstrong ML, Harrison DG (1990) Endothelium-dependent vascular relaxation is abnormal in the coronary microcirculation of atherosclerotic primates. Circulation 81:1586–1593
20. Sellke FW, Quillen JE, Brooks LA, Harrison DG (1990) Endothelial modulation of the coronary vasculature in vessels perfused via mature collaterals. Circulation 81:1938–1947
21. Simonetti I, Cooper SM, Harrison DG (1987) Dependence of native coronary collateral perfusion on normal zone resistance. Circulation 76 suppl IV:IV-327 (abstr.)
22. Simonetti I, Cooper SM, Harrison DG (1988) Effect of diltiazem on precollateral resistances. Circulation 78 suppl II:II-18 (abstr.)
23. Slotnik KL, Teigland JD (1951) Cardiac accidents following vasopressin injection (pitressin). JAMA 146:1126–1129

Authors' address:

David G. Harrison, M.D., University of Iowa, Iowa City, Iowa 52242, USA

α-Adrenergic Regulation of Myocardial Performance in the Exercising Dog: Evidence for Both Presynaptic α_1- and α_2-Adrenoceptors[*][†][‡]

B. D. Guth[1], E. Thaulow[2], G. Heusch[1], R. Seitelberger[3] and J. Ross, Jr.[4]

Seaweed Canyon Laboratory, Department of Medicine, University of California, La Jolla, California, USA

Summary

New evidence supporting both presynaptic α_1- and α_2-adrenoceptors playing a role in the regulation of myocardial contractile performance in the exercising dog is reviewed. Studies utilized chronically instrumented dogs having sonomicrometers for the measurement of regional wall thickening and transducers for the measurement of left ventricular and systemic hemodynamics. During steady state exercise, either the selective α_1-adrenoceptor blocker prazosin (80 µg/kg) or the selective α_2-adrenoceptor blocker idazoxan (80 µg/kg) was infused into the left atrium while exercise continued. Immediately following the administration of either α-adrenoceptor blocking agent, there were substantial increases in heart rate, left ventricular dP/dt and regional contractile function as assessed using sonomicrometers, and norepinephrine release by the myocardium increased substantially. β-adrenergic blockade prevented the heart rate and contractile effects of either α_1- or α_2-adrenoceptor blocker whereas norepinephrine release was further enhanced. These effects could not be attributed to baroreceptor unloading. In dogs studied under resting conditions with norepinephrine infusion to produce an increase in dP/dt similar to that observed during treadmill exercise, no sympathetic augmentation was observed following either α-blocker. Together, these studies provide evidence that both α_1- and α_2-adrenoceptors participate in the modulation of sympathetic neuronal norepinephrine release in the canine myocardium.

Introduction

The action of norepinephrine on the heart, whether released by sympathetic nerve terminals in the myocardium or arriving through the coronary circulation, is known to enhance inotropism and chronotropism by activating myocardial β-adrenoceptors. It is now well established that the direct effect of norepinephrine on the coronary vasculature is a vasoconstriction mediated by the activation of coronary α-adrenoceptors [1, 7, 13, 21]. However, the role for α-adrenergic coronary vasoconstriction under both physiological and pathophysiological conditions is still

Dr. Guth is the recipient of a research award from the Alexander von Humboldt-Stiftung, Jean-Paul Straße 12, D-5300 Bonn 2, FRG.

Current addresses: [1]Abteilung für Pathophysiologie, Zentrum für Innere Medizin der Universität Essen, Hufelandstraße 55, D-4300 Essen; [2]Medical Department B, Rikshospitalet, Pilestredet 32, Oslo 1, Norway; [3]II. Chirurgische Universitätsklinik, University of Vienna, Austria; [4]Division of Cardiology, Department of Medicine, M-013B, University of California, San Diego School of Medicine, La Jolla, CA, 92093. This study was supported by NIH grant HL-17682, Ischemic Heart Disease Specialized Center of Research (SCOR).

highly controversial (as can be seen in the variety of other contributions to this volume) since the metabolic vasodilation in response to activation of β-adrenoceptors is the dominant effect produced by the endogenous neurotransmitter norepinephrine. Norepinephrine has also been shown to activate α-adrenoceptors residing on sympathetic nerve terminals in the myocardium. Activation of such presynaptic α_2-adrenoceptors initiates a negative feedback mechanism resulting in inhibition of neurotransmitter release from the nerve terminal [17]. Consequently, the blockade of α-adrenoceptors using the non-selective blocking agent phentolamine in the exercising dog results in enhanced release of norepinephrine from the sympathetic nerve terminal and an augmented sympathetic response [6].

The α_2-adrenoceptors mediating this presynaptic negative feedback mechanism were so identified to differentiate them from α_1-adrenoceptors residing on the postsynaptic target organ which mediate vasoconstriction in the coronary circulation [9, 11]. Nevertheless, the presence of α_1-adrenoceptors on sympathetic nerve terminals mediating a negative feedback mechanism on neurotransmitter release has also been suggested [16].

The importance of this negative feedback mechanism on myocardial norepinephrine release, inotropic and chronotropic state during dynamic exercise was demonstrated by Heyndrickx et al. in conscious, exercising dogs [6]. The left ventricular hemodynamic response and coronary sinus catecholamine concentrations were measured before and after selective α_2-adrenergic blockade using yohimbine or non-selective α-adrenergic blockade using phentolamine administered systematically. Norepinephrine in the coronary sinus and left ventricular dP/dt and heart rate were significantly elevated after the administration of either α-blocking agent. The augmentation of heart rate and ventricular contractile function was completely blocked by β-adrenergic blockade, thus indicating that the effect required intact β-adrenoceptors and was not a direct effect of the blocking agents on the myocardium. However, in the study by Heyndrickx et al. it was concluded that the effect was mediated only by α_2-adrenoceptors because the selective α_1-adrenoceptor blocker prazosin did not appear to produce a similar sympathetic augmentation [6].

However, Gwirtz and coworkers observed that administration of the selective α_1-adrenoceptor blocker prazosin through a coronary artery in the exercising dog caused enhanced regional myocardial contractile function and blood flow [4]. This observation is consistent with the existence of an α_1-adrenoceptor-mediated negative feedback mechanism. However, this effect was not abolished by the intracoronary administration of the β-adrenoceptor blocker propranolol thereby leading these investigators to conclude that the mechanism of the increased contractile function during exercise was not the blockade of presynaptic α_1-receptors. Instead, they postulated that the observed increase in contractile function following prazosin administration was due only to the increased regional coronary blood flow subsequent to the blockade of postsynaptic α_1-adrenoceptors residing on the coronary vasculature [4]. Thus, the participation of α_1-adrenoceptors in the negative feedback of the myocardial sympathetic nerve terminal of the exercising dog is unresolved.

A series of studies was recently reported which was designed to clarify the potential role of α_1-adrenoceptors in the negative feedback mechanism of the myocardial sympathetic nerve terminal in dogs [3]. These will be reviewed and discussed in the context of the above-mentioned studies in order to draw some conclusions concerning the functional role of α-adrenoceptors as negative feedback mediators in the myocardium during dynamic exercise in the dog.

Methods

Experimental model

Dogs (20–30 kg) chronically instrumented for the measurement of left ventricular contractile function (sonomicrometers for left ventricular wall thickness) and hemodynamics (high-fidelity left-ventricular pressure (LVP) transducer and catheter in descending thoracic aorta) were utilized for these studies. A silastic catheter in the left atrium was used for drug infusion. Animals were trained to run on a motor-driven treadmill or to lay quietly on a table. Studies were performed no earlier than 1 week post-operatively when the dogs had fully recovered and could once again run on the treadmill comfortably. During studies, LVP, its first derivative (LV dP/dt), regional wall-thickness in the posterior and anterior left ventricular walls, and aortic pressure (catheter attached to external transducer for measurement of mean pressure only) were recorded on an oscillograph, as well as being converted to digital data (200 Hz) in real time for computer-assisted wave form analysis (CORDAT, Essen [15]). The contractile state of the myocardium was assessed both on a regional (sonomicrometers for percent systolic wall thickening and thickening velocity) and global ventricular level (LV dP/dt), while ventricular and systemic hemodynamics were measured. In four dogs, a silastic catheter was implanted into the coronary sinus for blood sampling for the determination of plasma norepinephrine.

α-Adrenoceptor blockers

The selective α_2-blocker idazoxan (2.0 mg (80 µg/kg), injected into the left atrium; supplied by Reckitt and Colman, Pharmaceutical Division, Hull, England), and the selective α_1-blocker prazosin (2.0 mg (80 µg/kg), injected into the left atrium) were, at the time of the study, the agents offering the highest degree of selectivity between α_1- and α_2-adrenoceptors. The selectivity of even a higher dose of prazosin (600 µg/kg or 1.2 mg/kg, iv) against the coronary vasoconstrictor effects of the selective α_1-agonists phenylephrine [2, 8] and methoxamine [5] was previously documented. Although idazoxan is less selective for α_2-adrenoceptors than is prazosin for α_1-adrenoceptors, it was the α_2-antagonist with the highest selectivity available at the time of the studies. The selectivity of blockade for both idazoxan and prazosin was assessed in four dogs having similar instrumentation to that described above and dosages were selected to produce a complete and selective blockade. This was assessed by determining the systemic pressor effects of the α_1-adrenoceptor agonist methoxamine (1.0 mg, diluted in 10 ml saline and injected into the left atrium as a bolus; Burroughs Wellcome Co., Research Triangle Park, North Carolina, USA) or the α_2-adrenoceptor agonist BHT 933 (1.0 mg, solubilized in 10 ml saline and injected into the left atrium as a bolus; supplied by Dr. Karl Thomae GmbH, Biberach, FRG) before and after administration of prazosin (80 µg/kg) or idazoxan (80 µg/kg). The dogs were studied after muscarinic blockade using atropine (60 µg/kg, iv) to prevent the profound reflex bradycardia normally observed after administration of either α-adrenoceptor agonist in the conscious dog. Methoxamine caused mean arterial blood pressure to increase by 66 mmHg, whereas after prazosin little (3 mmHg) pressor effect remained. Prazosin had no effect on the pressor effect of BHT 933 (35 mmHg vs 35 mmHg), thus demonstrating near-complete blockade of α_1-adrenoceptors together with a high degree of selectivity. Idazoxan reduced the pressor effect of BHT 933 from an average of 31 mmHg to only 12 mmHg, however, the pressor effect of methoxamine was also reduced slightly from 62 mmHg to 50 mmHg. Thus, it is clear that only prazosin produces complete blockade with the retention of selectivity, whereas idazoxan in the dosage used produced incomplete blockade with only partial selectivity. For the purposes of these studies, however, only the selectivity and effectiveness of prazosin is critical.

Administration of α-adrenoceptor blockers

The administration of the α-adrenergic blocking agents into a single coronary artery would be an attractive experimental approach for investigating the myocardial effects of α-adrenoceptor blocking agents in exercising dogs without systemic effects. Preliminary studies were conducted using dogs equipped with a coronary catheter so that α-blockers could be administered regionally, but this approach was quickly abandoned. Despite the successful use of this drug administration route by other laboratories [4, 20] in the exercising dog with normal coronaries, we were unable to restrict the action of the α-blocking agents to the area distal to the infusion site. Instead, we observed nearly simultaneous changes in regional contractile function (sonomicrometers) located in the area distal to the infusion site as well as in a control area well outside of the perfusion territory. These observations indicated a very rapid recirculation of the α-blockers during exercise. Since we were unable to produce a regional blockade using this method of drug administration, we opted for systemic administration of the α-blockers. Although this necessitated having to contend with systemic effects of the α-blockers, it has the important advantage of being able to use drug doses in which complete and selective α-adrenoceptor blockade can be clearly demonstrated (see above).

Protocols

The basic experimental protocol called for dogs to be studied first while standing quietly on the treadmill so that baseline hemodynamics and myocardial function could be measured. The dogs then ran for 3 min at a speed and treadmill inclination sufficient to elicit a heart rate greater than 200 beats per min. After 3 min of running, a second set of measurements was made during steady-state exercise. Then either prazosin or idazoxan was infused into the left atrium (dosages given above). After 3 min of additional exercise, a final set of data was obtained. The protocol was repeated on a separate day after the dogs received β-adrenoceptor blockade (1.0 mg/kg propranolol). In four dogs, arterial and coronary sinus plasma norepinephrine was measured at each experimental time point.

This experimental approach has a significant impact on the interpretation of the data and therefore deserves some discussion. An advantage of systemic administration of the α-blocking agents is that dosages can be easily administered and tested for efficacy, as noted above. One need not be concerned with restricting the agent to a small region of the myocardium and the problem of recirculation of the agent can be eliminated. Disadvantages of this approach include the fact that both $α_1$- and $α_2$-adrenoceptor blocking agents have systemic effects. Particularly, the $α_1$-adrenoceptor blocker prazosin produces systemic hypotension through the blockade of $α_1$-adrenoceptors residing in the peripheral circulation; $α_2$-adrenoceptor blockers have a similar, albeit smaller effect. Thus, systemic application of these agents requires that one must be able to cope with the resultant changes in systemic pressure. This is essential not only to match loading conditions on the left ventricle for the evaluation of contractile indices, but also due to the unloading of arterial baroreceptors which can activate the sympathetic nervous system independently from any myocardial α-adrenergic presynaptic mechanism. It should also be mentioned that prazosin has significant central effects in high doses which can lead to a central depression of the baroreflex [10]. Two separate experimental approaches were employed to address the problem of changes in systemic arterial pressure following the administration of the $α_1$-adrenoceptor blockade. First, in six dogs mean aortic pressure was not allowed to decrease following the prazosin administration by infusing angiotensin II (1–3 µg/min) through a left ventricular catheter, thereby offsetting the systemic vasodilation induced by prazosin. The infusion rate of angiotensin II was adjusted to maintain mean aortic blood pressure at the level observed during the steady-state run just prior to the administration of prazosin. Alternatively, in six additional dogs a valvuloplasty balloon (Mansfield, 2.3-cm balloon diameter) was implanted using aseptic technique into the aorta at the level of the diaphragm through the left carotid artery under fluoroscopy 1 day before the experimental protocol. Controlled inflation of the balloon was then used during running after prazosin administration to maintain the aortic pressure at the level observed before administration of prazosin.

Both experimental approaches were utilized since they each have certain advantages as well as disadvantages. For example, angiotensin II has been shown to facilitate norepinephrine release by a direct effect on the presynaptic sympathetic nerve endings in some experimental systems, which might account in part for the enhanced inotropic and chronotropic response during exercise in our dog model [3]. While the intraaortic balloon avoids this potential concern, to introduce the balloon one carotid artery had to be obstructed which left each dog with only the intraaortic and the carotid baroreceptors of the right side responsive to changes in arterial blood pressure. Fortunately for the interpretation of these data, the two preparations yielded identical results, thereby minimizing these concerns.

Presynaptic α-receptors

Results and discussion

In the first protocol, either idazoxan or prazosin was administered during treadmill exercise in the dogs. The dogs first ran for 3 min and a steady-state hemodynamic and contractile response was observed for the latter 2 min. After bolus injection of either α-blocker, the dogs continued running for an additional 3 min.

α₂-Blockade

The administration of the selective α_2-adrenoceptor blocker idazoxan resulted in significant enhancement of the myocardial chronotropic and inotropic state. This is consistent with the previous observations by Heyndrickx et al. and is indicative of a presynaptic α_2-adrenoceptor mediated negative feedback mechanism in the myocardium [6]. The response was characterized by a significant further increase in exercise heart rate (24 beats/min) and LV dP/dt (further 33% increase). Regional wall thickening was also enhanced, despite no changes in either end-diastolic LVP or mean arterial pressure. Peak LVP was increased by 9 mmHg and likely reflects the enhanced contractile state. It should be noted that the α_2-adrenoceptor blockade did not induce a systemic hypotension in this experimental model.

α₁-Blockade

The administration of prazosin during treadmill exercise also elicited a significant augmentation of chronotropism and inotropism as seen in Fig. 1a and summarized in Fig. 2. Exercise heart rate increased further after prazosin by 51 beats/min and dP/dt increased further by 36%. Regional systolic wall thickening increased significantly, although the ventricular loading conditions were altered by prazosin. Mean arterial pressure fell by 13 mmHg following prazosin, whereas end-diastolic LVP fell by 7 mmHg, indicative of the hypotensive effect of prazosin caused by the blockade of systemic α_1-adrenoceptors. The fact that peak LVP was unchanged by prazosin likely reflects the combined effect of the prazosin-induced vasodilation together with the inotropic stimulation. While these data are consistent with an α_1-adrenergic presynaptic negative feedback mechanism, the systemic hypotension also observed could be an adequate explanation for the large increase in sympathetic outflow to the myocardium, thereby making the studies in which arterial pressure could be supported essential to the interpretation of these data.

α₁-Adrenoceptor blockade with pressure match

Both angiotension II administration and adjustment of the intraaortic balloon provided well-matched arterial pressures during exercise before and after α_1-adrenoceptor blockade, and nearly identical results were observed. In both cases, even with identical arterial mean pressures, prazosin still elicited significant further augmentation of the chronotropic and inotropic response to treadmill exercise; shown in Fig. 1a is a study in which pressure was matched using an intraaortic balloon. Exercise heart rate further increased by 33 or 22 beats/min in the angiotensin II and aortic balloon groups, respectively, while exercise LV dP/dt increased further by 28% and 27%. It should be noted, however, that in both experimental models the LV end-diastolic pressures were significantly less following prazosin administration, despite matched peak LVP and mean aortic pressures. One may speculate that the reduced preload would only tend to minimize the observed increases in regional wall thickening or LV dP/dt.

At this point, the data strongly supported both α_1- and α_2-adrenoceptors being involved in the presynaptic modulation of norepinephrine release from sympathetic nerve terminals in the myocardium of dogs. These data are qualitatively similar to those of Gwirtz et al. in which a contractile function increase was observed following intracoronary prazosin infusion in exercising dogs [4], but contrast from the data of Heyndrickx et al. in which no α_1-adrenergic effect

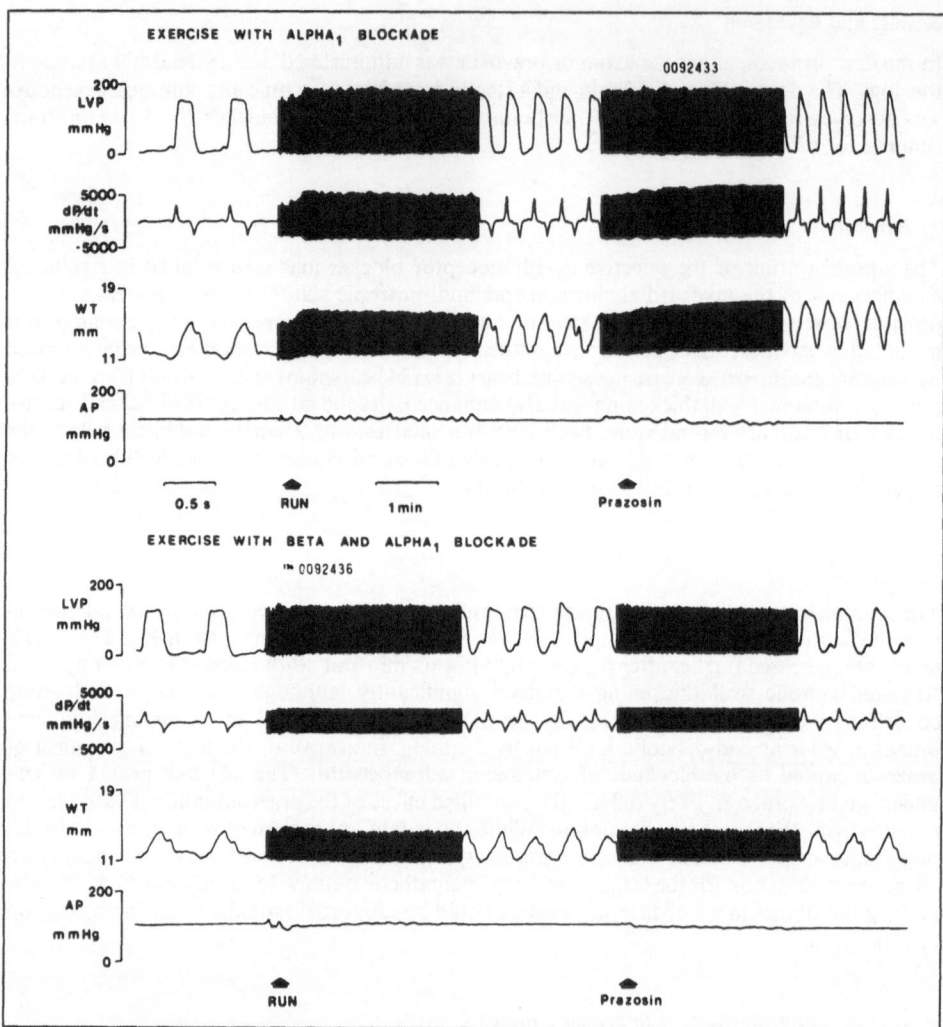

Fig. 1a. Original tracing from one study in which systemic arterial pressure was held constant following prazosin administration. With the onset of exercise there were significant increases in heart rate, LV dP/dt, and regional contractile function (increased wall thickening). Following the administration of prazosin, heart rate and contractility increased further.

Fig. 1b. Original tracing of the same dog but after administration of β-bloackade. Note the depressed response to exercise and the total lack of effect following prazosin administration. Arterial pressure was held constant following prazosin administration using the intraaortic balloon.

could be demonstrated [6]. Although the difference in the results cannot be exactly determined, it may be related to the substantial difference in the dosages of prazosin administered in the two studies. In the studies presented here, a dose of 80 μg/kg was used and administered as a bolus infusion [3]. This was shown to completely block the pressor effects of 1.0 mg methoxamine, thus indicating substantial α_1-blockade with retained selectivity since the pressor effects of BHT-933 were preserved. In the study by Heyndrickx et al. a prazosin dose of 500 μg/kg was used, thereby exceeding the present dosage by a factor of six-fold. Whereas the selectivity of

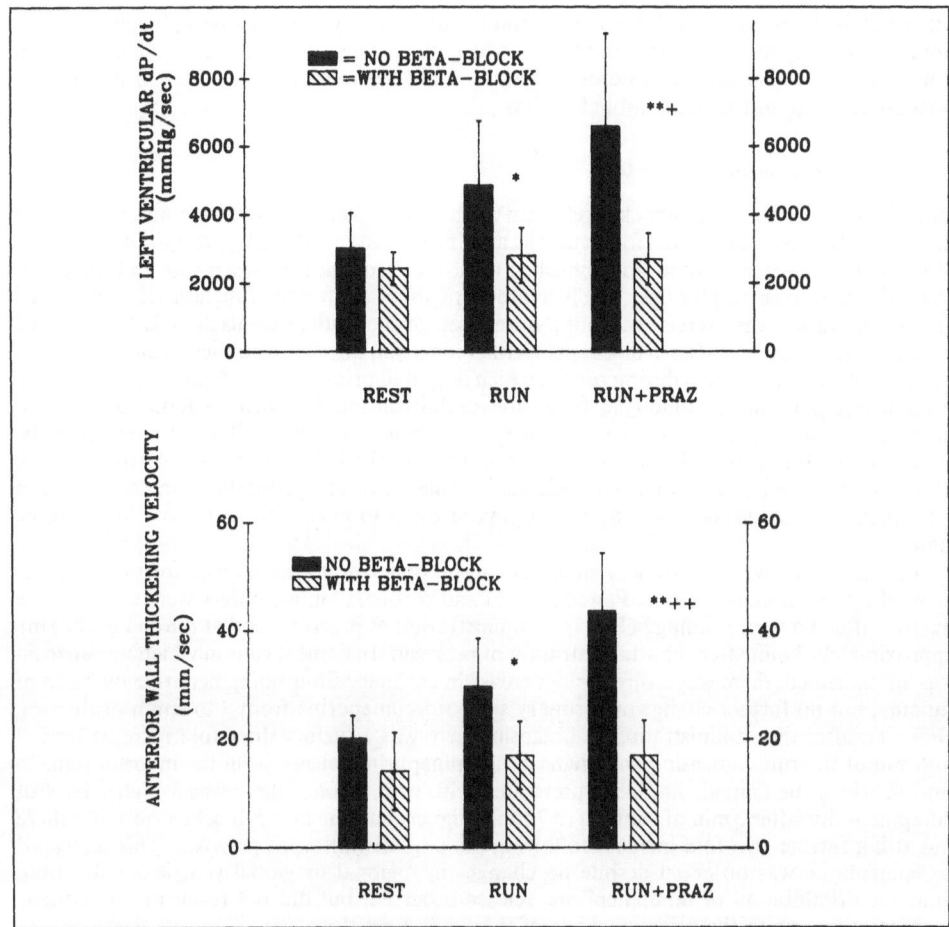

Fig. 2. Summary of global (LV dP/dt, above) and regional contractile indices (wall thickening velocity, below) following prazosin both before (filled bars) and after β-blockade (hatched bars). The significant augmentation of contractility observed after prazosin was completely eliminated by β-blockade.

prazosin is likely still intact with this dosage, one should be concerned about the possible central depressant effects that this drug exhibits [10]. For example, it is noteworthy that despite a substantial decrease in arterial pressure in the study by Heyndrickx et al. no increased heart rate or LV dP/dt was observed [6]. Such a large degree of hypotension would normally be countered with a significant sympathetic augmentation in the conscious dog due to barorecep-tor unloading, independent from any presynaptic effects prazosin may mediate. The failure to observe such reflex activation suggests the possibility of baroreceptor depression at this high dosage. Alternatively, the lack of a hypotension-induced tachycardia may be the result of prazosin's action on cholinergic nerve terminals in the myocardium eliciting augmented acetyl-choline release and bradycardia to offset any sympathetically mediated heart rate increase [12, 19].

A presynaptic α_1-adrenergic mechanism modulating norepinephrine release would be expected to be dependent upon intact β-adrenoceptors in the myocardium in order for the increased contractility and inotropism to be realized. Additionally, it is reasonable that a disinhibition of norepinephrine release from the sympathetic nerve terminal would only result ᵢin augmented heart rate and contractility if the nerve terminals are in an activated state. In

137

other words, if the sympathetic nerve terminal is not actively releasing norepinephrine, or is doing so at a very low level, then the removal (through presynaptic α-blockade) of a mechanism inhibiting the further release of norepinephrine should have little, if any, effect. Both of these possibilities were tested in additional studies.

Effect of β-adrenoceptor blockade

Runs in which either the α_1-blocker prazosin or the α_2-blocker idazoxan were administered as described above, were repeated following the administration of 1.0 mg/kg propranolol, IV (Fig. 1b). The β-adrenoceptor blockade abolished the α-blockade induced heart rate and LV dP/dt response to exercise in all dogs. The increase in regional contractile function was eliminated and, in contrast to the exercise without β-adrenoceptor blockade, end-diastolic LVP remained elevated during exercise after α-blockade. Neither idazoxan nor prazosin elicited any significant changes when administered during runs in which dogs had prior β-blockade although there was a tendency, particularly following prazosin administration, for mean arterial pressure to decrease, an indication of the blockade of systemic α-adrenoceptors. Thus, the sympathetic augmentation following either α_1- or α_2-adrenoceptor blockade is dependent upon the presence of functional β-adrenoceptors. This effectively rules out the possibility that either agent (idazoxan or prazosin) acts directly on the myocardium to affect contractile or chronotropic state.

In four dogs having a coronary sinus catheter, blood samples were taken for the measurement of plasma norepinephrine. Paired arterial and coronary sinus samples were taken before the run, after 3 min of running before the administration of prazosin, and at the end of the run, approximately 3 min after the administration of prazosin. In control runs in which no prazosin was administered, there was a three-fold increase in coronary sinus norepinephrine by 3 min of running, but no further change in coronary sinus norepinephrine from 3 to 6 min of the run. However, after the administration of prazosin, there was a further three-fold increase from 3 to 6 min of the run, indicating an enhanced norepinephrine release from the myocardium. In runs in which the animals had been pretreated with propranolol, there was a higher level of norepinephrine after 3 min of running (2.7 times higher than the non-β-blocked run), but there was still a further two-fold increase following the administration of prazosin. This increased norepinephrine was observed despite no changes in regional or global ventricular function; thus, the disinhibition of norepinephrine release occurred, but did not result in sympathetic augmentation due to the prior blockade of the β-adrenoceptors.

It is at this point that the observations of the present series of studies diverge from the observations of Gwirtz et al. [4]. In their study, the augmentation of regional contractile function and coronary artery blood flow following the intracoronary administration of prazosin was not blocked by the intracoronary administration of the β-adrenoceptor blocker propranolol. This is a key observation since a presynaptic mechanism would necessarily be dependent upon functional β-adrenoceptors. The failure of Gwirtz et al. to observe a prevention of this effect after β-adrenoceptor blockade lead these investigators to propose that the increased function was based entirely on an increased coronary blood flow and the alleviation of a preexistent subendocardial ischemia and contractile dysfunction occurring during normal exercise in the dog. The possible explanation of a "Gregg Effect" is not likely to be of significance in the intact heart [14] and this postulate was also rejected by Gwirtz and coworkers [4].

Unfortunately, there is currently no experimental evidence which can accommodate both the findings outlined in this review with the findings of Gwirtz et al. [4]. One can only speculate that the route of administration for both prazosin and propranolol may have lead to the discrepant conclusions. For instance, the injection of prazosin in the dose of 0.5 mg, as employed by Gwirtz et al., into the left circumflex coronary artery is likely to have had effects outside of the myocardium distal to the site of infusion. Indeed, the total dosage used by Guth et al. (reviewed here) was approximately 2.0 mg and resulted in complete systemic α_1-adrenoceptor blockade [3]. Thus, a dosage of 0.5 mg into the coronary, as used by Gwirtz et al. [4], likely produced at least partial blockade systemically including the presumably unblocked normal myocardium. Perhaps more importantly, with the administration of propranolol through the coronary catheter as performed by Gwirtz et al. a complete blockade on a first pass may be difficult to achieve. If a portion of β-adrenoceptors remained unblocked, it is possible that the augmentation of function and blood flow could still have occurred. The potential for divergent results based on the route of administration is further suggested by the fact that these investigators failed to observe enhanced contractile function after the relatively selective α_2-adrenoceptor blocker yohimbine was administered through a coronary artery in dogs exercising at a level comparable to that of dogs in the studies reviewed here [18]. Such an effect was marked both in the study by Heyndrickx et al. [6] and Guth et al., in which the α_2-blockade was given intravenously [3].

Finally, the blockade of a negative feedback loop should only result in enhanced output when the system is in an activated state. In the case of the current discussion, a negative feedback loop is achieved when norepinephrine released by the activated sympathetic nerve terminal in the myocardium binds with α_2-adrenoceptors or, as recently demonstrated, α_1-adrenoceptors residing on the nerve terminal itself. The subsequent inhibition of further norepinephrine release from the sympathetic nerve terminal to complete the negative feedback loop should only be operative when the sympathetic nerve terminal is actively releasing norepinephrine. This is the case during dynamic exercise when sympathetic discharge to the heart is high or in experimental settings where the sympathetic cardiac efferent nerves are stimulated. If β-adrenoceptor activation is the result of only circulating norepinephrine, then presynaptic α-adrenoceptor blockade should have no stimulatory effect, if this effect is through presynaptic disinhibition of norepinephrine release. Thus, in a final group of dogs, we tested the hypothesis that α_1-adrenoceptor or α_2-adrenoceptor blockade has no effect on global or regional contractile function or heart rate when administered to dogs resting quietly, but with adrenergic stimulation through intravenous norepinephrine infusion.

α-Adrenoceptor blockade during resting conditions with norepinephrine infusion

In five dogs, selective α-adrenoceptor blockade was performed during an intravenous infusion of norepinephrine while the dogs lay quietly on a table (total time approximately 30 min). The norepinephrine infusion rate ($2 \pm 1 \mu g/min$) was adjusted to achieve an increase in LV dP/dt similar to that observed in previous treadmill runs. Once a hemodynamic steady state was achieved (approximately 3 min) data were recorded. With continued norepinephrine infusion, the α_1-blocker prazosin or the α_2-blocker idazoxan (dosages as before) was infused into the left atrium. Once a new hemodynamic steady state was achieved (an additional 2 min) data were again recorded. The total infusion time was never greater than 7 min and the average total dose of norepinephrine given was $13 \mu g$. Significant activation of systemic α-adrenoceptors was apparent from a large increase in aortic pressure (Table 1) and all dogs became mildly excited during the norepinephrine infusion. With the administration of either α-blocker a substantial decrease in systemic diastolic and mean arterial pressures was observed, thereby indicating a significant blockade of systemic α-adrenoceptors in the peripheral circulation. However, unlike exercise, no augmentation of heart rate or LV dP/dt by prazosin was observed during norepinephrine infusion (Table 1). The remaining elevation of systemic pressure following α-blockade likely is the result of the intact β-adrenoceptor activation by norepinephrine as indicated by the enhanced contractile indices (LV dP/dt and regional function), whereas the lack of a tachycardia during norepinephrine infusion reflects the baroreceptor activation and reflex bradycardia due to the significantly elevated arterial pressures in the conscious dogs.

The results of this protocol strongly suggest that unless sympathetic stimulation is the result of norepinephrine released from the myocardial sympathetic nerve terminals, no sympathetic augmentation results from α_1- or α_2-presynaptic adrenoceptor blockade. It was assumed that the infused norepinephrine reached presynaptic α-adrenoceptors in the myocardium as well as the peripheral receptors mediating the systemic pressure increase. This experimental approach has

Table 1. Effect of selective α_1- or α_2-blockade during norepinephrine infusion.

	Control	NE INF	NE + BLOCK
α_1-blockade			
LVP (mmHg)	133 ± 6	222 ± 10*	170 ± 9*#
LVEDP (mmHg)	16.4 ± 2.3	30.8 ± 5.4*	22.9 ± 3.1*#
MAP (mmHg)	105 ± 8	182 ± 10*	134 ± 10*#
DAP (mmHg)	84 ± 17	150 ± 8*	108 ± 10*#
(+)LVdP/dt (mmHg/s)	2782 ± 474	5014 ± 612*	4744 ± 778*
HR (beats/min)	90 ± 26	104 ± 8	102 ± 21
%WT	26.0 ± 7.1	27.5 ± 3.9	36.1 ± 5.2*#
WTVel (mm/s)	12.4 ± 2.8	13.7 ± 1.5	18.9 ± 2.4*#
α_2-blockade			
LVP	130 ± 6	232 ± 21*	194 ± 9*#
LVEDP	12.6 ± 2.1	28.8 ± 3.3*	24.5 ± 2.5*
MAP	107 ± 14	185 ± 24*	155 ± 11*#
DAP	90 ± 21	153 ± 23*	128 ± 13*#
(+)LVdP/dt	2893 ± 533	4866 ± 706*	4857 ± 586*
HR	93 ± 25	106 ± 35	110 ± 37*
%WT	27.4 ± 7.0	25.9 ± 6.0	30.7 ± 7.5#
WTVel	12.8 ± 4.3	12.2 ± 3.6	15.6 ± 4.9*#

NE INF = infusion of norepinephrine; NE + BLOCK = effect of administration of α-blocker during infusion of norepinephrine; LVP = left ventricular pressure; LVEDP = end-diastolic LVP; MAP = mean aortic pressure; DAP = diastolic aortic pressure; (+)LVdP/dt = maximum first derivative of LVP; HR = heart rate;%WT = systolic thickening of the anterior wall; WTVel = thickening velocity of the anterior wall. Data are mean ± SD, N = 5, * = $P < 0.05$ vs Control, # = $P < 0.05$ vs NE INF.

been used previously to test for presynaptic mechanisms of angiotensin II on sympathetic nerve terminals [3].

CONCLUSIONS

The elucidation of a complex α-adrenergic negative feedback in an experimental model that itself is complex (the exercising, chronically instrumented dogs) is a difficult task. Clearly, the presynaptic α_1-adrenoceptor mediated negative feedback mechanism proposed here will require additional basic studies on the molecular level to substantiate its existence. Nevertheless, the experimental results reviewed herein indicate that this mechanism likely is active in the exercising dog and that α_1- or α_2-adrenoceptor blockade during dynamic exercise produces pronounced sympathetic augmentation. The remaining experimental inconsistencies are likely results of methodologic differences and will hopefully be resolved in the future.

The implications of the proposed α_1-adrenergic presynaptic inhibitory mechanism may be clinically important. For example, in patients receiving prazosin for the treatment of hypertension, the sympathetic response to exercise could be inappropriately augmented during exercise or stress which would be particularly worrisome for patients also having coronary artery disease and at risk of exercise-induced angina pectoris. The addition of β-adrenoceptor blockade in this subset of patients might be appropriate to prevent undesirable sympathetic outflow which could endanger the ischemic myocardium.

References

1. Andersson R, Holmberg S, Svedmyr N, Aberg G (1972) Adrenergic alpha- and beta-receptors in coronary vessels in man. Acta Med Scand 191:241–244
2. Chen DG, Dai X-Z, Zimmerman BG, Bache RJ (1988) Postsynaptic α_1- and α_2-adrenergic mechanisms in coronary vasoconstriction. J Cardiovasc Pharmacol 11:61–67
3. Guth BD, Thaulow E, Heusch G, Seitelberger R, Ross J Jr (1990) Myocardial effects of selective alpha-adrenoceptor blockade during exercise in dogs. Circ Res 66:1703–1712
4. Gwirtz PA, Overn SP, Mass HJ, Jones CE (1986) Alpha 1-adrenergic constriction limits coronary flow and cardiac function in running dogs. Am J Physiol 250:H1117–H1126
5. Heusch G, Deussen A, Schipke J, Thämer V (1984) α_1- and α_2-adrenoceptor-mediated vasoconstriction of large and small canine coronary arteries in vivo. J Cardiovasc Pharmacol 6:961–968
6. Heyndrickx GR, Vilaine JP, Moerman EJ, Leusen I (1984) Role of prejunctional alpha 2-adrenergic receptors in the regulation of myocardial performance during exercise in conscious dogs. Circ Res 54:683–693
7. Holtz J, Mayer E, Bassenge E (1977) Demonstration of alpha-adrenergic coronary control in different layers of canine myocardium by regional myocardial sympathectomy. Pfluegers Arch 372:187–194
8. Holtz J, Saeed M, Sommer O, Bassenge E (1982) Norepinephrine constricts the canine coronary bed via postsynaptic α_2-adrenoceptors. Eur J Pharmacol 82:199–202
9. Homcy CJ, Graham RM (1985) Molecular characterization of adrenergic receptors. Circ Res 56:635–650
10. McCall RB, Humphrey SJ (1981) Evidence for a central depressor action of postsynaptic alpha-1-adrenergic receptor antagonists. J Auton Nerv System 3:9–23
11. McGrath JC, Brown CM, Wilson VG (1983) Alpha-receptors: a critical review. Med Res Rev 9:407–533
12. McGrattan PA, Brown JH, Brown OM (1987) Parasympathetic effects on in vivo rat heart can be regulated through an alpha-1-adrenergic receptor. Circ Res 60:465–471
13. Mohrman DE, Feigl EO (1978) Competition between sympathetic vasoconstriction and metabolic vasodilation in the canine coronary circulation. Circ Res 42:79–86
14. Schulz R, Heusch G, Oudiz R, Guth BD (1989) Coronary pressure and flow have no independent effect on contractility in anesthetized swine. Faseb J 3:A405
15. Schulz R, Hücking G, Heusch G (1989) "CORDAT" – a new data acquisition and reduction program. Eur Heart J 10:309–(abstr.)
16. Starke K (1987) Presynaptic alpha-autoreceptors. Rev Physiol Biochem Pharmacol 107:73–146
17. Starke K, Göthert M, Kilbinger H (1989) Modulation of neurotransmitter release by presynaptic autoreceptors. Physiol Rev 69:864–989
18. Strader JR, Gwirtz PA, Jones CE (1988) Comparative effects of α_1- and α_2-adrenoceptors in modulation of coronary flow during exercise. J Pharmacol Exp Ther 246:772–778
19. Wetzel GT, Brown JH (1985) Presynaptic modulation of acetylcholine release from cardiac parasympathetic neurons. Am J Physiol 248:H33–H39
20. Young MA, Vatner DE, Knight DR, Graham RM, Homcy CJ, Vatner SF (1988) α-adrenergic vasoconstriction and receptor subtypes in large coronary arteries of calves. Am J Physiol 255:H1452–H1459
21. Zuberbuhler RC, Bohr DF (1965) Responses of coronary smooth muscle to catecholamines. Circ Res 16:431–440

Authors' address:

Brian D. Guth, Ph.D., Abteilung für Pathophysiologie, Zentrum für Innere Medizin der Universität Essen, Hufelandstraße 55, D-4300 Essen, FRG.

III. Adrenergic Coronary Vasoconstriction and Blood Flow Distribution In Ischemic Myocardium

Transmural Steal with Isoproterenol and Exercise in Poststenotic Myocardium

K. P. Gallagher

Seaweed Canyon Laboratory, School of Medicine, University of California, La Jolla, California and the Thoracic Surgery Research Laboratory, Departments of Surgery (Thoracic Section) and Physiology, University of Michigan Medical School, Ann Arbor, Michigan USA

Summary

Transmural coronary steal describes the phenomenon that can occur when coronary narrowing is severe enough to eliminate or nearly eliminate vasodilator reserve in the subendocardial layers. Because blood flow in a maximally vasodilated vascular bed is linearly dependent on perfusion pressure, additional reductions in perfusion pressure will decrease subendocardial blood flow. The subepicardial layers, operating on a different autoregulatory pressure-flow curve, may have vasodilator reserve available and display normal or even elevated blood flow when the subendocardium has reduced perfusion. Therefore, it appears as if subendocardial blood flow has been "stolen" by the subepicardial layers. Blood flow is not actually stolen but redistributed distal to a flow-limiting stenosis and the redistribution tends to favor the subepicardium because it can autoregulate to a lower pressure than the subendocardium.

Physiologic interventions such as exercise can alter myocardial oxygen requirements substantially. Vasodilator reserve will be utilized in those parts of the myocardium that have it available, in order to meet the augmented myocardial flow requirements associated with exercise. In poststenotic myocardium, however, decreased vascular resistance in subepicardial layers may reduce poststenotic perfusion pressure which will lead, in turn, to a decrease in blood flow to the subendocardial layers if they are maximally vasodilated. Because transmural systolic function (measured as wall thickening, for example) is largely dominated by changes in subendocardial perfusion, transmural steal during exercise may aggravate the level of dysfunction that occurs by augmenting the subendocardial flow deficit.

The concept of vascular "steal" was introduced by Reivich et al. [30] who coined the term subclavian steal to describe a situation in patients with subclavian arterial narrowing proximal to the origin of the vertebral artery. Reivich et al. observed neurologic symptoms consistent with reduced cerebral blood flow when the patients exercised an arm. Because vascular resistance decreased in the arm muscles during exercise, the limited amount of blood flow through the narrowed subclavian artery appeared to be redistributed or "stolen" from the vascular bed of the vertebral artery by the dilated skeletal muscle vascular bed.

The reductions in vertebral blood flow occurred because perfusion pressure distal to the flow restricting subclavian stenosis decreased during arm exercise. The poststenotic perfusion pressure decreased when the skeletal muscle vascular bed (supplied by the brachial artery) dilated. The vertebral bed was unable to reduce its vascular resistance further to maintain

This study was supported in part by National Institutes of Health Specialized Center of Research on Ischemic Heart Disease HL 17682 and NIH grant HL 32043.

perfusion because it was already maximally vasodilated. Vertebral blood flow decreased while blood flow increased in the parallel vascular circuit of the brachial artery which retained vasodilator reserve producing a pattern that looked as if one vascular bed had stolen blood flow from another [36]. Rowe [32] described how "coronary steal" could occur in the myocardium of patients with coronary artery disease. Myocardial blood flow in areas with exhausted or nearly exhausted vasodilator reserve (e.g., distal to severe coronary stenoses or minimally perfused by collateral vessels) could appear to be "stolen" and preferentially shunted elsewhere in circumstances that reduced vascular resistance in parallel vascular beds.

The main focus of this review is a variant of the phenomenon described by Rowe [32], known as transmural coronary steal. Transmural steal can occur when a major coronary artery is narrowed to the point that subendocardial vasodilator reserve is exhausted but subepicardial vasodilator reserve is still available. This situation is possible because the lower limit of autoregulating perfusion pressure [28] is higher for the subendocardial than subepicardial layers of the left ventricular wall [6, 10, 20, 23]. If the subendocardial vasculature is completely dilated to compensate for severe coronary narrowing, additional reductions in blood pressure will decrease subendocardial blood flow because that area's vasculature is operating on a linear portion of the pressure-flow curve. Consequently, when poststenotic perfusion pressure is reduced blood flow may appear to be "stolen" from the subendocardial layers by the subepicardial layers, the portion of the left ventricular wall which retains vasodilator reserve and can sustain normal or even increased levels of blood flow [18].

One way to demonstrate transmural steal in an experimental setting is to decrease coronary perfusion pressure with external constrictors or by cannulating the coronaries and altering coronary inflow rates or perfusion pressure, as numerous investigators have done over the last several years [1, 11, 12, 16, 18, 20, 23]. From the standpoint of pathophysiology, however, another means of producing transmural steal may be more important. Perfusion pressure distal to a critical stenosis will also decrease (thereby jeopardizing subendocardial perfusion) if vascular resistance is reduced in subepicardial layers which retain vasodilator reserve. Pharmacologic interventions that directly vasodilate subepicardial vasculature or physiologic manoeuvers that increase demand for blood flow and thereby produce subepicardial vasodilation can result in transmural steal because a reduction in subepicardial vascular resistance will decrease perfusion pressure distal to a flow-limiting coronary stenosis. Any reduction in perfusion pressure will lead to a decrease in subendocardial blood flow when the inner layers of the left ventricular wall no longer have autoregulatory vasodilator reserve available.

For example, in a study conducted several years ago, adenosine was infused distal to a critical coronary stenosis in open-chest, anesthetized dogs [11]. The adenosine did not change mean coronary blood flow (measured with an electromagnetic flowmeter) through the stenosis but it did reduce coronary perfusion pressure distal to the critical stenosis, indicating that vascular resistance must have changed to accommodate the same inflow at a lower pressure. The distribution of myocardial blood flow, measured with microspheres, showed that blood flow had increased in the subepicardium, but had decreased in the subendocardium consistent with the interpretation that vascular resistance had decreased in the subepicardial layers (enabling blood flow to increase); it had not changed in the subendocardial layers. Subendocardial blood flow decreased, paralleling the reduction in perfusion pressure. Similar observations have been made with other intracoronary and systemically administered vasodilators [3, 10, 18] including volatile anesthetics [4, 34].

Physiologic manoeuvers designed to augment myocardial work and oxygen demand have also been used to demonstrate transmural coronary steal [10]. For example, treadmill exercise in chronically instrumented dogs with acute stenoses [2] was associated with increased subepicardial blood flow and decreased subendocardial blood flow. Although reduced time in diastole no doubt contributed to the maldistribution of myocardial blood flow in these studies, the results supported the contention that augmentation of myocardial oxygen demand could produce subepicardial vasodilation and, thereby, induce transmural coronary steal.

To determine the potential consequences of transmural coronary steal in terms of regional myocardial function, we conducted two studies [13, 14]. In each study, sonomicrometers were implanted in chronically instrumented dogs to measure regional subendocardial segment tening and/or transmural wall thickening. Based on previous demonstrations that transmural mechanical function depended most strongly on deep myocardial perfusion [12, 31], we posed the hypothesis that subepicardial vasodilator reserve could be deleterious when myocardial oxygen demand was augmented in the presence of critical coronary stenosis. It was already established that increasing myocardial oxygen demands when coronary inflow was restricted led to regional mechanical dysfunction [25, 35]. We speculated that transmural steal could make the degree of dysfunction worse by augmenting the flow deficit in the crucial subendocardial layers of the left ventricular wall.

Ten to 14 days after surgery in the first study [13], the animals were studied while conscious to determine the effects of isoproterenol on the relationship between regional myocardial blood flow and function. Isoproterenol (0.20 μg/kg/min) was infused after producing a level of coronary stenosis in the circumflex artery with an adjustable hydraulic occluder. Two different degrees of stenosis were evaluated; one of them corresponded roughly to a critical stenosis which is the experimental setting that predisposes most strongly to development of transmural steal. Unfortunately, we did not measure coronary blood flow velocity in these studies [13] so that we could not determine whether we had successfully produced a critical (or hypercritical) stenosis until the study was completed and myocardial blood flow had been measured with microspheres. The stability of coronary stenosis was monitored by the steadiness of wall thickening impairment during the isoproterenol infusions.

During control conditions, myocardial blood flow increased uniformly to match the increased demand for blood flow associated with isoproterenol administration, which roughly doubled heart rate and peak positive dP/dt for the group as a whole. Global hemodynamics were similar during isoproterenol infusion with the level of coronary stenosis we identified as critical, but mean transmural blood flow in the ischemic region (measured with microspheres) was not significantly different from the baseline level $(0.82 \pm 0.16 \, \text{ml/min/g}$, mean $\pm$ SD, compared with $0.67 \pm 0.13 \, \text{ml/min/g}$, p = NS) and regional segment shortening and wall thickening were reduced dramatically. Examples of recorded tracings from an experiment similar to the ones undertaken as part of the study described by Gallagher et al. [13] are shown in Fig. 1. Because the dose of isoproterenol used in the example (Fig. 1) was somewhat lower than that used by Gallagher et al. [13], the hemodynamic changes are not as striking, but the wall thickness tracings demonstrate the effects of augmenting myocardial oxygen requirements when coronary inflow is restricted. Baseline recordings, shown on the left, illustrate the usual patterns of systolic wall thickening measured with sonomicrometers. The slow paper speed recordings in the center of the figure illustrate the rapidity of functional impairment that develops when coronary inflow is limited to baseline levels in this setting. For the group as a whole [13], percentage wall thickening in the flow-restricted area supplied by the stenosed circumflex artery was $9.2 \pm 3.8\%$, approximately a 75% reduction from systolic thickening measured during isoproterenol infusion without coronary stenosis $(39.6 \pm 11.9\%)$.

The transmural distribution of myocardial blood flow was altered in the experiments but the classical pattern of transmural steal was not apparent. Subepicardial blood flow increased significantly from a baseline value of $0.63 \pm 0.20 \, \text{ml/min/g}$ to $1.15 \pm 0.18 \, \text{ml/min/g}$ during isoproterenol infusion with coronary stenosis. To qualify as a transmural-steal situation, subendocardial blood flow should have decreased when subepicardial flow increased. Subendocardial perfusion, however, did not change significantly $(0.69 \pm 0.11$ to $0.61 \pm 0.09 \, \text{ml/min/g}$, p = NS). Although the data confirmed that restricted coronary inflow is unevenly distributed when myocardial oxygen demand is elevated, and it demonstrated that restriction of blood flow had serious effects on regional myocardial function, the results did not provide a clearcut answer regarding the functional consequences of transmural steal. Subendocardial blood flow did not increase to meet the augmented myocardial oxygen requirements produced by isoproterenol administration, but neither did it appear to be "stolen".

Accordingly, another study was conducted in chronically instrumented dogs with critical coronary stenosis [14]. In this group of animals, myocardial oxygen demand was increased by having the dogs run on a treadmill. The dogs were instrumented at sterile surgery a few weeks

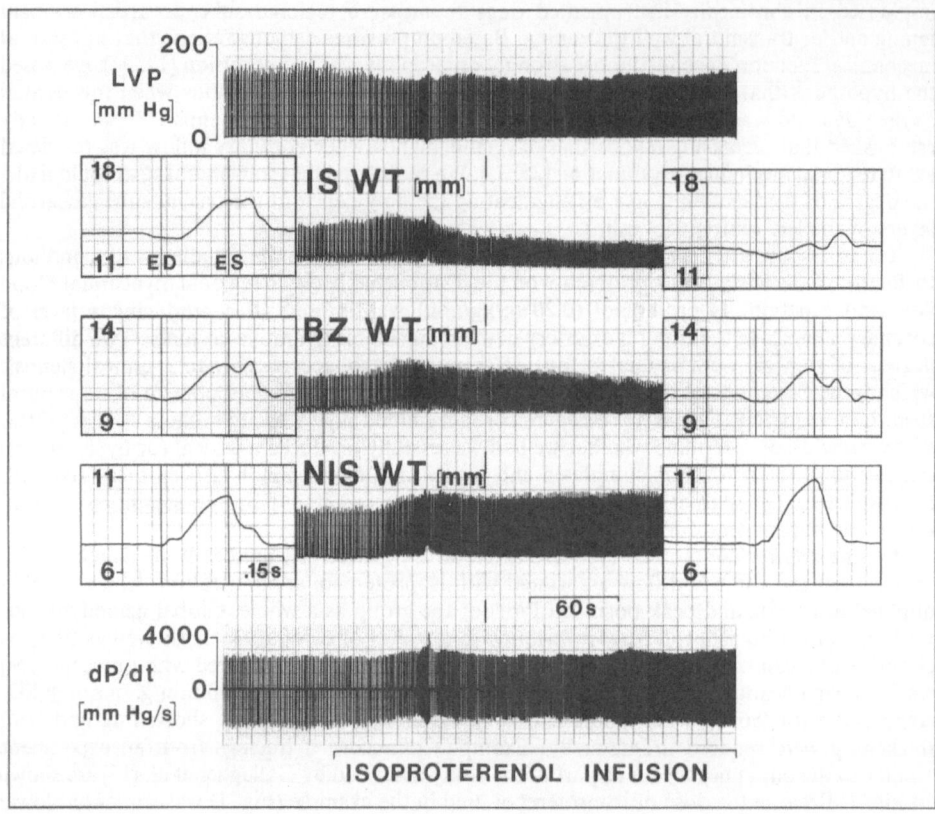

Fig. 1. Examples of analog tracings from a chronically instrumented dog with critical coronary stenosis of the circumflex artery. The critical stenosis was produced with an ameroid occluder that was implanted at sterile surgery approximately 2 weeks before the experiment shown here was performed. Isoproterenol (0.10 µg/kg/min) was infused intravenously to augment myocardial oxygen requirements similar to the study described by Gallagher et al. [13]. The global effects of the infusion are evident in the left ventricular pressure (LVP) and dP/dt tracings in the top and bottom rows, respectively. Wall-thickening in the ischemic area (IS WT) supplied by the ameroid-narrowed circumflex artery was substantially reduced, illustrating the effect of coronary flow restriction when myocardial oxygen requirements are elevated. Wall-thickening in the nonischemic area (NIS WT) increased. Wall-thickening in the border zone region at the margin of the ischemic area (BZ WT) was characterized by slight attenuation of systolic excursion.

before the experiments to enable measurement of regional wall thickening with sonomicrometers and determination of regional blood flow with microspheres. Flowprobes had not been implanted in the dogs who underwent isoproterenol infusions so there was no on-line monitor of stenosis severity available and this may have contributed to the confusing picture of transmural steal observed in the first study. It is possible that what we identified as critical stenosis in the isoproterenol study [13] may not have been so in all of the experiments. By implanting flowprobes in the dogs who would be exercised, a more controlled evaluation of the functional effects of transmural steal was possible. By using exercise as a means to increase myocardial oxygen requirements, a more physiologic stress was imposed to induce transmural steal.

The dogs were trained to run on an enclosed treadmill. The experiments were carried out with the dogs running at 10–13 km/h on a 5% grade. Producing critical stenosis proved to be a difficult proposition given the criterion we established that coronary blood flow velocity could

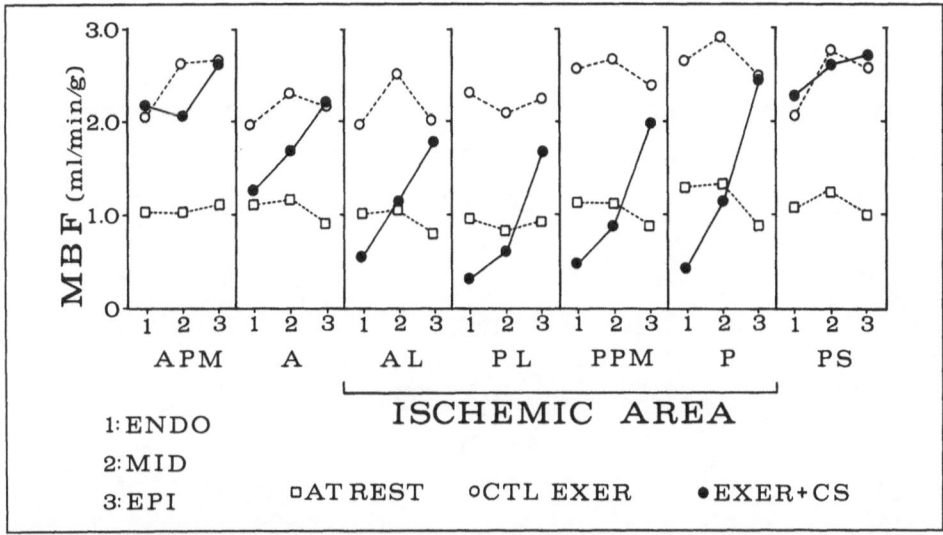

Fig. 2. Myocardial blood flow (MBF) data from one of the experiments in chronically instrumented dogs who underwent treadmill exercise while the circumflex artery was critically stenosed with an adjustable hydraulic occluder [14]. Data are shown from several contiguous transmural tissue samples spanning the ischemic area perfused by the circumflex artery. Blood flow was measured in three layers across each sample (1: ENDO, subendocardium; 2: MID, midmyocardium; 3: EPI, subepicardium) with microspheres. Data are shown during resting conditions (AT REST; dog standing on the treadmill), during control exercise without coronary stenosis (CTL EXER), and during exercise at the same speed and grade on the treadmill with critical coronary stenosis (EXER + CS). Blood flow was uniformly distributed in the samples at rest and during control exercise. During exercise plus critical stenosis, subendocardial blood flow decreased and subepicardial blood flow increased (compared with the baseline data obtained when the dog was at rest) in the ischemic area, providing a striking example of transmural coronary steal.

not deviate more than 25% from the baseline level after initiating exercise. It was frequently necessary to halt the study, reset the hydraulic occluder, and begin again after the dogs had recovered. In the control exercise experiments, heart rate increased from approximately 110 to 210 beats/min, left ventricular peak systolic pressure increased from an average of 132 to 169 mm Hg, and peak positive dP/dt increased from an average of 3629 to 6503 mm Hg/s. During exercise with critical stenosis, heart rates were similar but peak systolic pressure and peak positive dP/dt were significantly lower than during control exercise.

Blood flow data from one of the experiments are shown in Fig. 2 to demonstrate a particularly striking example of transmural flow redistribution during exercise with critical coronary stenosis. Blood flow in different tissue samples within and at the margins of the ischemic area are shown during baseline conditions, control exercise, and exercise with critical stenosis. Samples AL, PL, PPM, and P constituted the center of the ischemic area. During exercise plus critical stenosis, blood flow was substantially higher in the subepicardial than subendocardial layers. More importantly, subendocardial blood flow was lower than the baseline level of perfusion, supporting the contention that a transmural steal had developed secondary to subepicardial vasodilation and reduction in poststenotic perfusion pressure. For the group as a whole, subendocardial blood flow was significantly lower during exercise with critical stenosis (0.56 ± 0.26 ml/min/g) than during baseline conditions (1.11 ± 0.16/ml/min/g). Subepicardial blood flow increased during exercise with critical stenosis (1.84 ± 0.60 ml/min/g) compared with baseline (0.90 ± 0.12 ml/min/g) indicating that vascular resistance had decreased in the outer layers of the left ventricular wall. Consistent with the definition of transmural coronary steal, subendocardial flow had decreased and subepicardial flow had increased, appearing to do so at the expense of the subendocardium.

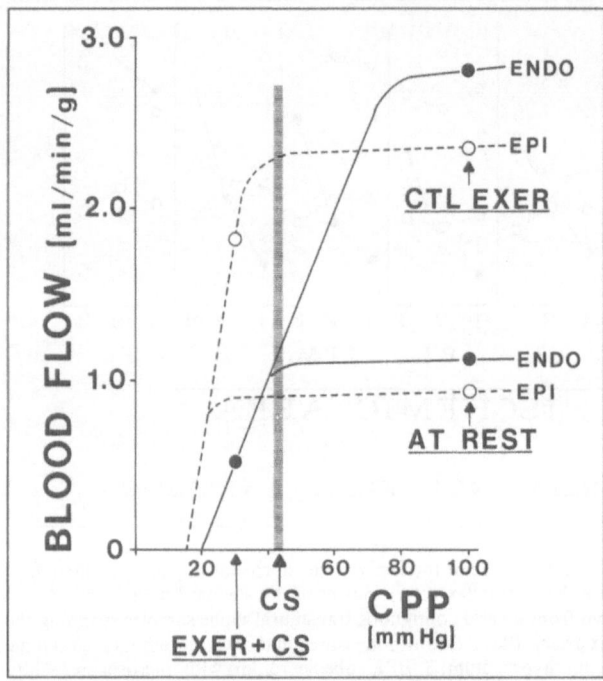

Fig. 3. Hypothetical subendocardial (ENDO) and subepicardial (EPI) pressure-flow curves to propose an explanation for the transmural steal observed in chronically instrumented dogs during exercise plus critical stenosis [14]. The values for coronary perfusion pressure (CPP) on the x-axis are hypothetical; the blood flow values shown with open and closed circles are average data from Gallagher et al. [14] for blood flow during conditions at rest (AT REST), during control exercise (CTL EXER), and exercise plus critical stenosis (EXER + CS). The pressure-flow curves connecting the points are speculative. The level of coronary perfusion pressure we propose corresponds to critical stenosis (CS) at rest is indicated with the cross-hatched line. During exercise plus critical stenosis, subendocardial blood flow decreased because perfusion pressure was "stolen" secondary to a reduction in subepicardial vascular resistance. Subepicardial blood flow was able to increase because it operates on a different pressure-flow curve than does the subendocardium.

We contend that subepicardial vasodilation led to a reduction of perfusion pressure distal to the stenosis and this, in turn, led to the decrease in subendocardial blood flow as shown in Fig. 3. Hypothetical autoregulatory pressure-flow curves are presented to suggest how the changes in myocardial blood flow distribution may have occurred. Superimposed on the curves are average subendocardial and subepicardial blood flow values from Gallagher et al. [14]. The solid lines correspond to subendocardial pressure-flow curves at rest (AT REST) and during conditions of augmented myocardial oxygen demand (CTL EXER). The dashed lines represent subepicardial pressure-flow curves. The plateaus of the curves indicate the amount of blood flow required for the level of work the heart was performing at rest (lower curves) or during treadmill exercise (upper curves). The inflexion points in each curve indicate the levels of coronary perfusion pressure (CPP) at which autoregulatory vasodilator reserve is exhausted. Consistent with previously reported data [10, 20, 23], the subepicardial inflexion points are at lower pressures than the subendocardial points.

We did not measure poststenotic pressure in these experiments, so the pressure levels shown in the graph (Fig. 3) are speculative. The values, however, are consistent with data reported by

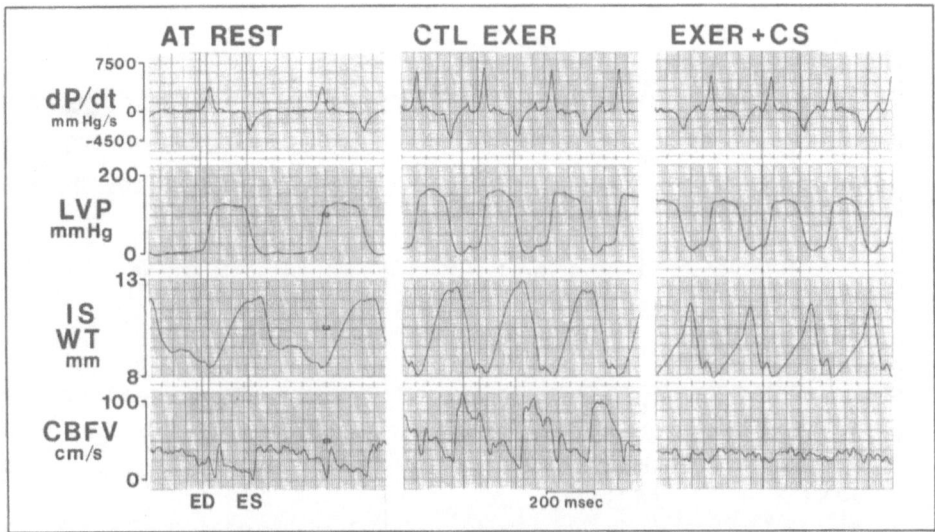

Fig. 4. Analog tracings of dP/dt, left ventricular pressure (LVP), ischemic wall thickness (IS WT), and coronary blood flow velocity (CBFV) from one of the experiments in chronically instrumented dogs undergoing treadmill exercise while the circumflex artery was critically stenosed with a hydraulic occluder [14]. Recorded tracings are shown while the dog stood on the treadmill before exercising (AT REST), during control exercise (CTL EXER) without coronary stenosis, and during exercise plus critical stenosis (EXER + CS). The main points to observe are the restriction of coronary blood flow velocity during exercise to the resting level by critical stenosis and the impairment of regional wall thickening during exercise plus critical stenosis.

other investigators who have measured poststenotic pressure in chronically instrumented dogs [2, 6, 17]. The pressure corresponding to critical stenosis (CS) at rest, for example, is similar to the value obtained by Canty [6] for the pressure level below which regional dysfunction develops in conscious, chronically instrumented dogs. We propose that imposition of critical stenosis at rest reduced poststenotic perfusion pressure to the point indicated by the cross-hatched line (labeled CS). Blood flow distribution would have remained normal and regional function would have been unaffected as long as myocardial oxygen requirements were unaltered.

During exercise with critical stenosis, however, regional dysfunction developed because coronary inflow was restricted far below the myocardial requirements for blood flow associated with exercise. The ischemia, we propose, was most pronounced in the subendocardial layers that were exposed not only to a restricted flow situation, but to a significant flow reduction from baseline because coronary perfusion pressure decreased secondary to subepicardial vasodilation. Subepicardial blood flow, operating on a different pressure-flow curve than the subendocardium, was capable of increasing. Subendocardial blood flow, already restricted to the lower portion of its pressure-flow curve by the critical stenosis, was reduced further. Decreased diastolic time associated with elevated heart rate and augmented left ventricular end-diastolic pressure may have contributed to the restriction of subendocardial blood flow, as well.

The functional consequences of restricted inflow during exercise are illustrated in Fig. 4. Examples of recorded tracings from one of the experiments are presented to demonstrate the striking change in regional systolic wall thickening during exercise with critical stenosis. During baseline conditions (standing on the treadmill) and control exercise, percentage wall thickening averaged 25 ± 6% and 32 ± 7%, respectively. During exercise with critical stenosis, percentage wall thickening was 9 ± 2%, a decrease of greater than 70% on the average compared with control exercise. Quite similar findings have been obtained in other studies in which dogs

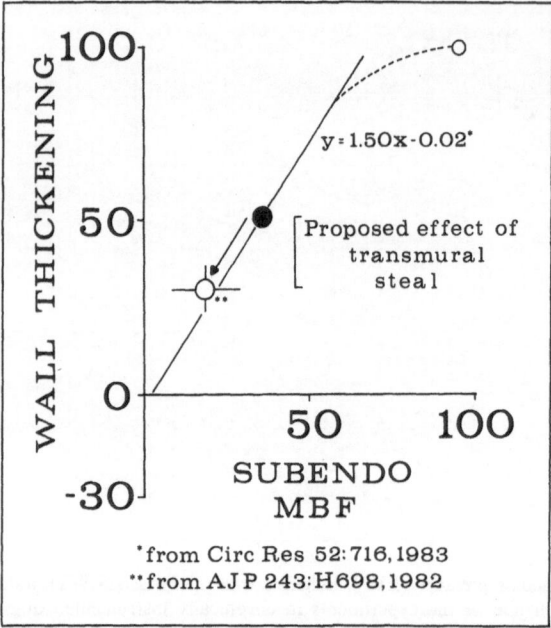

Fig. 5. Subendocardial flow-transmural function relationship to suggest how transmural steal may have aggravated regional dysfunction during exercise with critical stenosis. The regression line and equation were originally reported by Gallagher et al. [15]. Subendocardial blood flow (SUBENDO MBF) is presented as a percentage of blood flow in nonischemic myocardium; wall thickening is presented as a percentage of systolic thickening measured during control exercise without coronary stenosis. The open circle with standard deviation bars represents the average data from the critically stenosed dogs who underwent treadmill exercise [14]. The solid circle corresponds to the point on the regression line that would have characterized regional function if subendocardial blood flow had been restricted to the resting level during exercise with critical stenosis but had not been reduced further. We propose that the level of dysfunction we measured was worse because transmural steal occurred. When subepicardial resistance decreased in response to the positive inotropic and chronotropic changes associated with exercise, post-stenotic blood pressure decreased, driving subendocardial blood flow and wall thickening to a lower position on the flow-function regression line.

underwent treadmill exercise after the equivalent of critical stenosis had been produced with implanted ameroid occluders [19, 21, 26, 27].

The marked degree of regional dysfunction, we contend, was due not only to the restriction of total coronary inflow but to the maldistribution of blood flow. Subepicardial vasodilator reserve appeared to be available, accounting for the observation that subepicardial blood flow increased during exercise with critical stenosis (Fig. 3). Nonetheless, regional wall thickening was severely impaired, indicating that elevated subepicardial perfusion not only did little to sustain regional contractile function, but also suggesting that utilization of subepicardial vasodilator reserve had an adverse effect by producing a transmural steal. Had the transmural-steal effect not occurred, the reduction in subendocardial blood flow would have been less severe and the alteration in regional myocardial function would have been less dramatic. This proposal is schematically represented in Fig. 5, a graph relating changes in regional wall thickening to changes in subendocardial blood flow. The solid line corresponds to the linear equation describing this relationship for chronically instrumented dogs undergoing treadmill exercise with various levels of coronary stenosis which was reported by Gallagher et al. (see Fig. 5 of [15]). The average data for exercise with critical stenosis [14] are superimposed on the line with the open circle and standard deviation bars. Shown with the solid circle is the level of wall thickening that would have been observed had subendocardial blood flow not decreased due

152

to transmural steal. The reduction in wall thickening, we propose, would have been approximately 50% of the control exercise level, rather than the 70% reduction that was observed.

The fact that subepicardial blood flow increased raises another question that is difficult to answer. The analysis summarized above ascribes only detrimental effects to the changes in subepicardial perfusion. Had subepicardial blood flow failed to increase or even decreased during exercise with critical stenosis, it seems plausible that regional wall thickening might have been worse. It has been suggested that maintenance of subepicardial perfusion minimizes the impact of subendocardial ischemia by, at minimum, sustaining tension in the outer layers to avoid development of dyskinetic wall motion [12]. That elevation of subepicardial blood flow during exercise with critical stenosis exerted a similar salutary effect is a possibility that cannot be discounted. Given the dominant influence of subendocardial blood flow on transmural function, however, distinguishing a positive role for the subepicardium from the negative effects exerted by transmural steal secondary to subepicardial vasodilation will be a challenging proposition.

Recent studies have suggested that α-adrenergic vasoconstriction is most pronounced in the subepicardial layers, even with reduced coronary perfusion pressure [5, 29], leading to the conclusion that α-adrenergic vasoconstriction may exert an anti-transmural steal effect [7]. It has also been demonstrated that α-mediated subepicardial vasoconstriction helps maintain a normal distribution of blood flow across the ventricular wall during exercise with [8] or without [24] coronary inflow restriction due to experimental stenosis. It is possible, therefore, that in the absence of α-adrenergic effects, the degree of regional dysfunction we observed may have been worse. In Fig. 5, the position of the open symbol on the flow-function curve could represent the net result of both steal and anti-steal effects. The transmural steal effect would tend to drive subendocardial blood flow and transmural function downward on the curve, whereas the putative transmural anti-steal effect would tend to keep subendocardial blood flow higher than it would have been in the absence of subepicardial α-mediated vasoconstriction.

Although teleologically appealing, that there is an anti-transmural steal effect is not established with certainty. α-Adrenergic vasoconstriction in normal and ischemic myocardium is a controversial issue. Considerable evidence also exists indicating that α-adrenergic vasoconstriction is deleterious in ischemic myocardium [9, 22, 33], supporting the position that α-mediated effects aggravated rather than ameliorated the functional effects of transmural steal during exercise with coronary stenosis. Data obtained by Seitelberger et al. [33], for example, were obtained in chronically instrumented animals similar to those in which we evaluated the effects of critical stenosis during exercise [14]. Seitelberger et al. [33] showed that intracoronary administration of the $α_2$-blocker idazoxan (after propranolol administration) during treadmill exercise with flow-restricting coronary stenosis improved subendocardial perfusion and regional wall thickening substantially, indicating that α-adrenergic vasoconstriction exerts a deleterious effect. The extent to which the disparities in results can be attributed to differences in the specificity of various α-blocking and α-stimulating interventions remains a contentious issue [22]. Consequently, whether regional wall thickening is pushed down or pushed up the flow-function relation (Fig. 5) by α-adrenergic effects will require additional investigation.

References

1. Bach RJ, McHale PA, Greenfield JC (1977) Transmural myocardial perfusion during restricted coronary inflow in the awake dog. Am J Physiol 232:H645–H651
2. Ball RM, Bache RJ (1977) Transmural myocardial perfusion during restricted coronary inflow in the awake dog. Circ Res 38:60–66
3. Bache RJ, Tockman BA (1982) Effect of nitroglycerin and nifedipine on subendocardial perfusion in the presence of a flow-limiting coronary stenosis in the awake dog. Circ Res 50:678–687
4. Buffington CW (1986) Impaired systolic thickening associated with halothane in the presence of a coronary stenosis is mediated by changes in hemodynamics. Anesthesiology 64:632–640
5. Buffington CW, Feigl EO (1963) Effect of coronary artery pressure on transmural distribution of adrenergic coronary vasoconstriction in the dog. Circ Res 53:613–621

6. Canty JM Jr (1988) Coronary pressure-function and steady-state pressure-flow relations during autoregulation in the unanesthetized dog. Circ Res 63:821–836
7. Chiarello M, Ribeiro LGT, Davis MA, Maroko PR (1977) "Reverse coronary steal" induced by coronary vasoconstriction following coronary artery occlusion in dogs. Circulation 56:809–815
8. Chilian WM, Ackell PH (1988) Transmural differences in sympathetic coronary constriction during exercise in the presence of coronary stenosis. Circ Res 62:216–225
9. Deussen A, Heusch G, Thämer V (1985) Alpha-2 adrenoceptor-mediated coronary vasoconstriction persists after exhaustion of coronary dilator reserve. Eur J Pharmacol 115:147–153
10. Feigl EO (1983) Coronary physiology. Physiol Reviews 63:1–205
11. Gallagher KP, Folts JD, Shebuski RJ, Rankin JHG, Rowe GG (1980a) Subepicardial vasodilator reserve in the presence of critical coronary stenosis in dogs. Am J Cardiol 46:67–73
12. Gallagher KP, Kumada T, Koziol JA, McKown MD, Kemper WS, Ross J Jr (1980b) Significance of regional wall thickening abnormalities relative to transmural myocardial perfusion in anesthetized dogs. Circulation 62:1266–1274
13. Gallagher KP, Kumada T, Battler A, Kemper WS, Ross J Jr (1982) Isoproterenol-induced myocardial dysfunction in dogs with coronary stenosis. Am J Physiol 242:H260–H267
14. Gallagher KP, Osakada G, Matsuzaki M, Kemper WS, Ross J Jr (1982) Myocardial blood flow and function with critical coronary stenosis in exercising dogs. Am J Physiol 243:H698–H707
15. Gallagher KP, Matsuzaki M, Osakada G, Kemper WS, Ross J Jr (1983) Effect of exercise on the relationship between myocardial blood flow and systolic wall thickening in dogs with acute coronary stenosis. Circ Res 52:716–729
16. Gallagher KP, Stirling MC, Choy M, Szpunar CA, Gerren RA, Botham MJ, Lemmer JH (1985) Dissociation between epicardial and transmural function during acute myocardial ischemia. Circulation 71:1279–1291
17. Gould KL (1978) Pressure-flow characteristics of coronary stenoses in unsedated dogs at rest and during coronary vasodilation. Circ Res 43:242–253
18. Gross GJ, Warltier DC (1981) Coronary steal in four models of single or multiple vessel obstruction in dogs. Am J Cardiol 48:84–92
19. Guth BD, Heusch G, Seitelberger R, Ross J Jr (1987) Elimination of exercise-induced regional myocardial dysfunction by a bradycardic agent in dogs with chronic coronary artery stenosis. Circulation 75:661–669
20. Guyton RA, McClenathan JH, Newman GE, Michaelis LL (1977) Significance of subendocardial S–T segment elevation caused by coronary stenosis in the dog. Am J Cardiol 40:373–380
21. Heusch G, Guth BD, Seitelberger R, Ross J Jr (1987) Attenuation of exercise-induced myocardial ischemia in dogs with recruitment of coronary vasodilator reserve by nifedipine. Circulation 75:482–490
22. Heusch G (1990) Alpha-adrenergic mechanisms in myocardial ischemia. Circulation 81:1–13
23. Hoffman JIE (1984) Maximal coronary flow and the concept of coronary vascular reserve. Circulation 70:153–159
24. Huang AH, Feigl EO (1988) Adrenergic coronary vasoconstriction helps maintain uniform transmural blood flow distribution during exercise. Circ Res 62:286–298
25. Kumada T, Gallagher KP, Shirato K, McKown D, Miller M, Kemper WS, White F, Ross J Jr (1980) Reduction of exercise-induced regional myocardial dysfunction by propranolol: Studies in a canine model of chronic coronary artery stenosis. Circ Res 46:190–200
26. Matsuzaki M, Gallagher KP, Patritti J, Tajimi T, Kemper WS, Ross J Jr (1983) Effects of a calcium entry blocker (diltiazem) on regional myocardial flow and function during exercise in conscious dogs. Circulation 69:801–814
27. Matsuzaki M, Guth B, Tajimi T, Kemper WS, Ross J Jr (1984) Effect of the combination of diltiazem and atenolol on exercise-induced regional myocardial ischemia in conscious dogs. Circulation 72:233–243
28. Mosher P, Ross J Jr, McFate PA, Shaw RF (1964) Control of coronary blood flow by an autoregulatory mechanism. Circ Res 14:250–259
29. Nathan HJ, Feigl EO (1986) Adrenergic vasoconstriction lessens transmural steal during coronary hypoperfusion. Am J Physiol 250:H645–H653
30. Reivich M, Holling HE, Roberts BE (1961) Reversal of blood flow through the vertebral artery and its effect on the cerebral circulation. New Engl J Med 265:878–885
31. Roan PG, Buja M, Izquierdo C, Hoshimi H, Saffer S, Willerson JT (1981) Interrelationships between regional left ventricular function, coronary blood flow, and myocellular necrosis during the initial 24 hours and 1 week after experimental coronary occlusion in awake, unsedated dogs. Circ Res 49:31–40

32. Rowe GG (1970) Inequalities of myocardial perfusion in coronary artery disease ("coronary steal"). Circulation 42:193–194
33. Seitelberger R, Guth BD, Heusch G, Lee J-D, Katayama K, Ross J Jr (1988) Intracoronary alpha-2 adrenergic blockade attenuates ischemia in conscious dogs during exercise. Circ Res 62:436–442
34. Tatekawa S, Traber KB, Hantler CB, Tait AR, Gallagher KP, Knight PR (1987) Effects of isoflurane on myocardial blood flow, function, and oxygen consumption in the presence of critical coronary stenosis in dogs. Anesth Analg 66:1073–1082
35. Tomoike H, Franklin D, McKown D, Kemper WS, Guberek M, Ross J Jr (1978) Regional myocardial dysfunction and hemodynamic abnormalities during strenuous exercise in dogs with limited coronary flow. Circ Res 42:487–496
36. Wolf GL, Wilson WJ (1974) Vasodilator reserve, parallel vascular beds, and significant stenosis; a review for the angiographer. CRC Crit Rev Clin Radiol Nucl Med 5:1–41

Author's address:

Kim P. Gallagher, Ph.D., Thoracic Surgery Research Laboratory, B560 MSRB II, Box 0686, University of Michigan, Ann Arbor, MI 48109 USA.

Mechanisms of Benefit in the Ischemic Myocardium due to Heart Rate Reduction

B. D. Guth,[1] C. Indolfi,[2] G. Heusch,[1] R. Seitelberger[3] and J. Ross, Jr.[4]

Seaweed Canyon Laboratory, Department of Medicine, University of California, La Jolla, California, USA

Summary

The studies reviewed here examine the role of heart rate reduction in the beneficial effect observed following β-adrenoceptor blockade during exercise-induced ischemia in conscious dogs. To further study the effects of heart rate reduction on regional blood flow in an ischemic bed without collateral circulation, anesthetized swine with controlled coronary perfusion were also studied. Measurements of regional myocardial blood flow (microspheres) and contractile function (sonomicrometers) during steady state exercise in dogs with chronic coronary artery stenosis indicated the existence of severe regional contractile dysfunction and subendocardial ischemia. The administration of β-adrenoceptor blockade (1.0 mg/kg atenolol) improved regional contractile function when heart rate was reduced from 220 to 165 beats/min. Atrial pacing during exercise to prevent the bradycardia following β-adrenoceptor blockade eliminated the improved regional function and blood flow. Thus, the beneficial effect of β-blockade was only apparent when exercise heart rate was reduced. In anesthetized swine with constant inflow coronary perfusion, two levels of coronary hypoperfusion were examined at heart rates of 91 beats/min or 55 beats/min. Bradycardia was produced using the bradycardic agent UL-FS 49 (0.3 mg/kg). Regional contractile function and subendocardial blood flow were markedly improved at the lower heart rate for either level of reduced coronary perfusion, indicating a redistribution of blood flow towards the subendocardium. The improvement in contractile function was larger than predicted on the basis of the improvement in blood flow per min to the subendocardium. Independent relationships between regional contractile function and the subendocardial blood flow per min were observed for each heart rate. Thus, the studies in conscious exercising dogs indicated that heart rate reduction is an essential mechanism for the improvement of ischemic regional myocardial contractile function during exercise by β-blockade. This is likely the result of the marked improvement in subendocardial blood flow per beat which accompanies the reduced heart rate; regional myocardial blood flow per beat appears to be a predictor of regional contractile function during ischemia both at rest and during exercise.

Introduction

The predominant effect of sympathetic activation on the heart is an increase in heart rate and contractility, both being mediated through the activation of β-adrenoceptors in the myocar-

---

Dr. Guth is the recipient of a research award from the Alexander von Humboldt-Stiftung, Bonn, FRG.

Current addresses: [1]Abteilung für Pathophysiologie, Zentrum für Innere Medizin der Universität Essen, FRG; [2]Clinica Medica, Cattedra di Cardiologia, Facolta di Medicina, Napoli, Italy; [3]II. Chirurgische Universitätsklinik, University of Vienna, Austria; [4]Division of Cardiology, Department of Medicine, University of California, San Diego, School of Medicine, La Jolla, California, USA.

This study was supported by NIH grant HL-17682, Ischemic Heart Disease Specialized Center of Research (SCOR).

dium. It has been recognized for some time that such sympathetic effects and the resultant increase in myocardial oxygen requirements can further compromise ischemic myocardium [1, 3, 6, 7, 14, 16, 19, 20]. Consequently, β-adrenoceptor blocking drugs have been first-line agents for the treatment of patients with ischemic heart disease [2, 5, 17, 23, 31].

The beneficial effects of cardioselective β-adrenoceptor blockade during exercise-induced ischemia were demonstrated in dogs with a chronic coronary artery stenosis by Matsuzaki et al. [22]. Improved regional myocardial function in the ischemic myocardium was associated with an increase in subendocardial blood flow and an improvement of the endocardial to epicardial blood flow ratio. Exercise with β-blockade was characterized by reductions of heart rate, arterial pressure, and contractile state of the nonischemic myocardium. From this study it was not clear how much of the improvement observed could be attributed to the reduction of exercise heart rate and how much to the reduction of exercise contractile state or afterload reduction. Of these possible mechanisms, reducing the frequency of myocardial contraction has several advantages for the ischemic myocardium. First, a reduced heart rate provides increased diastolic duration; blood flow to the subendocardium occurs predominantly during diastole due to the compression of the coronary vasculature during systole. Since the contractile function of the left ventricle is closely related to blood flow to the subendocardium [9, 10, 29], the increased diastolic duration per se may result in enhanced blood flow to the ischemic subendocardium and, hence, increased contractile function. Secondly, the ischemic subendocardium is subjected to coronary "steal" due to the metabolic vasodilation of adjacent (lateral and transmural) non-ischemic myocardium [8, 30]. Reduction of heart rate will reduce the oxygen requirements of these adjacent areas, lessen metabolic vasodilation, and thus could potentially reduce any coronary "steal" [14, 18]. Heart rate reduction, therefore, could allow for both an increase in absolute blood flow per minute to the ischemic perfusion bed, as well as provide an improvement in the distribution of blood within the ischemic vascular bed. Finally, the oxygen requirement of the ischemic myocardium itself may be reduced by the reduced heart rate. Thus, as far as ischemia is a mismatch between oxygen supply (reduced) and oxygen requirements (enhanced during exercise), then such an imbalance would be lessened by the reduced heart rate if the oxygen requirements of the ischemic subendocardium are reduced due to fewer contractions per minute and the reduced contractile state, despite the finding that the contraction of each beat is improved after β-blockade [22].

The purpose of these studies described was to determine the role of heart rate reduction in the beneficial effect of β-adrenergic blockade during exercise-induced ischemia. The results of these studies indicated that not only was heart rate reduction a mechanism of benefit during β-blockade, but β-blockade without its usual heart rate reduction did not benefit ischemic blood flow or contractile function [12]. Due to the apparent benefit of heart rate reduction on the ischemic myocardium during exercise, further studies were performed in an anesthetized swine model. Carefully controlled studies investigated the effects of heart rate on the distribution of blood flow within the ischemic myocardium in a model in which coronary collaterals play no role [25]. These studies demonstrated that the relationship between regional subendocardial blood flow per minute and contractile function is heart-rate dependent. Thus, the amount of blood flow per beat measured in the acutely ischemic subendocardium could be closely related to the level of regional contractile function [18].

Methods

Studies on exercise-induced ischemia

The role of heart rate reduction in the beneficial effect of β-adrenoceptor blockade was assessed in a canine model of single vessel chronic coronary artery stenosis. This model has been used for testing the efficacy of numerous pharmacologic agents for treating exercise-induced ischemia [13, 21, 22]. Briefly, dogs were chronically instrumented for the assessment of regional myocardial contractile function using sonomicrometers measuring wall thickness in the potentially ischemic region and a control area. Ventricular hemodynamics were measured using a high-fidelity pressure transducer implanted in the left ventricle. Catheters were implanted in the aorta and left atrium for the microsphere method of measuring regional myocardial blood flow [4]. Electrodes were attached to the left atrium for pacing the heart. The left circumflex coronary artery was fitted with a Doppler flow probe, an ameroid constrictor, and a pneumatic occlusive cuff. The ameroid constrictor was designed to produce gradual constriction of the artery over a period of 2–3 weeks. The dogs were allowed at least 2 weeks of recovery prior to any studies. Studies were conducted when the ameroid constrictor had produced a coronary stenosis sufficient to produce regional contractile dysfunction during a brief test run.

Dogs were trained prior to surgery to run on a motor-driven treadmill with minimal restraint at an external workload sufficient to produce a steady-state heart rate of over 200 beats/min. Experimental runs lasted for approximately 6 min so that the measurement of regional blood flow with microspheres could be performed in the steady-state following 4 min of running. The amount of regional ischemia induced, as well as the resultant contractile dysfunction, have been shown to be highly reproducible in runs conducted 3 h apart [21]. Therefore, in this study two identical runs were performed 3 h apart and myocardial blood flow and function were measured. Fifteen min prior to the second run, dogs received a cardioselective β-adrenoceptor blocker (1.0 mg/kg atenolol) through the left atrial catheter. Two min into the run, after an initial steady-state was apparent, the heart was paced to a rate identical to that observed in the first, control run. Microspheres were injected during the subsequent steady-state to measure regional myocardial blood flow at the matched heart rate. Thus, the role of heart rate reduction was assessed by comparing two identical runs, one with and one without β-blockade, but at identical heart rates produced by atrial pacing [12].

Studies in anesthetized swine

Anesthetized swine were used to study the effect of heart rate on the relationship between blood flow and function in a model without coronary collaterals and in which coronary flow can be precisely controlled. Swine (30 kg) anesthetized with isoflurane (1–2% with oxygen) were instrumented with sonomicrometers measuring left ventricular wall thickness (control and ischemic zones) and atrial pacing wires. The left anterior descending coronary artery was cannulated and perfused with blood from the carotid artery. Coronary perfusion was controlled by a roller pump and coronary perfusion pressure was measured from a side arm on the cannula near the tip (characteristics of this system have been described in detail previously [26]).

In these studies [18], coronary blood flow to the left anterior descending artery was reduced to produce two different levels of regional contractile dysfunction; one was described as "moderate" ischemia (function equal to 18% of control) and the other as "severe" ischemia (mild dyskinesia). The effect of heart rate on the distribution of blood flow within the ischemic vascular bed, together with the contractile function, was studied first at a heart rate paced just above the intrinsic heart rate of the experimental preparation (91 beats/min), and then at a very low sinus heart rate produced by administration of the bradycardic agent UL-FS 49 (55 beats/min) (UL-FS 49 was a gift from Karl Thomae, Biberach/Riss, FRG). UL-FS 49 is a bradycardic agent which has been shown to have no effect on ventricular contractility independent from its effect on the heart rate [11, 18]. Thus, the effect of heart rate on a known amount of coronary blood flow could be assessed in terms of blood flow distribution across the wall and regional contractile function.

Results and discussion

Heart rate reduction during exercise-induced ischemia

The effect of β-blockade on regional contractile function during exercise-induced ischemia when heart rate was controlled is summarized in Fig. 1. During the first exercise bout, regional contractile function in the ischemic region was severely reduced from 27.8% to 7.2%, whereas

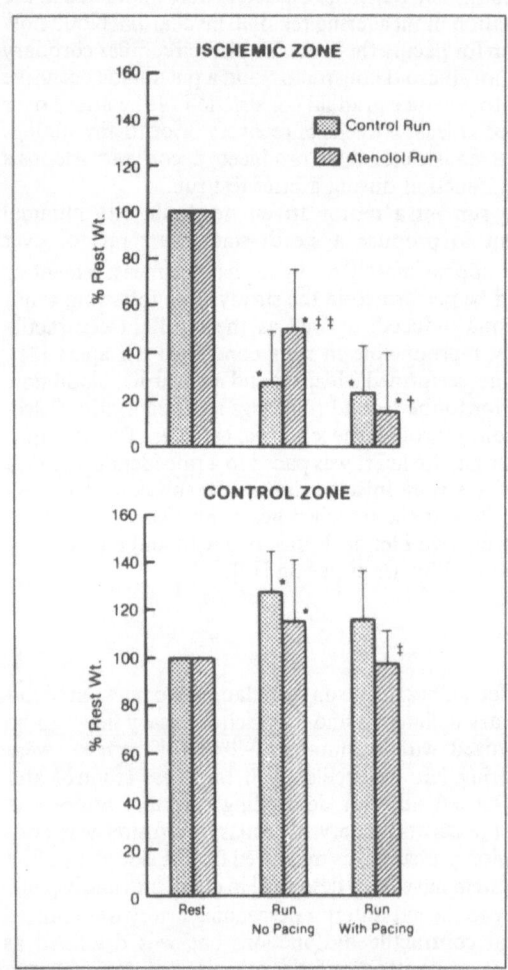

Fig. 1. Comparison of systolic wall thickening (expressed as percent of resting control value) in the posterior ischemic zone, and anterior control zone during the control run (stippled bars) and the run with atenolol (hatched bars). %WT in ischemic zone did not decrease during exercise before pacing as much as during the control run, verifying the beneficial effect of β-blockade when heart rate is decreased. With atrial pacing, %WT decreased significantly so that %WT tended to be less than that observed in the control run at the matched heart rate. * = p < 0.05 vs rest; † = p < 0.05 vs run no pacing; ‡ = p < 0.05, ‡‡ = p < 0.01 vs control run. (Reproduced from [12] with the permission of the American Heart Association.)

function increased in the control region (32.4% to 37.4%). Heart rate increased to 218 beats/min during this run and dP/dt increased by 61%. During the initial portion of the second run with β-adrenoceptor blockade, heart rate was reduced from 218 to 166 beats/min and dP/dt was unchanged from the resting, pre-run condition. Regional contractile function, expressed as the percent change in wall thickness during systole, was modestly improved at the reduced heart rate (11.4% vs 8.3%). With subsequent atrial pacing during this same run, heart rate was successfully matched to the first run without β-blockade, but left ventricular dP/dt was still markedly attenuated (3,642 vs 4,730 mm Hg/s). Contractile function in the ischemic region fell with atrial pacing and was consistently less than that observed in the first run (4.1% vs 7.2%).

Heart rate reduction during ischemia

During the run with β-blockade, the contractile function of the non-ischemic wall was also reduced compared to the control run (26.0% vs 37.4%).

During the control ischemic run, blood flow to the control subendocardium increased by nearly two-fold [1.14 vs 2.18 (ml/min)/g]. During the second run with β-blockade and matched heart rate, there was no significant change in regional myocardial blood flow to the control region, but subendocardial blood flow tended to be reduced compared to the control run [1.92 (ml/min)/g].

Blood flow to the ischemic region of the heart was markedly changed during exercise compared to the resting condition. Subepicardial blood flow increased significantly [0.88 vs 1.96 (ml/min)/g], whereas subendocardial blood flow actually decreased from 1.07 to 0.72 (ml/min)/g. During the run with β-blockade and atrial pacing to match heart rate, subepicardial blood flow was unchanged but subendocardial blood flow tended to be further decreased compared to the control run [0.56 (ml/min)/g]. Thus, when the bradycardia of β-blockade was prevented by atrial pacing, no beneficial effect of β-blockade on either regional myocardial blood flow or function was apparent [12].

Beneficial mechanisms of β-blockade during exercise

Due to the complexity of the experimental model used, the studies in the exercising dog with regional myocardial ischemia do not lend themselves to a precise definition of mechanisms. Instead, these studies provide an opportunity to study an integrated and complex hemodynamic response, such as exercise, in a clinically relevant situation. The results above indicate the importance of heart rate reduction to the beneficial effect of β-blockade during exercise-induced ischemia. The effect of β-blockade on regional ischemic blood flow and function, when heart rate reduction was prevented, was no longer beneficial in terms of either the perfusion of the ischemic region or its contractile function. It is possible that β-blockade without heart rate reduction unmasks α-adrenoceptor mediated vasoconstriction which has been shown to worsen ischemia during exercise in dogs [27].

The possible mechanisms involved in the improvement in regional contractile function and blood flow when heart rate is reduced can only be speculated from the results of these studies. Of importance is likely the prolonged diastolic interval at reduced exercise heart rates which permits increased subendocardial blood flow per min to the ischemic region, as observed by Matsuzaki et al. [22]. When no heart rate reduction was permitted following β-blockade, no improvement in either subendocardial or transmural blood flow to the ischemic region was observed. Secondly, the reduced number of contractions per min and the negative inotropic effect in the non-ischemic myocardium could lessen any coronary "steal". Coronary "steal" occurs when metabolic vasodilation of one region of a perfusion bed causes a decrease in perfusion pressure which decreases blood flow to another region of the bed that is pressure-dependent. Thus, reducing the number of contractions per min should reduce the work of the outer layer of the myocardium and thereby lessen the metabolic vasodilation there. A reversal of transmural "steal" following β-blockade was reported by Matsuzaki et al. when heart rate was reduced [22], whereas no reversal of coronary "steal" was observed when heart rate was paced to prevent the β-blockade-induced heart rate reduction [12]. These mechanisms may contribute to an increase in subendocardial perfusion which, in turn, can result in improved regional contractile function.

What about the potential benefit of reducing the oxygen requirements of the ischemic myocardium through heart rate reduction and the negative inotropic effect of β-adrenoceptor blockade? This potential mechanism of benefit may not actually apply to the acutely ischemic myocardium, as recently proposed by Ross [24]. In the exercising dog with ischemia, contractile function (and presumably oxygen consumption) of an ischemic wall is enhanced with the onset of exercise [22]. However, the contractile function s oon begins to decrease in the ischemic region as a consequence of the blood flow deficit; a proportional decrease in the oxygen consumption of the ischemic wall would be expected assuming little change in the myocardial oxygen extraction. Thus, the oxygen requirement of an acutely ischemic and dysfunctional wall is reduced due to the limited coronary inflow. Ross proposed that the regional myocardial function is reduced in proportion to the blood flow deficit since it can consume only the amount of oxygen it receives, and the ability to increase oxygen extraction is limited. He described this situation as a steady-state "perfusion-contraction matching" since the regional contractile function is reduced during steady-state ischemia to "match" the blood flow available [24]. This has interesting implications for studies involving heart rate reduction and β-blockade. It could be reasonably assumed that β-blockade reduced oxygen requirements of the myocardium during exercise due to heart rate reduction and a reduced contractility. However, in the ischemic region the opposite effect on oxygen consumption per beat likely occurs. Thus, if myocardial blood flow to the ischemic region is increased (e.g., β-blockade with reduced heart rate), the oxygen consumption per beat must increase, as indicated by the increased contractile function. In this situation the matching between blood flow and function in the steady-state is preserved, but at a higher level of function made possible by the increase in blood flow, particularly when expressed as subendocardial flow per beat [24].

Although β-adrenoceptor blockade may result in enhanced subendocardial blood flow (provided bradycardia occurs), the β-blockade may limit the ability of the ischemic myocardium to respond to increases in blood flow with increased contractile function. This may be a distinct disadvantage for patients in which ischemia impairs left ventricular pump function. If the benefit of reduced heart rate can be achieved without the blockade of β-adrenoceptors, a highly favorable effect might be expected. This was shown to be the case in a study in which the bradycardic agent UL-FS 49 was used in the above-described model of exercise-induced ischemia in which a more pronounced increase in contractile function of the ischemic region was achieved without depression of normal regions during exercise [11].

The chronically instrumented dog having single-vessel coronary stenosis has a highly collateralized coronary vasculature [25, 28]. Therefore, no differentiation can be made between changes in antegrade coronary inflow through the stenotic artery and changes in collateral blood flow. Thus, it was not known to what extent the increased subendocardial flow in the ischemic region was due to an overall increase in blood flow to the perfusion bed (through collaterals or through the stenotic artery) or merely to a more favorable redistribution of the available blood flow towards the subendocardium. Since arterial pressure was lower after β-blockade, the pressure gradient for collateral perfusion was not likely increased. However, it was suggested that both mechanisms were possible [22]. To determine the extent of transmural redistribution caused by bradycardia during ischemia with constant coronary inflow, the studies in the anesthetized swine model were performed.

Studies on the role of heart rate reduction in swine

An original tracing from one study in which heart rate was reduced from 90 to 52 beats/min is presented in Fig. 2. In the group of swine in which moderate ischemia was produced, coronary blood flow in the left anterior descending artery was reduced from 64 ml/min to 19 ml/min and regional wall thickening during systole fell from 25% to 6% when heart rate was 91 beats/min. After a complete return to control conditions, heart rate was reduced to 54 beats/min with UL-FS 49. An identical hypoperfusion was then performed, but this time regional function was not changed from the control measurement despite the reduction in blood flow to 19 ml/min (Fig. 2). Since left ventricular systolic pressure was also reduced at this lower heart rate and might, in part, explain the remarkable improvement in regional contractile function, a subset of animals was studied in which aortic constriction was performed to enable a comparison of regional function and blood flow at matched systolic loads. In this set of studies, regional wall thickening fell from 33% to 6% in the control ischemia but to only 23% in the second ischemia at the reduced heart rate with matched systolic pressures. Thus, only a small portion of the profound increase in regional function could be attributed to the reduction in systolic pressure [18].

A comparison of regional myocardial blood flow at the same coronary inflow but at different heart rates indicated substantial changes in the transmural distribution of blood flow

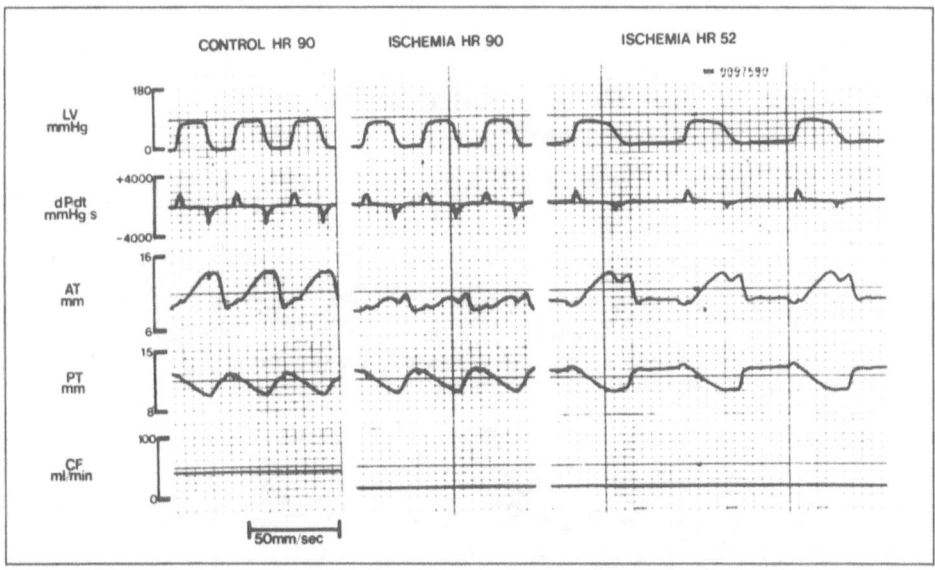

Fig. 2. Original recordings illustrating hemodynamic and regional dimension changes during ischemia without and with heart rate reduction. Tracings show left ventricular pressure (LV), wall thickness in the ischemic anterior myocardium (AT), segmental shortening in the posterior control area (PT), and mean coronary flow (CF). During constant reduced coronary inflow, bradycardia induced a marked improvement in the regional ischemic function. (Reproduced from [18] with the permission of the American Heart Association.)

in the ischemic perfusion territory. During the first hypoperfusion period with a heart rate of 91 beats/min, subendocardial blood flow was 0.24 (ml/min)/g, whereas during the hypoperfusion at a heart rate of 54 beats/min, subendocardial blood flow was 0.42 (ml/min)/g; endo/epi ratio improved from 0.55 to 0.86. Thus, the reduced heart rate significantly enhanced regional subendocardial blood flow in the ischemic region. Similar effects were observed during the "severe" ischemia with an improvement of subendocardial blood flow from 0.14 to 0.21 (ml/min)/g and endo/epi ratio from 0.46 to 0.75. In this series of studies with "severe" ischemia, regional contractile function during hypoperfusion was improved from − 5% (systolic wall thinning) to 13%; wall thickening was 24% and 25% during the two respective control periods.

During ischemia, the regional contractile function has been shown to be directly related to the amount of regional myocardial blood flow (transmural or subendocardial) [9, 10, 29]. In these studies in anesthetized swine, plotting of the regional wall function data (%WT) vs the subendocardial blood flow per minute [(ml/min)/g] indicated that separate relationships occurred, depending on the heart rate (Fig. 3). Thus, when blood flow vs function was plotted for swine with a heart rate of 55 beats/min and compared to a group of swine in which heart rate was 122 beats/min [26], two distinct relationships could be seen (Fig. 3). However, when blood flow to the ischemic subendocardium was calculated on the basis of flow per beat (i.e., ([ml/min]/g)/heart rate), the data could be well described by a single function [18]. These results support the idea of a matching between subendocardial blood flow per beat and regional contraction during various degrees of ischemia and are similar to previous such analyses done in the exercising dog subjected to acute regional ischemia [24].

The effects of heart rate slowing on the hemodynamics of the left ventricle having regional myocardial ischemia are not all favorable for the alleviation of ischemia. During each hypoperfusion, there was an increase in the left ventricular end-diastolic pressure. During ischemia at the reduced heart rate, end-diastolic left ventricular pressure was even further increased.

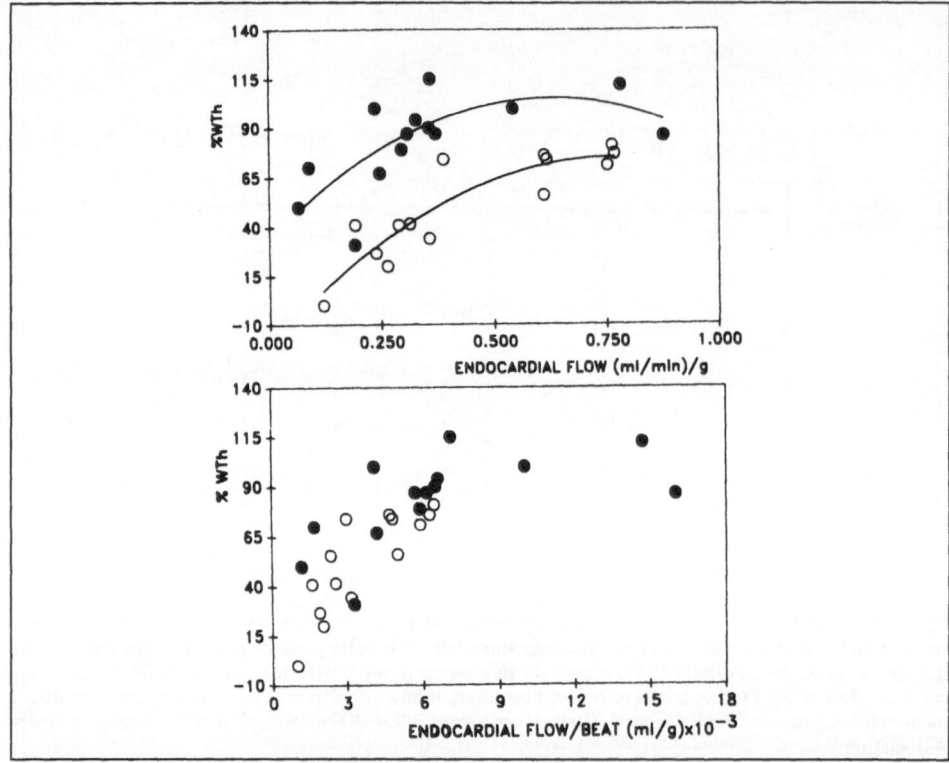

Fig. 3. Top panel: Plot of relation between subendocardial blood flow per minute and percent wall thickening (%WTh, % of control) compared at heart rates of 122 beats/min (open circles) and 55 beats/min (closed circles). An increase in subendocardial blood flow per minute was not sufficient to explain the striking increase in function, and independent relationships between blood flow per minute and function were found for each heart rate. Lines are second-order polynomial regressions for each group of data. Bottom panel: When the subendocardial blood flow was normalized for heart rate (endocardial flow per beat) data from heart rate of 122 beats/min (open circles) and from heart rate of 55 beats/min (closed circles) can be described by a single relation. Thus, the subendocardial blood flow per beat predicts wall function independently of heart rate. (Reproduced from [18] with the permission of the American Heart Association).

Elevated end-diastolic pressure would be expected to further compromise perfusion of the ischemic subendocardium since it is a primary determinant of diastolic wall stress. Despite the elevated end-diastolic pressure, both subendocardial blood flow and function were remarkably improved during the bradycardia. A similar observation was made when heart rate was reduced using UL-FS 49 in exercising dogs with regional ischemia despite extremely high end-diastolic left ventricular pressure [11]. Thus, the overall effect of reduced heart rate on the ischemic subendocardium and regional contractile function is beneficial despite the possibility that the elevated end-diastolic left ventricular pressure per se may not be advantageous.

The studies in anesthetized swine provide strong evidence that in a perfusion bed without coronary collaterals, reduction of heart rate results in a markedly favorable redistribution of blood flow within the ischemic bed that is associated with enhanced contractile function.

Conclusions

The role of heart rate reduction for mediating the beneficial effect of β-adrenoceptor blockade in exercising dogs was examined and found not only to be an important component of the benefit, but in addition without heart rate reduction, no beneficial effect could be demonstrated. Although the reduction of inotropic state by β-blockade may reduce coronary steal in the ischemic perfusion bed by reducing the oxygen demands of non-ischemic myocardium, this effect appears to be of minor significance compared to the heart rate reduction. Studies in anesthetized swine with controlled coronary perfusion indicate that a favorable redistribution of blood flow to the ischemic subendocardium occurs with heart rate reduction. Moreover, the subendocardial blood flow per beat appears to be a primary determinant of the contractile function of the ischemic myocardium during hypoperfusion.

Reduction of heart rate by means other than β-adrenoceptor blockade may have advantages in certain clinical settings, such as exercise-induced ischemia. In conditions where it may be advisable to maintain the inotropic state of the myocardium, bradycardia could still benefit the ischemic myocardium. It must be acknowledged that even with β-adrenoceptors intact, bradycardia may still cause a decrease in arterial pressure through a negative force-frequency effect such that additional support of systemic pressure might be required in some settings, such as during acute ischemia at rest.

References

1. Becker L (1976) Effect of tachycardia on left ventricular blood flow distribution during coronary occlusion. Am J Physiol 230:1072-1077
2. Boyle DMcC, Barber JM, McIlmoyle EL, Evans AE, Cran G, Elwood JH, Shanks RG (1983) Effect of very early intervention with metoprolol on myocardial infarct size. Br Heart J 49:229-233
3. Buffington CW (1985) Hemodynamic determinants of ischemic myocardial dysfunction in the presence of coronary stenosis in dogs. Anesthesiology 63:651-662
4. Domenech RJ, Hoffman JIE, Noble MIM, Saunders KB, Henson JR, Subijanto S (1969) Total and regional coronary blood flow measured by radioactive microspheres in conscious and anesthetized dogs. Circ Res 25:581-596
5. Epstein SE, Braunwald E (1966) Beta-adrenergic receptor blocking drugs. Mechanisms of action and clinical applications. New Engl J Med 275:1106-1112
6. Fedor JM, Rembert JC, McIntosh DM, Greenfield Jr JC (1980) Effects of exercise- and pacing-induced tachycardia on coronary collateral flow in the awake dog. Circ Res 46:214-220
7. Forrester JS, Helfant RH, Pasternac A, Most AS, Kemp HG, Gorlin R (1971) Atrial pacing in coronary heart disease. Effect on hemodynamics, metabolism and coronary circulation. Am J Cardiol 27:237-243
8. Gallagher KP, Folts JD, Shebuski RJ, Rankin JHG, Rowe GG (1980) Subepicardial vasodilator reserve in the presence of critical coronary stenosis in dogs. Am J Cardiol 46:67-73
9. Gallagher KP, Kumada T, Koziol JA, McKown MD, Kemper WS, Ross Jr J (1980) Significance of regional wall thickening abnormalities relative to transmural myocardial perfusion in anesthetized dogs. Circulation 62:1266-1274
10. Gallagher KP, Matsuzaki M, Koziol JA, Kemper WS, Ross Jr J (1984) Regional myocardial perfusion and wall thickening during ischemia in conscious dogs. Am J Physiol 247:H727-H738
11. Guth BD, Heusch G, Seitelberger R, Ross Jr J (1987) Elimination of exercise-induced regional myocardial dysfunction by a bradycardic agent in dogs with chronic coronary stenosis. Circulation 75:661-669
12. Guth BD, Heusch G, Seitelberger R, Ross Jr J (1987) Mechanism of beneficial effect of beta-adrenergic blockade on exercise-induced myocardial ischemia in conscious dogs. Circ Res 60:738-746
13. Guth BD, Tajimi T, Seitelberger R, Lee JD, Matsuzaki M, Ross Jr J (1986) Experimental exercise-induced ischemia: Drug therapy can eliminate regional dysfunction and oxygen supply-demand imbalance. J Am Coll Cardiol 7:1036-1046
14. Heusch G, Yoshimoto N (1983) Effects of heart rate and perfusion pressure on segmental coronary resistances and collateral perfusion. Pfluegers Arch 397:284-289
15. Heusch G, Yoshimoto N, Heegemann H, Thämer V (1983) Interaction of methoxamine with compensatory vasodilation distal to coronary stenoses. Drug Res 33:1647-1650
16. Heusch G, Yoshimoto N, Müller-Ruchholtz ER (1982) Effects of heart rate on hemodynamic severity of coronary artery stenosis in the dog. Basic Res Cardiol 77:562-573
17. Hjalmarson A (1984) The Göteborg metoprolol trial in acute myocardial infarction. Am J Cardiol 53 Suppl.: 1D-2D

18. Indolfi C, Guth BD, Miura T, Miyazaki S, Schulz R, Ross Jr J (1989) Mechanisms of improved ischemic regional dysfunction by bradycardia. Studies on UL-FS 49 in swine. Circulation 80:983–993
19. Lederman SN, Wenger TL, Harrell FE, Strauss HC (1987) Effects of different paced heart rates on canine coronary occlusion and reperfusion arrhythmias. Am Heart J 113:1365–1369
20. Maseri A, L'Abbate A, Pesola A, Michelassi C, Marzilli M, de Nes M (1977) Regional myocardial perfusion in patients with atherosclerotic coronary artery disease, at rest and during angina pectoris induced by tachycardia. Circ Res 55:423–433
21. Matsuzaki M, Gallagher KP, Patritti J, Tajimi T, Kemper WS, White FC, Ross Jr J (1984) Effects of a calcium-entry blocker (diltiazem) on regional myocardial flow and function during exercise in conscious dogs. Circulation 69:801–814
22. Matsuzaki M, Patritti J, Tajimi T, Miller M, Kemper WS, Ross Jr J (1984) Effects of β-blockade on regional myocardial flow and function during exercise. Am J Physiol 247:H52–H60
23. MIAMI Trial Research Group (1985) General discussion. Am J Cardiol 56:55G–57G
24. Ross Jr J (1989) Mechanisms of regional ischemia and antianginal drug action during exercise. Prog Cardiovasc Dis 31:455–466
25. Schaper W, Schaper J, Xhonneux R, Vandesteene R (1969) The morphology of intercoronary anastomoses in chronic coronary artery occlusion. Cardiovasc Res 3:315–323
26. Schulz R, Miyazaki S, Miller M, Thaulow E, Heusch G, Ross Jr J, Guth BD (1989) Consequences of regional inotropic stimulation of ischemic myocardium on regional myocardial blood flow and function in anesthetized swine. Circ Res 64:1116–1126
27. Seitelberger R, Guth BD, Heusch G, Lee JD, Katayama K, Ross Jr J (1988) Intracoronary alpha 2-adrenergic receptor blockade attenuates ischemia in conscious dogs during exercise. Circ Res 62:436–442
28. Tomoike H, Franklin D, Kemper WS, McKown D, Ross Jr J (1981) Functional evaluation of coronary collateral development in conscious dogs. Am J Physiol 241:H519–H524
29. Vatner SF (1980) Correlation between acute reductions in myocardial blood flow and function in conscious dogs. Circ Res 47:201–207
30. Wilcken DEL, Paoloni HJ, Eikens E (1971) Evidence for intravenous dipyridamole (Persantin) producing a "coronary steal" effect in the ischemic myocardium. Aust N Z J Med 1:8–14
31. Yusuf S, Sleight P, Rossi P, Ramsdale D, Peto R, Furze L, Sterry H, Pearson M, Motwani R, Parish S, Gray R, Bennett D, Bray C (1983) Reduction in infarct size, arrhythmias and chest pain by early intravenous beta blockade in suspected acute myocardial infarction. Circulation 67 suppl I:I32–I41

Authors' address:

Brian D. Guth, Ph.D., Abteilung für Pathophysiologie, Zentrum für Innere Medizin der Universität Essen, Hufelandstraße 55, D-4300 Essen, FRG.

Adrenergic Control of Transmural Coronary Blood Flow

E. O. Feigl

Department of Physiology and Biophysics, University of Washington, Seattle, USA

Summary

Tachycardia and an increase in myocardial metabolism result from the sympathetic activation that occurs during baroreceptor reflexes, emotion, and exercise. Paradoxically, a concomitant adrenergic α-receptor-mediated coronary vasoconstriction competes with the local metabolic coronary vasodilation that occurs during these conditions, and thereby limits metabolic hyperemia. Measurements of transmural blood flow in α-receptor blocked and α-receptor intact regions of the left ventricle during exercise demonstrate that adrenergic vasoconstriction helps maintain blood flow to the vulnerable subendocardium during tachycardia. This may be the explanation as to why paradoxical adrenergic coronary vasoconstriction has evolved. During controlled conditions of constant coronary flow, an anti-transmural steal effect due to adrenergic vasoconstriction in the subepicardium can be demonstrated during ischemic conditions. These observations demonstrate unexpected beneficial effects of adrenergic coronary vasoconstriction during tachycardia and cardiovascular stress.

Introduction

The focus of this symposium is the unexpected finding of Heusch and Deussen [24] that the imposition of an epicardial coronary artery stenosis *potentiates* adrenergic coronary vasoconstriction elicited by electrical stimulation of efferent sympathetic nerves to the heart. These authors emphasize that the adrenergic vasoconstriction during partial ischemia is primarily mediated via α_2- rather than α_1-receptors [12, 24]. This α_2-receptor-mediated vasoconstriction distal to a coronary artery stenosis has also been demonstrated during ischemia produced by exercise in Ross's laboratory [44]. Chilian and Ackell [9] and Bache's group [32] have also found evidence for sympathetic coronary vasoconstriction in ischemic myocardium distal to a stenosis during exercise, although potentiation of an α_2-receptor effect was not observed [32].

Myocardial ischemia is a powerful stimulus to coronary vasodilation, and prior to Heusch and Deussen's [24] report it was assumed that ischemic vasodilation would overwhelm adrenergic coronary vasoconstriction when the ischemia became severe enough. The Heusch and Deussen finding that ischemia potentiates, rather than attenuates, adrenergic coronary constriction is an unexpected puzzle in coronary physiology. A related unexpected finding is that coronary vasodilator reserve is not exhausted during ischemic hypoperfusion at a perfusion pressure of 40 mm Hg or less and that pharmacological dilation with adenosine increases coronary flow under these conditions [1, 8, 42, 48].

Heusch [25] has extended the observation of adrenergic coronary vasoconstriction during ischemia to a hypothesis on the reflex genesis of myocardial infarction, utilizing Malliani and co-workers' [36] demonstration of a sympathosympathetic reflex originating from and returning to the heart. Heusch suggests that myocardial ischemia excites sympathetic afferent fibers

Supported by NIH grant HL 16910.

ventricular wall that synapse in the spinal cord with sympathetic efferent fibers that return to the heart and activate coronary α_2-receptors to cause constriction that will worsen the ischemia. This is a regenerative, positive loop, reflex mechanism, whereby moderate ischemia will result in ever increasing ischemia. In addition, the potentiation of adrenergic vasoconstriction by ischemia will add gain to such a system, so that one might expect moderate ischemia to spiral inevitably to infarction by this reflex mechanism. Whether such a mechanism operates during clinical myocardial infarction remains to be demonstrated.

Intriguing as these observations during ischemia are, there is an additional paradox that underlies these findings: What function does adrenergic coronary vasoconstriction serve in normal coronary physiology? Sympathetic activation during cardiovascular stress results in tachycardia, increased contractility, and thus, an augmented myocardial oxygen consumption. Therefore, a concomitant adrenergic coronary vasoconstriction that limits flow when oxygen delivery must increase seems paradoxical [16]. Although the *net* effect of sympathetic activation to the heart is almost always an increase in coronary blood flow due to the augmented metabolism, the adrenergic constrictor effect competes with the metabolic vasodilation and retards the functional vasodilation by about 20–30%, resulting in an increased oxygen extraction by the heart [37].

The cardiac inotropic and chronotropic effects of sympathetic activation are mediated via β-receptors and can be blunted by β-receptor blocking drugs. Therefore, it is possible to "unmask" α-adrenergic coronary constriction with β-blockade. Following β-blockade, the usual response to sympathetic activation is a net decrease in coronary blood flow, thus α-constriction is "unmasked," and may subsequently be prevented with an α-receptor blocking agent.

α-Receptor-mediated coronary vasoconstriction appears to be present whenever there is a generalized sympathetic activation in response to cardiovascular stress, as during exercise [2, 3, 9–11, 21, 22, 26, 27, 29, 32, 33, 40, 44, 46]. Adrenergic coronary vasoconstriction is part of the baroreceptor reflex sympathetic response to carotid sinus hypotension [13–15, 37, 43] and psychological stress [4]. A marked decrease in blood flow through a stenotic coronary artery, similar to the Heusch findings, has been observed in the post-anger state by Verrier and co-workers [47].

There is also evidence of adrenergic coronary constriction in humans, although the measurements are often indirect. Bicycle or handgrip exercise results in constriction of stenotic segments of coronary arteries [6, 7, 18]. Estimates of coronary blood flow with radioactive rubidium or coronary sinus thermodilution have demonstrated unexpected decreases in coronary blood flow in some atherosclerotic patients during handgrip exercise [34] or during generalized sympathetic activation in a cold pressor test [17, 30, 35, 38, 39].

Transmural flow during exercise

The Seattle laboratory has investigated the reason that paradoxical adrenergic coronary vasoconstriction is part of the generalized sympathetic activation that occurs during exercise. The hypothesis is that adrenergic vasoconstriction helps maintain a uniform transmural blood flow in the left ventricle during the tachycardia of exercise. The hemodynamic problem is that the tissue pressure developed during systole in the wall of the left ventricle compresses coronary vessels more in the inner than in the outer layers of the ventricle [23]. This has the effect of restricting flow to the inner layer of the ventricle during systole so that flow to the subendocardium occurs only during diastole. During tachycardia the diastolic proportion of the cardiac cycle declines, yet the myocardial oxygen consumption may increase many fold. The question is how a dramatic increase in blood flow per minute, as required by the increase in metabolism, can be sustained with less diastolic time per minute. The hypothesis is that a concomitant vasoconstrictor influence adapts the coronary vascular impedance during tachycardia so that flow to the vulnerable inner layer of the left ventricle is maintained.

Dogs were prepared with catheters in the left atrium, coronary sinus, and aorta for the measurement of regional blood flow with radioactive microspheres, myocardial oxygen con-

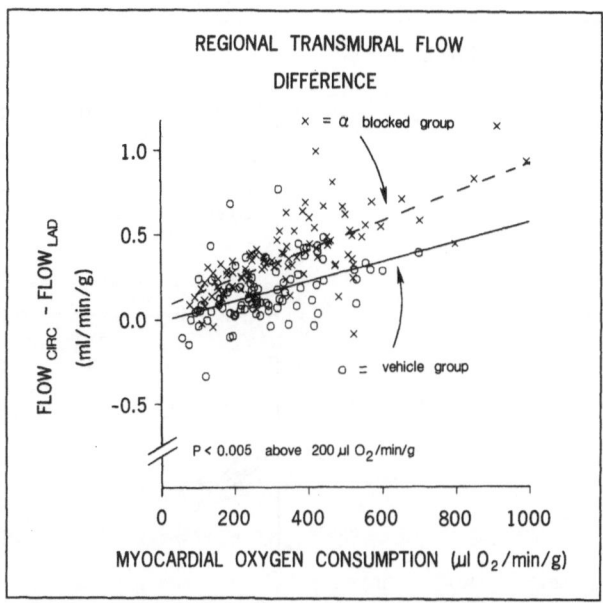

Fig. 1. Paired differences in regional transmural flow between the circumflex (α-blocked or vehicle-treated) region and the anterior descending (control) region in the α-blocked and vehicle groups. Lines show the mean differences calculated for each group by multiple linear regression, and individual points show the residual variance within dogs plotted around the mean regression lines. The slopes of the two regression lines were not significantly different from each other, but the regional difference in flow was significantly greater ($p < 0.005$) in the α-blocked group than in the vehicle group, when myocardial oxygen consumption was greater than $200 \, \mu l \, O_2/min/g$. (From [29].)

sumption, blood pressure, and heart rate. The left circumflex coronary artery vascular bed was selectively treated with saline vehicle or the combined α_1- and α_2-receptor blocking agent phenoxybenzamine. Paired comparisons were made between the untreated left anterior descending coronary artery bed and the treated (vehicle or α-blockade) circumflex bed during rest and graded treadmill exercise.

Average transmural blood flow tended to be greater in the circumflex region than in the anterior descending region, even when the treatment was only the control vehicle, suggesting a small normal regional difference. When the circumflex region was selectively α-blocked with phenoxybenzamine, a significant difference between blood flow in the two regions was observed (Fig. 1). Average transmural blood flow was greater with α-receptor blockade, as compared to vehicle control, when myocardial oxygen consumption exceeded $200 \, \mu l/min$ per g during exercise. This confirms the numerous studies where α-receptor coronary vasoconstriction was observed during exercise.

The inner/outer blood flow ratio in the free wall of the left ventricle was also changed by α-receptor blockade during exercise. The inner/outer flow ratio normally tends to fall with increasing levels of exercise and myocardial oxygen consumption, as illustrated in the untreated anterior descending region and vehicle-treated circumflex region data in Fig. 2. The inner/outer flow ratio is normally somewhat larger in the circumflex region than in the anterior descending region [49], probably because of different radii of curvature and stress distributions in the two regions. When α-receptors were selectively blocked in the circumflex region, the normal relationship was reversed and the inner/outer flow ratio was *less* than in the anterior descending region (Fig. 3). This indicates that coronary α-receptor activation helps maintain a more uniform transmural distribution of blood flow during exercise.

169

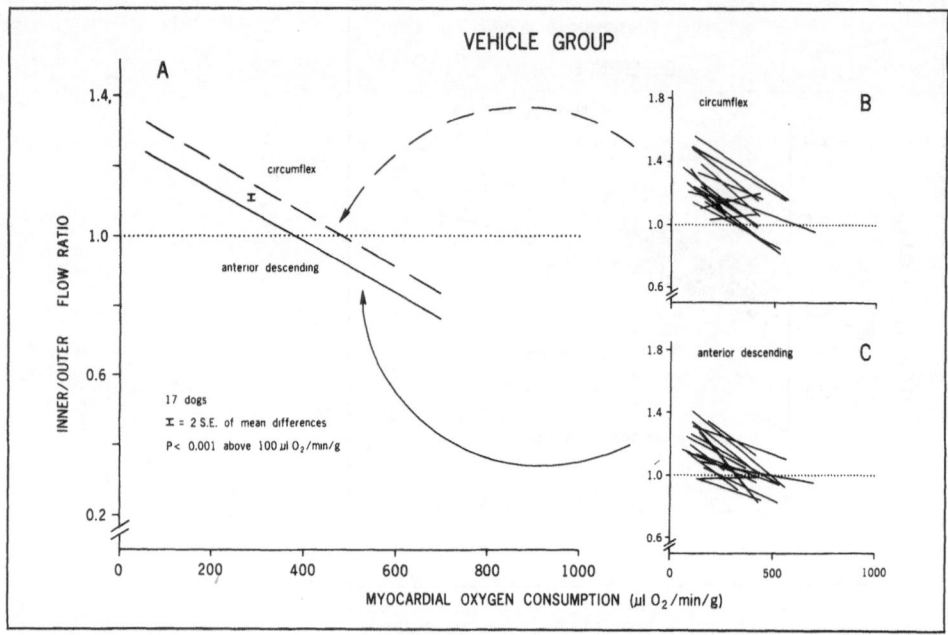

Fig. 2. Inner/outer flow ratio vs. myocardial oxygen consumption in the vehicle group. A: Lines summarize observations in the control (anterior descending) region and in the vehicle-treated (circumflex) region according to multiple linear regression analysis. The slopes of the two lines were not significantly different ($0.8 < p < 0.9$). The inner/outer ratio in the circumflex region was significantly higher than in the paired anterior descending region when myocardial oxygen consumption was $100\,\mu l\,O_2/min/g$ or more ($p < 0.001$). B and C: Simple regression lines calculated for the control (anterior descending) and vehicle-treated (circumflex) regions of each dog illustrate the variability among dogs. (From [29].)

Since the inner/outer ratio may be altered by flow changes in either the inner or outer layer of the left ventricle, an additional comparison was made. The paired differences in subendocardial flow in the untreated anterior descending region and treated circumflex region are given in Fig. 4. These data indicate that inner layer flow (ml/min per g) was better maintained at high levels of exercise with α-receptors intact (vehicle) than with α-receptor blockade. This higher inner layer flow with α-receptors intact was observed despite the prejunctional and postjunctional vasodilating effects of combined α_1- and α_2-receptor blockade increasing mean transmural flow (Fig. 1).

In summary, these data indicate that adrenergic coronary vasoconstriction helps maintain a uniform distribution of blood flow across the left ventricular wall during exercise. This observation may answer the paradox of why there is a relative adrenergic coronary vasoconstriction during exercise.

Transmural steal

A vascular steal occurs when perfusion pressure for a vascular bed (in which flow is pressure-dependent) is lowered by vasodilation in a parallel vascular bed, both beds being distal to a flow- and pressure-limiting stenosis. If the pressure distal to the stenosis at the branch point between the two vascular beds is below the range where autoregulation maintains flow relatively constant, then a drop in pressure will result in a large decrease in flow. The lower limit of autoregulation of blood flow in the outer layer of the left ventricle is about 40 mm Hg, but it is approximately 70 mm Hg in the inner layer of the left ventricle [20]. The conditions for a

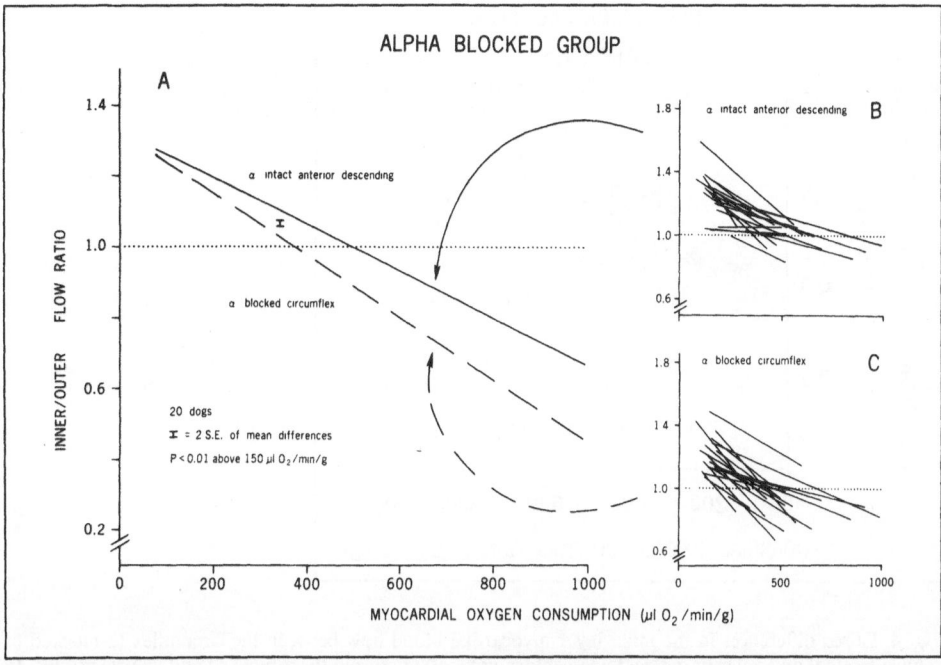

Fig. 3. Inner/outer flow ratio vs. myocardial oxygen consumption in the α-blocked group. A: Lines summarize observations in the α-receptor intact (anterior descending) region and in the α-receptor blocked (circumflex) region according to multiple linear regression analysis. The slopes of the two lines were different (p < 0.05). The inner/outer flow ratio of the α-blocked region was below the paired α-intact region when myocardial oxygen consumption was 150 μl O₂/min/g or more (p < 0.01). B and C: Simple regression lines calculated for the α-intact (anterior descending) and α-blocked (circumflex) regions of each dog illustrate the variability among dogs. (From [29].)

potential transmural steal are present when a stenosis in an epicardial coronary artery lowers the pressure at the branch point between the inner and outer layers below about 70 mm Hg. Under these conditions, vasodilation in the outer layer of the left ventricle will lower the perfusion pressure for the inner layer and thus "steal" flow from the subendocardium. The situation is exacerbated by the nonlinear pressure-flow characteristics of a stenosis. The pressure drop (ΔP) across a stenosis is an exponential function of the flow through the stenosis [5, 19]. At low resting flows the pressure drop across a stenosis may only be about 5 mm Hg. Therefore, with a normal aortic pressure of 90 mm Hg, the pressure distal to the stenosis is 85 mm Hg and is within the autoregulatory range for both the inner and outer layers of the left ventricle. A doubling of flow through the stenosis may increase the pressure drop to 20 mm Hg, which would lower the distal pressure to 70 mm Hg if the aortic pressure remains at 90 mm Hg. This is about the lower limit of autoregulation for the inner layer, and any further drop in pressure will result in a decrease in subendocardial blood flow. If, under these conditions, there is a vasodilation in the outer layer of the left ventricle, flow through the stenosis will increase with a further exponential exacerbation of the pressure drop across the stenosis. The fall in pressure distal to the stenosis will decrease flow in the pressure-dependent (little autoregulation) inner layer, thus vasodilation in the outer layer will "steal" flow from the inner layer. The key element is that vasodilation in the outer layer decreases perfusion pressure for the inner layer, where flow is pressure-dependent.

The hypothesis to be tested was if adrenergic vasoconstriction in the outer layer could preserve flow to the vulnerable inner layer of the left ventricle by an anti-steal mechanism [41].

171

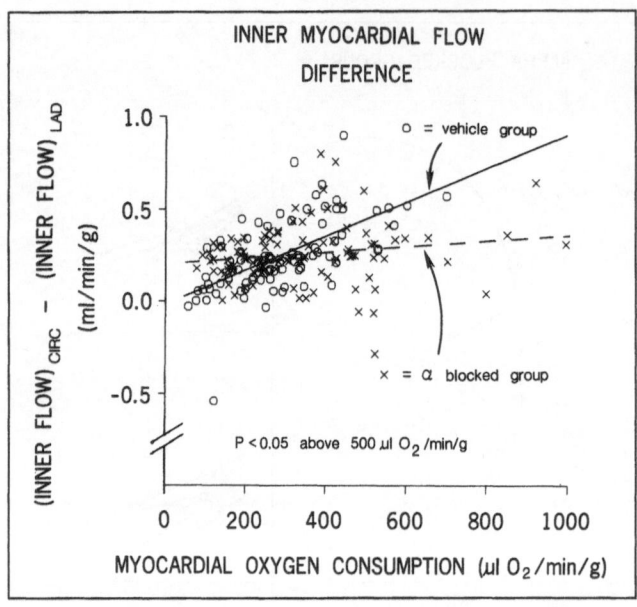

Fig. 4. Paired differences in the inner layer myocardial blood flow between the circumflex (α-blocked or vehicle-treated) and anterior descending (α-intact, control) regions in the α-blocked and vehicle groups. In each group, the anterior descending region served as the paired, α-intact, control region. Lines indicate the mean differences calculated for each group by multiple linear regression, and individual points show the residual variance within dogs, plotted around the regression lines. The slopes of the two regression lines were significantly different ($p < 0.01$) and the difference in the α-blocked group was significantly less than in the vehicle group when myocardial oxygen consumption was $500\,\mu l\,O_2/min/g$ or greater ($p < 0.05$). Thus, inner myocardial blood flow in the circumflex region was better maintained when α-receptors were intact (vehicle group) than when they were blocked (α-blocked group), despite the prejunctional and postjunctional vasodilating effects of α-blockade. (From [29].)

The left anterior descending and circumflex regions of the left ventricular free wall were compared. In seven experiments, the anterior descending vascular bed was regionally α_1- and α_2-blocked with phenoxybenzamine, and in seven experiments the circumflex vascular bed was regionally α-blocked. The two regions were separately pump-perfused at the same flow rate (ml/min per g of myocardium) with constant (controlled) flow pumps. Constant flow perfusion is hydraulically equivalent to a stenosis, since vasodilation results in a decrease in pressure and a potential steal. Technically, it is easier to match pumps than it is to match nonlinear stenosis characteristics in two regions. Propranolol was administered to blunt β-receptor-mediated effects, and α-receptor activation in both regions was achieved with an intracoronary infusion of norepinephrine. Microspheres were used to measure transmural blood flow as total coronary flow was reduced from 100% to 80, 70, 60, and 50% of normal in an equal manner in both regions. When total coronary flow was reduced sufficiently to cause maldistribution across the ventricular wall and ischemia, the inner layer blood flow was greater in the α-receptor intact region than in the α-blocked region (Fig. 5). The total transmural flow was equal in both regions by experimental design, but at ischemic perfusion levels the inner layer flow was greater with α-receptor-mediated vasoconstriction than without.

In summary, these data indicate that α-receptor-mediated coronary vasoconstriction in the outer layer of the left ventricle has an unexpected beneficial anti-transmural steal effect when the hemodynamic conditions for a transmural steal are present.

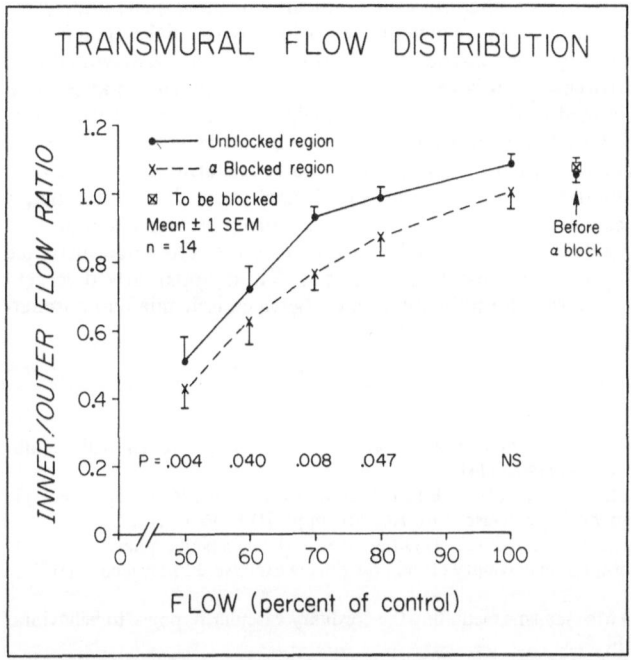

Fig. 5. Inner-to-outer myocardial blood flow ratios in control and phenoxybenzamine-treated regions of the left ventricle during intracoronary norepinephrine infusion in β-blocked dogs are shown. At normal flows both before and after (100%) administration of phenoxybenzamine there was no significant difference between the two regions. At all four levels of hypoperfusion (80%, 70%, 60%, and 50% of control flow) α-receptor activation prevented a redistribution of blood flow away from inner to outer layers of myocardium. Standard error bars indicate variability among animals, while p values are for two-tailed, paired *t*-test of the difference between α-blocked and unblocked regions. (From [41].)

Discussion

Numerous studies demonstrating α-receptor-mediated coronary vasoconstriction during exercise were cited in the introduction. The reason for this paradoxical adrenergic vasoconstriction may be explained by the exercise study of Huang and Feigl [29], described above, where α-vasoconstriction helps maintain a more uniform transmural blood flow during the tachycardia of exercise. The hemodynamic mechanism for this beneficial effect was not demonstrated by the experiment. Since there was no coronary artery stenosis in the exercise study, a simple anti-steal mechanism seems unlikely. It is possible that adrenergic vasoconstriction results in a decreased intramural vascular capacitance, such that systolic compression is lessened and diastolic reexpansion is facilitated. That is, the adrenergic action on coronary vascular smooth muscle may "tune" the coronary input impedance during tachycardia to lessen the wasted oscillatory retrograde and antegrade flow that occurs successively during systole and diastole [28, 31, 45].

Although the experimental conditions were not the same, the results of Nathan and Feigl [41] are not in agreement with the work of Heusch and coworkers. No evidence for potentiation of α-receptor-mediated vasoconstriction was observed during ischemic hypoperfusion; however, the experiment was not designed to test the Heusch findings. The Seitelberger et al. [44] observation of detrimental adrenergic coronary vasoconstriction distal to a stenosis during exercise has not been confirmed by similar experiments of Chilian and Ackell [9]. Bache's group [32] did not confirm Seitelberger et al.'s [44] finding of α_2-mediated detrimental coronary vasoconstriction distal to a stenosis during exercise, but did observe an unpotentiated α_1-

mediated constriction. These apparently conflicting results indicate our incomplete understanding of adrenergic coronary vasoconstriction – particularly during myocardial ischemia.

In conclusion, adrenergic coronary vasoconstriction is part of the generalized sympathetic activation that occurs during baroreceptor reflexes, emotion, and exercise. In the absence of coronary stenosis, this α-receptor-mediated vasoconstriction helps maintain a uniform transmural distribution of coronary blood flow during exercise tachycardia. This may be the explanation as to why this paradoxical vasoconstriction has evolved. During controlled conditions of constant coronary blood flow, an anti-transmural steal effect due to adrenergic vasoconstriction in the subepicardium can be demonstrated during ischemic conditions. However, there are conflicting results from different laboratories as to the action of α-mediated coronary vasoconstriction during ischemia distal to a stenosis. These apparently divergent results indicate the need for further research on the interaction between ischemia and adrenergic coronary vasoconstriction.

References

1. Aversano T, Becker LC (1985) Persistence of coronary vasodilator reserve despite functionally significant flow reduction. Am J Physiol 248:H403–H411
2. Bache RJ, Dai X-Z, Herzog CA, Schwartz JS (1987) Effects of nonselective and selective α_1-adrenergic blockade on coronary blood flow during exercise. Circ Res 61(Suppl. II):II-36–II-41
3. Bache RJ, Homans DC, Schwartz JS, Dai X-Z (1988) Differences in the effects of alpha-1 adrenergic blockade with prazosin and indoramin on coronary blood flow during exercise. J Pharmacol Exp Ther 245:232–237
4. Billman GE, Randall DC (1981) Mechanisms mediating the coronary vascular response to behavioral stress in the dog. Circ Res 48:214–223
5. Brown BG, Bolson EL, Dodge HT (1984) Dynamic mechanisms in human coronary stenosis. Circulation 70:917–922
6. Brown BG, Josephson MA, Petersen RB, Pierce CD, Wong M, Hecht HS, Bolson E, Dodge HT (1981) Intravenous dipyridamole combined with isometric handgrip for near maximal acute increase in coronary flow in patients with coronary artery disease. Am J Cardiol 48:1077–1085
7. Brown BG, Lee AB, Bolson EL, Dodge HT (1984) Reflex constriction of significant coronary stenosis as a mechanism contributing to ischemic left ventricular dysfunction during isometric exercise. Circulation 70:18–24
8. Canty JM, Klocke FJ (1985) Reduced regional myocardial perfusion in the presence of pharmacologic vasodilator reserve. Circulation 71:370–377
9. Chilian WM, Ackell PH (1988) Transmural differences in sympathetic coronary constriction during exercise in the presence of coronary stenosis. Circ Res 62:216–225
10. Chilian WM, Harrison DG, Haws CW, Snyder WD, Marcus ML (1986) Adrenergic coronary tone during submaximal exercise in the dog is produced by circulating catecholamines. Evidence for adrenergic denervation supersensitivity in the myocardium but not in coronary vessels. Circ Res 58:68–82
11. Dai X-Z, Herzog CA, Schwartz JS, Bache RJ (1986) Coronary blood flow during exercise following nonselective and selective α_1-adrenergic blockade with indoramin. J Cardiovasc Pharmacol 8:574–581
12. Deussen A, Heusch G, Thämer V (1985) α_2-Adrenoceptor-mediated coronary vasoconstriction persists after exhaustion of coronary dilator reserve. Eur J Pharmacol 115:147–153
13. DiSalvo J, Parker PE, Scott JB, Haddy FJ (1971) Carotid baroceptor influence on coronary vascular resistance in the anesthetized dog. Am J Physiol 221:156–160
14. Ely SW, Sawyer DC, Anderson DL, Scott JB (1981) Carotid sinus reflex vasoconstriction in right coronary circulation of dog and pig. Am J Physiol 241:H149–H154
15. Feigl EO (1968) Carotid sinus reflex control of coronary blood flow. Circ Res 23:223–237
16. Feigl EO (1987) The paradox of adrenergic coronary vasoconstriction. Circulation 76:737–745
17. Feldman RL, Whittle JL, Marx JD, Pepine CJ, Conti CR (1982) Regional coronary hemodynamic responses to cold stimulation in patients without variant angina. Am J Cardiol 49:665–673
18. Gage JE, Hess OM, Murakami T, Ritter M, Grimm J, Krayenbuehl HP (1986) Vasoconstriction of stenotic coronary arteries during dynamic exercise in patients with classic angina pectoris – reversibility by nitroglycerin. Circulation 73:865–876
19. Gould KL (1985) Quantification of coronary artery stenosis in vivo. Circ Res 57:341–353
20. Guyton RA, McClenathan JH, Newman GE, Michaelis LL (1977) Significance of subendocardial S–T

segment elevation caused by coronary stenosis in dog – epicardial S–T segment depression, local ischemia and subsequent necrosis. Am J Cardiol 40:373–380

21. Gwirtz PA, Overn SP, Mass HJ, Jones CE (1986) α_1-Adrenergic constriction limits coronary flow and cardiac function in running dogs. Am J Physiol 250:H1117–H1126

22. Gwirtz PA, Stone HL (1981) Coronary blood flow and myocardial oxygen consumption after *alpha* adrenergic blockade during submaximal exercise. J Pharmacol Exp Ther 217:92–98

23. Heineman FW, Grayson J (1985) Transmural distribution of intramyocardial pressure measured by micropipette technique. Am J Physiol 249:H1216–H1223

24. Heusch G, Deussen A (1983) The effects of cardiac sympathetic nerve stimulation on perfusion of stenotic coronary arteries in the dog. Circ Res 53:8–15

25. Heusch G, Deussen A, Thämer V (1985) Cardiac sympathetic nerve activity and progressive vasoconstriction distal to coronary stenoses: feed-back aggravation of myocardial ischemia. J Auton Nerv Syst 13:311–326

26. Heyndrickx GR, Muylaert P, Pannier JL (1982) α-Adrenergic control of oxygen delivery to myocardium during exercise in conscious dogs. Am J Physiol 242:H805–H809

27. Heyndrickx GR, Vilaine JP, Moerman EJ, Leusen I (1984) Role of prejunctional α_2-adrenergic receptors in the regulation of myocardial performance during exercise in conscious dogs. Circ Res 54:683–693

28. Hoffman JIE, Baer RW, Hanley FL, Messina LM (1985) Regulation of transmural myocardial blood flow. J Biomech Eng 107:2–9

29. Huang AH, Feigl EO (1988) Adrenergic coronary vasoconstriction helps maintain uniform transmural blood flow distribution during exercise. Circ Res 62:286–298

30. Kern MJ, Ganz P, Horowitz JD, Gaspar J, Barry WH, Lorell BH, Grossman W, Mudge GH (1983) Potentiation of coronary vasoconstriction by beta-adrenergic blockade in patients with coronary artery disease. Circulation 67:1178–1185

31. Kresh JY, Fox M, Brockman SK, Noordergraaf A (1990) Model-based analysis of transmural vessel impedance and myocardial circulation dynamics. Am J Physiol 258:H262–H276

32. Laxson DD, Dai X-Z, Homans DC, Bache RJ (1989) The role of α_1- and α_2-adrenergic receptors in mediation of coronary vasoconstriction in hypoperfused ischemic myocardium during exercise. Circ Res 65:1688–1697

33. Liang IYS, Stone HL (1982) Effect of exercise conditioning on coronary resistance. J Appl Physiol 53:631–636

34. Lowe DK, Rothbaum DA, McHenry PL, Corya BC, Knoebel SB (1975) Myocardial blood flow response to isometric (handgrip) and treadmill exercise in coronary artery disease. Circulation 51:126–131

35. Malacoff RF, Mudge GH, Holman L, Idoine J, Bifolck L, Cohn PF (1983) Effect of the cold pressor test on regional myocardial blood flow in patients with coronary artery disease. Am Heart J 106:78–84

36. Malliani A, Schwartz PJ, Zanchetti A (1969) A sympathetic reflex elicited by experimental coronary occlusion. Am J Physiol 217:703–709

37. Mohrman DE, Feigl EO (1978) Competition between sympathetic vasoconstriction and metabolic vasodilation in the canine coronary circulation. Circ Res 42:79–86

38. Mudge GH Jr, Goldberg S, Gunther S, Mann T, Grossman W (1979) Comparison of metabolic and vasoconstrictor stimuli on coronary vascular resistance in man. Circulation 59:544–550

39. Mudge GH Jr, Grossman W, Mills RM Jr, Lesch M, Braunwald E (1976) Reflex increase in coronary vascular resistance in patients with ischemic heart disease. N Engl J Med 295:1333–1337

40. Murray PA, Vatner SF (1979) α-Adrenoceptor attenuation of the coronary vascular response to severe exercise in the conscious dog. Circ Res 45:654–660

41. Nathan HJ, Feigl EO (1986) Adrenergic vasoconstriction lessens transmural steal during coronary hypoperfusion. Am J Physiol 250:H645–H653

42. Pantely GA, Bristow JD, Swenson LJ, Ladley HD, Johnson WB, Anselone CG (1985) Incomplete coronary vasodilation during myocardial ischemia in swine. Am J Physiol 249:H638–H647

43. Powell JR, Feigl EO (1979) Carotid sinus reflex coronary vasoconstriction during controlled myocardial oxygen metabolism in the dog. Circ Res 44:44–51

44. Seitelberger R, Guth BD, Heusch G, Lee J-D, Katayama K, Ross J Jr (1988) Intracoronary α_2-adrenergic receptor blockade attenuates ischemia in conscious dogs during exercise. Circ Res 62:436–442

45. Spaan JAE (1985) Coronary diastolic pressure-flow relation and zero flow pressure explained on the basis of intramyocardial compliance. Circ Res 56:293–309

46. Strader JR, Gwirtz PA, Jones CE (1988) Comparative effects of alpha-1 and alpha-2 adrenoceptors in modulation of coronary flow during exercise. J Pharmacol Exp Ther 246:772–778

47. Verrier RL, Hagestad EL, Lown B (1987) Delayed myocardial ischemia induced by anger. Circulation 75:249–254

48. Warltier DC, Gross GJ, Brooks HL (1981) Pharmacologic- vs. ischemia-induced coronary artery vasodilation. Am J Physiol 240:H767–H774
49. Wüsten B, Flameng W, Schaper W (1974) The distribution of myocardial flow. Part I: Effects of experimental coronary occlusion. Basic Res Cardiol 69:422–434

Author's address:

Eric O. Feigl, M.D., Dept. of Physiology & Biophysics SJ-40, University of Washington Medical School, Seattle, WA 98195, U.S.A.

α_1-Adrenergic Coronary Constriction during Exercise and Ischemia

C. E. Jones and P. A. Gwirtz*

Department of Physiology, Texas College of Osteopathic Medicine, Fort Worth, Texas, USA

Summary

This paper reviews work primarily from our laboratories, examining an α_1-adrenergic receptor-mediated coronary constriction during exercise and myocardial ischemia in dogs. It was demonstrated that in the quiescent conscious dog, the coronary circulation is devoid of an α_1-coronary constriction. Furthermore, it was shown by the intracoronary injection of selective agonists that both α_1- and α_2-receptor subtypes are present in coronary vessels. However, during exercise or ischemia only the selective α_1-antagonist prazosin caused an increase in coronary inflow, indicating that only α_1-receptors were activated. During both conditions, the increase in flow caused by α_1-blockade was associated with an increased contractile function in subendocardium. In experiments on anesthetized dogs, it was shown that prazosin caused an equal increase in perfusion of subepicardial and subendocardial layers during stellate ganglion stimulation. However, contractile function was increased only in subendocardium. It was proposed that only in deeper muscle layers does an α_1-coronary constriction impose a flow-limitation on contractile function. Finally, recent results indicate that myocardial ischemia, produced either by partial coronary stenosis or by maintenance of coronary inflow at the resting level during exercise, may initiate a vicious cycle with a further increase in α_1-adrenergic coronary constriction. Abolition of this positive feedback mechanism may partially explain the anti-infarction effects of chronic ventricular sympathectomy, as previously observed in our laboratories.

Introduction

True regulation of coronary blood flow in accordance with the nutrient need of the myocardium is vested in local mechanisms that are independent of external nervous or hormonal influences. The local mechanisms regulating coronary blood flow are tightly coupled to oxygen availability. Thus, the oxygen need of myocardium is well correlated with its external work load, and coronary blood flow is well correlated with this oxygen need. The local mechanisms regulating coronary blood flow are not completely defined, but a large body of evidence suggests that the local "metabolic signal" involves vasoactive metabolites, such as adenosine, released from myocardial myocytes in direct proportion to myocardial work or to the balance between myocardial oxygen supply and oxygen demand.

Many external factors are superimposed on local regulatory mechanisms and can substantially modulate coronary blood flow. One such modulatory influence is the sympathetic

Work reported from the authors' laboratories was supported by NIH Grant Numbers HL-29232 and HL-34172, as well as by grants from the American Heart Association and its Texas affiliate, the American Osteopathic Association, and the American Health Assistance Foundation.

nervous system. Thus, sympathetic nerves are closely associated with the large coronary conduit vessels as well as with the smaller coronary resistance vessels [12]. In addition, in all species of animals studied the coronary vascular wall contains adrenergic receptors and is responsive to adrenergic agonists. The direct response of the coronary circulation to adrenergic stimulation, whether through neurally released or circulating catecholamines, is one of vaso-constriction. During sympathetic stimulation of the heart, myocardial β-adrenergic receptor activation leads to an increased cardiac activity with a resulting increase in myocardial oxygen need. As a consequence, the coronary circulation undergoes vasodilation through local mechanisms. However, it has been clearly demonstrated that the direct adrenergic coronary constrictor effect may compete with and limit this local metabolic vasodilation [42].

There remains intense confusion regarding the receptor subtype which predominates in eliciting coronary constriction during various conditions of adrenergic stimulation. As a brief overview, some researchers using exogenous agonists have provided evidence for the existence of both subtypes in large coronary arteries and small resistance vessels [50]. However, others have reported that during cardiac sympathetic nerve stimulation following severe coronary artery stenosis, the α_2-subtype mediated a sustained coronary constriction in smaller downstream vessels [23]. Likewise, work from this same laboratory suggested that the α_1-subtype predominates in large coronary arteries, while the α_2-subtype predominates in resistance vessels [24]. Finally, at the other extreme, research by us and others has indicated that the α_1-subtype is primarily responsible for the observed coronary adrenergic constriction during various conditions such as exercise and ischemia [11, 20, 37, 39]. The distinction between the roles of α_1- and α_2-receptor subtypes in mediating adrenergic coronary constriction is of obvious physiological and pharmacological importance, but the results of a multitude of animal studies remain dilemmatic.

Many aspects of adrenergic mechanisms in the coronary vasculature are not at present understood, and many seemingly controversial reports currently exist. Indeed, the precise physiological function served by sympathetic modulation of myocardial perfusion is not clear at the present time. However, it does seem clear from current evidence that sympathetic vasoconstriction of the coronary vessels may, under certain conditions, impose a pathological impediment to normal coronary flow regulation. The present paper will review the data from our laboratories and others showing an important influence of sympathetic coronary vasoconstriction, especially that mediated by α_1-adrenergic receptor activation, during various pathological and physiological conditions. Emphasis will be placed on the influence of such vasoconstriction during sympathetic stimulation of the heart associated with exercise and myocardial ischemia. This paper is not intended to suggest that only the α_1-receptor subtype is involved in adrenergic coronary constriction under these conditions, but it is aimed at a review of results indicating a potentially important role of this particular receptor subtype.

α-Adrenergic coronary constriction at rest and during sympathetic stimulation of the heart

Evidence indicates that in the conscious, quiescent subject, sympathetic stimulation of the heart and coronary vasculature is low or nonexistent. Thus, Chilian et al. [9] used topically applied phenol to chemically sympathectomize a portion of the left ventricle, while innervation to the remainder of the ventricle remained intact. Regional left ventricular blood flow was measured while the dogs were in the conscious resting state. No differences in perfusion of the sympathectomized and innervated regions were observed, and the endocardial-to-epicardial blood flow ratio was not different between the two regions. We have observed that in conscious instrumented dogs trained to lie quietly, intracoronary administration of the selective α_1-adrenergic antagonist prazosin, at a dose which abolished the constrictor response to phenylephrine, caused no observable changes in coronary inflow [20]. Results of a typical experiment are shown in Fig. 1. It may be noted that not only was coronary inflow unaffected by α_1-adrenergic blockade in the resting dog, but there was no change in any of the measured variables. In view of the above evidence, we feel that in the unstressed, conscious subject the coronary circulation is without a significant adrenergic constriction. Although some inves-

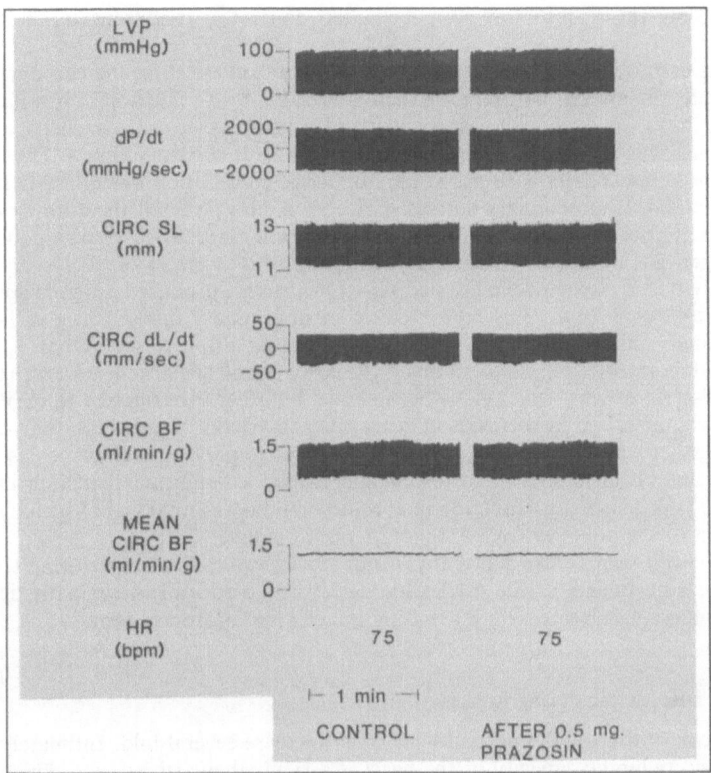

Fig. 1. Tracings from a resting, conscious dog before and immediately following intracoronary injection of the α_1-adrenergic antagonist prazosin. Note that at rest, prazosin had no effects on any recorded variable. Abbreviations: LVP = left ventricular pressure, dP/dt = rate of left ventricular pressure change, CIRC SL = segment length in the circumflex perfusion territory, CIRC dL/dt = rate of change of circumflex territory segment length, CIRC BF = circumflex blood flow, HR = heart rate. Data are from Gwirtz et al. [21].

tigators have reported an adrenergic coronary constriction at rest, we feel these contradictory results may be due to a certain degree of anxiety in the animals used in those studies. It has been reported that psychological stress may be associated with sympathetic stimulation of the heart and coronary circulation [4].

Even though there may be little sympathetic constriction of the coronary circulation at rest, a multitude of factors which cause increased sympathetic stimulation of the heart are associated with increased adrenergic coronary constriction. For example, Brachfield et al. [5] demonstrated that in the simple, anesthetized open-chest dog preparation, transecting the pericoronary nerves led to a 13% increase in coronary blood flow. That an adrenergic coronary constriction exists following anesthesia and opening the chest has also been demonstrated by us [32]. While examining the effects of adrenergic coronary vasoconstriction on coronary pressure-flow autoregulation in the open-chest, anesthetized dog, we noted that at a normal coronary perfusion pressure of 100 mm Hg, intracoronary administration of prazosin led to a significant increase in coronary inflow. Interestingly, we also noted that although both the α_1-agonist methoxamine and the α_2-agonist clonidine caused a transient reduction in coronary inflow, the α_2-antagonist yohimbine, unlike prazosin, did not lead to an increased coronary inflow. These results suggested that although both α-receptor subtypes were present in the

179

coronary vasculature, only the α_1-receptors were activated under the conditions of these experiments.

A variety of specific cardiovascular reflexes also have a significant effect on the coronary vasculature. For example, the arterial baroreflex, which is initiated by a reduction in arterial blood pressure and which is associated with vasoconstriction in many vascular beds, elicits an adrenergic coronary constriction [13]. The chemoreflex may also have coronary effects. Thus, during arterial hypoxemia, researchers have noted an adrenergic constriction much like that seen with the baroreflex [44]. In preliminary studies, Grice et al. [16] perfused the coronary circulation of dogs with normoxic blood while producing systemic hypoxemia by respiring the animals with low oxygen gas mixtures. Following β-adrenergic blockade to prevent reflex increases in cardiac activity and oxygen demand, peripheral hypoxemia produced a significant reduction in coronary inflow of 20%. This reflex reduction in coronary inflow was totally abolished by intracoronary administration of prazosin, indicating the involvement of α_1-receptors. Work in our laboratories has also examined the coronary effects of reflexes arising from disturbances of visceral organs, which have been shown by other investigators to elicit cardiovascular reflexes characterized by increases in heart rate and arterial blood pressure. We observed that in the anesthetized dog following cardiac β-blockade to prevent a reflex increase in cardiac activity, mechanical distention of the esophagus caused a significant reduction in coronary inflow of 18% [14]. This reflex reduction in coronary inflow was totally abolished by coronary administration of the nonselective α-antagonist phentolamine.

From the above review, it seems clear that many, if not most, perturbations which lead to generalized sympathetic adjustments of the cardiovascular system also lead to sympathetic stimulation of the coronary vasculature with a resulting increase in constrictor tone.

α_1-Adrenergic coronary constriction during exercise

During exercise, the work output of the normal heart may increase several-fold. Intimately involved in this increased pumping capacity of the heart is a sympathetic stimulation which increases both heart rate and the inotropic state of the myocardium. The increased work output of the heart causes increased oxygen and blood flow demands. Due to local metabolic influences on the coronary circulation, coronary resistance vessels undergo a substantial dilation, and coronary blood flow may increase by as much as five-fold or more during maximal exercise. Because of the substantial increase in coronary blood flow observed during exercise, and because coronary flow is closely correlated with myocardial oxygen consumption, it has long been felt that adrenergic constrictor influences on the coronary circulation during exercise are relatively unimportant and may be overridden by the more powerful local metabolic mechanisms. However, when it is realized that during exercise adrenergic constrictor influences may increase concomitantly with the local metabolic dilatory influences, it becomes easier to predict that in this condition an adrenergic coronary constrictor tone may still exist.

Effects of α_1-vasoconstriction on coronary flow, oxygen delivery, and myocardial contractile function

Results from several laboratories have indeed indicated that an α-adrenergic coronary constriction limits the increase in coronary blood flow during moderate to severe exercise. However, many of the earlier studies employed systemic injection of a nonselective α-antagonist. Murray and Vatner [45] used free-running dogs and noted that following systemic injection of phentolamine, late diastolic coronary resistance was reduced to a value of 30% less than that observed in the same dogs without α-blockade. Similarly, Gwirtz and Stone [21] injected systemic phentolamine while dogs were running on a motor-driven treadmill at a rate of 6.4 km/hr on a 16% grade. They observed that α-adrenergic blockade caused a 42% increase in coronary blood flow during exercise. Interestingly, they also observed an increase in the dP/dt max of the left ventricle following α-blockade. Heyndrickx et al. [26] also employed systemic phentolamine in dogs exercising on a treadmill. They also observed that α-blockade during exercise was associated with a significant increase in coronary blood flow, as well as a significant increase in left ventricular dP/dt max. Furthermore, Heyndrickx and coworkers measured myocardial oxygen delivery and consumption and observed that without α-blockade, there was a decreased myocardial oxygen delivery-to-oxygen consumption ratio. They suggested that myocardial oxygen demand increased out of proportion to oxygen supply during normal exercise. However, after α-blockade with phentolamine, they observed that the changes in oxygen delivery were more proportionate to the changes in oxygen consumption. They concluded that an α-adrenergic coronary constriction actually limits oxygen delivery to the myocardium during exercise in the normal dog.

While the above studies strongly imply that an α-adrenergic coronary constriction limits the increase in coronary blood flow and oxygen delivery during exercise, certain questions remained. First, systemic administration of adrenergic antagonists will not only affect the coronary circulation but will also affect other peripheral vascular beds. These effects may lead to circulatory reflexes which may secondarily affect the coronary circulation. Second, the use of nonselective α-antagonists does not provide insight into the α-receptor subtypes involved in the coronary constriction.

To overcome these difficulties, Gwirtz et al. [20] conducted experiments in which the selective α_1-antagonist prazosin was administered directly into a coronary artery during exercise in dogs. Coronary blood flow was measured with a Doppler flow probe implanted around the circumflex artery, and a small catheter was implanted in the circumflex artery distal to the flow probe. Global left ventricular performance was measured with a solid-state transducer implanted through the ventricular apex, and regional myocardial function was monitored with two pairs of piezoelectric crystals implanted in the perfusion territory of the left anterior descending artery, as well as in the perfusion territory of the circumflex artery. A coronary sinus catheter was also implanted to permit measurement of left ventricular oxygen extraction. The animals were subjected to a modified Bruce treadmill regimen lasting 20 min and culminating at a submaximal exercise level of 6.4 km/h and 16% grade. At this level of exercise, prazosin (0.5 mg) was injected into the circumflex catheter. Figure 2 shows the results of a typical experiment. Note that exercise was associated with increases in peak left ventricular pressure, dP/dt max, the rate of myocardial segment length shortening, heart rate, and circumflex blood flow. The injection of prazosin during exercise resulted in a further increase in circumflex inflow, and this increase in flow was followed by an increase in the rate of segmental shortening in the circumflex perfusion territory, as well as in dP/dt max of the left ventricle. In 15 dogs, the mean increases in coronary blood flow, rate of segmental shortening in the circumflex perfusion territory, and dP/dt max were 21%, 37%, and 21%, respectively. These increases were statistically significant. There was no change in segmental shortening in the left anterior descending perfusion territory, indicating that the dose of prazosin administered did not recirculate in such a way that it affected the anterior ventricle. Associated with the changes described above, it was observed that the administration of prazosin during exercise caused a 25% increase in left ventricular oxygen consumption. These results indicated an important role of α_1-receptors in mediating a coronary constrictor tone in exercise with a resulting limitation of the increase in myocardial oxygen delivery and a possible limitation of the increase in cardiac contractile function.

The precise basis for the increased contractile performance associated with the prazosin-induced increase in coronary flow is not certain. This effect cannot be attributed to increased adrenergic stimulation of myocardial β-receptors since in the doses employed, prazosin does not cause a substantial increase in norepinephrine release from sympathetic nerve terminals. Furthermore, the increased performance cannot be attributed to inhibition of postjunctional myocardial α-adrenergic receptors, since if these receptors are of functional importance, their inhibition would be expected to cause myocardial depression. Also, the increased myocardial performance following prazosin was not due to a direct inotropic effect of the agent, but was dependent on a high sympathetic drive to the heart since prazosin did not elicit an increase in contractile performance in the resting dog (see Fig. 1). It is attractive to propose that the

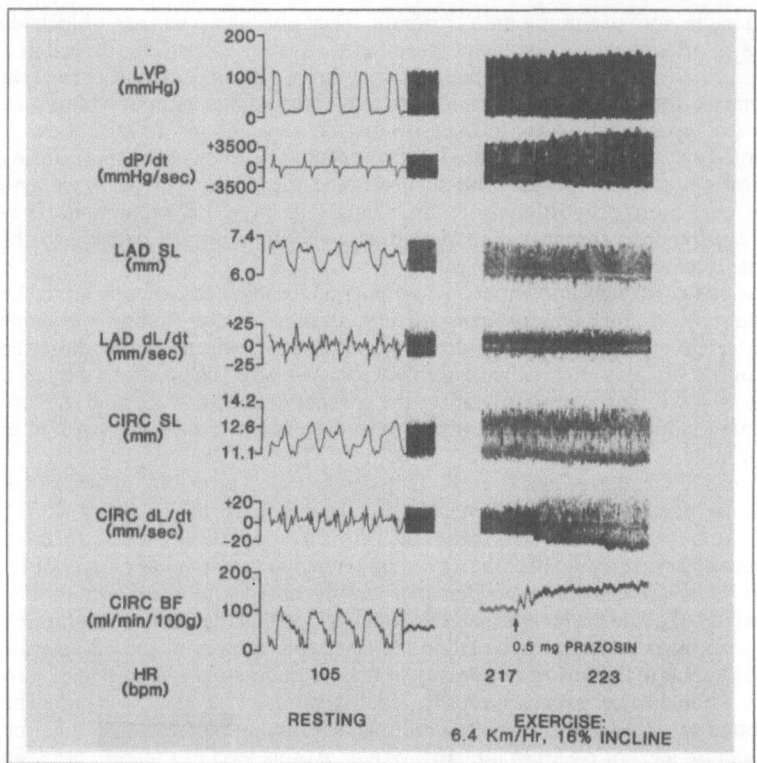

Fig. 2. Tracings showing effects of an intracircumflex injection of prazosin (0.5 mg) during submaximal exercise in the dog. Note that prazosin injection during exercise caused a substantial increase in coronary flow which was associated with an increase in systolic dL/dt max in the circumflex perfusion territory and in left ventricular dP/dt max. Abbreviations: LAD SL = segment length in the left anterior descending perfusion territory, other abbreviations as in Fig. 1. Data are from Gwirtz et al. [21].

prazosin-induced increase in contractile performance of the myocardium was a direct result of the increase in blood flow. Indeed, recent experiments compared the effects of intracoronary prazosin with those caused by intracoronary administration of the vasodilator adenosine [18]. When adenosine was administered at a dose which mimicked the increase in coronary flow caused by prazosin in each of six exercising dogs, the resulting increases in rate of myocardial segmental shortening and dP/dt max were not different from those seen with prazosin. These results confirmed the proposal that the increased contractile performance associated with the increase in flow was, indeed, the result of the increase in flow and was independent of the means by which flow was increased.

Commensurate with these observations is the implication that during moderate to heavy exercise, contractile performance, at least in the deeper myocardial layers (in which the segment length crystals were implanted), becomes somewhat flow-limited, and a contributing factor to this flow limitation may be an α_1-adrenergic coronary constrictor tone. Abolition of this vasoconstriction permits an increase in myocardial contractile function. However, it must also be recognized that the increased myocardial function secondary to the prazosin-induced increase in flow may be explained by the so-called "Gregg phenomenon" by which, even in nonischemic myocardium, an increase in coronary flow or pressure elicits an increased myocardial contraction. At the present time, use of the Gregg phenomenon to explain the increase in myocardial performance associated with an increase in coronary flow is feasible since there is only little evidence that the myocardium becomes overtly ischemic during moderate to severe exercise. For example, coronary venous pO_2 is substantially reduced in exercise, indicating that the increase in coronary flow is not sufficient to meet the increase in myocardial oxygen demand. It is also relevant that Barnard [3] reported that in normal human subjects undergoing strenuous exercise without a prior warm-up period, there was electrocardiographical evidence of myocardial ischemia. Finally, even in the absence of overt ischemia, it is possible that contractile function in the more vulnerable subendocardial layers may become flow-limited.

In the study by Gwirtz et al. [20] described above, it was proposed that an adrenergic vasoconstriction was mediated by the α_1-receptor subtype. Additional studies were performed to more systematically examine this proposal [49]. In these studies, dogs were instrumented as described above and were subjected to the same treadmill exercise regimen. At the peak level of exercise, the response to intracoronary injection of the selective α_1-adrenergic receptor antagonist prazosin was compared to that of the selective α_2-adrenergic antagonist yohimbine. Prazosin caused a significant increase in coronary flow of 28%, while yohimbine caused no change in coronary inflow. Results of a typical experiment are illustrated in Fig. 3. In this case, it can be seen that yohimbine caused no change in any recorded variable, while prazosin injected in the same animal caused a substantial increase in coronary blood flow which was associated with increases in the rate of shortening in the posterior ventricle, as well as in ventricular dP/dt max. Thus, in our laboratories the adrenergic coronary constriction in exercise seemed to be due primarily to α_1-receptors. Results similar to those above have recently been reported by Dai et al. [11]. They noted that following α_1-blockade with prazosin, coronary blood flow during exercise was significantly greater in comparison to control, while α_2-blockade with idazoxan did not alter the coronary response to exercise. They also noted that combined α_1- and α_2-adrenergic blockade was not more effective in increasing blood flow during exercise than was α_1-adrenergic blockade alone. The finding of an α-adrenoceptor-mediated limitation of flow and contractile function in the normal heart during moderate exercise was surprising. In the normal heart, this effect is only slight and does not lead to overt ischemia. However, under certain conditions, this vasoconstrictor tone may become more important. Thus, it is notable that the coronary circulation becomes more responsive to α-agonists after exercise training [22]. Furthermore, during exercise with pre-existing coronary artery obstruction the α-constrictor tone may exert an especially important limitation on myocardial flow and function.

That the α_1-receptor subtype is involved in exerting an adrenergic coronary constriction in exercise and ischemia has been reported by Laxson et al. [37]. These researchers occluded the circumflex artery in dogs until distal pressure fell to 40 mm Hg. During exercise, intracoronary prazosin increased subendocardial flow by 87% and mean transmural flow by 49%. A significant increase in segmental shortening in subendocardium also occurred. Selective α_2-blockade with intracoronary idazoxan was without effects. Results of experiments by Seitelberger et al. [48] are in agreement with those of Laxson and coworkers in regard to the existence of an α-adrenergic coronary constriction in exercise with coronary stenosis but are in contrast regarding the receptor subtype involved. These researchers observed that during treadmill exercise in dogs with a severe coronary stenosis and myocardial β-blockade, intracoronary infusion of the α_2-antagonist idazoxan increased subendocardial perfusion in ischemic myocardium by approximately 150%, and increased ventricular wall thickening in that region by approximately 100%. They concluded that an α-receptor mediated coronary constriction can impede myocardial flow and function during exercise with coronary stenosis and that this vasoconstriction is due to activation of the α_2-receptor subtype. These results compared to those from our laboratories and others indicating noninvolvement of α_2-receptors in coronary constriction seen in exercise with or without coronary stenosis emphasize the current confusion regarding the specific contributions of both α-receptor subtypes.

Role of neurogenic and circulating catecholamines

While there seems to be little doubt that an α-adrenergic coronary constriction exists during exercise, and while much evidence indicates that this coronary constriction is mediated by

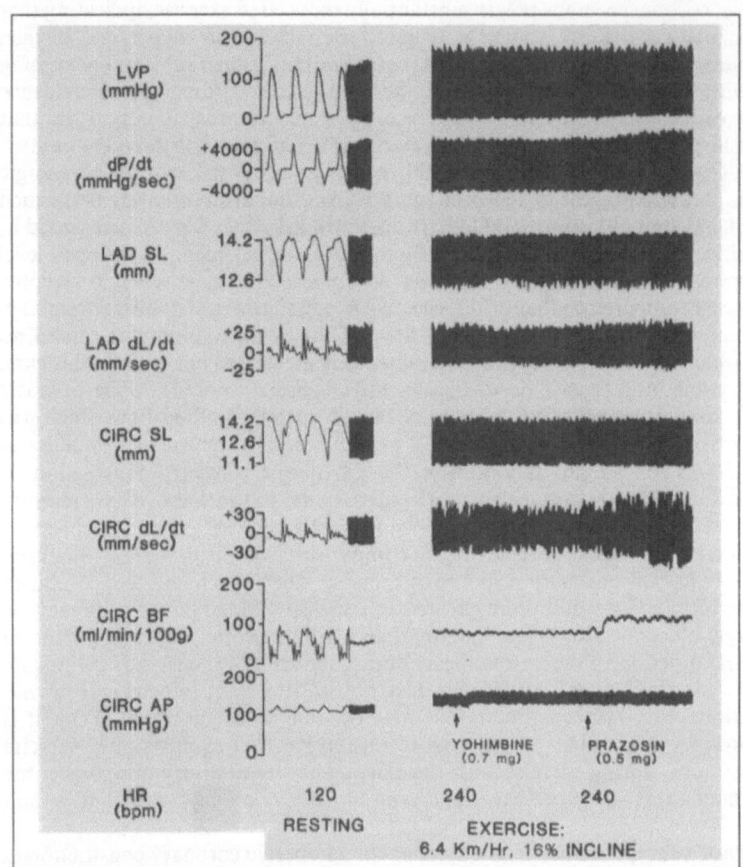

Fig. 3. Tracings showing the responses to intracircumflex injection of the α_2-receptor antagonist yohimbine (0.7 mg) and the α_1-receptor antagonist prazosin (0.5 mg) during submaximal exercise in the dog. Note that yohimbine caused no effects on CIRC BF, while prazosin caused a substantial increase in CIRC BF. The prazosin-induced increase in coronary flow was associated with an increase in systolic dL/dt max in the circumflex perfusion territory as well as in left ventricular dP/dt. Abbreviations: CAP = circumflex arterial pressure, other abbreviations as in Figs. 1 and 2. Data are from Strader et al. [49].

α_1-adrenergic receptors, there is some controversy regarding whether the coronary α-receptors are activated by neurally released norepinephrine or by circulating catecholamines. Chilian et al. [10] chemically sympathectomized a portion of the left ventricle using phenol, while leaving the innervation of the remainder of the left ventricle intact. Using the tracer microsphere technique, they found no difference in myocardial perfusion between the innervated and the sympathectomized regions during exercise sufficient to increase heart rate to 180–200 beats/min. Furthermore, they observed that following nonselective α-adrenergic blockade in conjunction with β-adrenergic blockade, coronary resistance in both regions was significantly less than during β-blockade alone. These results imply that the α-adrenergic coronary constriction was entirely due to circulating catecholamines. However, experiments performed in our laboratories have provided contradictory results [17]. In these experiments, one group of animals underwent intrapericardial surgical sympathectomy of the ventricles, leaving supraventricular innervation intact. Another group of dogs served as sham-operated controls. Both groups of dogs were exercised on a treadmill for 20 min to a level of 16 km/h and 16% incline, sufficient to increase heart rate to 180–220 beats/min. At this level of exercise prazosin was injected into an indwelling circumflex catheter. In the sham-operated animals prazosin elicited a statistically significant 26% increase in coronary inflow, but in the sympathectomized group prazosin elicited only a 5% increase in coronary flow, which was not significant. Thus, our data indicated that the adrenergic coronary constriction existing during exercise is neurogenic since abolition of the sympathetic nerves to the heart almost completely abolished the α_1-adrenergic coronary constrictor tone, and since previous results from our laboratory using the same model could not find evidence for a role of α_2-receptors. The basis of the discrepancy between these two studies is not readily apparent. However, it is noted that in the study of Chilian et al., the α-antagonist phentolamine was injected systemically and caused approximately a 20 mm Hg reduction in mean arterial pressure. It is reasonable to suggest that this reduction in coronary perfusion pressure may have led to the observed reduction in coronary resistance, even in the sympathectomized regions of the heart, due to the known coronary pressure-flow autoregulation phenomenon.

Transmural distribution of coronary adrenergic constriction during exercise

Few studies have examined the transmural distribution of the coronary adrenergic constrictor tone during exercise. Huang and Feigl [28] proposed that the adrenergic coronary constriction may be more intense in the subepicardial muscle layers, such that it diverts blood flow to the more vulnerable deeper layers during exercise. These investigators tested this hypothesis in dogs by selectively blocking α-receptors in one myocardial region, using the nonselective antagonist phenoxybenzamine, and measuring regional myocardial perfusion during treadmill exercise. They observed that average transmural flow was less in the unblocked region than in the blocked region, as would be expected from the results of other investigators. However, they also observed that the ratio of subendocardial-to-subepicardial perfusion was better maintained in the region in which α-receptors were intact. Although the mechanism for the apparent preservation of uniform transmural flow by α-adrenergic coronary constriction is uncertain, it may provide a physiological basis for this constriction. But there is no evidence for a nonuniform distribution of α_1-adrenergic receptors in left ventricular coronary vessels. Thus, Saffitz [47], using receptor binding techniques in cats, found no difference in the density of α_1-receptors between subepicardial and subendocardial coronary arterioles. Recognizing the absence of data indicating a nonuniform distribution of α-receptors across the ventricular wall, Huang and Feigl suggested the possibility that the endogenous vasodilator adenosine is released preferentially from myocardial myocytes in the deeper layers, where contractile function may be more flow-limited. Since adenosine is known to attenuate release of norepinephrine from sympathetic nerve terminals, it is then feasible that neuronal release of norepinephrine during exercise is less in subendocardial layers than in subepicardial layers, resulting in a greater adrenergic vasoconstriction in subepicardial layers. A similar nonuniform adrenergic constriction across the ventricular wall has also been reported by Chilian and Ackell [8] during exercise following coronary stenosis.

We have also examined the transmural nature of an α_1-coronary constriction during sympathetic stimulation of the heart. In two groups of dogs, we examined the effects of intracoronary prazosin on left ventricular transmural contractile function and perfusion during stellate ganglion stimulation [19]. In one group piezoelectric crystals were implanted in subepicardial and subendocardial layers of the circumflex perfusion territory for measurement of segmental shortening and rate of shortening in these regions. During stellate ganglion stimulation that produced changes in cardiac performance like those seen in exercise, intracoronary prazosin resulted only in a further increase in contractile function in the deeper layers. In a second group, the effect of prazosin on perfusion in subepicardial and subendocardial layers during stellate ganglion stimulation was examined. Notably, prazosin caused an equal and significant increase in perfusion in both layers by about 34%. We feel that the increase in contractile function caused by prazosin only in the deeper layers may be due to the fact that, even though adrenergic vasoconstriction was uniform across the entire ventricular wall, only in the deeper, more ischemia-prone layers did this vasoconstriction limit contractile function. Also, it may be noted that the prazosin-induced increase in perfusion in both epicardial and endocardial layers coupled with an increased contractile function only in the deeper layers suggests that the increased contractile function was not secondary to the Gregg phenomenon. In regard to transmural distribution of an α-constriction, it should also be recalled that Laxson et al. [37] saw an 87% increase in subendocardial perfusion following selective α_1-blockade in exercising dogs following partial coronary stenosis.

While the proposal of Huang and Feigl [28] provides a logical physiological basis for an adrenergic coronary constriction in exercise and other conditions, this hypothesis requires further exploration.

α_1-Adrenergic coronary constriction during myocardial ischemia

Oxygen availability to myocardium is determined both by the rate of coronary blood flow and by the oxygen extraction from coronary blood. Over a wide range of oxygen demand, an adrenergic coronary constriction may not be an impediment to oxygen availability since any resulting reduction in coronary perfusion may be balanced by an increase in oxygen extraction. But if a significant adrenergic constriction still exists under conditions in which the extraction is very high (i.e., a minimal extraction reserve), such adrenergic coronary constriction may then serve as an impediment to myocardial oxygen availability. During myocardial ischemia, such a situation may exist.

Effects of an α_1-adrenergic coronary constriction on myocardial blood flow and oxygenation during ischemia

The myocardium becomes ischemic during severe systemic hypotension. Initial experiments from our laboratories examined whether or not an α-adrenergic coronary constriction existed under this condition [31], especially since it may be predicted that metabolic vasodilatory signals would be very high. In 17 anesthetized dogs, the effects of the nonselective α-antagonist phenoxybenzamine infused into the main left coronary artery on coronary blood flow, left ventricular oxygen extraction, and oxygen consumption were determined both under normotensive conditions and during hypotension produced by hemorrhage. At all coronary perfusion pressures from 90 to 35 mm Hg, phenoxybenzamine resulted in substantial increases in left coronary inflow. The mean increase in flow was approximately 40%. However, at normotensive pressures (greater than 80 mm Hg) the increase in coronary flow was largely balanced by a reduction in oxygen extraction such that myocardial oxygen consumption increased by only 13%. This increase in oxygen consumption may reflect the prejunctional action of phenoxybenzamine. Interestingly, at hypotensive pressures of less than approximately 60 mm Hg, oxygen extraction prior to phenoxybenzamine infusion had increased from its normotensive value of 13 ml/100 ml of blood to values as high as 18 ml/100 ml of blood. Under these conditions, it was observed that the increase in coronary flow caused by α-blockade was not accompanied by substantial reductions in oxygen extraction, and oxygen consumption increased by 38%. It was proposed that during severe hemorrhagic hypotension, the heart becomes ischemic, and an α-adrenergic coronary constriction contributes to this ischemia by limiting coronary dilation and myocardial oxygenation.

In hemorrhagic hypotension, an intense sympathetic stimulation of the heart and coronary circulation may occur, largely because of peripheral reflexes associated with systemic hypotension. Therefore, we wished to examine whether an α-adrenergic coronary constriction could be demonstrated in the open-chest dog during myocardial ischemia produced by selective coronary hypotension. In addition, we wished to delineate the specific receptor subtypes involved. For these reasons, we performed experiments in which a subclavian-to-main left coronary shunt was employed, and coronary perfusion pressure was servo-controlled [33]. After control data were obtained, coronary perfusion pressure was abruptly reduced to

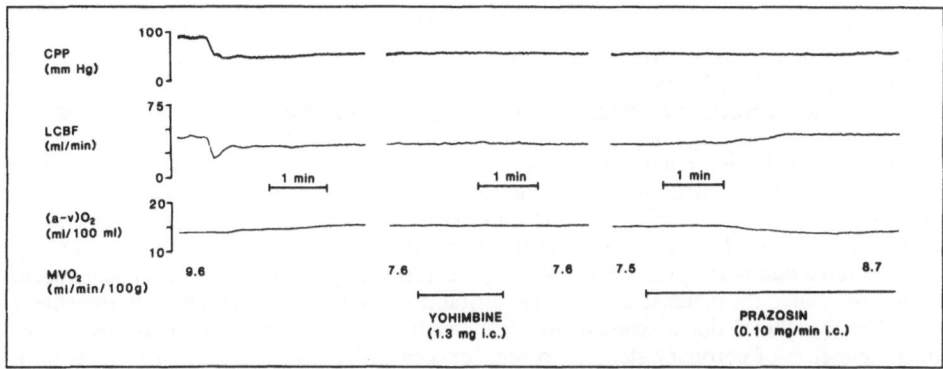

Fig. 4. Tracings showing effects of yohimbine and prazosin in the anesthetized dog following a reduction in left CPP to 50 mm Hg. The dog had been subjected to cardiac β-adrenergic blockade by intracoronary injection of propranolol. Note that during coronary hypoperfusion yohimbine injected into the main left coronary artery caused no effects. However, prazosin caused a substantial increase in LCBF and in $M\dot{V}O_2$. Abbreviations: CPP = coronary perfusion pressure in the main left coronary artery, LCBF = left coronary blood flow, $(a-v)O_2$ = left ventricular oxygen extraction, $M\dot{V}O_2$ = left ventricular oxygen consumption. Data are from Liang and Jones [39].

50 mm Hg, and the effects of prazosin were compared to those of the nonselective-antagonist phenoxybenzamine both with and without prior β-blockade with propranolol. In all cases, the reduction in coronary perfusion pressure caused reductions in left coronary inflow and myocardial oxygen consumption of approximately 35% and 20%, respectively. Intracoronary prazosin in the absence of β-blockade caused increases in coronary inflow and oxygen consumption of 22% and 15%, respectively, and these prazosin-induced changes were not affected by prior administration of propranolol. However, intracoronary phenoxybenzamine without prior β-blockade caused increases in coronary flow and oxygen consumption of 31% and 30%, respectively, which were significantly greater than the changes elicited by prazosin. The greater effects of phenoxybenzamine could potentially have been due to a prejunctional α_2-adrenergic blockade with an increase in norepinephrine release and a resulting metabolic dilation. Or the greater effects could have been due to a postjunctional α_2-blockade additive to the α_1-blockade, resulting in a greater coronary dilation than with α_1-blockade alone. However, since the changes in coronary inflow and oxygen consumption caused by phenoxybenzamine in the presence of β-blockade were not significantly different from those observed with prazosin, it became clear that there was no added coronary dilation secondary to postjunctional α_2-antagonism. These results suggested that in the anesthetized, open-chest dog a coronary α-adrenergic constriction was present during selective and severe coronary hypotension. This adrenergic constriction was sufficient to limit myocardial oxygenation and was mediated primarily by the α_1-receptor subtype.

To more directly confirm the receptor subtype involved, additional experiments were performed in which the effects of prazosin were compared to the effects of the α_2-receptor antagonist yohimbine [39]. The experimental model was identical to that described above. It was observed that in the presence of β-blockade, prazosin elicited effects similar to those outlined above, while intracoronary yohimbine was entirely without effects. A typical experiment using both yohimbine and prazosin is illustrated in Fig. 4. It is clear that yohimbine caused no discernible effects during coronary hypotension, while prazosin administered in the same animal caused a sizeable increase in left coronary blood flow and myocardial oxygen consumption. Thus, during exercise or coronary hypotension, we have been able to demonstrate an α-adrenergic coronary constriction mediated by the α_1-receptor subtype. In contrast to the work of some others, we have been unable to demonstrate a role for postjunctional α_2-receptors. From our results, it was difficult to discern whether the coronary vessels involved in the α_1-constriction were large vessels, true arterioles, or both. However, from the magnitude

187

of the flow increases resulting from α_1-adrenergic blockade, it is apparent that the constriction existing during coronary hypoperfusion in the open-chest, anesthetized dog was sufficient to impose a serious impediment to myocardial perfusion.

A coronary adrenergic vasoconstriction existing in myocardial ischemia has been proposed by a number of investigators. Thus, Mudge et al. [43] demonstrated that in humans with coronary artery disease, reflex coronary adrenergic constriction was sufficient to cause anginal pain, and could be abolished by the nonselective antagonist phentolamine. Likewise, Hillis and Braunwald [27] and Maseri et al. [41] implicated an adrenergic coronary constriction in coronary vasospasm resulting in myocardial infarction. Furthermore, Buffington and Feigl [7] demonstrated that in the presence of coronary stenosis in dogs, α-blockade with phenoxybenzamine attenuated the increase in coronary resistance caused by administration of norepinephrine. They concluded that a sympathetic α-adrenergic coronary constriction may operate even in the presence of coronary stenosis to limit oxygen delivery to the heart. Brown et al. [6] observed a reflex coronary constriction during sympathetic stimulation of the heart elicited by hand-grip exercise in humans with significant coronary artery stenosis. Aubry et al. [2] reported that in anesthetized, open-chest dogs, α_1-adrenergic blockade with prazosin reduced the ventricular arrhythmias resulting from coronary occlusion. This action of prazosin was associated with a reduction in the ischemia-induced rise in ventricular filling pressure and an increased coronary perfusion.

While experiments in our laboratories appear to clearly indicate that the adrenergic coronary constriction existing in myocardial ischemia is mediated by α_1-receptor subtypes, and while our data are in agreement with those of certain other investigators, there are also reports that the adrenergic coronary constriction is mediated primarily by α_2-receptors. Thus, Heusch and Deussen [23] noted that during stellate ganglion stimulation following severe coronary stenosis in dogs, an adrenergic coronary constriction was alleviated by the specific α_2-antagonist rauwolscine, while prazosin was without effects. Likewise, Kopia et al. [34] suggested a role for both α_1- and α_2-adrenergic receptors in causing coronary constriction following coronary stenosis with α_2-receptors playing the dominant role. The basis for these discrepancies is not readily apparent.

Transmural distribution of adrenergic coronary constriction in myocardial ischemia

As pointed out earlier, it has been proposed that during exercise a coronary adrenergic constriction exists predominantly in the subepicardial layers and operates to preserve flow to the subendocardial layers. A similar mechanism has been proposed to operate during coronary hypoperfusion by Nathan and Feigl [46]. In open-chest, anesthetized dogs, two regions of the left ventricle were separately perfused, and in one region α-receptors were blocked with the nonselective antagonist phenoxybenzamine. Norepinephrine was infused into both regions of the ventricle as coronary inflow to both regions was reduced progressively from 100% to 50% of normal. Regional flows in each region were measured using tracer microspheres. It was observed that at low coronary flow levels, subendocardial perfusion was greater in the ventricular region in which α-receptors were intact. These results are in concert with those of Giudicelli et al. [15] who reported a greater α-adrenergic vasoconstriction in the subepicardium during sympathetic nerve stimulation of β-blocked, nonischemic hearts. Furthermore, Johannsen et al. [29] reported that during maximal adenosine-induced coronary dilation following β-blockade in dogs, a condition which may simulate the high metabolic vasodilatory signal thought to exist in myocardial ischemia, sympathetic nerve stimulation caused vasoconstriction in the subepicardium only.

While the concept of a physiological, protective role for an adrenergic coronary constriction in myocardial ischemia is extremely interesting it is not without contradiction. Thus, as pointed out earlier, Saffitz [47] found that α_1-receptors were equally distributed in coronary vessels throughout the ventricular wall of the cat. Furthermore, our data indicate that during stellate ganglion stimulation of the nonischemic heart, intracoronary prazosin caused an equal increase in coronary perfusion in both subendocardial and subepicardial regions [19]. And Laxson et al.

[37] saw an 87% increase in subendocardial perfusion caused by prazosin in the exercising dog with coronary stenosis, indicating a substantial α_1-mediated vasoconstriction in this region.

The contradictory results regarding the transmural distribution of an α-adrenergic coronary constriction may be largely due to the different experimental models and protocols employed. It is plausible that with mild myocardial ischemia and mild sympathetic stimulation of the heart, only the small coronary resistance vessels are under the potential influence of an adrenergic constriction. If, in such a case, some factor prevents α-adrenergic receptor activation in subendocardial resistance vessels (perhaps a higher production of adenosine in subendocardium, as suggested by Huang and Feigl [28]), the conditions would exist to promote flow to subendocardial layers at the expense of subepicardial layers. However, with more severe myocardial ischemia and more intense sympathetic stimulation, larger coronary conduit vessels may undergo vasoconstriction that would impede flow transmurally. In addition, a more intense sympathetic stimulation may result in constriction of resistance vessels in deeper muscle layers. It is also established that local ischemia may cause a nerve impulse-independent release of norepinephrine from sympathetic nerve terminals [1]. Such a condition would then promote norepinephrine release and adrenergic constriction in subendocardium. It is apparent that this important issue requires further study.

Ischemia-induced adrenergic coronary constriction: a possible positive feedback

In addition to peripheral reflexes which may result from myocardial ischemia with cardiac pump failure, there are two mechanisms by which myocardial ischemia itself, once initiated, may directly stimulate an adrenergic coronary constriction and, thereby, exacerbate the existing ischemia. First, it is established that ischemia itself may cause a localized release of norepinephrine from sympathetic nerve terminals [1]. Such localized release of norepinephrine may be expected to stimulate myocardial β-adrenergic receptors, possibly causing a regional increase in myocardial oxygen demand. The increase in norepinephrine release may also be expected to stimulate vascular α-adrenergic receptors causing an increase in coronary constrictor tone opposing local metabolic vasodilation. Second, it has been reported that conditions existing in myocardial ischemia may initiate a spinal sympathetic reflex causing an increase in sympathetic stimulation of the heart [40]. The increase in sympathetic stimulation resulting from this "cardiocardiac reflex" may have a coronary component by which adrenergic constriction of the coronary vasculature is increased.

Experiments in our laboratories indeed suggest that severe myocardial ischemia may directly initiate adrenergic coronary constriction [38]. Eight dogs were instrumented for measurement of circumflex inflow velocity, and a hydraulic occluder was placed around the artery distal to the probe. A small catheter was implanted distal to the occluder. Regional contractile function in the posterior left ventricle was measured by implantation of piezoelectric crystals in the subendocardium, and arterial pressure was measured with a small catheter implanted in the aorta. With these animals lying quietly in the conscious state, intracoronary injection of prazosin (0.5 mg) caused no effects in any measured variable. These results indicated that in the resting quiescent state, the coronary circulation was devoid of significant sympathetic stimulation. After control measurements, the coronary occluder was inflated sufficiently to cause a 21% reduction in regional segmental shortening. This ischemia-induced reduction in posterior contractile function was not severe enough to cause a reduction in global ventricular function such that arterial pressure was unaffected. Following partial coronary occlusion, the same dose of prazosin caused a significant increase in circumflex inflow of 36%, and segmental shortening increased to a value not different from the preocclusion value. This α_1-adrenergic coronary constriction which was present following partial coronary stenosis, but not before stenosis, could not be explained by the arterial baroreflex since no change in arterial pressure was observed. Furthermore, the partial stenosis caused no discernible discomfort or anxiety in the animals. It is attractive to speculate that the adrenergic coronary constriction induced by partial coronary stenosis was due to one of the potential positive feedback mechanisms described above. Notably, the results of these experiments also clearly indicated that an α_1-coronary constriction was present in the conscious dog even in myocardial ischemia sufficiently severe to cause a reduction in contractile function.

The results of our experiments in the awake dog, as described above, are compatible with more complete experiments performed in the anesthetized dog by Heusch et al. [25]. These researchers noted that severe circumflex artery stenosis for 20 min was associated with an increased firing of postganglionic sympathetic nerves and approximately a 30% increase in poststenotic coronary resistance, which resulted in a net lactate production by ischemic myocardium. Arterial pressure was unchanged, indicating noninvolvement of the baroreflex. The selective α_2-antagonist rauwolscine or the calcium channel antagonist nifedipine prevented the progressive increase in poststenotic resistance and the lactate production, but not the increase in nerve activity. On the other hand, spinal anesthesia of segments C7-T6 prevented the increase in nerve activity, as well as the increases in coronary resistance and lactate production. Thus, these results strongly indicate that a positive feedback mechanism exists once severe myocardial ischemia is initiated and is at least, in large measure, due to the neural cardiocardiac reflex as described by Malliani et al. [40]. In addition, the work of Heusch et al. [25] suggests a major role of α_2-receptors in mediating the reflex vasoconstriction, whereas our results also suggest a major role of α_1-receptors.

A second set of preliminary experiments from our laboratories (unpublished) also indicate that a similar positive feedback mechanism may occur in ischemia caused by exercise with coronary stenosis. Dogs were instrumented as described above. When the dogs were treadmill exercised at the level of 6.4 km/h and 8% grade, coronary inflow increased by approximately 30%, and posterior ventricular rate of segmental shortening increased by approximately 50%. Heart rate increased to about 150 beats/min. Intracoronary injection of prazosin (0.5 mg) under these conditions resulted in a further increase in coronary inflow of 8–10%. These experiments were repeated in the same dogs on separate days. However, in these experiments, coronary inflow was held at the resting level during treadmill exercise by inflation of the circumflex occluder. Under these conditions, injection of the same dose of prazosin caused a substantially larger increase in circumflex inflow of 80–100%. Thus, the results of these preliminary experiments were again compatible with the concept that myocardial ischemia exacerbated an α_1-adrenergic constriction of the coronary vasculature. The experiments did not provide insight into whether this effect involved a cardiocardiac reflex, a local ischemia-induced release of norepinephrine, or both. Nevertheless, it is attractive to speculate that abolition of such a positive feedback mechanism partially explains the anti-infarction effects of chronic ventricular sympathectomy during coronary artery occlusion [30]. Thus, following long-term sympathectomy, not only are neural pathways interrupted, but the norepinephrine content of the ventricles is depleted.

Conclusions

From the research reviewed in this paper, particularly that from our laboratories, the following important points appear to emerge:

First, an adrenergic coronary constriction is negligible in the normal, conscious subject at rest. However, factors which increase sympathetic stimulation of the cardiovascular system also increase an adrenergic constrictor tone of coronary vessels.

Second, during exercise and myocardial ischemia, an adrenergic coronary constrictor tone may limit a coronary dilation and, thereby, limit the increase in myocardial oxygen delivery and contractile function.

Third, both α_1- and α_2-adrenergic receptors are present in the coronary vasculature. However, under the conditions of our experiments the vasoconstriction seen in exercise and myocardial ischemia is mediated primarily by the α_1-receptor subtype. It is noted that a controversy exists in this regard since several other laboratories have observed an important role of α_2-receptors, especially in myocardial ischemia.

Fourth, evidence suggests that the adrenergic coronary constriction associated with sympathetic stimulation of the heart is uniform across the left ventricular wall. However, the resulting limitation in contractile function is seen predominantly in subendocardium.

Fifth, severe myocardial ischemia, once initiated, may directly increase adrenergic coronary constriction, such that conditions exist for a vicious cycle to become operative.

Acknowledgements: We wish to thank graduate students Andre' Meintjes and Nancy Longlet, as well as postdoctoral trainees Drs. Pamela Gayheart and Marilyn Brandt for their valuable review of this manuscript. We are also grateful to Mrs. Wendy Ching for typing it.

References

1. Abrahamsson T, Almgren O, Carlsson L (1985) Ischemia-induced local release of myocardial noradrenaline. J Cardiovasc Pharmacol 7:S19–S22
2. Aubry ML, Davey MJ, Petch B (1985) Cardioprotective and antidysrhythmic effects of α_1-adrenoceptor blockade during myocardial ischaemia and reperfusion in dogs. J Cardiovasc Pharmacol 7:S93–S102
3. Barnard RJ, Gardner GW, Diaco NV, MacAlpen RN, Kattus AA (1973) Cardiovascular responses to sudden strenuous exercise – heart rate, blood pressure, and ECG. J Appl Physiol 34:833–837
4. Billman GE, Randall DC (1981) Mechanisms mediating the coronary vascular response to behavioral stress in the dog. Circ Res 48:214–222
5. Brachfield N, Monroe RG, Gorlin R (1960) Effect of pericoronary denervation on coronary hemodynamics. Am J Physiol 199:174–178
6. Brown BG, Lee AB, Bolson EL, Dodge HT (1984) Reflex constriction of significant coronary stenosis as a mechanism contributing to ischemic left ventricular dysfunction during isometric exercise. Circulation 70:18–24
7. Buffington CW, Feigl EO (1981) Adrenergic coronary vasoconstriction in the presence of coronary stenosis in the dog. Circ Res 48:416–423
8. Chilian WM, Ackell PH (1986) Sympathetic coronary tone during exercise prevents transmural steal in the presence of stenosis. Fed Proc 45:533 (abstr.)
9. Chilian WM, Boatwright RB, Shoji T, Griggs DM, Jr (1981) Evidence against significant resting sympathetic coronary vasoconstrictor tone in the conscious dog. Circ Res 49:866–876
10. Chilian WM, Harrison DC, Haws CW, Snyder WD, Marcus ML (1986) Adrenergic coronary tone during submaximal exercise in the dog is produced by circulating catecholamines. Circ Res 58:68–82
11. Dai X, Sublett E, Lindstrom P, Schwartz JS, Homans DC, Bache RJ (1989) Coronary flow during exercise after α_1- and α_2-adrenergic blockade. Am J Physiol 256:H1148–H1155
12. Dolezel S, Gerova M, Jero J, Sladek T, Vasku J (1978) Adrenergic innervation of the coronary arteries and the myocardium. Acta Anat 100:306–316
13. Feigl EO (1968) Carotid sinus reflex control of coronary blood flow. Circ Res 23:223–237
14. Gayheart PA, Gwirtz PA, Longlet NJ, Bravenec JS, Jones CE (1990) An α-adrenergic coronary constriction during esophageal distention in the dog. FASEB J 4:A852 (abstr.)
15. Giudicelli JF, Berdeaux A, Tato F, Garnier M (1980) Left stellate stimulation: regional myocardial flows and ischemic injury in dogs. Am J Physiol 239:H359–H364
16. Grice DP, Watanabe N, Yonekura S, Williams AG, Jones CE, Downey HF (1987) Alpha-one adrenoceptor mediated coronary vasoconstriction during systemic hypoxia. Physiologist 30(4):188 (abstr.)
17. Gwirtz PA, Brandt MA, Meintjes AF, Jones CE (1990) Neuronally released catecholamines are primarily responsible for the coronary constriction tone noted during exercise. FASEB J 4:1072 (abstr.)
18. Gwirtz PA, Dodd-o JM, Brandt MA, Jones CE (1990) Augmentation of coronary flow improves myocardial function in exercise. J Cardiovasc Pharmacol 15:752–758
19. Gwirtz PA, Jones CE, Dodd-o JM, Hamrick ML, Downey HF, Williams AG (1989) A transmural alpha$_1$-adrenergic constriction during stellate stimulation. FASEB J 3:A895 (abstr.)
20. Gwirtz PA, Overn SP, Mass HJ, Jones CE (1986) α_1-Adrenergic constriction limits coronary flow and cardiac function in running dogs. Am J Physiol 250:H1117–H1126
21. Gwirtz PA, Stone HL (1981) Coronary blood flow and myocardial oxygen consumption after alpha-adrenergic blockade during submaximal exercise. J Pharmacol Exp Ther 217:92–98
22. Gwirtz PA, Stone HL (1984) Coronary vascular response to adrenergic stimulation in exercise-conditioned dogs. J Appl Physiol 57:315–320
23. Heusch G, Deussen A (1983) The effects of cardiac sympathetic nerve stimulation on perfusion of stenotic coronary arteries in the dog. Circ Res 53:8–15
24. Heusch G, Deussen A, Schipke J, Thämer V (1984) α_1- and α_2-adrenoceptor mediated vasoconstriction of large and small canine coronary arteries in vivo. J Cardiovasc Pharmacol 6:961–968

25. Heusch G, Deussen A, Thämer V (1985) Cardiac sympathetic nerve activity and progressive vasoconstriction distal to coronary stenoses: feed-back aggravation of myocardial ischemia. J Auton Nerv Syst 13:311–326
26. Heyndrickx GR, Muylaert P, Pannier JL (1982) α-Adrenergic control of oxygen delivery to myocardium during exercise in conscious dogs. Am J Physiol 242:H805–H809
27. Hillis LD, Braunwald E (1978) Coronary-artery spasm. N Engl J Med 299:695–702
28. Huang AH, Feigl EO (1988) Adrenergic coronary vasoconstriction helps maintain uniform transmural blood flow distribution during exercise. Circ Res 62:286–298
29. Johannsen UJ, Mark AL, Marcus ML (1982) Responsiveness to cardiac sympathetic nerve stimulation during maximal coronary dilation produced by adenosine. Circ Res 50:510–517
30. Jones CE, Beck LY, DuPont E, Barnes GE (1978) Effects of coronary ligation on the chronically sympathectomized dog ventricle. Am J Physiol 235:H429–H434
31. Jones CE, Farrell TA, Ator R (1983) Evidence that a coronary α-adrenergic tone limits myocardial blood flow and oxygenation in acute hemorrhagic hypotension. Circ Shock 11:329–340
32. Jones CE, Liang IYS, Gwirtz PA (1987) Effects of α-adrenergic blockade on coronary autoregulation in dogs. Am J Physiol 253:H365–H372
33. Jones CE, Liang IYS, Maulsby MR (1986) Cardiac and coronary effects of prazosin and phenoxybenzamine during coronary hypotension. J Pharmacol Exp Ther 236:204–211
34. Kopia GA, Kipaciewicz LJ, Ruffolo RR, Jr (1986) Alpha adrenoceptor regulation of coronary artery blood flow in normal and stenotic canine coronary arteries. J Pharmacol Exp Ther 239:641–647
35. Langer SZ (1984) Presynaptic regulation of the release of catecholamines. Pharmacol Rev 32:337–362
36. Langer SZ, Hicks PE (1984) Alpha-adrenoreceptor subtypes in blood vessels: physiology and pharmacology. J Cardiovasc Pharmacol 6:S547–558
37. Laxson DD, Dai XZ, Homans DC, Bache RJ (1989) The role of α_1- and α_2-adrenergic receptors in mediation of coronary vasoconstriction in hypoperfused ischemic myocardium during exercise. Circ Res 65:1688–1697
38. Liang IYS, Elsken CS, Carroll KA, Maulsby MA, Jones CE (1987) Coronary sympathetic constriction impedes myocardial function during partial coronary occlusion. Fed Proc 46:832 (abstr.)
39. Liang IYS, Jones CE (1985) α_1-Adrenergic blockade increases coronary blood flow during coronary hypoperfusion. Am J Physiol 249:H1070–H1077
40. Malliani A, Peterson DD, Bishop VS, Brown AM (1972) Spinal sympathetic cardiocardiac reflexes. Circ Res 30:158–166
41. Maseri A, L'Abbate A, Baroldi G, Chierchia S, Marzilli M, Ballestra AM, Severi S, Parodi O, Biagini A, Distante A, Pesola A (1978) Coronary vasospasm as a possible cause of myocardial infarction. N Engl J Med 299:1271–1277
42. Mohrman DE, Feigl EO (1978) Competition between sympathetic vasoconstriction and metabolic vasodilation in the canine coronary circulation. Circ Res 42:79–86
43. Mudge GH Jr, Grossman W, Mills RM Jr, Lesch M, Braunwald E (1976) Reflex increase in coronary vascular resistance in patients with ischemic heart disease. N Engl J Med 295:1333–1337
44. Murray PA, Lavalle M, Vatner SF (1984) Alpha adrenergic-mediated reduction in coronary blood flow secondary to carotid chemoreceptor reflex activation in conscious dogs. Circ Res 54:96–106
45. Murray PA, Vatner SF (1979) α-Adrenoceptor attenuation of the coronary vascular response to severe exercise in the conscious dog. Circ Res 45:654–660
46. Nathan HJ, Feigl EO (1986) Adrenergic vasoconstriction lessens transmural steal during coronary hypoperfusion. Am J Physiol 250:H645–H653
47. Saffitz JE (1989) Distribution of α_1-adrenergic receptors in myocytic regions and vasculature of feline myocardium. Am J Physiol 257:H162–H169
48. Seitelberger R, Guth BD, Heusch G, Lee J, Katayama K, Ross J, Jr (1988) Intracoronary α_2-adrenergic receptor blockade attenuates ischemia in conscious dogs during exercise. Circ Res 62:436–442
49. Strader JR, Gwirtz PA, Jones CE (1988) Comparative effects of alpha$_1$- and alpha$_2$-adrenoceptors in modulation of coronary flow during exercise. J Pharmacol Exp Ther 246:772–778
50. Woodman OL, Vatner SF (1987) Coronary vasoconstriction mediated by α_1- and α_2-adrenoceptors in conscious dogs. Am J Physiol 253:H388–H393
51. Young MA, Vatner DE, Knight DR, Graham RM, Homcy CJ, Vatner SF (1988) α-adrenergic vasoconstriction and receptor subtypes in large coronary arteries of calves. Am J Physiol 255:H1452–H1459

Authors' address:

Dr. Carl E. Jones, Professor and Chairman, Department of Physiology, Texas College of Osteopathic Medicine, 3500 Camp Bowie Blvd., Fort Worth, Texas 76107-2690, USA.

Contribution of Postsynaptic α_2-Adrenoceptors to Reflex Sympathetic Constriction of Stenotic Coronary Vessels

A. Deussen, P. Busch, J. Schipke, V. Thämer and G. Heusch*

Zentrum für Physiologie der Heinrich-Heine-Universität Düsseldorf, FRG

Summary

Increases in the activity of efferent cardiac sympathetic nerves by $35 \pm 9\%$ were induced by 60 s bilateral occlusion of the common carotid arteries (BCO) in anesthetized dogs. Under control conditions the reflex rise in sympathetic nerve activity enhanced left ventricular pressure (115 ± 4 mm Hg) by 47% and regional myocardial oxygen consumption (9.7 ± 1.1 ml/min·100 g) by 56%. Simultaneously, end-diastolic circumflex coronary resistance (0.99 ± 0.11 mm Hg·min·100 g/ml) decreased by 16%. After exhaustion of coronary dilator reserve by production of a severe coronary stenosis, BCO enhanced left ventricular pressure (107 ± 4 mm Hg) by 49%, oxygen consumption of the poststenotic area (7.6 ± 0.8 ml/min·100 g) increased by 21%, and circumflex coronary resistance (0.54 ± 0.05 mm Hg·min·100 g/ml) also increased by 19%. The reflex increase in coronary resistance during BCO was abolished after infusion of the α_2-adrenoceptor antagonist rauwolscine (0.2 mg/kg i.v.). Administration of rauwolscine, however, did not prevent the reflex increase of left ventricular pressure and regional myocardial oxygen consumption. Comparable increases in poststenotic coronary resistance during BCO were found in dogs which either received propranolol (2 mg/kg i.v.) or in which the reflex rise in mean aortic pressure was limited to 13 ± 3 mm Hg. In both experimental groups, rauwolscine also effectively prevented the BCO-induced rise in coronary resistance. In contrast, the reflex increase of total peripheral resistance was not significantly reduced by rauwolscine, but was blunted after additional administration of the selective α_1-adrenoceptor antagonist prazosin (1.2 mg/kg i.v.). We conclude that: 1) Poststenotic coronary vasoconstriction occurs during shortlasting increases in efferent cardiac sympathetic discharge within the physiological range. 2) This increase in poststenotic coronary resistance is significantly reduced after administration of the α_2-adrenoceptor antagonist rauwolscine. 3) In contrast to poststenotic coronary resistance, functionally innervated α_2-adrenoceptors are of minimal importance for the reflex increase in total peripheral resistance.

Introduction

It is well established that sympathetic activation of α-adrenoceptors effectively controls coronary blood flow and myocardial oxygen extraction (for review see [10]). This α-adrenergic vasoconstriction competes with a vasodilation that is mediated by activation of coronary β-adrenoceptors and the metabolic consequences of myocardial β-adrenergic activation [10, 16, 25, 36]. Under control conditions the β-adrenoceptor-mediated and, in particular, the metabolic dilation prevail. However, in the presence of compensatory coronary dilation as it may occur distal to a severe stenosis, metabolic dilation is no longer present, while α-adrenoceptor-induced vasoconstriction persists [3, 27]. The increase in coronary resistance distal to a severe stenosis

---

*Prof. Heusch's present address is: Abt. Pathophysiologie, Zentrum Innere Medizin, Universität Essen, Hufelandstr. 55, 4300 Essen.

may even result in net lactate production and regional contractile dysfunction indicative of myocardial ischemia [17]; (for review see [16]).

A few studies have attributed the mediation of coronary vasoconstriction to the activation of α_1-adrenoceptors [13, 41]. Other studies from independent laboratories including our own suggest that peripheral coronary constriction is mediated by the α_2-adrenoceptor subtype [19, 24, 39, 40]. Very recently, the coexistence of both α_1- as well as α_2-adrenoceptor-mediated coronary vasoconstriction was demonstrated [45]. A major fraction of peripheral α-adrenoceptors controlling the perfusion of various organ beds are known to be of the α_1-subtype (for overview see [37]). It is thus conceivable that selective α_2-adrenoceptor antagonists may reduce coronary vasoconstriction, but permit systemic blood pressure and, thereby, coronary perfusion pressure to still increase during a general activation of the sympathetic nervous system. Therefore, aims of the present study were: 1) To evaluate whether increases in sympathetic nerve activity within the physiologic range cause significant α-coronary vasoconstriction after coronary dilator reserve has been exhausted. 2) To test whether blockade of postsynaptic vascular α_2-adrenoceptors can significantly attenuate the coronary vasoconstriction, but leave the rise of total peripheral vascular resistance unaffected.

For this purpose experiments were performed in anesthetized, thoracotomized, and vagotomized dogs. Since α-coronary vasoconstriction is part of the baroreceptor reflex [9], bilateral carotid occlusion was used to stimulate efferent sympathetic nerve activity. In addition, this protocol permits to study the effects of a general increase in sympathetic activity and to compare them with those effects induced by electrical stimulation of cardiac sympathetic nerves [17]. Possible interactions of pressure autoregulation [10, 28] and presynaptic α_2-adrenoceptors [21, 38] were examined in separate sets of experiments.

Methods

General preparation

Mongrel dogs (n = 27) with a body mass of 22–35 kg were anesthetized intravenously with α-D-chloralose (50 mg/kg) and urethane (250 mg/kg). Artificial respiration was maintained through a tracheal tube with a respirator (874052, Braun-Melsungen). Arterial pO_2, pCO_2 and pH were repeatedly measured (BMS 2 MK 2, Radiometer Copenhagen) and maintained within normal limits [11]. Arterial pCO_2 was adjusted by the ventilation, arterial pO_2 could be raised by supplementing the inspired air with oxygen, and arterial pH was regulated by intravenous infusion of sodium bicarbonate solution. Rectal temperature was kept close to 38 °C by means of a heating pad. The spleen was removed to prevent increases of hematocrit during activation of the sympathetic nervous system [29, 42]. In addition, the vagus nerves were cut bilaterally in the cervical region to eliminate possible cholinergic effects on heart rate and coronary circulation [14, 26, 43].

Left ventricular pressure and pressure in the ascending aorta were recorded with catheter-tip-manometers (PC 350, Millar). Left ventricular dP/dt and heart rate were derived from the ventricular pressure signal. A left thoracotomy was performed and the heart was suspended in a pericardial cradle. To measure cardiac output an electromagnetic flowprobe (SP 7515, Statham) was placed around the ascending aorta. Another flowprobe was used to measure circumflex coronary blood flow. Flowmeters (SP 2202, Statham) were electrically calibrated and "zero" of coronary flow was adjusted during shortlasting occlusions of the circumflex coronary artery at the end of bilateral carotid occlusions (see below). A small branch of the circumflex coronary artery was cannulated with a polyethylene catheter (1 mm o.d.) for measurement of peripheral epicardial coronary pressure (P 23 ID, Statham). The hemodynamic data were continuously recorded on a polygraph (R 612, Beckman).

To measure regional myocardial oxygen consumption, a 5F Goodale-Lubin catheter was advanced into a coronary vein, draining the myocardium supplied by the circumflex coronary artery. Oxygen content was determined in blood samples taken from the left femoral artery and the cannulated circumflex vein (Lex-O_2-Con TL, Lexington Instruments). As previously reported [20], by using the described catheterization technique blood samples can be accepted as representative for the circumflex-perfused myocardium. In addition, coronary venous pressure was measured with a pressure transducer (P 23 BB, Statham) by means of this catheter.

Rauwolscine experiments (n = 13)

Under control conditions the coronary dilator reserve was quantified by determination of the reactive hyperemia repayment [6] following a 15 s coronary occlusion. Then, approximately 10 min later a 60 s bilateral carotid occlusion was performed. In order to produce a stenosis a circumferential constrictor device (3 mm diameter, 4 mm length) was placed on the left circumflex coronary artery distal to the flowprobe and proximal to the cannulated circumflex branch. The constrictor device was adjusted to constrict the vessel until coronary blood flow was decreased by approximately 50%. Under this condition the reactive hyperemic response was almost completely abolished. In the presence of the coronary stenosis, a second bilateral carotid occlusion was performed. Finally, both common carotid arteries were occluded after administration of the selective α_2-adrenoceptor blocker rauwolscine-hydrochloride (0.2 mg/kg i.v., Roth, Karlsruhe) with the stenosis left unchanged.

To allow a comparison of the effects of 60 s bilateral carotid occlusion on efferent cardiac sympathetic nerve activity with those effects induced by a 20 min coronary hypoperfusion (previously reported in [20]), sympathetic discharge was measured in this experimental group. Signals obtained from a multi-fiber preparation of the left ventrolateral cervical cardiac nerve were amplified by a low level preamplifier (Tektronix 122), displayed on an oscilloscope (Tektronix 565), and then passed through a rate meter that counted impulses with amplitudes between two predetermined levels. Spikes per second were continuously recorded together with the hemodynamic parameters. Zero level of nerve activity was determined by ganglionic blockade (6 mg/kg hexamethonium i.v., Sigma, Munich) at the end of the experiments. Since cardiac sympathetic nerve activity was evaluated from multifiber preparations, no absolute values are reported. Control sympathetic nerve activity at the start of each experiment was defined as 100%. Changes were expressed in percent of control activity.

Rauwolscine-propranolol experiments (n = 5)

Before and after production of a severe coronary stenosis the effects of 60 s bilateral carotid occlusion were tested. Then the stenosis was released and the dogs received 2 mg/kg propranolol i.v. (Rhein-Pharma, Plankstadt). After treatment with propranolol the following experimental protocol was identical to that used in the rauwolscine group: bilateral carotid occlusions were performed before production of a stenosis, in the presence of a severe stenosis, and after additional administration of rauwolscine (0.2 mg/kg i.v.).

Pressure-controlled experiments (n = 9)

The left femoral artery was cannulated with a large bore tube (5 mm i.d.) and connected to a pressurized reservoir, the pressure of which was adjusted to mean arterial blood pressure. During bilateral carotid occlusion the pressure was allowed to increase by approximately 15 mm Hg. Coagulation and clotting of blood cells were prevented by 4 mg/kg heparin and 2 mg/kg aspirin (aspisol, Bayer, Wuppertal), respectively. Bilateral carotid occlusions were performed before and after coronary dilator reserve had been exhausted by the severe coronary stenosis, as well as after additional administration of rauwolscine (0.2 mg/kg i.v.).

Calculations and statistics

Blood samples and measurements of hemodynamic parameters were taken under steady-state conditions 10 s before and 40–60 s after bilateral carotid occlusion. End-diastolic peripheral coronary resistance was calculated according to Ohm's law from the pressure gradient between peripheral coronary pressure and coronary venous pressure and the left circumflex coronary blood flow. Oxygen consumption was calculated according to Fick's principle from the arterio-venous difference in oxygen content and coronary blood flow. After each experiment the poststenotic perfusion area of the left circumflex coronary artery was visualized by injection of calcium sulphate solution via the coronary catheter, then excised and weighed. Coronary flow, calculated peripheral coronary resistance, as well as myocardial oxygen consumption were related to 100 g of myocardium.

Data given in this study are means $\pm$ SEM. Changes during BCO were compared with control values using two-tailed Student's t-test for paired observations. Significance was considered to be present when p was less than 0.05.

Results

Under control conditions reactive hyperemia repayment following a 15 s complete occlusion of the left circumflex coronary artery was $410 \pm 40\%$ (n = 27). Production of a stenosis decreased mean coronary blood flow by $38 \pm 5\%$ (n = 27) and exhausted coronary dilator reserve almost completely; reactive hyperemia repayment following a 15 s coronary occlusion in the presence of a stenosis was $10 \pm 10\%$ (range 0–70%; n = 27). There was no significant difference in the reactive hyperemic response or in the degree of restriction of coronary blood flow by the stenosis between the three experimental groups (Tables 1–3).

Rauwolscine experiments

With coronary dilator reserve present, bilateral carotid occlusion (BCO) increased cardiac sympathetic nerve activity by $35 \pm 9\%$ (p < 0.05, n = 10). Simultaneously, the following changes in the hemodynamic and metabolic parameters occurred: mean aortic pressure (+51%), peak left ventricular pressure (+47%), dP/dt_{max} (+48%), and regional myocardial oxygen consumption (+58%), respectively (Table 1). Calculated end-diastolic coronary resistance decreased by 16% (Fig. 1).

While total cardiac performance remained unchanged following production of a severe coronary stenosis, regional myocardial oxygen consumption was slightly depressed (Table 1). As depicted in Fig. 2, in the presence of a severe coronary stenosis BCO induced increases in aortic pressure and cardiac performance comparable to those observed in the absence of a coronary stenosis (Table 1). Although peripheral coronary pressure considerably increased, coronary flow rose by a lesser degree (Fig. 2). Thus, end-diastolic poststenotic coronary resistance significantly (20%) increased (Fig. 1). In parallel, oxygen consumption in the circumflex-perfused myocardium increased only moderately (Table 1).

Administration of rauwolscine induced an increase in left ventricular dP/dt_{max} (Table 1) and a transient decrease in aortic pressure. However, within 10–15 min following application of rauwolscine, sympathetic nerve activity increased by $16 \pm 8\%$ (p < 0.05, n = 10). Hence, at the end of this period mean aortic pressure did not significantly differ from the pre-rauwolscine value. The other parameters remained unchanged (Table 1, Fig. 1).

Following rauwolscine, BCO caused a slightly lesser increase of sympathetic nerve activity ($21 \pm 5\%$ vs $33 \pm 5\%$ before rauwolscine). This difference was not statistically significant. During BCO peripheral coronary pressure and coronary blood flow increased by 31% and 37%, respectively (Table 1). Thus, end-diastolic poststenotic coronary resistance remained unchanged (Fig. 1). Regional myocardial oxygen consumption increased by 33% (Table 1).

Propranolol-rauwolscine experiments

The effects of BCO on hemodynamic parameters determined prior to infusion of propranolol were comparable to those reported for the rauwolscine experiments (data not shown). The

Table 1. Effects of 60 s bilateral carotid occlusion (BCO) on cardiac performance and perfusion of coronary arteries. BCO was performed under control conditions (no stenosis), with a severe stenosis, and after additional administration of rauwolscine (0.2 mg/kg i.v.).

	No stenosis		Severe stenosis		Severe stenosis + rauwolscine	
	Control	BCO	Control	BCO	Control	BCO
Left ventricular pressure (mm Hg)	115 ± 4	169 ± 8*	107 ± 4	159 ± 8*	105 ± 5	148 ± 9*
Left ventricular end-diastolic pressure (mm Hg)	4.6 ± 0.9	5.0 ± 0.8	4.8 ± 1.0	4.9 ± 1.2	4.8 ± 0.9	4.8 ± 0.9
Left ventricular dP/dt_{max} (mm Hg/s)	2000 ± 100	2960 ± 230*	1860 ± 110	2740 ± 250*	2690 ± 260	3670 ± 370†
Heart rate (bpm)	168 ± 5	180 ± 6†	170 ± 4	182 ± 5†	171 ± 6	184 ± 5†
Mean aortic pressure (mm Hg)	105 ± 4	159 ± 7*	98 ± 4	150 ± 8*	93 ± 5	129 ± 8*
Mean peripheral coronary pressure (mm Hg)	103 ± 4	157 ± 7*	53 ± 3	86 ± 8*	48 ± 5	63 ± 8*
Mean coronary blood flow (ml/min per 100 g)	69 ± 6	135 ± 17*	49 ± 3	69 ± 6*	51 ± 5	70 ± 8*
Regional myocardial oxygen consumption (ml/min per 100 g)	9.7 ± 1.1	15.1 ± 1.4*	7.6 ± 0.8	9.2 ± 1.4*	7.9 ± 0.8	10.8 ± 1.8*

Data are mean ± SEM, n = 13.
† $p < 0.05$, * $p < 0.01$ BCO vs Control.

197

Table 2. Effects of bilateral carotid occlusion (BCO) on cardiac performance and perfusion of coronary arteries in the presence of propranolol (2 mg/kg i.v.). BCO was performed under control conditions (no stenosis), with a severe stenosis, and after additional administration of rauwolscine (0.2 mg/kg i.v.).

	No stenosis		Severe stenosis		Severe stenosis + rauwolscine	
	Control	BCO	Control	BCO	Control	BCO
Left ventricular pressure (mm Hg)	107 ± 8	150 ± 9*	113 ± 7	162 ± 8*	117 ± 4	174 ± 3*
Left ventricular end-diastolic pressure (mm Hg)	2.6 ± 0.9	6.8 ± 1.2*	2.6 ± 0.6	6.8 ± 0.7*	3.2 ± 0.5	9.4 ± 1.4*
Left ventricular dP/dt_{max} (mm Hg/s)	1620 ± 120	1940 ± 90*	1580 ± 180	2030 ± 140*	1620 ± 150	2220 ± 120*
Heart rate (bpm)	121 ± 7	125 ± 8	119 ± 8	120 ± 8	120 ± 8	122 ± 9
Mean aortic pressure (mm Hg)	98 ± 9	141 ± 9*	104 ± 8	151 ± 8*	108 ± 4	161 ± 4*
Mean peripheral coronary pressure (mm Hg)	97 ± 9	141 ± 9*	47 ± 6	80 ± 9*	51 ± 3	86 ± 5*
Mean coronary blood flow (ml/min per 100 g)	54 ± 5	73 ± 10*	30 ± 7	45 ± 11*	32 ± 5	58 ± 8*

Data are mean ± SEM, n = 5.
*$p < 0.01$ BCO vs Control.

Table 3. Effects of bilateral carotid occlusion (BCO) on cardiac performance and perfusion of coronary arteries when the rise of aortic pressure was limited. BCO was performed under control conditions (no stenosis), with a severe stenosis, and after additional administration of rauwolscine (0.2 mg/kg i.v.).

	No stenosis		Severe stenosis		Severe stenosis + rauwolscine	
	Control	BCO	Control	BCO	Control	BCO
Left ventricular pressure (mm Hg)	106 ± 4	119 ± 4*	100 ± 3	112 ± 3*	99 ± 5	115 ± 6*
Left ventricular end-diastolic pressure (mm Hg)	4.4 ± 1.0	4.8 ± 0.8	4.2 ± 0.8	4.2 ± 0.9	4.4 ± 0.8	4.6 ± 1.0
Left ventricular dP/dt$_{max}$ (mm Hg/s)	1970 ± 70	2170 ± 110*	1780 ± 60	1980 ± 90*	2320 ± 290	2710 ± 340*
Heart rate (bpm)	165 ± 8	179 ± 10†	168 ± 7	184 ± 7†	175 ± 10	185 ± 12
Mean aortic pressure (mm Hg)	97 ± 4	108 ± 4*	91 ± 4	102 ± 4*	85 ± 5	100 ± 5*
Mean peripheral coronary pressure (mm Hg)	95 ± 4	107 ± 4*	47 ± 3	57 ± 3*	42 ± 3	49 ± 4*
Mean coronary blood flow (ml/min per 100 g)	67 ± 8	80 ± 10†	40 ± 5	39 ± 6	33 ± 2	40 ± 3†

Data are mean ± SEM, n = 9.
†p < 0.05, *p < 0.01 BCO vs Control.

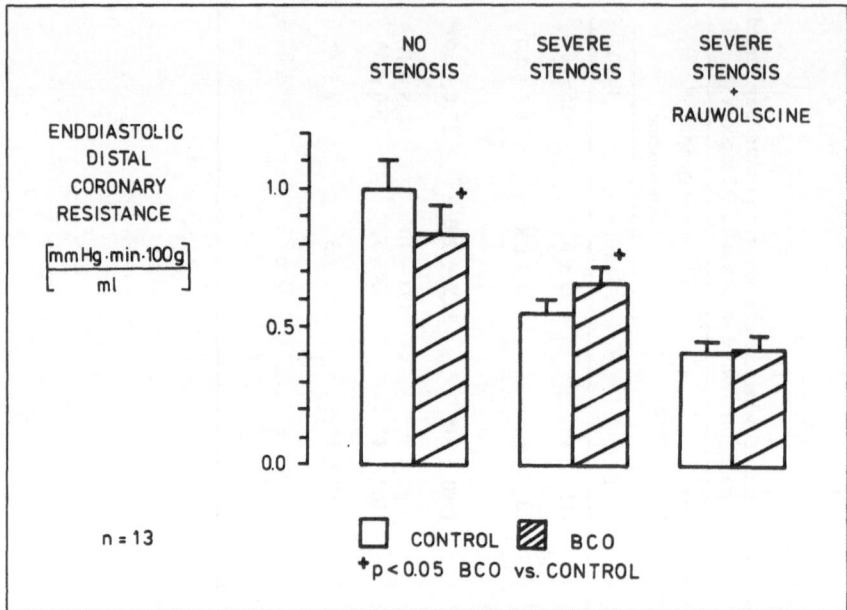

Fig. 1. Effect of 60 s bilateral carotid occlusion (BCO) on calculated end-diastolic distal coronary resistance. BCO was performed under control conditions (no stenosis), with a severe coronary stenosis, and after additional administration of rauwolscine (0.2 mg/kg i.v.).

mean increase in end-diastolic poststenotic coronary resistance prior to β-blockade was 16% (n = 5) compared to 20% (n = 13) in the rauwolscine experiments. Administration of propranolol (2 mg/kg i.v.) decreased heart rate by 24%, left ventricular dP/dt_{max} by 34%, and coronary blood flow by 15%. The other hemodynamic parameters were not significantly affected.

As shown in Table 2, in the presence of propranolol, reactions of the hemodynamic parameters to BCO were comparable under all conditions tested (no stenosis, severe stenosis, severe stenosis + rauwolscine). Peak left ventricular pressure increased by 45%, left ventricular end-diastolic pressure by 175%, left ventricular dP/dt_{max} by 29%, and mean aortic pressure by 47% (Table 2). In the absence, as well as in the presence of a severe coronary stenosis, end-diastolic coronary resistance significantly increased by 18% and 22%, respectively (Fig. 3). Infusion of rauwolscine prevented the increase of coronary resistance during BCO almost completely, although a small (4%), but still significant increase in end-diastolic poststenotic coronary resistance remained present (Fig. 3). The residual increase in resistance during BCO was blunted after additional infusion of prazosin (resistance during BCO -4 ± 6%, n = 3).

To estimate the contribution of peripheral α_2-adrenoceptors to the reflex rise in aortic blood pressure, cardiac output was measured in this experimental group. During BCO, cardiac output either remained constant (no treatment) or fell (β-adrenoceptor blockade, combined β- and α_2-adrenoceptor blockade). Calculated total peripheral resistance significantly increased by 54%, 79%, and 76%, respectively (Fig. 4). Additional infusion of the selective α_1-adrenoceptor antagonist prazosin (1.2 mg/kg i.v.) greatly abolished this increase of total peripheral resistance (Fig. 4).

Pressure-controlled experiments

In this experimental group, the rise of mean aortic pressure during BCO was limited to

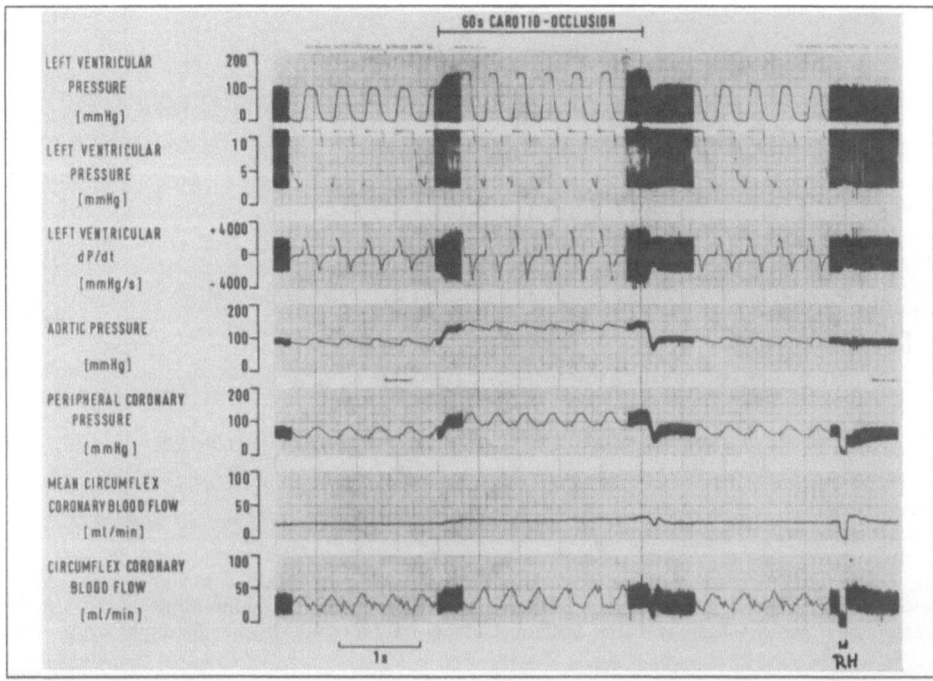

Fig. 2. Original recording showing the effects of a 60 s bilateral carotid occlusion on hemodynamic parameters in the presence of a severe coronary stenosis on the left circumflex coronary artery. While peripheral coronary perfusion pressure is notably increased, coronary blood flow is only slightly enhanced. "RH" indicates the 15 s coronary occlusion for determination of the reactive hyperemic response.

13 ± 3 mm Hg. This allowed peak left ventricular pressure and left ventricular dP/dt_{max} to increase by 14% and 13%, respectively. Regional myocardial oxygen consumption was not significantly affected by BCO (changes less than 8%, data not shown). In the absence of a stenosis, coronary blood flow increased to a comparable extent as coronary perfusion pressure (19% and 12%, respectively) (Table 3). Thus, end-diastolic coronary resistance remained unchanged (Fig. 5). However, after exhaustion of coronary dilator reserve by means of a stenosis, an increase in coronary perfusion pressure of 21% did not further enhance coronary blood flow (-2%; Table 3). Thus, end-diastolic poststenotic coronary resistance increased by 18% (Fig. 5). This resistance increase was blunted after rauwolscine (Table 3, Fig. 5).

Discussion

The present study provides evidence that a shortlasting increase in efferent cardiac sympathetic nerve activity within the physiologic range may increase end-diastolic poststenotic coronary resistance via postsynaptic vascular α_2-adrenoceptors. When poststenotic coronary resistance and cardiac afterload increase simultaneously, the increase in myocardial oxygen consumption becomes limited. In contrast to coronary resistance, total peripheral resistance is preferentially under control of α_1-adrenoceptors.

Neurogenic coronary vasoconstriction is part of the carotid sinus reflex [9]. This reflex vasoconstriction occurs independently of changes in myocardial oxygen consumption [35]. The present study demonstrates this vasoconstriction, even without previous blockade of cardiac β-adrenoceptors, when coronary dilator reserve is exhausted by a coronary stenosis. In the

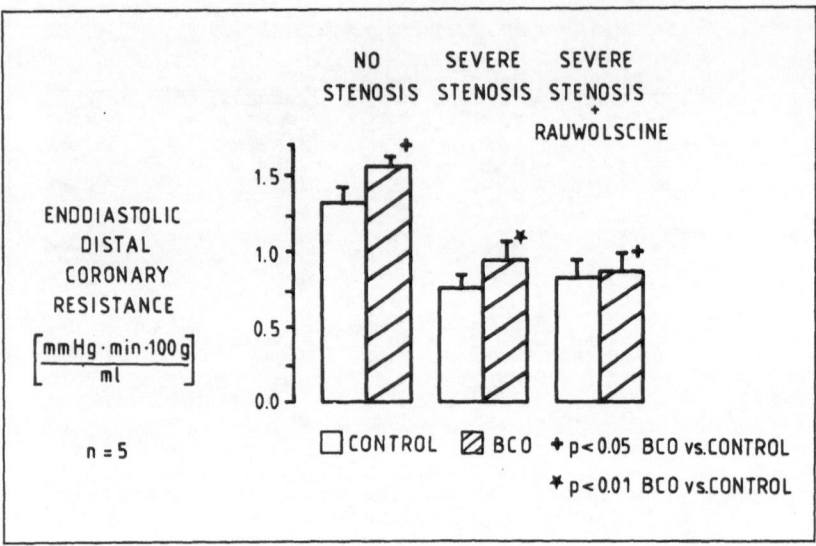

Fig. 3. Effects of 60 s bilateral carotid occlusion (BCO) on end-diastolic distal coronary resistance in the presence of propranolol (0.2 mg/kg i.v.). BCO was induced under control conditions (no stenosis), in the presence of a severe stenosis, and after additional administration of rauwolscine (0.2 mg/kg i.v.).

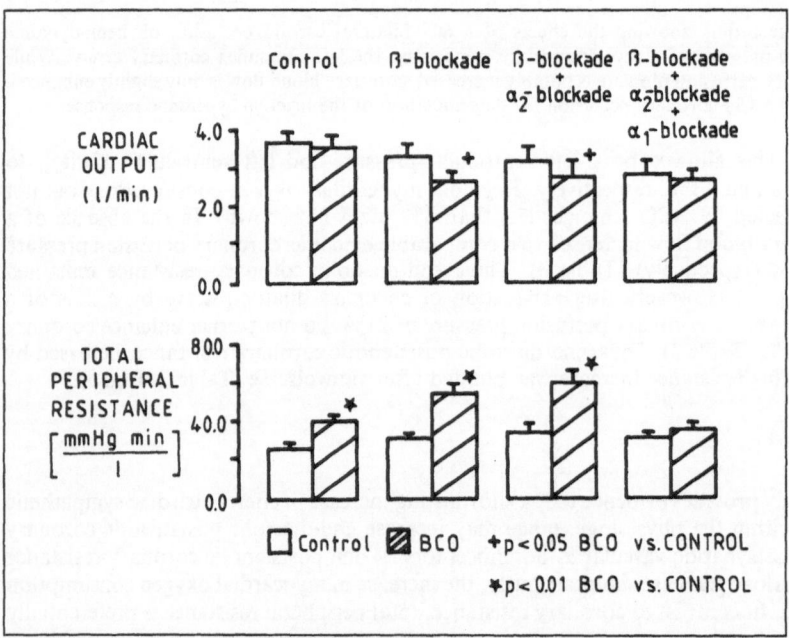

Fig. 4. Effects of 60 s bilateral carotid occlusion (BCO) on cardiac output and total peripheral resistance under control conditions (n = 5), with β-blockade (propranolol 2 mg/kg i.v., n = 5), with combined β- and α_2-blockade (rauwolscine 0.2 mg/kg i.v., n = 5), and with combined β-, α_2-, and α_1-blockade (prazosin 1.2 mg/kg i.v., n = 3).

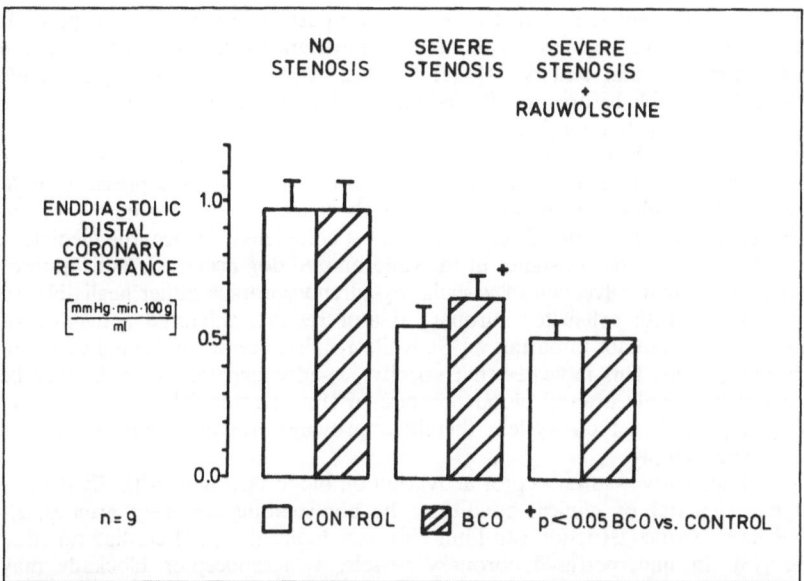

Fig. 5. Effects of 60 s bilateral carotid occlusion (BCO) on end-diastolic distal coronary resistance. The increase in aortic pressure during BCO was limited to 13 ± 3 mm Hg. BCO was performed under control conditions (no stenosis), with a severe stenosis, and after additional administration of rauwolscine (0.2 mg/kg i.v.).

above studies the reflex increase of coronary resistance during BCO ranged between 15% and 22% (with coronary dilator reserve intact or exhausted). In contrast, Woodman [44] reported a 70% increase of late diastolic resistance in non-stenotic coronary vessels during bilateral carotid occlusion (BCO).

BCO considerably increased systemic arterial blood pressure (Tables 1, 2). To relate the increase in poststenotic coronary resistance to an adrenergic constrictor mechanism, influences of extravascular compression, as well as autoregulation of coronary resistance must be excluded. The effects of an increased left ventricular pressure on coronary resistance were minimized by calculating end-diastolic resistances [7]. Although end-diastolic ventricular pressure remained unchanged during BCO in the rauwolscine group (Table 1), end-diastolic poststenotic coronary resistance significantly increased (Fig. 1), suggesting that this rise occurred independently of extravascular compression. However, end-diastolic left ventricular pressure rose significantly in the β-blocked dogs during BCO. Hence, additional experiments were performed in which the increase of left ventricular afterload was limited. This procedure also prevented the increase in left ventricular preload while the increase in end-diastolic poststenotic coronary resistance during BCO persisted (Fig. 5), indicating that the rise in coronary resistance was largely independent of extravascular compression. In addition, this result provides evidence that vascular autoregulation following an increase in coronary perfusion pressure [10, 28] was no major determinant of the enhanced coronary resistance under these conditions. The attribution of the increase in poststenotic coronary resistance to a direct adrenergic mechanism, rather than to extravascular compression or autoregulation, is finally emphasized by the effectiveness of rauwolscine in preventing the increase in end-diastolic poststenotic coronary resistance.

α_2-Adrenoceptors have been demonstrated at pre- as well as at postsynaptic sites [15, 21, 23, 38]. Presynaptic α_2-adrenoceptors mediate a feed-back inhibition of neuronal norepinephrine release, and blockade of these receptors may result in metabolic coronary vasodilation secondary to postsynaptic myocardial β-adrenoceptor activation [21, 38]. Since rauwolscine still prevented most of the reflex coronary vasoconstriction after pretreatment with propranolol, this constriction seems to be directly mediated via postsynaptic α_2-adrenoceptors. It is unlikely that the prevention of the coronary constriction by rauwolscine was due to an action on neurons in the cardiovascular centers [31]. While reflex increases in efferent sympathetic nerve activity were comparable before and after administration of rauwolscine, the changes in coronary resistance were largely abolished.

Rauwolscine diminished the reflex rise in aortic pressure by only 25% in the rauwolscine group (no propranolol) and had no effect on the pressure response in the presence of β-adrenoceptor blockade. Additional administration of the selective α_1-antagonist prazosin prevented the increase of total peripheral resistance almost completely. Hence, the mediation of the increase in total peripheral resistance in the vagotomized dog occurs via activation of α_1-adrenoceptors, while the involvement of vascular α_2-adrenoceptors is rather negligible. In the coronary circulation with exhausted coronary dilator reserve, enhanced activation of α_1-adrenoceptors may be of minor importance [18], while α_2-adrenoceptor-mediated vasoconstriction clearly persists [8]. This indicates that selective α_2-adrenoceptor blockade may be advantageous, since it permits arterial blood pressure to still increase following a general activation of the sympathetic nervous system, thereby improving coronary perfusion pressure and myocardial oxygen supply.

The net effect of coronary α-adrenoceptor activation on blood flow may critically depend on the actual experimental or clinical condition. In non-stenotic coronary arteries α_1-adrenoceptor-mediated vasoconstriction can limit coronary blood flow and cardiac function during exercise [13]. In underperfused coronary vessels, α_1-adrenoceptor blockade may improve resting blood flow, if coronary perfusion pressure is kept constant [30]. However, with freely variable coronary perfusion pressure, activation of α_1-adrenoceptors by the selective agonist methoxamine has been shown to favor a redistribution of flow into underperfused areas [4, 22]. This beneficial effect of selective α_1-stimulation was later explained by the observation that methoxamine exerts a stronger constrictor effect on the resistance of normally perfused than of underperfused coronary vessels [18].

Recently, Nathan and Feigl [34] reported experiments in which they found an accentuation of subendocardial ischemia following treatment with the α-adrenoceptor antagonist phenoxybenzamine. Chilian and Ackell [5] provided evidence that, distal to a flow-limiting stenosis, α-adrenergic vasoconstriction in the outer layers of the left ventricle facilitates a redistribution of flow toward the subendocardium. This beneficial effect of coronary α-adrenoceptors may potentially counteract the global flow reduction to the peripheral vascular bed of a constricted coronary artery [17]. It should be noted, however, that in recent studies by Seitelberger et al. [40] the reduction of subendocardial blood flow and systolic wall thickening in ischemic myocardium of dogs during exercise was significantly attenuated by the selective α_2-antagonist idazoxan.

Reflex-induced myocardial ischemia was claimed to be of clinical importance. In patients with coronary artery disease, an increase in plasma catecholamines [33] as well as an increase in coronary resistance may occur during the cold pressor test [32, 12]. In some of these patients angina is precipitated [32]. Very recently, in patients with stable exertional angina, infusion of phentolamine into the most severely stenotic coronary arteries significantly improved the ischemic response toward ergometric exercise [1]. The existence of prazosin-resistant α-adrenoceptors in human coronary vessels has meanwhile been recognized [2]. With respect to the results of the present study it seems likely that these coronary α-adrenoceptors may contribute to coronary vasoconstriction during reflex increases of sympathetic nerve activity.

References

1. Berkenboom GM, Abramowicz M, Vandermoten P, Degre SG (1986) Role of alpha-adrenergic coronary tone in exercise-induced angina pectoris. Am J Cardiol 57:195–198
2. Berkenboom GM, Fontaine J, Desmet JM, Degre SG (1987) Comparison of the effect of beta adrenergic antagonists with different ancillary properties on isolated canine and human coronary arteries. Cardiovasc Res 21:299–304

3. Buffington CW, Feigl EO (1981) Adrenergic coronary vasoconstriction in the presence of coronary stenosis in the dog. Circ Res 48:416–423
4. Chiariello M, Ribeiro LGT, Davis MA, Maroko PR (1977) "Reverse Coronary Steal" induced by coronary vasoconstriction following coronary artery occlusion in dogs. Circulation 56:809–815
5. Chilian WM, Ackell PH (1988) Transmural differences in sympathetic coronary constriction during exercise in the presence of coronary stenosis. Circ Res 62:216–225
6. Coffman JD, Gregg DE (1960) Reactive hyperemia characteristics of the myocardium. Am J Physiol 199:1143–1149
7. Denison AB, Bardhanabaedya S, Green HD (1956) Adrenergic drugs and blockade on coronary arterioles and myocardial contraction. Circ Res 4:653–658
8. Deussen A, Heusch G, Thämer V (1985) α_2-Adrenoceptor-mediated coronary vasoconstriction persists after exhaustion of coronary dilator reserve. Eur J Pharmacol 115:147–153
9. Feigl EO (1968) Carotid sinus reflex control of coronary blood flow. Circ Res 23:223–237
10. Feigl EO (1983) Coronary physiology. Physiol Rev 63:1–205
11. Feigl EO, D'Alecy LG (1972) Normal arterial blood pH, oxygen, and carbon-dioxide tensions in unanesthetized dogs. J Appl Physiol 32:152–153
12. Gunther S, Green L, Muller JE, Mudge GH, Grossman W (1979) Inappropriate coronary vasoconstriction in patients with coronary artery disease: A role for nifedipine? Am J Cardiol 44:793–797
13. Gwirtz PA, Overn SP, Mass HJ, Jones CE (1986) α_1-Adrenergic constriction limits coronary flow and cardiac function in runnings dogs. Am J Physiol 250:H1117–H1126
14. Hackett JG, Abboud FM, Mark AL, Schmid PG, Heistad DD (1972) Coronary vascular response to stimulation of chemoreceptors and baroreceptors. Evidence for reflex activation of vagal cholinergic innervation. Circ Res 31:8–17
15. Hamilton CA, Reid JL (1982) A postsynaptic location of alpha$_2$-adrenoceptors in vascular smooth muscle: in vivo studies in the conscious rabbit. Cardiovasc Res 16:11–15
16. Heusch G (1990) α-Adrenergic mechanisms in myocardial ischemia. Circulation 81:1–13
17. Heusch G, Deussen A (1983) The effects of cardiac sympathetic nerve stimulation on perfusion of stenotic coronary arteries in the dog. Circ Res 53:8–15
18. Heusch G, Yoshimoto N, Heegeman H, Thämer V (1983) Interaction of methoxamine with compensatory vasodilation distal to coronary stenoses. Drug 33:1647–1650
19. Heusch G, Deussen A, Schipke J, Thämer V (1984) α_1- and α_2-adrenoceptor-mediated vasoconstriction of large and small canine coronary arteries in vivo. J Cardiovasc Pharmacol 6:961–968
20. Heusch G, Deussen A, Thämer V (1985) Cardiac sympathetic nerve activity and progressive vasoconstriction distal to coronary stenoses: feed-back aggravation of myocardial ischemia. J Auton Nerv Syst 13:311–326
21. Heyndrickx GR, Vilaine JP, Moerman EJ, Leusen I (1984) Role of prejunctional α_2-adrenergic receptors in the regulation of myocardial performance during exercise in conscious dogs. Circ Res 54:683–693
22. Hirshfield JW, Borer JS, Goldstein RE, Barrett MJ, Epstein SE (1974) Reduction in severity and extent of myocardial infarction when nitroglycerin and methoxamine are administered during coronary occlusion. Circulation 49:291–297
23. Hoffman BB, Lefkowitz RJ (1980) Alpha-adrenergic receptor subtypes. N Engl J Med 302:1390–1396
24. Holtz J, Saeed M, Sommer O, Bassenge E (1982) Norepinephrine constricts the canine coronary bed via postsynaptic α_2-adrenoceptors. Eur J Pharmacol 82:199–202
25. Imai S, Otorii T, Takeda K, Katano Y (1975) Coronary vasodilatation and adrenergic receptors in the dog heart and coronary. Jap J Pharmacol 25:423–432
26. Ito BR, Feigl EO (1985) Carotid baroreceptor reflex coronary vasodilation in the dog. Circ Res 56:486–495
27. Johannsen UJ, Mark AL, Marcus ML (1982) Responsiveness to cardiac sympathetic nerve stimulation during maximal coronary dilation produced by adenosine. Circ Res 50:510–517
28. Kirchheim H (1976) Systemic arterial baroreceptor reflexes. Physiol Rev 56:100–176
29. Kramer K, Luft UC (1951) Mobilization of red cells and oxygen form the spleen in severe hypoxia. Am J Physiol 165:215–228
30. Liang IYS, Jones CE (1985) α_1-Adrenergic blockade increases coronary blood flow during coronary hypoperfusion. Am J Physiol 249:H1070–H1077
31. McCall RB, Schuette MR, Humphrey SJ, Lahti RA, Bahrsun C (1983) Evidence for a central sympathoexcitatory action of alpha-2 adrenergic antagonists. J Pharmacol Exp Ther 224:501–507
32. Mudge GH, Grossman W, Mills RM, Lesch M, Braunwald E (1976) Reflex increase in coronary vascular resistance in patients with ischemic heart disease. N Engl J Med 295:1333–1337
33. Mueller HS, Rao PS, Rao PB, Gory DJ, Mudd JG, Ayres SM (1982) Enhanced transcardiac 1-norepi-

nephrine response during cold pressor test in obstructive coronary artery disease. Am J Cardiol 50:1223–1228

34. Nathan HJ, Feigl EO (1986) Adrenergic vasoconstriction lessens transmural steal during coronary hypoperfusion. Am J Physiol 250:H645–H653
35. Powell JR, Feigl EO (1979) Carotid sinus reflex coronary vasoconstriction during controlled myocardial oxygen metabolism in the dog. Circ Res 44:44–51
36. Ross G (1976) Adrenergic responses of coronary vessels. Circ Res 39:461–465
37. Ruffolo RR (1987) The α_1-adrenergic receptors, Humana Press, Clifton
38. Saeed M, Sommer O, Holtz J, Bassenge E (1982) α-Adrenoceptor blockade by phentolamine causes β-adrenergic vasodilation by increased catecholamine release due to presynaptic α-blockade. J Cardiovasc Pharmacol 4:44–52
39. Saeed M, Holtz J, Elsner D, Bassenge E (1985) Sympathetic control of myocardial oxygen balance in dogs mediated by activation of coronary vascular α_2-adrenoceptors. J Cardiovasc Pharmacol 7:167–173
40. Seitelberger R, Guth BD, Heusch G, Lee JD, Katayama K, Ross Jr J (1988) Intracoronary α_2-adrenergic receptor blockade attenuates ischemia in conscious dogs during exercise. Circ Res 62:436–442
41. Thomas JX, Jones CE, Randall WC (1984) Neural modulation of coronary blood flow. In: Randall WC (ed) Nervous control of cardiovascular function. New York, Oxford Univ. Press, pp 178–198
42. Vatner SF, Higgins CB, Millard RW, Franklin D (1974) Role of the spleen in the peripheral vascular response to severe exercise in untethered dogs. Cardiovasc Res 8:276–282
43. Vatner SF, McRitchie RJ (1975) Interaction of the chemoreflex and the pulmonary inflation reflex in the regulation of coronary circulation in conscious dogs. Circ Res 37:664–673
44. Woodman OL (1987) The role of α_1- and α_2-adrenoceptors in the coronary vasoconstrictor responses to neurally released and exogenous noradrenaline in the dog. Naunyn-Schmiedeberg's Arch Pharmacol 336:161–168
45. Woodman OL, Vatner SF (1987) Coronary vasoconstriction mediated by α_1- and α_2-adrenoceptors in conscious dogs. Am J Physiol 253:H388–H393

Authors' address:

Priv. Doz. Dr. A. Deussen, Zentrum für Physiologie der Heinrich-Heine-Universität Düsseldorf, Moorenstr. 5, 4 Düsseldorf 1, FRG.

α_2-Adrenergic Coronary Constriction in Ischemic Myocardium during Exercise

R. Seitelberger[1], B. D. Guth[2], G. Heusch[2] and J. Ross, Jr.*

Seaweed Canyon Laboratory, Division of Cardiology, Department of Medicine, University of California, San Diego, School of Medicine, La Jolla, California, USA

Summary

The effect of either selective α_1- or α_2-adrenoceptor blockade on ischemic myocardial blood flow and function was examined in β-blocked dogs trained to run on a motor-driven treadmill. The animals were instrumented with sonomicrometers for the assessment of regional systolic wall thickening (%WTh) of the left ventricle. For drug infusion, an intracoronary catheter was implanted in the circumflex coronary artery and a hydraulic cuff was placed proximally around the artery. Following systemic β-blockade with 0.8 mg/kg propranolol, an acute stenosis of the circumflex coronary artery inflated during exercise induced severe dysfunction in the posterior wall. Intracoronary infusion of 80 μg/kg of the selective α_2-adrenoceptor blocking agent idazoxan improved posterior wall (PW)-%WTh from 5.1 $\pm$ 1.6 to 10.8 $\pm$ 2.8% (p < 0.06) and regional myocardial blood flow (radiolabelled microspheres) in the subendocardium of the posterior wall from 0.17 $\pm$ 0.05 to 0.45 $\pm$ 0.30 (ml/min)/g (p < 0.05). No increases in regional blood flow and regional myocardial function were observed after infusion of the selective α_1-adrenoceptor blocking agent prazosin (20 μg/kg) under the same experimental conditions.

It is concluded that during severe ischemia, significant postjunctional α_2-adrenoceptor mediated coronary vasoconstriction exists. Regional α_2-adrenoceptor blockade, but not α_1-adrenoceptor blockade is effective in reducing regional ischemia and dysfunction by attenuating sympathetic vasoconstriction in the conscious dog.

Introduction

During sympathetic activation induced by exercise or stress, humoral and neuronal release of norepinephrine activates myocardial and coronary α- as well as β-adrenoceptors. Whereas stimulation of β-adrenoceptors increases myocardial performance and oxygen demand under nonischemic conditions, the stimulation of α-adrenoceptors competes with coronary vasodilation caused by local metabolic mechanisms and increases coronary resistance [10, 14, 21]. Whether α-adrenergic coronary constriction is powerful enough to limit coronary blood flow also under ischemic conditions, when endogenous coronary dilator reserve is exhausted, remains controversial [8, 16, 23]. Another important question is which α-adrenoceptor subtype, α_1 or α_2, is responsible for sympathetic coronary vasoconstriction during ischemia.

Studies in anesthetized and conscious dogs demonstrated that in the presence of severe coronary stenoses, sympathetic coronary vasoconstriction mediated by α_2-adrenoceptors can overcome local metabolic vasodilation and aggravate myocardial ischemia during cardiac

Current Addresses: [1]2nd Department of Surgery, University of Vienna, Austria; [2]Department of Pathophysiology, Center of Internal Medicine, University of Essen, FRG.

sympathetic nerve stimulation [9] or exercise [27]. However, studies by Liang and Jones (19) and Jones et al. [15] in anesthetized dogs, and Laxson et al. [18] in exercising dogs reported that only selective α_1-adrenoceptor blockade has beneficial effects on the myocardium during moderate hypoperfusion. The goal of the present paper is to review evidence for the role of α_1- and α_2-adrenoceptor-mediated coronary vasoconstriction during severe myocardial ischemia and activation of the sympathetic nervous system by a normal physiological stimulus. These data include those from a previously published study on the anti-ischemic effects of the selective α_2-adrenoceptor blocker idazoxan during exercise [27], as well as of additional experiments with the selective α_1-adrenoceptor blocker prazosin in the same experimental model.

Methods

Animal model

Mongrel dogs (24–33 kg) were trained to run on a motor-driven treadmill before undergoing surgery under general anesthesia. A left lateral thoracotomy was performed and the pericardium opened widely. A miniature pressure transducer (Konigsberg P7) and a Tygon fluid-filled catheter were inserted through a stab wound in the apex of the left ventricle to measure left ventricular pressure. The transducer was calibrated by direct measurement of left ventricular pressure through the fluid-filled catheter (Statham P23D6), zero pressure reference being taken at the estimated level of the right atrium. A silicone rubber catheter (1.57 mm i.d.) was inserted into the descending aorta for blood withdrawal and measurement of arterial pressure. An additional catheter was placed into the left atrium through the atrial appendage for the injection of microspheres.

The proximal left circumflex coronary artery was dissected free and a heparin-filled Tygon catheter (0.5 mm i.d.) was inserted near its origin. The catheter was maintained in position by sewing a Dacron patch attached to the catheter to the surrounding tissue. Special care was taken to place the tip of the intracoronary catheter proximal to the origin of the first main branch of the circumflex artery. A hydraulic cuff was then placed proximal to the coronary catheter.

For the measurement of regional wall thickening, two pairs of miniature ultrasonic crystals were implanted in the left ventricular wall using standard techniques [24]. One pair was implanted in the posterior wall supplied by the circumflex coronary artery, and the other was positioned in the anterior wall within the distribution area of the left anterior descending artery. The pericardium was left open and all wires and tubes were passed subcutaneously to the back of the dog and brought through the skin between the scapulae. The pneumothorax was evacuated through a chest tube in the 6th intercostal space. Figure 1 illustrates a completely instrumented dog heart.

Measurement of regional myocardial blood flow

Myocardial blood flow was determined using radionuclide-labelled microspheres by the reference withdrawal method [6, 12]. Two of the following radionuclide labels were used randomly in each dog: ^{141}Ce, ^{113}Sn, ^{103}Ru, ^{95}Nb, or ^{46}Sc (15 μm diameter, New England Nuclear, Boston). For each measurement, 1–2 ml containing 6–10 million microspheres (suspended in 10% dextran with Tween-80) were injected within approximately 5 s into the left atrial catheter and then flushed with warm saline. The arterial reference sample was withdrawn from the aortic catheter (8 ml/min over a period of 90 s), starting at least 5 s before the microsphere injection.

At the end of the study, the animals were killed with an overdose of barbiturate, and the heart was removed and placed in 10% formalin for 5 days. The atria, right ventricle, and blood vessels were then removed, and the left ventricular surface was cleaned of epicardial fat. The left ventricle was sliced perpendicular to the long axis and then sectioned as described previously. Data are reported only from transmural tissue samples containing the ischemic and

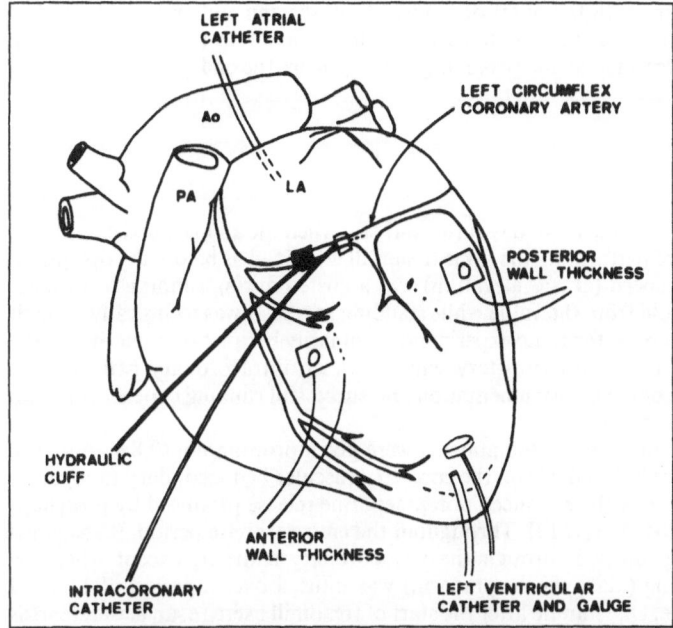

Fig. 1. Schematic illustration of the surgical instrumentation performed in dog hearts. PA = pulmonary artery; Ao = aorta; LA = left atrium.

control crystal pairs. However, blood flow was examined in the entire circumference of the left ventricle at the level of the ischemic crystal pair to assure that the crystals were placed within the central zone of ischemia.

Each transmural plug was subdivided into endocardial, midwall, and epicardial pieces, weighed, and placed in glass counting vials. The radioactivity of tissue and arterial blood samples was counted in a Packard Autogamma Spectrometer (model 5912) with a multichannel analyzer at the energy peak of each radionuclide [6]. The number of counts per energy window was determined and corrected for background activity and overlapping counts from accompanying isotopes by solving simultaneous equations using a matrix inversion technique [25]. Myocardial blood flow is reported as milliliters per min per gram of myocardium ((ml/min)/g). To help ensure that each sample contained no less than 400 microspheres, no sample weighed less than 0.5 gram.

The number of spheres in each tissue sample was calculated using the number of counts per min per sphere that had been previously measured for each microsphere label employed. Tissue samples from the ischemic zone later underwent histological examination to assess whether small emboli from the intracoronary catheter induced myocardial necrosis. No dog included in this study showed tissue damage between the sonomicrometers, and in each study reported, the sonomicrometers were appropriately positioned across the wall.

Data analysis

Hemodynamic data were recorded on a Brush forced-ink recorder and magnetic tape. Data recorded on the magnetic tape were used for subsequent digitization, beat averaging, and analysis by computer (PDP11/03). Fifteen consecutive beats were averaged for each observation. Parameters measured and analyzed included peak left ventricular (LV) pressure and end-diastolic pressure (LVEDP), maximum positive left ventricular dP/dt (max dP/dt), heart rate, mean arterial pressure, and anterior and posterior wall thickness (WTh) at end-diastole (defined as the time coincident with the onset of a positive dP/dt) and end-systole (defined as the point of maximal systolic thickening within 20 ms before maximum (−) LV dP/dt [17]. Systolic wall thickening is presented as the percentage change from the end-diastolic thickness (%WTh).

Study protocol

All experiments were performed at least 10 days after surgery, when the animals had completely recovered and were capable of performing the same treadmill workload as before the operation. For each dog, the treadmill speed (at 5% elevation) was adjusted during running to increase the heart rate by 1.6- to 1.8-fold from the value while standing at rest. It was technically difficult to achieve fully instrumented dogs for these experiments, primarily because of a relatively high incidence of thrombosis of the coronary artery when both an intracoronary catheter and hydraulic cuff were placed. Complete instrumentation and successful running experiments with ischemia were accomplished in five dogs.

Ten min before the start of exercise, the animals were given propranolol (0.8 mg/kg i.v.). This was done to prevent metabolic effects on the coronary vascular bed secondary to myocardial β-adrenoceptor activation by the enhanced norepinephrine release produced by presynaptic α_2- or α_1-adrenoceptor blockade [7, 13]. Throughout the entire exercise period, 0.9% NaCl (2 ml/min) was continuously infused through the intracoronary catheter, except when the intracoronary α-blocking drug (idazoxan or prazosin) was infused over a period of 1 min at exactly the same infusion rate. One minute after the start of treadmill exercise, an acute stenosis of the circumflex coronary artery was set by partially inflating and then adjusting the hydraulic cuff to achieve an average decrease of 75% in systolic thickening of the posterior wall. When all recorded systemic hemodynamic measurements, and both the anterior and posterior %WTh had reached a steady state for at least 1 min, no further adjustments of the cuff were made and the withdrawal of the arterial reference blood sample was started.

Microspheres were then injected to assess regional myocardial blood flow during acute coronary stenosis. Immediately after the withdrawal period of 90 s, the selective α_2-adrenoceptor blocking agent idazoxan (80 μg/kg, n = 5) or the selective α_1-adrenoceptor blocking agent prazosin (20 μg/kg, n = 3) were infused by the intracoronary route over 1 min. When %WTh of the posterior wall had reached a new steady state lasting at least 30 s (at most 4 min after the beginning of the intracoronary drug infusion), withdrawal of the second reference blood sample was commenced and microspheres were injected, with a steady-state maintained for the next 90 s. After the end of the withdrawal period (90 s), the hydraulic cuff was deflated and the run terminated.

Control experiments with ischemia, but without administration of any α-adrenoceptor blocking drug were performed in three dogs. In these experiments, the animals continued running on the treadmill for at least 8 min after the acute coronary stenosis was set, a period equivalent to the total running time when microsphere injections and drug administration were performed during ischemia.

Statistical analysis

All data are reported as mean ± SD. Serial changes in hemodynamic variables and regional wall thickness values were analyzed by a repeated measure analysis of variance, and when a significant overall effect was observed, single mean values were compared with Tukey's test. Since blood flow measurements were made in the steady state at only two time points (the control microsphere injection during running and the microsphere injection after selective α-adrenoceptor-blockade), a paired nonparametric test was utilized (Mann-Whitney test). Due to the low number of experiments (n = 3), data with prazosin are presented as mean ± SD without statistical analysis.

Results

Figure 2 shows an original tracing of a complete experiment with infusion of the selective α_2-adrenoceptor blocker idazoxan. Because of the prior β-adrenoceptor blockade with propranolol, the increases in regional and global myocardial function and heart rate were relatively small immediately after the onset of exercise (Table 1). The production of an acute coronary stenosis reduced left ventricular pressure and maximum ($+$) LV dP/dt, but the additional small increase in heart rate was not significant.

After the coronary stenosis was set, %WTh of the posterior wall fell by an average of 75.7%. Intracoronary infusion of the selective α_2-adrenoceptor blocking agent idazoxan increased %WTh of the posterior wall in every dog, the average increasing from 5.1 to 10.8%, but it had no effect on %WTh of the anterior wall (Table 1). Hemodynamic variables were not significantly changed after the drug infusion. Changes in regional myocardial blood flow are shown in Table 2. After acute circumflex stenosis, blood flow to the anterior wall showed a normal blood flow distribution. However, there was a maldistribution of transmural blood flow in the posterior wall, with the lowest flow in the subendocardium. Intracoronary infusion of idazoxan markedly improved blood flow to the subendocardium and midmyocardium of the posterior wall in all five dogs (Table 2). Average blood flow to the subendocardium increased from 0.17 to 0.45 (ml/min)/g. Drug infusion had no effect on subepicardial blood flow, and the endocardial/epicardial flow ratio increased in the posterior wall (Table 2).

In experiments with running, after β-adrenoceptor blockade and subsequent intracoronary infusion of the selective α_1-adrenoceptor blocking agent prazosin (n = 3), the decrease in posterior wall %WTh due to acute coronary stenosis during the run was 82.3%. Changes in systemic hemodynamic parameters after setting of the acute coronary stenosis were similar to the experiments with idazoxan. However, the intracoronary infusion of prazosin (20 μg/kg) over 1 min did not improve posterior wall function in any experiment (Table 1). Accordingly, no increase in regional myocardial blood flow was detected after prazosin infusion (Table 2).

In control experiments (n = 3) with running, after β-adrenoceptor blockade, but without intracoronary infusion of α-adrenoceptor blocking drugs, inflation and setting of the hydraulic cuff caused a decrease in posterior %WTh by 84.8% under steady-state conditions (2 min after the stenosis was set), and the run was then continued for more than 8 min without further adjustment of the cuff. During this period, %WTh of the posterior wall did not improve over time in any of the three animals, but rather showed a tendency toward further decreases averaging -3.5 ± 1.6%, $-1.6\% \pm 3.1$%, -2.8 ± 2.4%, and -1.9 ± 3.4% at 2, 4, 6, and 8 min, respectively. Hemodynamic variables remained unchanged throughout this exercise period.

Discussion

The role of adrenergic coronary vasoconstriction during myocardial hypoperfusion has been investigated in various experiments and clinical models. In 1967, Bassenge et al. [1] reported a shift from metabolic coronary dilation to coronary constriction during intracoronary norepinephrine infusion and progressive reduction of coronary perfusion pressure. Buffington and Feigl [2] first identified coronary α-adrenoceptors as the mediator of sympathetic coronary vasoconstriction in dogs with moderate coronary stenosis, and Mudge et al. [22] suggested an adrenergically triggered increase in coronary resistance in patients with ischemic heart disease during the cold pressor test. Studies from Heusch et al. [9] in anesthetized dogs with electrical cardiac sympathetic nerve stimulation demonstrated that metabolic coronary dilation is progressively reversed to α-adrenergic coronary constriction with increasing severity of coronary stenosis. In addition, they identified postsynaptic α_2-adrenoceptors as the predominant mediators of sympathetic coronary vasoconstriction induced by sympathetic nerve stimulation during myocardial ischemia [9]. The goal of the present study was to examine the question of whether or not sympathetic coronary vasoconstriction mediated by α_2- and/or α_1-adrenoceptors also affects ischemic myocardium during the physiological stress of exercise. The results of

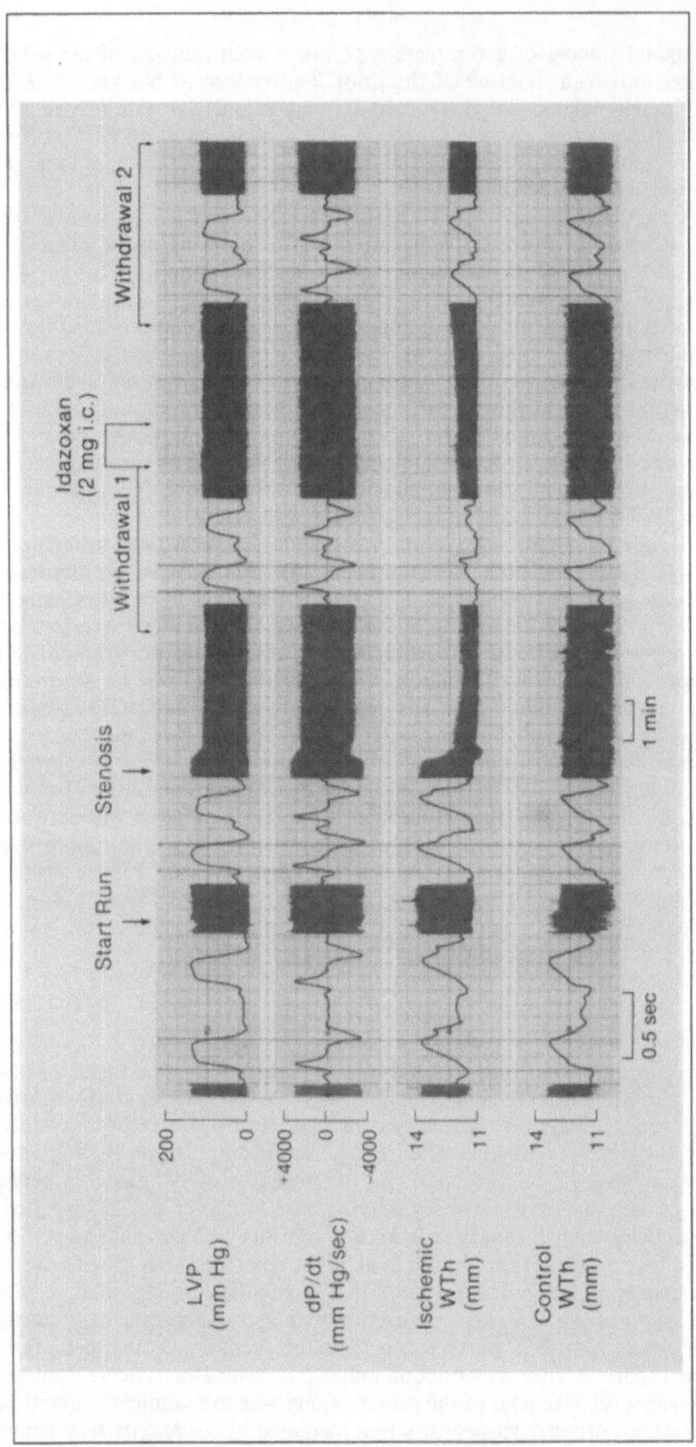

Fig. 2. Original recording from one study demonstrating the improvement of ischemic regional wall thickening in a conscious dog during treadmill exercise by intracoronary infusion of the selective α_2-adrenoceptor blocker idazoxan distal to the coronary stenosis.
The beginning of exercise (Start Run) and the commencement of the coronary stenosis (Stenosis) are noted. Arterial sampling periods for the microsphere technique are indicated as Withdrawal 1 (before drug infusion) and Withdrawal 2 (after Idazoxan infusion). LVP = left ventricular pressure; dP/dt = the first derivative of LVP; Ischemic or Control WTh = wall thickness in the ischemic or control region.

Table 1. Influence of intracoronary infusion of either the selective α_2-adrenoceptor blocker idazoxan or the selective α_1-adrenoceptor blocker prazosin on hemodynamics and regional myocardial function during treadmill exercise with acute coronary stenosis

		Rest		During run		
		Before propranolol	After propranolol	Before stenosis	Stenosis-control	Stenosis-treatment
HR (1/min)	Idazoxan	104 ± 22	105 ± 17	160 ± 14++	165 ± 20	167 ± 22
	Prazosin	106 ± 10	94 ± 5	160 ± 8	174 ± 18	177 ± 22
LVP max (mm Hg)	Idazoxan	138 ± 9	149 ± 9*	150 ± 28	127 ± 27#	127 ± 22#
	Prazosin	136 ± 26	151 ± 37	132 ± 19	116 ± 19	105 ± 10
(+)LV dP/dt (mm Hg/s)	Idazoxan	3610 ± 338	2697 ± 72**	3068 ± 570+	2593 ± 528#	2704 ± 628#
	Prazosin	3658 ± 543	2705 ± 213	2693 ± 652	2457 ± 632	2204 ± 684
LVEDP (mm Hg)	Idazoxan	13.7 ± 5.9	17.5 ± 7.4	28.0 ± 2.9++	28.5 ± 6.1	26.6 ± 4.1
	Prazosin	8.1 ± 4.8	9.4 ± 5.1	25.4 ± 7.3	24.3 ± 6.9	17.6 ± 7.4
AP-mean (mm Hg)	Idazoxan	114 ± 11	122 ± 14*	136 ± 15+	121 ± 14#	120 ± 14#
	Prazosin	103 ± 12	106 ± 13	124 ± 15	112 ± 15	102 ± 15
PW - %WTh	Idazoxan	22.5 ± 4.0	17.6 ± 2.9**	21.0 ± 1.8+	5.1 ± 1.6##	10.6 ± 2.2# ●
	Prazosin	24.1 ± 4.1	19.8 ± 3.2	22.0 ± 4.8	3.9 ± 4.3	1.8 ± 5.3
AW - %WTh	Idazoxan	20.4 ± 8.4	14.3 ± 6.1**	15.6 ± 5.2+	15.6 ± 5.8	15.4 ± 5.6
	Prazosin	19.3 ± 5.5	14.6 ± 3.2	15.8 ± 4.3	15.3 ± 5.4	14.4 ± 3.8

HR = heart rate; LVP max = maximum systolic left ventricular pressure; (+) LV dP/dt = maximum first derivative of LVP; LVEDP = left ventricular end-diastolic pressure; AP-mean = mean aortic pressure; PW = posterior wall; %WTh = percent systolic wall thickening; AW = anterior wall. Data are mean ± SD, n = 5 (idazoxan), n = 3 (prazosin, no statistical analysis); *p < 0.05, **p < 0.01 rest after vs before propranolol; +p < 0.05, ++p < 0.01 run before stenosis vs rest after propranolol; # p < 0.05, ## p < 0.01 run after stenosis vs before stenosis; ● p < 0.05 run after idazoxan (control) vs after idazoxan.

Table 2. Myocardial blood flow in (ml/min)/g during exercise with acute coronary stenosis before and after intracoronary idazoxan or prazosin

		Ischemic area		Control area	
		Control	Treatment	Control	Treatment
ENDO	Idazoxan	0.170 ± 0.05	0.452 ± 0.30*	0.975 ± 0.24	1.160 ± 0.21
	Prazosin	0.180 ± 0.16	0.109 ± 0.04	0.899 ± 0.05	0.877 ± 0.15
MID	Idazoxan	0.421 ± 0.08	0.822 ± 0.21*	0.985 ± 0.09	1.157 ± 0.28
	Prazosin	0.359 ± 0.25	0.214 ± 0.18	0.976 ± 0.10	1.017 ± 0.27
EPI	Idazoxan	0.831 ± 0.21	0.783 ± 0.15	0.839 ± 0.48	0.900 ± 0.34
	Prazosin	0.675 ± 0.32	0.548 ± 0.34	0.855 ± 0.10	0.924 ± 0.25
TRANSMURAL	Idazoxan	0.474 ± 0.08	0.685 ± 0.17	0.933 ± 0.10	1.073 ± 0.26
	Prazosin	0.405 ± 0.25	0.291 ± 0.18	0.920 ± 0.06	0.939 ± 0.23
ENDO/EPI	Idazoxan	0.219 ± 0.09	0.595 ± 0.41*	1.177 ± 0.37	1.219 ± 0.47
	Prazosin	0.221 ± 0.15	0.233 ± 0.08	1.027 ± 0.15	0.967 ± 0.10

ENDO, MID, EPI: myocardial blood flow in subendocardial, midwall, and subepicardial regions, respectively; TRANSMURAL: transmural myocardial blood flow; ENDO/EPI: ratio of subendocardial to subepicardial blood flow. Data are mean ± SD, n = 5 (idazoxan), n = 3 (prazosin, no statistical analysis); *p < 0.05.

this study clearly demonstrate that acute regional α_2-adrenoceptor blockade can diminish regional myocardial dysfunction and increase myocardial blood flow to the ischemic zone.

The experiments were designed to allow the assessment of regional effects of α-adrenoceptor blockers by intracoronary infusion, thereby using doses without significant effects on systemic hemodynamic parameters. All studies were performed after β-adrenergic blockade in order to unmask and define the extent of α-adrenoceptor-mediated coronary vasoconstriction during ischemia and to prevent any secondary effects of enhanced norepinephrine release caused by presynaptic α_1- or α_2-adrenoceptor blockade during exercise [7, 13]. The prolonged exercise period of about 12 min required a steady-state condition in order to assess the drug effect on the ischemic coronary vasculature and myocardium. Conceivably, ischemia and continued exercise could have triggered the recruitment of collateral vessels, thereby inducing an increase in blood flow and function in the underperfused area independent from any α-adrenergic effect. However, in control runs with a comparable degree of regional myocardial dysfunction after acute coronary stenosis, no gradual improvement in regional function was detected during an observation period of more than 8 min. Thus, the almost immediate increase in myocardial function after idazoxan infusion was not likely due to spontaneous changes.

The increases in blood flow after idazoxan occurred in the subendocardium and midwall without significant changes in the subepicardium. This may either be due to the fact that α_2-adrenoceptors mainly regulate vasoconstriction of small resistance vessels [10] or to the recent observation that the significance of sympathetic vasoconstriction increases with the severity of ischemia [11], which is most prominent in the subendocardium.

In contrast to all previously discussed studies that indicate a deleterious effect of α-adrenergic coronary constriction during myocardial ischemia, Nathan and Feigl [23] reported that α-adrenergic coronary constriction exerts a favorable effect on ischemic myocardium by preventing a transmural redistribution of blood flow away from the ischemic subendocardium. In anesthetized, β-blocked dogs with constant inflow coronary hypoperfusion, they observed an increase in interpolated subendocardial to subepicardial blood flow ratios in a control as compared with a phenoxybenzamine-treated region during intracoronary norepinephrine infusion. However, the use of constant inflow coronary hypoperfusion may have prevented the most significant beneficial effect of α-adrenoceptor blockade, that is, an increase in total coronary blood flow. In addition, the authors did not present absolute data for subendocardial blood flow or measures of regional myocardial function in order to substantiate the aggravation of ischemia in the phenoxybenzamine-treated region. Since α_1- but not α_2-adrenergic coronary constriction is attenuated by ischemia [5], the use of phenoxybenzamine in this investigation is particularly problematic, because it has been shown to only partially block α_2-adrenoceptors [4].

In recent studies it was concluded that α_1- but not α_2-adrenoceptor-mediated coronary vasoconstriction is reponsible for sympathetic coronary vasoconstriction during hypoperfusion. Liang and Jones [19] and Jones et al. [15] employed a constant coronary hypoperfusion by limiting coronary perfusion pressure to 50 mm Hg. In this setting, α_1-adrenoceptor blockade with prazosin, but not α_2-adrenoceptor blockade with yohimbine or idazoxan increased coronary flow. Since ischemia in their study was neither severe nor transmural, prazosin could have removed any α_1-adrenergic tone to the subepicardium and thereby increased total coronary inflow. The relatively mild level of ischemia and sympathetic activation may explain the fact that idazoxan or yohimbine had no effect on coronary hypoperfusion in this study.

Laxson et al. [18] recently reported that in chronically instrumented dogs the intracoronary infusion of the selective α_1-adrenoceptor blocker prazosin, but not the selective α_2-adrenoceptor blocker idazoxan increased transmural blood flow during treadmill exercise in the presence of an acute coronary stenosis that decreased distal perfusion pressure to 40 mm Hg. In addition to the relatively mild degree of coronary hypoperfusion (endocardial flow > 0.6 (ml/min)/g), several other reasons may explain the results of Laxson et al. [18] as compared to the present investigation. Laxson et al. [18] did not use systemic β-blockade in their study. Since both presynaptic α_2- and also α_1-adrenoreceptor blockade result in increased norepinephrine release from sympathetic nerve terminals [7], myocardial β-adrenoceptors are stimulated and induce a further increase in regional myocardial oxygen demand [7, 13] even in the acutely ischemic myocardium [26]. In consequence, metabolic coronary vasomotion secondary to myocardial β-adrenoceptor activation cannot be distinguished from α-adrenergic coronary vasomotion. Another reason for the relative weak anti-ischemic effect of idazoxan in the study by Laxson et al. [18] may be the administration of a substantially lower dose of idazoxan ($10 \mu g/kg$) in their study as compared to our study (average dose: $80 \mu g/kg$). Since the dosage applied in our study did not alter arterial pressure during exercise and β-adrenoceptor blockade, we can still exclude that idazoxan-induced changes in coronary perfusion pressure may have influenced our results.

In contrast to the selective α_2-adrenoceptor blocker idazoxan, the selective α_1-adrenoceptor blocker prazosin in the present study induced no increase, but rather a further decrease in myocardial blood flow and myocardial function in the underperfused posterior wall during exercise in our study. This was probably caused by the slight decrease in arterial and, as a consequence, coronary perfusion pressure, which is a critical determinant for poststenotic coronary flow and which was artificially maintained constant in the study by Laxon et al. [18]. However, the slight decrease in peripheral pressure after prazosin indicates that even peripheral α_1-adrenoceptors were partially blocked, supporting the conclusion that the administered dose was sufficient to block any intracoronary α_1-adrenoceptors. It has to be mentioned, however, that the small number of only three experiments with prazosin does not allow any statistical analysis of this subset of data.

It should also be pointed out that under non-ischemic conditions in dogs, α_2-adrenoceptors contribute approximately 75% to the increase in coronary resistance during sympathetic activation [10] or norepinephrine infusion [14, 28]. During ischemia, vasoconstrictor responses to α_1-adrenoceptor agonists appear to be progressively diminished, whereas the vasoconstrictor response to α_2-adrenoceptor agonists persists even in the presence of a severe coronary stenosis [5]. Whereas the transmural distribution of coronary α_1- and α_2-adrenoceptors is uniform in nonischemic myocardium [3], the preferential α_2-adrenergic coronary constriction in the most ischemic inner myocardial layers may be related to the attenuation of α_1- but not α_2-adrenergic coronary constriction by ischemia, potentially because of the different sensitivity of α_1- and α_2-adrenoceptors to acidosis [20].

These results clearly demonstrate that coronary vasoconstrictor tone exists distal to flow limiting arterial stenoses during exercise, and even in the presence of severe myocardial dysfunction. Blood flow and regional myocardial function to the region of hypoperfused myocardium was enhanced by selective α_2-adrenoceptor blockade with idazoxan, but not α_1-adrenoceptor blockade with prazosin. These results indicate that α_2-adrenoceptor mediated coronary vasoconstriction aggravates myocardial hypoperfusion distal to coronary stenosis during exercise.

References

1. Bassenge E, Walter P, Doutheil U (1967) Wirkungsumkehr der adrenergischen Coronargefäßreaktion in Abhängigkeit vom Koronargefäßtonus. Pflügers Arch 297:146–155
2. Buffington CW, Feigl EO (1981) Adrenergic coronary vasoconstriction in the presence of coronary stenosis in the dog. Circ Res 48:416–423
3. Chen DG, Dai X-Z, Zimmerman BG, Bache RJ (1988) Postsynaptic α_1- and α_2-adrenergic mechanism in coronary vasoconstriction. J Cardiovasc Pharmacol 11:61–67
4. Constantine JW, Lebel W (1980) Complete blockade by phenoxybenzamine of α_1- but not α_2-vascular receptors in dogs and the effect of propranolol. Naunyn Schmiedebergs Arch Pharmacol 314:149–156
5. Deussen A, Heusch G, Thämer V (1985) α_2-adrenoceptor-mediated coronary vasoconstriction persists after exhaustion of coronary dilator reserve. Eur J Pharmacol 115:147–153
6. Gallagher KP, Matsuzaki M, Osakada G, Kemper WS, Ross J Jr (1983) Effect of exercise on the relationship between myocardial blood flow and systolic wall thickening in dogs with acute coronary stenosis. Circ Res 52:716–729
7. Guth BD, Thaulow E, Heusch G, Seitelberger R, Ross J Jr (1990) Myocardial effects of selective α-adrenoceptor blockade during exercise in dogs. Circ Res 66:1703–1712
8. Heusch G (1990) α-adrenergic mechanisms in myocardial ischemia. Circulation 81:1–13
9. Heusch G, Deussen A (1983) The effects of cardiac sympathetic nerve stimulation on perfusion of stenotic coronary arteries in the dog. Circ Res 53:8–15
10. Heusch G, Deussen A, Schipke J, Thämer V (1984) α_1- and α_2-adrenoceptor-mediated vasoconstriction of large and small canine coronary arteries in vivo. J Cardiovasc Pharmacol 6:961–968
11. Heusch G, Deussen A, Thämer V (1985) Cardiac sympathetic nerve activity and progressive vasoconstriction distal to coronary stenosis: Feed-back aggravation of myocardial ischemia. J Auton Nerv Syst 13:311–326
12. Heymann MA, Payne BD, Hoffman JIE, Rudolph AM (1977) Blood flow measurements with radionuclide labelled particles. Prog Cardiovasc Dis 20:55–79
13. Heyndrickx GR, Vilaine JP, Moerman EJ, Leusen I (1984) Role of prejunctional α_2-adrenergic receptors in the regulation of myocardial performance during exercise in conscious dogs. Circ Res 54:683–693
14. Holtz J, Saeed M, Sommer O, Bassenge E (1982) Norepinephrine constricts the canine coronary bed via postsynaptic α_2-adrenoceptors. Eur J Pharmacol 82:199–202
15. Jones CE, Liang IYS, Maulsby MR (1986) Cardiac and coronary effects of prazosin and phenoxybenzamine during coronary hypotension. J Pharmacol Exp Ther 236:204–211
16. Kitakaze M, Hori M, Gotoh K, Sato H, Iwakura K, Kitabatake A, Inoue M, Kamada T (1989) Beneficial effects of α_2-adrenoceptor activity on ischemic myocardium during coronary hypoperfusion in dogs. Circ Res 65:1632–1645
17. Kumada T, Karliner JS, Pouleur H, Gallagher K, Shirato K, Ross J Jr (1979) Effects of coronary occlusion on early ventricular diastolic events in conscious dogs. Am J Physiol 237:H542–H549
18. Laxson DD, Dai XZ, Homans DC, Bache RJ (1989) The role of α_1- and α_2-adrenergic receptors in mediation of coronary vasoconstriction in hypoperfused ischemic myocardium during exercise. Circ Res 65:1688–1697
19. Liang IY, Jones CE (1985) α_1-adrenergic blockade increases coronary blood flow during coronary hypoperfusion. Am J Physiol 249:H1070–H1077
20. McGrath JC (1982) Evidence for more than one type of postjunctional α-adrenoceptor. Biochem Pharmacol 32:467–484
21. Mohrman DE, Feigl EO (1978) Competition between sympathetic vasoconstriction and metabolic vasodilation in the canine coronary circulation. Circ Res 42:79–86
22. Mudge GH Jr, Grossman W, Mills RM Jr, Lesch M, Braunwald E (1976) Reflex increase in coronary vascular resistance in patients with ischemic heart disease. N Engl J Med 295:1333–1337
23. Nathan HJ, Feigl EO (1986) Adrenergic vasoconstriction lessens transmural steal during coronary hypoperfusion. Am J Physiol 250:H645–H653

24. Sasayama S, Franklin D, Ross J Jr, Kemper WS, McKnown D (1976) Dynamic changes in left ventricular wall thickness and their use in analyzing cardiac function in the conscious dog. Am J Cardiol 38:870–879
25. Schosser R, Arfors KE, Messmer K (1979) MIC-II – A program for the determination of cardiac output, arteriovenous shunt and regional blood flow using the radioactive microsphere method. Comput Programs Biomed 9:19–38
26. Schulz R, Miyazaki S, Miller M, Thaulow E, Heusch G, Ross J Jr (1989) Consequences of regional inotropic stimulation of ischemic myocardium on regional myocardial blood flow and function in anesthetized swine. Circ Res 64:1116–1126
27. Seitelberger R, Guth BD, Heusch G, Lee JD, Katayama K, Ross J Jr (1988) Intracoronary α_2-adrenergic receptor blockade attenuates ischemia in conscious dogs during exercise. Circ Res 62:436–442
28. Woodman OL, Vatner SF (1987) Coronary vasoconstriction mediated by α_1- and α_2-adrenoceptors in conscious dogs. Am J Physiol 253:H388–H393

Authors' address:

Dr. Rainald Seitelberger, II. Department of Surgery, University of Vienna, Spitalgasse 23, 1090 Vienna, Austria.

Prevention of α-Adrenergic Coronary Constriction by Calcium-Antagonists

G. Heusch, A. Deussen and B. D. Guth

*Abteilung Pathophysiologie, Universitätsklinikum Essen, FRG; Abteilung Herz- und Kreislaufphysiologie, Universität Düsseldorf, FRG; Division of Cardiology, University of California, San Diego, USA

Summary

This manuscript reviews the experimental evidence for a functional antagonism of Ca-antagonists against α-adrenoceptor-mediated increases in coronary vasomotor tone. In studies on anesthetized dogs, intravenous nifedipine effectively prevented the α_1-adrenoceptor-mediated increase in epicardial coronary resistance, as well as the increase in end-diastolic resistance mediated by both α_1- and α_2-adrenoceptors during cardiac sympathetic nerve stimulation. Both intracoronary and intravenous administration of nifedipine also prevented the α_2-adrenoceptor-mediated increase in coronary resistance distal to severe stenoses, as well as the resulting ischemic dysfunction and net lactate production during cardiac sympathetic nerve stimulation. Felodipine was equally effective as nifedipine in preventing an α_2-adrenoceptor-mediated increase in coronary resistance and the resulting contractile dysfunction distal to severe coronary stenoses. α_1- and α_2-Adrenergic coronary constriction also contribute to the severity of myocardial ischemia in conscious dogs during treadmill exercise. Again, nifedipine improved regional myocardial blood flow and attenuated regional contractile dysfunction during exercise-induced ischemia in conscious dogs with a chronic coronary stenosis. This beneficial effect of nifedipine was attributed to a recruitment of coronary dilator reserve and not to a reduction in heart rate or afterload. In conclusion, there is solid experimental evidence for a functional antagonism of Ca-antagonists against α-adrenergic coronary constriction and its contribution to myocardial ischemia.

Introduction

α-Adrenergic coronary vasoconstriction effectively competes with metabolic coronary vasodilation during the sympathetic activation induced by exercise or excitement. Even under normal conditions when a substantial coronary dilator reserve is present, α-adrenergic constriction acts to limit metabolic coronary dilation by about 30%, such that myocardial oxygen extraction increases together with coronary blood flow during sympathetic activation to match oxygen supply to the increased myocardial oxygen demand [12, 34, 36].

Segmental and transmural distribution of coronary vascular α_1- and α_2-adrenoceptors

Recent studies have shown that there are at least two types of postsynaptic vascular α-adrenoceptors, i.e., α_1- and α_2-adrenoceptors, in the coronary circulation of dogs [6, 23, 27, 51],

*The studies reported in this manuscript were supported by grants of the German Research Foundation (SFB 30 Kardiologie Düsseldorf und He 1320/1-1, 2-1, 3-1 and 3-2) and the National Heart, Lung and Blood Institute, Bethesda (HL-17682).

calves [52], guinea pigs [9], monkeys, and also humans [48]. These vascular α-adrenoceptors are not homogeneously distributed throughout the coronary vascular bed. The α-adrenoceptor-mediated constriction of epicardial coronary arteries is less marked than that of the coronary resistance vessels [23, 28]. In dogs, the epicardial coronary vasoconstriction is exclusively mediated by α_1-adrenoceptors [23, 40, 48], whereas in calves [52] and humans [48] α_2-adrenoceptors may also be involved in epicardial coronary vasoconstriction. The vasoconstriction of the coronary resistive vessels in dogs is predominantly mediated by α_2-adrenoceptors [23, 27, 45]. Coronary collaterals are not at all responsive to α-adrenoceptor activation [16–18]. Thus, collateral blood flow during sympathetic activation is mainly determined by the driving pressure gradient across the collaterals between the autoregulating donor vessels and the ischemic recipient vessels [5]. The constrictor response of the coronary circulation upon direct activation of α_1- and α_2-adrenoceptors is transmurally homogeneous, with no transmural redistribution of blood flow occurring in nonischemic myocardium [6]. However, during coronary hypoperfusion which results in transmurally inhomogeneous ischemia, α_1-adrenergic coronary constriction is attenuated [26], whereas α_2-adrenergic coronary constriction is not [10] – potentially because of the different sensitivity of α_1- and α_2-adrenoceptors to acidosis [33]. Therefore, during ischemia α_2-adrenergic coronary vasoconstriction may predominate in the most ischemic inner myocardial layers [44].

α-Adrenergic coronary vasoconstriction in myocardial ischemia

α-Adrenergic coronary vasoconstriction can contribute to the initiation and aggravation of myocardial ischemia during sympathetic activation (for review see [20]). When coronary dilator reserve is exhausted distal to severe coronary stenoses in anesthetized dogs, cardiac sympathetic nerve stimulation induces α_2-adrenergic coronary vasoconstriction which is powerful enough to precipitate poststenotic myocardial ischemia, as evidenced by regional net lactate production, regional contractile dysfunction, and the initiation of malignant arrhythmias [21]. Also, in conscious dogs during the sympathetic activation of treadmill exercise, α_2-adrenergic coronary constriction contributes to the severity of poststenotic myocardial ischemia. Conversely, selective α_2-blockade particularly improves subendocardial blood flow and attenuates contractile dysfunction [44]. A contribution of α_1-adrenergic coronary constrictor tone to the severity of myocardial ischemia has also been reported in studies on open-chest dogs with a presumably high resting sympathetic tone [29], as well as in conscious dogs during exercise-induced myocardial ischemia [15].

In the clinical setting, α-adrenergic coronary vasoconstriction contributes to the narrowing of epicardial dynamic stenoses during isometric [3, 4] and dynamic exercise [13]. α-Adrenergic coronary vasoconstriction was also suggested to contribute to the increase in coronary vascular resistance in patients with stable angina during the cold pressor test [35] and during exercise [1, 14].

Ca-antagonists and α-adrenergic coronary vasoconstriction

Calcium antagonists are established drugs for the treatment of ischemic heart disease [2, 19, 37]. The beneficial effects of Ca-antagonists on ischemic myocardium may be due to their favorable influence on systemic hemodynamics through a reduction of heart rate and afterload, as well as to a decrease in myocardial inotropism and, thus, oxygen demand. In addition, Ca-antagonists reduce coronary vasomotor tone and, thereby, increase oxygen supply. Finally, Ca-antagonists are reported to have additional (currently poorly understood) cardioprotective effects on ischemic metabolism. The relative contribution of these beneficial effects to the therapeutic action of Ca-antagonists varies between diltiazem-type, verapamil-type, and nifedipine-type Ca-antagonists. Whereas diltiazem also exerts negative chronotropic effects [31] and verapamil also exerts negative chronotropic and inotropic effects [38], the therapeutic effects of dihydropyridine Ca-antagonists of the nifedipine-type are preferably attributed to their vasodilator action, in particular their coronary vasodilator action.

Ca-antagonists prevent α-adrenergic coronary constriction

The main purpose of the present review is to summarize the experimental evidence suggesting that the beneficial effects of dihydropyridine-Ca-antagonists are largely due to the attenuation of α-adrenergic coronary constrictor tone.

Nifedipine functionally antagonizes α_1-adrenoceptor-mediated epicardial and α_2-adrenoceptor-mediated resistive vessel constriction

Eighteen mongrel dogs were anesthetized with chloralose-urethane and artifically ventilated. After a left thoracotomy and pericardiotomy, the proximal left circumflex coronary artery was isolated, cannulated, and perfused at constant pressure from a windkessel supplied with blood flow from both femoral arteries by a roller pump. Coronary perfusion pressure was measured at the tip of the cannula through a small internal tube connected to a Statham pressure transducer. Coronary blood flow was measured within the extracorporeal perfusion circuit with an electromagnetic flowmeter. The vasomotion of the large epicardial coronary arterial segment 1–2 cm distal to the perfusion cannula tip was continuously recorded with ultrasonic transit-time technique [23, 42]. After postmortem measurement of length and mass of the epicardial segment, at which the sonomicrometers were located, the internal changing diameter could be calculated from the measured external diameter [39]. Thus, the internal cross-sectional area and changes in epicardial coronary resistance (% changes in the fourth power of the vessel radius) could also be calculated. The vasomotion of the small resistive coronary vessels was determined by calculation of end-diastolic resistance from coronary perfusion pressure, coronary venous pressure, and coronary blood flow which was normalized to 100 g of perfused tissue. To prevent reflex vagal effects secondary to changes in blood pressure and cardiac dimensions, a bilateral cervical vagotomy was performed. To prevent an interaction with β-adrenergic coronary vasomotion and metabolic coronary vasomotion secondary to myocardial β-adrenoceptor activation, propranolol (2 mg/kg i.v.) was administered. The left ventrolateral cervical cardiac sympathetic nerve was electrically stimulated at 20 Hz with 5 ms pulses of 1–5 V.

Cardiac sympathetic nerve stimulation increased epicardial coronary arterial resistance by 11.1 ± 3.8 (SD)% and end-diastolic coronary resistance by $31.8 \pm 12.7\%$ (Fig. 1). The increase in epicardial resistance was prevented by the selective α_1-antagonist prazosin (n = 6, 1.2 mg/kg i.v.), which still permitted an increase in end-diastolic resistance by $23.0 \pm 13.6\%$. The increase in end-diastolic resistance was prevented by the selective α_2-antagonist rauwolscine (n = 6, 0.2 mg/kg i.v.), which still permitted an increase in epicardial resistance by $11.9 \pm 10.1\%$. Apparently, in this in situ canine heart preparation, epicardial vasoconstriction is exclusively mediated by α_1-adrenoceptors. The vasoconstriction of the resistance vessels is predominantly due to α_2-adrenoceptors; however, α_1-adrenoceptors are also involved. Our observation that only about 26% of the increase in end-diastolic resistance during cardiac sympathetic nerve stimulation can be attributed to the activation of α_1-adrenoceptors is in agreement with previous findings by Holtz et al. [27] using intracoronary infusion of norepinephrine. Prazosin only attenuated the increase in end-diastolic resistance during sympathetic stimulation, leaving the α_2-adrenoceptor mediated part of vasoconstriction unchanged, whereas rauwolscine completely prevented the increase in end-diastolic resistance. These findings probably reflect the better selectivity of prazosin for α_1-adrenoceptors as compared to that of rauwolscine for α_2-adrenoceptors [33].

The dihydropyridine Ca-antagonist nifedipine (n = 6, 20 μg/kg i.v.) completely prevented both the α_1-adrenoceptor mediated increase in epicardial coronary resistance and the increase in end-diastolic resistance which was mediated by α_1- und α_2-adrenoceptors (Fig. 1). A dilation of both large epicardial coronary arteries and coronary resistance vessels by nifedipine has previously been demonstrated in conscious dogs [50] and humans [43]. Our data suggest that the large and small vessel dilation by nifedipine in normal coronary arteries may, at least in part, be due to a functional antagonism at α_1- as well as α_2-adrenoceptors which both upon activation trigger an influx of Ca^{2+} [40, 46, 49].

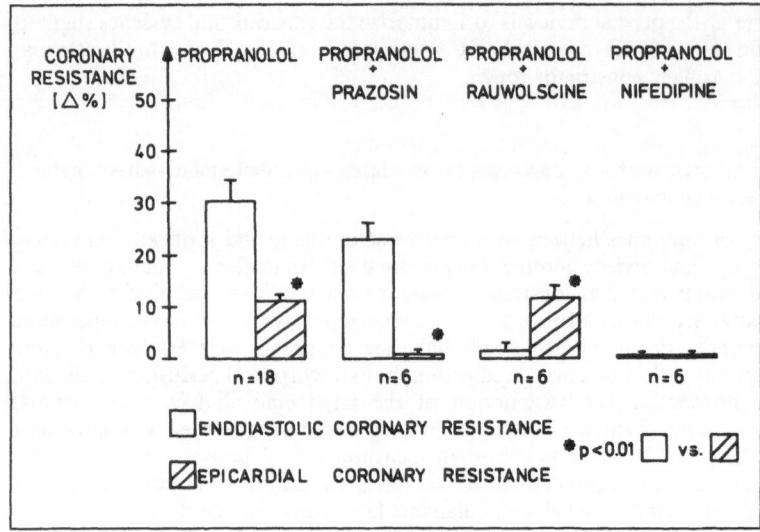

Fig. 1. Effects of cardiac sympathetic nerve stimulation (CSNS) on end-diastolic and epicardial coronary resistance. After β-blockade with propranolol (2 mg/kg iv) CSNS increased both end-diastolic and epicardial coronary resistance. After selective α_1-adrenoceptor blockade with prazosin (1.2 mg/kg iv) CSNS still increased end-diastolic resistance, although to a lesser extent, whereas epicardial resistance was unchanged. After selective α_2-adrenoceptor blockade by rauwolscine (0.2 mg/kg iv), CSNS still increased epicardial coronary resistance, whereas end-diastolic resistance remained unchanged. Nifedipine (20 μg/kg iv) prevented the increases in both end-diastolic and epicardial coronary resistance during CSNS. Bars indicate 1 SEM. (From [23] by permission of Raven Press.)

Nifedipine prevents α_2-adrenergic coronary vasoconstriction and the resulting myocardial ischemia distal to severe coronary stenoses

α_2-Adrenergic coronary vasoconstriction during cardiac sympathetic nerve stimulation is not attenuated during myocardial ischemia. In contrast, it significantly contributes to the initiation of regional myocardial ischemia (net lactate production, regional contractile dysfunction) [21]. We, therefore, tested whether the dihydropyridine Ca-antagonist nifedipine would also antagonize α_2-adrenergic coronary vasoconstriction in ischemic myocardium [22].

Eleven mongrel dogs were anesthetized with chloralose-urethane and artifically ventilated. After a left thoracotomy and pericardiotomy the left circumflex coronary artery was equipped with an electromagnetic flowprobe for measurement of coronary blood flow and a constrictor device for the production of coronary stenoses. A small side branch of the left circumflex coronary artery distal to the site of the constrictor was cannulated for measurement of coronary perfusion pressure and drug infusion. End-diastolic coronary resistance was calculated from coronary perfusion pressure, coronary venous pressure, and coronary blood flow, and normalized to 100 g of perfused tissue. Regional subendocardial segment shortening of the circumflex-perfused myocardium was analyzed using sonomicrometry. Cardiac sympathetic nerve stimulation was performed, as described above, after bilateral cervical vagotomy.

Cardiac sympathetic nerve stimulation decreased end-diastolic coronary resistance of intact coronary arteries with a substantial coronary dilator reserve (Fig. 2). However, after production of a severe coronary stenosis on the proximal left circumflex coronary artery that exhausted the endogenous dilator reserve of the poststenotic vascular bed (lack of postocclusive reactive hyperemia), cardiac sympathetic nerve stimulation increased poststenotic end-diastolic coronary resistance (Fig. 2). This increase in poststenotic coronary resistance resulted in regional net lactate production and a decrease in systolic segment shortening from 8.4 ± 2.3%

Ca-antagonists prevent α-adrenergic coronary constriction

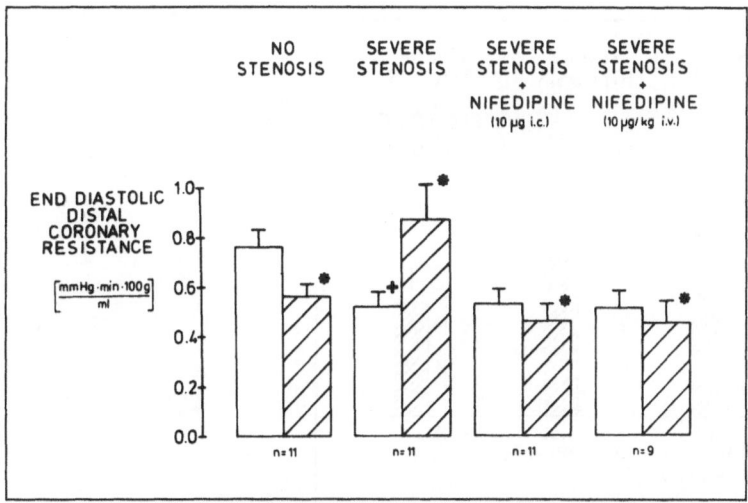

Fig. 2. Effects of cardiac sympathetic nerve stimulation (CSNS) on the end-diastolic resistance of intact and severely stenotic coronary arteries. Whereas the resistance of intact coronary arteries is decreased, the resistance distal to a severe stenosis is increased during CSNS. This increase in poststenotic resistance is prevented by intracoronary (ic) as well as by intravenous (iv) administration of nifedipine. Control: open columns. CSNS: hatched columns. *p < 0.05 CSNS vs control. +p < 0.05 severe stenosis vs no stenosis. Bars indicate 1 SEM. (From [22] by permission of Raven press.)

to $7.0 \pm 2.3\%$. Both intracoronary administration of $10 \, \mu g$ nifedipine into the poststenotic coronary vascular bed and systemic administration of $10 \, \mu g/kg$ i.v. nifedipine prevented the increase in poststenotic end-diastolic resistance (Fig. 2) and the resulting net lactate production and deterioration of regional contractile function during cardiac sympathetic nerve stimulation. The equal effectiveness of intracoronary and systemic nifedipine in the present experiment indicates that the beneficial action of nifedipine on poststenotic myocardium during sympathetic activation is based on the prevention of α_2-adrenergic coronary vasoconstriction rather than a decrease in afterload or an increase in collateral blood flow from nonischemic myocardial areas.

Felodipine prevents α_2-adrenergic coronary vasoconstriction and the resulting myocardial ischemia distal to severe coronary stenoses

In another series, we tested whether the new dihydropyridine Ca-antagonist felodipine, which exhibits a high degree of vascular selectivity [30], would also functionally antagonize α_2-adrenergic coronary vasoconstriction distal to severe coronary stenoses and prevent the initiation of poststenotic myocardial ischemia [11].

In this series, ten beagle dogs were anesthetized with enflurane and nitrous oxide and artificially ventilated. After a left thoracotomy and pericardiotomy the proximal left circumflex coronary artery was equipped with an electromagnetic flowprobe for measurement of coronary blood flow and a constrictor device. A small side-branch of the left circumflex coronary artery distal to the site of the constrictor was cannulated for measurement of coronary perfusion pressure and drug infusion. Mean coronary resistance was calculated from coronary perfusion pressure and coronary blood flow normalized to 100 g of perfused tissue. Regional myocardial function was determined as systolic wall thickening in the control anterior myocardium and the circumflex-perfused posterolateral myocardium using sonomicrometry [41]. α_2-Adrenergic coronary vasoconstriction was induced by intracoronary infusion of the selective α_2-agonist BHT 933 (200 μg).

Intracoronary infusion of BHT 933 into intact coronary arteries increased coronary resis-

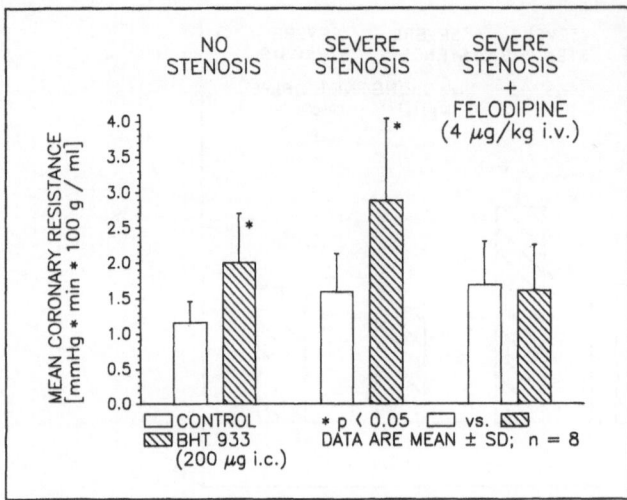

Fig. 3. Effects of the selective α_2-agonist BHT 933 on the coronary resistance of intact and severely stenotic coronary arteries. BHT 933 increased coronary resistance in the absence and presence of a severe stenosis. The BHT 933-induced increase in resistance of severely stenotic coronary arteries was prevented by felodipine. Bars indicate 1 SD. (From [11] by permission of Kluwer Press.)

tance (Fig. 3), but did not alter regional contractile function (systolic wall thickening $14.2 \pm 2.8\%$ vs $14.1 \pm 2.7\%$). After production of a severe coronary stenosis that exhausted the endogenous dilator reserve of the poststenotic vascular bed, as indicated by a decrease in systolic wall thickening to $11.9 \pm 2.7\%$, intracoronary infusion of BHT 933 again increased coronary resistance (Fig. 3), but also precipitated myocardial ischemia, as evidenced by a further decrease in systolic wall thickening from $11.9 \pm 2.7\%$ to $8.2 \pm 3.1\%$. Pretreatment with felodipine ($4 \mu g/kg$ iv) prevented the BHT 933 – induced increase in coronary resistance (Fig. 3) and the resulting ischemic dysfunction of the poststenotic myocardium (systolic wall thickening $12.1 \pm 3.0\%$ vs $12.6 \pm 2.9\%$).

Thus, felodipine is equally effective as nifedipine in preventing the poststenotic coronary vasoconstriction and the resulting myocardial ischemia induced by activation of coronary vascular α_2-adrenoceptors.

Recruitment of coronary dilator reserve and attenuation of exercise-induced myocardial ischemia in conscious dogs by nifedipine

To establish the relevance of the functional antagonism exerted by dihydropyridine Ca-antagonists against α_2-adrenergic coronary vasoconstriction during more physiological situations of sympathetic activation, we studied the effect of nifedipine in chronically instrumented conscious dogs with a chronic left circumflex coronary obstruction during exercise-induced myocardial ischemia. We had previously shown that α_2-adrenergic coronary vasoconstriction also contributes to the severity of poststenotic myocardial ischemia during treadmill exercise [44].

In the present study [25], nine dogs were anesthetized and instrumented under sterile conditions. Left ventricular pressure and derived indices were measured with a micromanometer introduced into the left ventricle through the apical dimple. Regional wall thickness was measured in the anterior control region and the circumflex-perfused posterolateral region using sonomicrometry. Regional myocardial blood flow was measured with tracer microspheres injected into the left atrium, and a reference sample was withdrawn from the descending thoracic aorta. The proximal left circumflex coronary artery was equipped with an ameroid constrictor

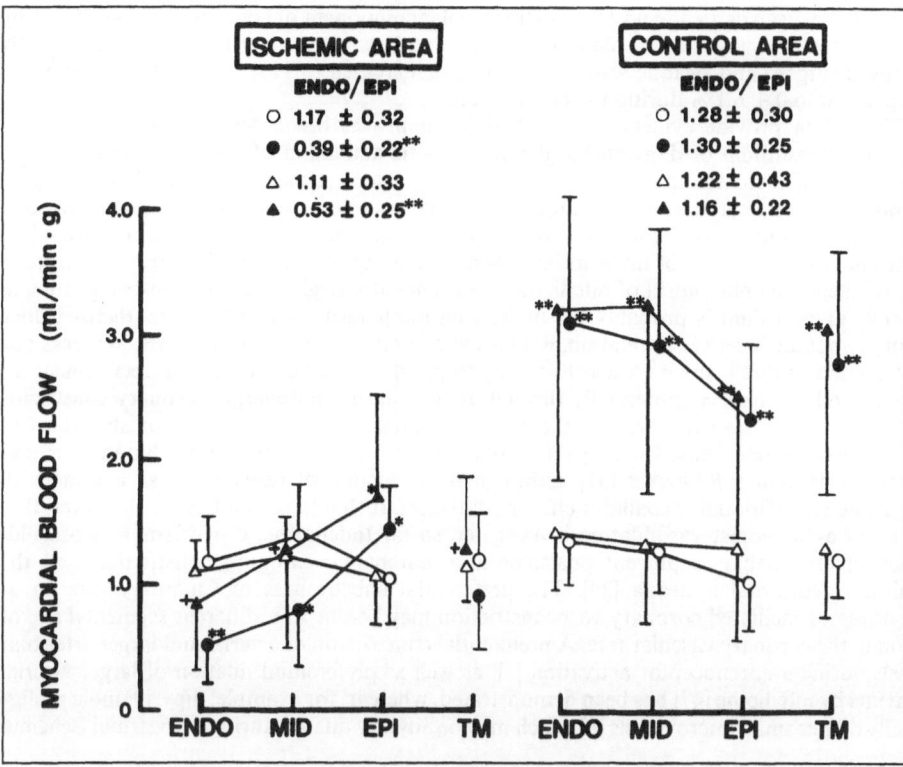

Fig. 4. Transmural distribution of regional myocardial blood flow in the ischemic and control regions at rest and during exercise. At rest (○), blood flow distribution in the ischemic region was not different from that in the control area. During the control treadmill exercise (●), however, myocardial blood flow in the control region increased in all transmural layers, whereas in the ischemic region a marked transmural gradient with a decrease in subendocardial blood flow and an increase in subepicardial blood flow developed. Nifedipine did not change myocardial blood flow at rest (△) in either region. However, the decreases in blood flow to the subendocardium and the midmyocardium of the ischemic region during exercise (▲) were significantly attenuated. ENDO = subendocardium, MID = midmyocardium, EPI = subepicardium, TM = mean transmural myocardial blood flow. Bars indicate 1 SD. *p < 0.05, **p < 0.01 exercise vs rest. +p < 0.05 nifedipine vs control. (From [25] by permission of the American Heart Association.)

which gradually occluded the coronary artery over a 2–3-week period, allowing sufficient time for the collateral circulation to develop. The developing collateral circulation was able to compensate for the increasing obstruction of the proximal left circumflex coronary artery and to maintain a normal regional myocardial blood flow and contractile function at rest. However, severe regional myocardial ischemia was precipitated during submaximal treadmill exercise, as indicated by a decrease in regional myocardial blood flow, particularly in the subendocardial and midmyocardial layers (Fig. 4), and a decrease in systolic wall thickening from 24.3 ± 5.8% at rest to 6.0 ± 6.1% during steady-state exercise. The dogs were again subjected to an identical treadmill exercise after a 3 h recovery period; reproducible changes in systemic hemodynamics, as well as in regional myocardial blood flow and function, are observed after 3 h in the absence of any treatment [31]. Intravenous nifedipine (10 μg/kg i.v.) did not change systemic hemodynamics and regional myocardial blood flow and function at rest. Also, nifedipine did not attenuate the increases in heart rate and blood pressure, i.e., the major determinants of myocardial oxygen demand and of the extravascular component of coronary resistance, during exercise. However, myocardial blood flow to the most ischemic inner layers was significantly improved (Fig. 4) and systolic wall thickening was increased to $11.4 \pm 7.8\%$ by nifedipine as compared to $6.0 \pm 6.1\%$ during the control run.

These data provide evidence that nifedipine improves blood flow and function in the ischemic myocardium of dogs during the sympathetic activation of exercise. The lack of an attenuation of increases in heart rate and blood pressure strongly suggests that nifedipine recruits a dilator reserve by a reduction in coronary vasomotor tone in the ischemic myocardium. Since α_2-adrenergic coronary constrictor tone has previously been demonstrated to contribute to the severity of myocardial ischemia in conscious dogs during treadmill exercise [44], a functional antagonism of nifedipine against α_2-adrenergic coronary constrictor tone in ischemic myocardium is probably the underlying mechanism, not only in anesthetized dogs during sympathetic nerve stimulation, but also in conscious dogs during exercise. Whereas our own studies utilized the vasoselective dihydropyridine Ca-antagonists, a recruitment of coronary dilator reserve, potentially through reduction of α-adrenergic coronary constrictor tone, is probably also involved in the beneficial effects of other Ca-antagonists. The Ca-antagonist diltiazem caused a larger improvement in ischemic myocardial blood flow and function [31] than a β-blocker [32], although the reduction of heart rate was less marked, suggesting an additional vasodilator effect of diltiazem. It should be emphasized, however, that the non-Ca-antagonist vasodilators investigated so far (adenosine, dipyridamole, isosorbid-dinitrate) are unable to prevent poststenotic α_2-adrenergic coronary constriction and the resulting myocardial ischemia [24]. The preferential effectiveness of Ca-antagonists on α-adrenoceptor mediated coronary vasoconstriction may be due to a different segmental site of action in the coronary vascular tree. A preferential constriction of arterial and larger arteriolar vessels during α-adrenoceptor activation [7], as well as preferential dilation of larger arterial segments by nifedipine [47] has been demonstrated, whereas, for example, dipyridamole preferentially dilates small microvessels [8] which may be anyway dilated during myocardial ischemia [47].

In conclusion, Ca-antagonists can functionally antagonize α_1- and α_2-adrenergic coronary vasoconstriction. The α_2-adrenergic coronary constrictor mechanisms contributing to the precipitation of poststenotic myocardial ischemia are effectively antagonized by nifedipine and felodipine. Attenuation of exercise-induced myocardial ischemia through recruitment of coronary dilator reserve in ischemic myocardium was also achieved by nifedipine in conscious dogs during the physiological sympathetic activation of treadmill exercise.

Acknowledgements: The technical assistance of Ms. Bethina Blank, Ms. Margaret Hill, Ms. Denice Jio, and Ms. Barbara Patzer, as well as the secretarial help of Ms. Maria Kranke are appreciated.

References

1. Berkenboom GM, Abramowicz M, Vandermoten P, Degre SG (1986) Role of alpha-adrenergic coronary tone in exercise-induced angina pectoris. Am J Cardiol 57:195–198
2. Braunwald E (1982) Calcium-channel blockers: Pharmacologic considerations. Am Heart J 104:665–671
3. Brown BG, Bolson EL, Dodge HT (1984) Dynamic mechanisms in human coronary stenosis. Circulation 70:917–922
4. Brown BG, Lee AB, Bolson EL, Dodge HT (1984) Reflex constriction of significant coronary stenosis as a mechanism contributing to ischemic left ventricular dysfunction during isometric exercise. Circulation 70:18–24
5. Busch P, Deussen A, Heusch G (1988) Sympathetic effects on segmental coronary resistances and their role in coronary collateral perfusion. J Appl Cardiol 3:145–160
6. Chen DG, Dai X-Z, Zimmerman BG, Bache RJ (1988) Postsynaptic α_1- and α_2-adrenergic mechanisms in coronary vasoconstriction. J Cardiovasc Pharmacol 11:61–67
7. Chilian WM, Layne SM, Eastham CL, Marcus ML (1989) Heterogeneous microvascular coronary α-adrenergic vasoconstriction. Circ Res 64:376–388

8. Chilian WM, Layne SM, Klausner EC, Eastham CL, Marcus ML (1989) Redistribution of coronary microvascular resistance produced by dipyridamole. Am J Physiol 256:H383–H390

9. Decker N, Schwartz PJ (1985) Postjunctional alpha1- and alpha2-adrenoceptors in the coronaries of the perfused guinea-pig heart. J Pharmacol Exp Ther 232:251–257

10. Deussen A, Heusch G, Thämer V (1985) Alpha 2-adrenoceptor-mediated coronary vasoconstriction persists after exhaustion of coronary dilator reserve. Eur J Pharmacol 115:147–153

11. Ehring T, Heusch G (1990) Felodipine prevents the poststenotic myocardial ischemia induced by α_2-adrenergic coronary constriction. Cardiovasc Drugs Ther 4: 443–450

12. Feigl EO (1975) Control of myocardial oxygen tension by sympathetic coronary vasoconstriction in the dog. Circ Res 37:88–95

13. Gage JE, Hess OM, Murakami T, Ritter M, Grimm J, Krayenbuehl HP (1986) Vasoconstriction of stenotic coronary arteries during dynamic exercise in patients with classic angina pectoris: reversibility by nitroglycerin. Circulation 73:865–876

14. Gould L, Reddy GV, Gombrecht RF (1973) Oral phentolamine in angina pectoris. Jap Heart J 14:393–397

15. Guth BD, Heusch G, Indolfi C, Thaulow E, Ross Jr J (1989) Alpha1-adrenergic blockade reduces exercise-induced ischemia and contractile dysfunction in conscious dogs. Circulation 80 suppl II:II-98 (abstr)

16. Harrison DG, Chilian WM, Marcus ML (1986) Absence of functioning alpha-adrenergic receptors in mature canine coronary collaterals. Circ Res 59:133–142

17. Hautamaa PV, Dai X-Z, Homans DC, Bache RJ (1989) Vasomotor activity of moderately well-developed canine coronary collateral circulation. Am J Physiol 256:H890–H897

18. Hautamaa PV, Dai XZ, Homans DC, Robb JF, Bache RJ (1987) Vasomotor properties of immature canine coronary collateral circulation. Am J Physiol 252:H1105–H1111

19. Henry PD (1980) Comparative pharmacology of calcium antagonists: nifedipine, verapamil and diltiazem. Am J Cardiol 46:1047–1058

20. Heusch G (1990) α-Adrenergic mechanisms in myocardial ischemia. Circulation 81:1–13

21. Heusch G, Deussen A (1983) The effects of cardiac sympathetic nerve stimulation on the perfusion of stenotic coronary arteries in the dog. Circ Res 53:8–15

22. Heusch G, Deussen A (1984) Nifedipine prevents sympathetic vasoconstriction distal to severe coronary stenoses. J Cardiovasc Pharmacol 6:378–383

23. Heusch G, Deussen A, Schipke J, Thämer V (1984) α_1- and α_2-Adrenoceptor-mediated vasoconstriction of large and small canine coronary arteries in vivo. J Cardiovasc Pharmacol 6:961–968

24. Heusch G, Deussen A, Schipke J, Thämer V (1986) Adenosine, dipyridamole and isosorbiddinitrate are ineffective to prevent the sympathetic initiation of poststenotic myocardial ischemia. Drug Res 36:1045–1048

25. Heusch G, Guth BD, Seitelberger R, Ross Jr J (1987) Attenuation of exercise-induced myocardial ischemia in dogs with recruitment of coronary vasodilator reserve by nifedipine. Circulation 75:482–490

26. Heusch G, Yoshimoto N, Heegemann H, Thämer V (1983) Interaction of methoxamine with compensatory vasodilation distal to coronary stenoses. Drug Res 33:1647–1650

27. Holtz J, Saeed M, Sommer O, Bassenge E (1982) Norepinephrine constricts the canine coronary bed via postsynaptic α_2-adrenoceptors. Eur J Pharmacol 82:199–202

28. Kelley KO, Feigl EO (1978) Segmental alpha-receptor-mediated vasoconstriction in the canine coronary circulation. Circ Res 43:908–917

29. Liang IYS, Jones CE (1985) Alpha 1-adrenergic blockade increases coronary blood flow during coronary hypoperfusion. Am J Physiol 249:H1070–H1077

30. Ljung B (1985) Vascular selectivity of felodipine. Drugs 29:46–58

31. Matsuzaki M, Gallagher KP, Patritti J, Tajimi T, Kemper WS, White FC, Ross Jr J (1984) Effects of a calcium-entry blocker (diltiazem) on regional myocardial flow and function during exercise in conscious dogs. Circulation 69:801–814

32. Matsuzaki M, Patritti J, Tajimi T, Miller M, Kemper WS, Ross Jr J (1984) Effects of β-blockade on regional myocardial flow and function during exercise. Am J Physiol 247:H52–H60

33. McGrath JC (1982) Evidence for more than one type of postjunctional alpha-adrenoceptor. Biochem Pharmacol 31:467–484

34. Mohrman DE, Feigl EO (1978) Competition between sympathetic vasoconstriction and metabolic vasodilation in the canine coronary circulation. Circ Res 42:79–86

35. Mudge GH, Grossman W, Mills Jr RM, Lesch M, Braunwald E (1976) Reflex increase in coronary vascular resistance in patients with ischemic heart disease. N Engl J Med 295:1333–1337

36. Murray PA, Vatner SF (1979) α-Adrenoceptor attenuation of coronary vascular response to severe exercise in the conscious dog. Circ Res 45:654–660
37. Opie LH (1984) Calcium Antagonists and Cardiovascular Disease. Raven Press
38. Osakada G, Kumada T, Gallagher KP, Kemper WS, Ross Jr J (1981) Reduction of exercise-induced ischemic regional myocardial dysfunction by verapamil in conscious dogs. Am Heart J 101:707–712
39. Pagani M, Baig H, Sherman A, Manders WT, Quinn P, Patrick T, Franklin D, Vatner SF (1978) Measurement of multiple simultaneous small dimensions and study of arterial pressure dimension relations in conscious animals. Am J Physiol 235:H610–H617
40. Rimele TJ, Rooke TW, Aarhus LL, Vanhoutte PM (1983) Alpha-1 adrenoceptors and calcium in isolated canine coronary arteries. J Pharmacol Exp Ther 226:668–672
41. Schipke J, Heusch G, Schulz R, Thämer V (1987) An easy and quick implantation procedure for the measurement of myocardial wall thickness using sonomicrometry. Basic Res Cardiol 82:411–414
42. Schipke J, Oswald S, Deussen A, Heusch G (1984) Ein verbessertes Ultraschallverfahren zur Messung des Durchmessers epikardialer Koronararterien. Biomed Tech 29:3–5
43. Schulz W, Krauss G, Kaltenbach M, Kober G (1981) Einfluß von intrakoronarem und intravenösem Nifedipine auf die allgemeine und lokale Gefäßweite von epikardialen Koronararterien bei stabiler Angina pectoris- ein antianginöser Wirkaspekt? Z Kardiol 70:809–815
44. Seitelberger R, Guth BD, Heusch G, Lee JD, Katayama K, Ross Jr J (1988) Intracoronary alpha 2-adrenergic receptor blockade attenuates ischemia in conscious dogs during exercise. Circ Res 62:436–442
45. Seitelberger R, Guth BD, Lee JD, Katayama K, Heusch G, Ross Jr J (1986) Alpha1 and alpha2-receptor stimulation in conscious dogs increase coronary resistance but not myocardial function. J Am Coll Cardiol 7 suppl A:81A (abstr)
46. Sharma AD, Saffitz JE, Lee BI, Sobel BE, Corr PB (1983) Alpha adrenergic-mediated accumulation of calcium in reperfused myocardium. J Clin Invest 72:802–818
47. Tillmanns H, Neumann FJ, Parekh N, Dussel R, Zimmermann R, Dorigo O, Tiefenbacher C, Steinhausen M (1988) Unterschiedliche Wirkung von Koronardilatatoren auf die terminale Strombahn des ischämischen Myokards. Z Kardiol 77 (suppl I):147 (abstr)
48. Toda N (1986) Alpha-adrenoceptor subtypes and diltiazem actions in isolated human coronary arteries. Am J Physiol 250:H718–H724
49. van Meel JCA, Timmermans PBMWM, van Zwieten PA (1983) Alpha 1- and alpha 2-adrenoceptor stimulation in the isolated perfused hindquarters of the rat: in vitro model. J Cardiovasc Pharmacol 5:580–585
50. Vatner SF, Hintze TH (1982) Effects of a calcium-channel antagonist on large and small coronary arteries in conscious dogs. Circulation 66:579–588
51. Woodman OL, Vatner SF (1987) Coronary vasoconstriction mediated by α_1- and α_2-adrenoceptors in conscious dogs. Am J Physiol 253:H388–H393
52. Young MA, Vatner DE, Knight DR, Graham RM, Homcy CJ, Vatner SF (1988) α-Adrenergic vasoconstriction and receptor subtypes in large coronary arteries of calves. Am J Physiol 255:H1452–H1459

Authors' address:

Prof. Dr. Gerd Heusch, Abteilung für Pathophysiologie, Zentrum für Innere Medizin, Universitätsklinikum Essen, Hufelandstraße 55, 4300 Essen, FRG.

Effects of Selective Cardiac Denervation on Collateral Blood Flow after Coronary Artery Occlusion in Conscious Dogs

Y-T. Shen, D. R. Knight, J. X. Thomas, Jr. and S. F. Vatner

Department of Medicine, Harvard Medical School, Brigham & Women's Hospital, and the New England Regional Primate Research Center, Southborough, Massachusetts, USA

Summary

The extent to which cardiac nerves regulate responses to myocardial ischemia remains controversial. Our data in conscious dogs indicate that neither selective posterior left ventricular (LV) wall denervation nor selective ventricular denervation, leaving the atria intact, modifies the effects of coronary artery occlusion (for 24 h) on regional myocardial function and infarct size as compared to normally innervated dogs. Since hemodynamic changes were similar among the three groups after coronary artery occlusion, it is possible to speculate that responses of collateral blood flow to the ischemic zone were also not modified by chronic selective cardiac denervation. To address this, individual samples were selected and included in either the infarcted (TTC negative) or salvaged (TTC positive) group. The infarcted and salvaged samples were paired according to blood flow levels of 0.1–0.2, 0.2–0.3, or 0.3–0.4 ml/min/g at either 5 min, 1 h, 3 h, or 6 h after coronary artery occlusion. The results demonstrated similar patterns of myocardial blood flow in tissue samples within the area at risk after coronary artery occlusion in the animals, regardless of whether the ischemic zone was innervated or denervated. While blood flow rose in ischemic tissue that ultimately was salvaged, and tended not to rise over the 24 h monitoring period in tissue samples that became necrotic, no differences could be discerned on the basis of intact or absent innervation of the ischemic zone. Thus, chronic absence of cardiac nerves does not affect regulation of ischemic zone blood flow following coronary artery occlusion in conscious dogs.

Introduction

Following coronary artery occlusion, the extent to which cardiac nerves modify responses to myocardial ischemia remains controversial. Several studies demonstrated beneficial effects of either total cardiac denervation or selective ventricular denervation on collateral blood flow [6, 14, 16, 17, 31], recovery of regional myocardial function [35], development of arrhythmias [7], and extent of necrosis [2, 14, 15] after coronary artery occlusion. In contrast, the studies from our laboratory in conscious dogs indicated that total cardiac denervation exerts an adverse effect on myocardial infarct size which may be related to the sustained elevation in LV end-diastolic pressure that can compromise coronary collateral blood flow [25]. Further studies focused on the conscious model with either selective posterior left ventricular (LV) wall denervation or selective total ventricular denervation, leaving the atria intact [33]. In these models LV end-diastolic pressure did not rise more than was observed in intact dogs following coronary artery occlusion. We found that neither selective posterior LV wall denervation nor total ventricular denervation improved regional myocardial function, blood flow or reduced

This work was supported in part by US Public Health Service Grants, HL 33065 and RR 00168.

infarct size after permanent or brief coronary artery occlusion [33]. The objective of this article is to review these findings, focusing on a new analysis of myocardial blood flow, to determine whether chronic elimination of cardiac nerves affected regulation of collateral blood flow after coronary artery occlusion. To accomplish this goal, data from prior studies [33, 34] were reanalyzed to examine the potential differences in transmural blood flow in the ischemic zone, following coronary artery occlusion in the presence and absence of selective chronic cardiac denervation.

Methods

The details of the methodology have been presented previously [25, 33, 34] and, accordingly, will be reviewed only briefly. Twenty-five adult dogs were anesthetized with sodium pentobarbital (30 mg/kg, i.v.). Through a left thoracotomy, Tygon catheters were implanted in the aorta and left atrium and a miniature pressure gauge was implanted in the left ventricle. An occluder was implanted around the left circumflex coronary artery. Wall thickness transducers were placed in the anterior and in the posterior LV regions. In nine dogs, posterior LV wall denervation was accomplished with a combination of surgical and chemical techniques, as described previously [3, 23, 26, 29]. Ventricular denervation was performed in six dogs by using either a combination of surgical and chemical techniques, or the surgical technique alone [6, 21, 29]. In 10 dogs, no surgical or chemical denervation of the heart was utilized; they constituted the intact group.

Hemodynamic data were recorded prior to occlusion and continuously throughout the first 3 h of occlusion and again for a 1 h period at 24 h after occlusion. Regional myocardial blood flow was measured by the radioactive microsphere technique [5, 25, 33]. After the first injection of microspheres the snare occluder was tightened. At 3–5 min after complete coronary artery occlusion, a second injection of microspheres was given followed by additional injections of microspheres at 1, 3, 6 and 24 h.

The dogs were anesthetized with sodium pentobarbital (30 mg/kv, i.v.) at 24 h after coronary artery occlusion and the heart was removed after an overdose of pentobarbital and placed on a perfusion apparatus [24, 25, 33]. The ascending aorta was perfused with Evan's blue dye and the left circumflex coronary artery was perfused with saline. After completion of cardiac perfusion, the left ventricle was divided into 6–7 rings. Color slides of both sides of each ring were taken for subsequent evaluation of the size of the perfusion bed of the left circumflex coronary artery. After this procedure, the rings were incubated in 1.0% triphenyl tetrazolium chloride (TTC) to reveal the viable tissue and necrotic myocardium [10, 25, 33]. The slides were projected on a digitizing tablet at similar magnification for measurement of area at risk and infarct size. Each ring was cut into small sections and further subdivided into endocardial, midmyocardial, and epicardial regions, making a total of 200–300 pieces from each left ventricle. Myocardial samples from the area at risk and the non-ischemic zone were counted in a gamma counter. In this way the pattern of blood flow was determined for the entire left ventricle. In addition, so as to eliminate variations in regional myocardial blood flow related to "microsphere loss" from the ischemic tissue, a correction factor was utilized [20, 27, 28]. Samples that were infarcted or salvaged using the TTC criteria were dissected from the area at risk for retrospective blood flow analysis. At each time point after coronary artery occlusion, i.e., 5 min, 1 h, 3 h, and 6 h, all samples within the range of blood flow (0.1–0.2, 0.2–0.3, and 0.3–0.4 ml/min/g) were analyzed. The total number of samples analyzed was 954. However, the total number of data points analyzed was 1505, as some of the samples were used for more than one analysis. The epicardial, midmyocardial, and endocardial distribution of these data points in infarcted tissue was 108, 282, and 399, while the data points in salvaged tissue were 370, 221 and 125, respectively. As expected, the majority of infarcted samples were derived from the endocardial region, whereas the majority of salvaged samples were derived from the epicardial region.

The data were analyzed using analysis of variance with Student-Newman-Keuls test (1) and student's t-test. Significance was accepted when the p value was less than 0.05. All values are reported as mean $\pm$ SEM.

Collateral blood flow after coronary occlusion

Results

Global and regional myocardial function

The effects of coronary artery occlusion on LV systolic pressure, LV end-diastolic pressure, LV dP/dt, mean arterial pressure, systolic wall thickening in the ischemic zone, and heart rate have been reported previously [33] and are summarized in Table 1. LV end-diastolic pressure and heart rate increased early after coronary artery occlusion, while LV systolic pressure, LV dP/dt, and mean arterial pressure did not change significantly. Systolic wall thickening in the posterior wall was diminished completely over the 24 h monitoring period. At 24 h after coronary artery occlusion, LV systolic pressure, LV dP/dt, and mean arterial pressure were decreased. There were no differences in any of these measurements among the three groups studied.

Infarct size and plasma creatine kinase

There were no differences in areas at risk and infarct size weights among the three groups, as reported previously [33]. Infarct size, expressed as percent of the area at risk, was similar among the intact (44.9 ± 7.1%), posterior LV wall denervated (47.6 ± 6.4%), and ventricular denervated (48.3 ± 8.4%) groups (Fig. 1a). The relationship between infarct size/area at risk and collateral (epicardial) blood flow was also analyzed (Fig. 1b). These relationships were not different among the three groups. After coronary artery occlusion CK values peaked to similar levels in the intact (1402 ± 335 IU/ml), posterior LV wall denervated (1489 ± 308 IU/ml), and ventricular denervated (2083 ± 470 IU/ml) groups, respectively (Fig. 1c). The time to peak CK levels was also similar: 10.9 ± 0.8 h, 10.6 ± 1.0 h, and 11.0 ± 0.4 h among the three groups.

Regional myocardial blood flow

The patterns of epicardial and endocardial blood flow in infarcted and salvaged tissues for equal blood flow values at 5 min after coronary artery occlusion in a 0.1–0.2 ml/min/g range in conscious dogs with intact nerves, posterior LV wall denervation, and ventricular denervation are shown in Fig. 2. There were major differences in responses of blood flow in salvaged tissue as compared with responses in infarcted tissue at 1 h, 3 h, 6 h and 24 h after coronary occlusion. There were no differences among the three groups, as reported previously [34].

In order to examine more closely the role of cardiac nerves in regulating blood flow to ischemic myocardium, a new analysis was conducted from the data reported previously [34]. Infarcted and salvaged tissue samples with blood flows within the three ranges (0.1–0.2, 0.2–0.3 and 0.3–0.4 ml/min/g) were compared for equivalent blood flow values at 5 min and 1 h (Fig. 3), and 3 h and 6 h (Fig. 4) after coronary artery occlusion in the three groups. The tissue samples from salvaged myocardium or infarcted myocardium were also compared for equal reductions in blood flow at 5 min and 1 h, at 1 h and 3 h, and at 3 h and 6 h after coronary artery occlusion in the three groups (Fig. 5). It is clear that at each time point analyzed, blood flow rose significantly more in the salvaged tissue samples than was observed in infarcted tissue samples. However, there were no differences either in salvaged or infarcted tissue among the three groups in any of the three blood flow ranges studied.

Discussion

The extent to which cardiac nerves affect responses to coronary artery occlusion in terms of alterations in collateral blood flow, and regional myocardial function, development of arrhythmias, and ultimate necrosis of myocardial tissue at risk remains controversial. The majority of prior studies have been conducted in acutely prepared anesthetized animals and have demonstrated markedly beneficial effects of either total cardiac denervation or selective ventricular denervation on collateral blood flow [6, 14, 16, 17, 31], recovery of regional myocardial function

Table 1. Hemodynamic effects of coronary artery occlusion

	Baseline	Time after coronary artery occlusion			
		3–5 min	1 h	3 h	24 h
LV Systolic Pressure (mmHg)					
Intact	121 ± 3.6	112 ± 1.9	109 ± 4.4	111 ± 5.3	101 ± 3.9*
Posterior denervated	125 ± 3.8	121 ± 4.0	122 ± 5.4	123 ± 4.6	96 ± 4.4*
Ventricular denervated	127 ± 3.0	122 ± 8.6	121 ± 9.3	115 ± 5.5	98 ± 6.3*
LV End-diastolic Pressure (mmHg)					
Intact	7.0 ± 0.5	10.9 ± 0.7*	10.1 ± 1.0	8.6 ± 0.7	8.6 ± 0.7
Posterior denervated	6.8 ± 0.8	11.0 ± 1.1*	9.3 ± 1.5	9.4 ± 1.3	7.7 ± 1.4
Ventricular denervated	7.6 ± 0.9	12.3 ± 1.7*	8.9 ± 1.6	9.0 ± 1.7	8.6 ± 1.5
LV dP/dt (mmHg/s)					
Intact	3359 ± 156	3383 ± 122	3171 ± 218	2904 ± 226	2302 ± 151*
Posterior denervated	3656 ± 188	3440 ± 189	3514 ± 166	3215 ± 237	2359 ± 202*
Ventricular denervated	3449 ± 287	3618 ± 458	3392 ± 453	2899 ± 138	2336 ± 117*
Mean Arterial Pressure (mmHg)					
Intact	98 ± 2.8	103 ± 2.5	98 ± 3.6	100 ± 6.4	84 ± 4.0
Posterior denervated	99 ± 3.5	105 ± 4.4	103 ± 4.2	103 ± 3.0	82 ± 4.4
Ventricular denervated	102 ± 4.3	103 ± 2.0	101 ± 4.6	95 ± 3.4	87 ± 4.6
Heart Rate (beats/min)					
Intact	93 ± 3.9	142 ± 5.4*	137 ± 8.1*	129 ± 7.8*	167 ± 10.5*
Posterior denervated	93 ± 2.3	150 ± 6.6*	138 ± 5.6*	134 ± 5.2*	153 ± 4.0*
Ventricular denervated	94 ± 5.9	147 ± 12.8*	136 ± 11.6*	131 ± 9.5*	158 ± 13.7*
Ischemic Zone					
Systolic wall thickening (mm)					
Intact	3.01 ± 0.24	0.12 ± 0.27*	0.33 ± 0.31*	0.42 ± 0.31*	0.24 ± 0.29*
Posterior denervated	2.38 ± 0.21	0.04 ± 0.12*	0.04 ± 0.13*	0.21 ± 0.19*	0.14 ± 0.20*
Ventricular denervated	2.39 ± 0.38	0.05 ± 0.07*	0.26 ± 0.14*	0.27 ± 0.13*	0.27 ± 0.16*

* Different from baseline, $p < 0.05$.

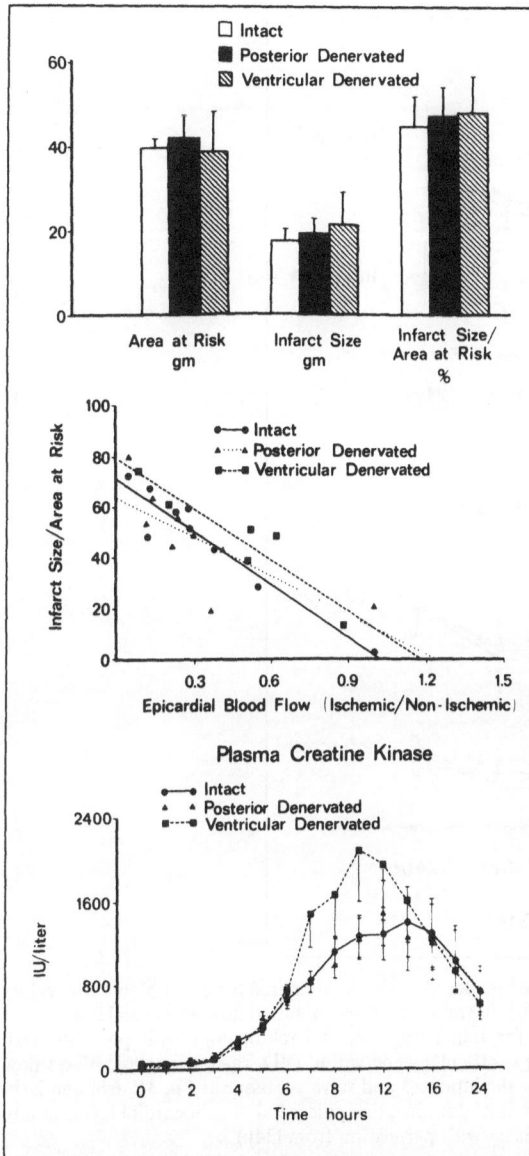

Fig. 1. The effects of coronary artery occlusion on infarct size (top panel), relationship between infarct size/area at risk and collateral blood flow (middle panel), and creatine kinase (bottom panel) are shown for intact dogs, dogs with posterior wall denervation, and dogs with ventricular denervation. Coronary artery occlusion did not result in any statistical difference in any of these parameters among the three groups. (Reproduced with permission from [33].)

[35], development of arrhythmias [7], and extent of necrosis [2, 14, 15] following coronary artery occlusion. In contrast, a prior study from our laboratory conducted in conscious dogs with total cardiac denervation found that coronary artery occlusion resulted in increased infarct size and lower levels of collateral coronary flow associated with greater elevation in LV end-diastolic pressure as compared to those in intact dogs [25]. It is conceivable that, in dogs with total cardiac denervation, the greater increase in LV end-diastolic pressure following coronary artery occlusion might be due to the lack of atrial innervation. Our subsequent investigations focused on the effects of coronary artery occlusion in conscious dogs with selective posterior LV wall denervation and total ventricular denervation [33]. In contrast to results in dogs with total cardiac denervation, responses of global LV function after coronary artery occlusion were similar in intact dogs and dogs with selective regional cardiac denervation. Systolic wall

233

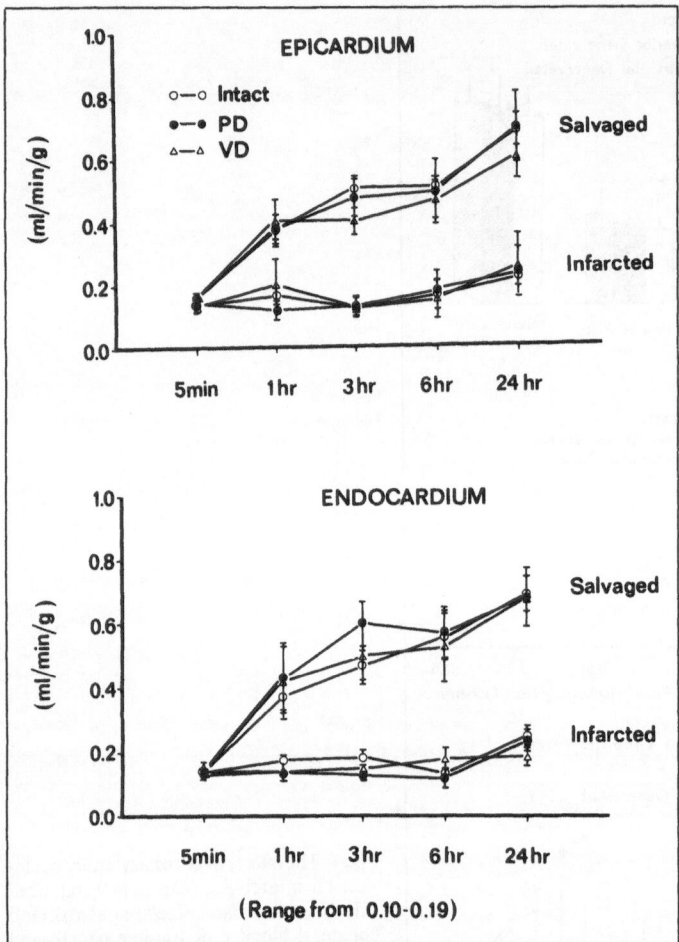

Fig. 2. Tissue samples from salvaged or infarcted myocardium were compared for equal blood flow values at 5 min after coronary artery occlusion in the 0.1–0.2 ml/min/g range of blood flow; epicardial layers (top) and endocardial layers (bottom) are shown for intact dogs (open circles), dogs with posterior wall denervation (PD, solid circles), and dogs with ventricular denervation (VD, open triangles). Blood flow values were compared in those three groups for the infarcted and salvaged tissue at 1 h, 3 h, 6 h, and 24 h. There were no differences among the three groups of dogs in either epicardial or endocardial layers at any time after coronary artery occlusion. (Reproduced with permission from [34].)

thickening in the ischemic zone was replaced by no wall thickening or wall thinning and did not recover over the 24 h monitoring period in any of the groups studied, indicating that selective denervation did not improve regional myocardial function. An almost identical infarct size, expressed as % of area at risk, was observed among intact and denervated groups. Furthermore, even when the infarcts were normalized for collateral blood flow, coronary artery occlusion did not result in any statistical difference in infarct size among the three groups. The total amount of CK released and time-to-peak CK values after coronary artery occlusion were also not different, indicating that selective cardiac denervation did not substantially reduce or delay myocardial damage after coronary artery occlusion.

It should be recognized that in most prior studies the data analyzed for myocardial function or myocardial blood flow reflected average information from an entire ischemic zone, which is

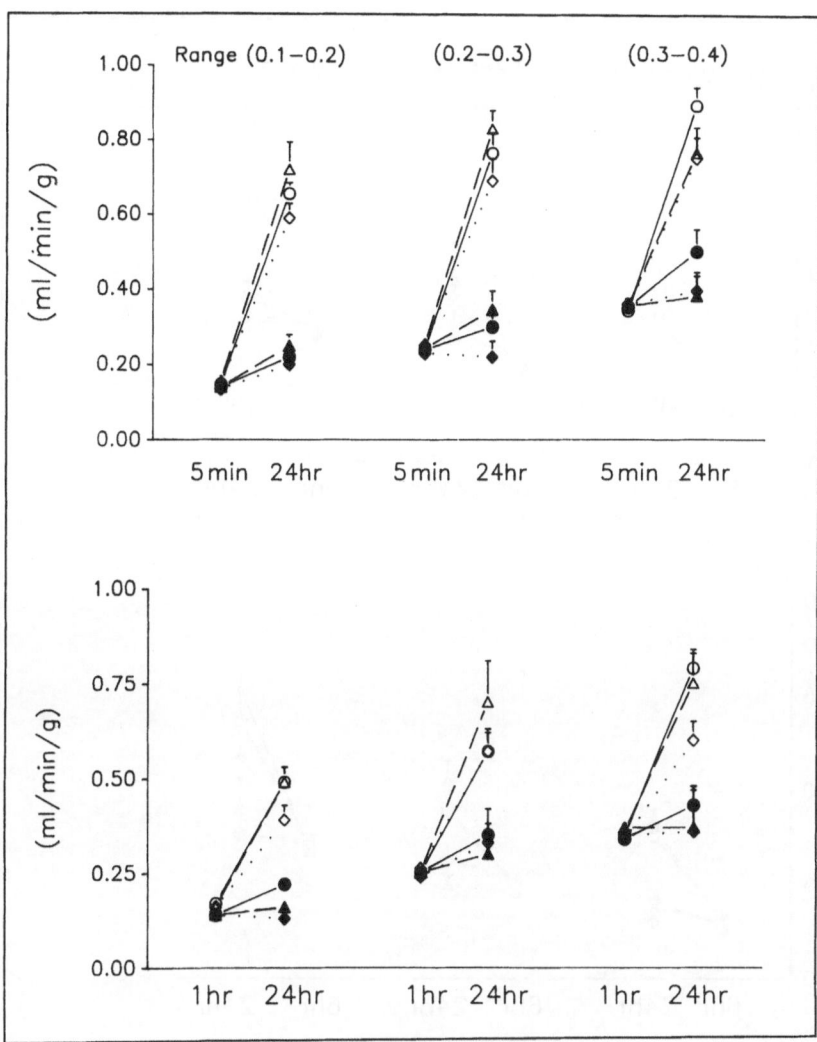

Fig. 3. Tissue samples from salvaged (open symbols) or infarcted (closed symbols) myocardium compared for equal reductions in blood flow at 5 min (top panel) and 1 h (bottom panel) after coronary artery occlusion are shown for intact dogs (solid line), dogs with posterior wall denervation (dashed line), and dogs with ventricular denervation (dotted line). The samples were subdivided into 0.1–0.2, 0.2–0.3, and 0.3–0.4 ml/min/g in ranges of blood flow. Similar blood flow patterns in either infarcted or salvaged tissue were observed among the three groups of dogs. In infarcted tissue there was no change or only a modest increase in blood flow over the subsequent 24 h and 23 h. In contrast, in the tissue samples that were salvaged, there was a significantly greater increase in blood flow over the following 24 h and 23 h periods in all of the three blood flow ranges.

known to exhibit considerable heterogeneity of blood flow, as has been noted by several investigators [4, 8, 11, 22, 30, 32]. Therefore, a more systematic analysis of transmural blood flow distribution focusing on small, discrete areas within the area at risk that either were infarcted or salvaged would be important for understanding the role of cardiac nerves in regulation of regional blood flow to ischemic myocardium. To conduct this type of analysis, the discrete tissue samples within the area at risk were divided into primarily infarcted or

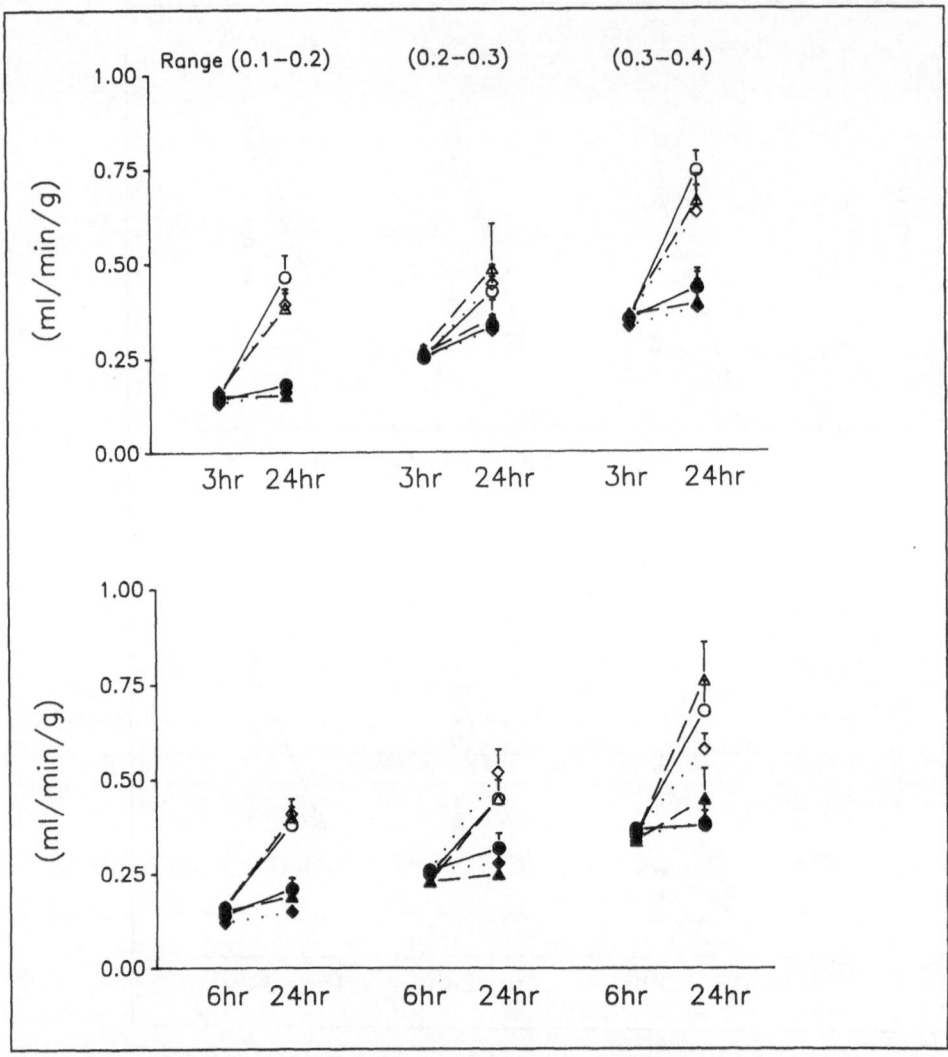

Fig. 4. Tissue samples from salvaged (open symbols) or infarcted (closed symbols) myocardium compared for equal reductions in blood flow at 3 h (top panel) and 6 h (bottom panel) after coronary artery occlusion are shown for intact dogs (solid line), dogs with posterior wall denervation (dashed line), and dogs with ventricular denervation (dotted line). The samples were subdivided into 0.1–0.2, 0.2–0.3, and 0.3–0.4 ml/min/g in ranges of blood flow. Similar blood flow patterns in either infarcted or salvaged tissue were observed among the three groups of dogs. In infarcted tissue there was no change or only a modest increase in blood flow over the subsequent 21 h and 18 h. In contrast, in the tissue samples that were salvaged, there was a significantly greater increase in blood flow over the following 21 h and 18 h period in all of the three blood flow ranges.

salvaged myocardium on the basis of TTC staining. As shown in Figs. 2–5, there was a progressive rise in blood flow to the salvaged myocardium. In contrast, blood flow to infarcted myocardium changed only slightly. It is clear that either in the epicardium or endocardium, there were no differences at 1 h, 3 h, 6 h, and 24 h after coronary occlusion among intact, posterior LV wall denervated and ventricular denervated groups (Fig. 2). In addition, blood

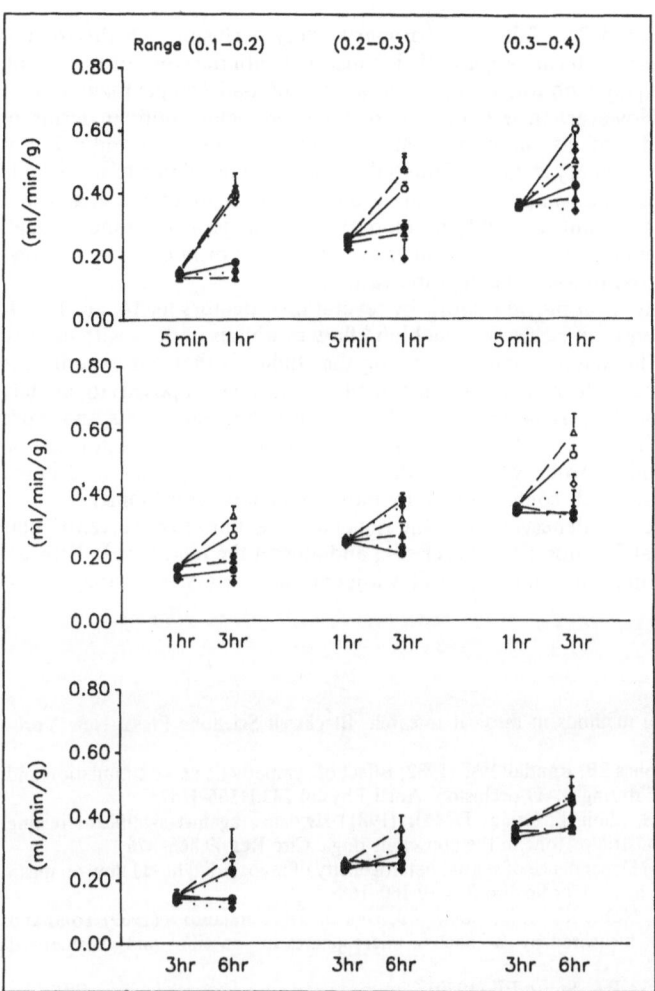

Fig. 5. Tissue samples from salvaged (open symbols) or infarcted (closed symbols) myocardium were compared at 5 min and 1 h (top panel), at 1 h and 3 h (middle panel), and at 3 h and 6 h (bottom panel) after coronary artery occlusion in intact dogs (solid line), dogs with posterior wall denervation (dashed line), and dogs with ventricular denervation (dotted line). The samples were subdivided into 0.1–0.2, 0.2–0.3, and 0.3–0.4 ml/min/g at 5 min, 1 h, and 3 h after coronary artery occlusion. At each time point the blood flow rose more in the salvaged tissue samples than in infarcted tissue samples. However, there were no differences among the three groups of dogs.

flow responses in infarcted and salvaged tissue samples were also compared at equal reductions in blood flow at 5 min and 1 h (Fig. 3), and 3 h and 6 h (Fig. 4) after coronary artery occlusion in the three groups. The similar patterns of blood flow to ischemic tissue at all time points after coronary artery occlusion among intact and denervated groups indicate that elimination of cardiac nerves did not modify ischemic zone blood flow. This is consistent with the infarct size results and CK appearance in blood. Thus, the evidence from our studies clearly indicates no difference in ischemic zone blood flow following coronary artery occlusion, whether or not cardiac nerves are intact. These results are apparently inconsistent with the work of Feigl et al. [9], and Heusch and Deussen [13], who found important aspects of α-adrenergic regulation of ischemic zone blood flow. However, their studies were relatively acute, both in terms of applying sympathetic blockade and also in terms of monitoring blood flow after induction of ischemia that was induced by coronary stenosis. Our studies were conducted in a more chronic setting, with neural denervation and complete coronary artery occlusion, and they suggest that sympathetic nerves are not important in modifying blood flow to the ischemic zone through collateral channels. In this connection, work by Harrison et al. [12] is of interest, as it shows that collateral vessels are devoid of α-adrenergic innervation.

Our results differ markedly from previous work by several investigators [6, 14, 16, 17, 31] that showed important differences in ischemic zone blood flow responses in dogs with cardiac or ventricular denervation. The major differences among the studies is that our experiments were carried out in conscious, chronically instrumented animals as opposed to acutely prepared, anesthetized animals [19]. However, the differences between our results and more recent data from Jones et al. [18] are less clear, since their experiments were also conducted on conscious animals. In that study, coronary artery catheters were chronically implanted in the animals for a 2-month period, which could potentially induce collateral development.

In conclusion, neither selective denervation of the ischemic zone nor selective ventricular denervation improved regional function and infarct size, and altered the relationship between collateral blood flow distribution and infarction following coronary artery occlusion in conscious dogs.

References

1. Armitage P (1973) Statistical methods in medical research. Blackwell Scientific Press, New York, pp 116–126
2. Barber MJ, Thomas JX Jr, Jones SB, Randall WC (1982) Effect of sympathetic nerve stimulation and cardiac denervation on MBF during LAD occlusion. Am J Physiol 243:H566–H574
3. Chilian WM, Boatwright TS, Shoji T, Griggs DM Jr (1981) Evidence against significant resting sympathetic coronary vasoconstrictor tone in the conscious dogs. Circ Res 49:866–876
4. Conway RS, Weiss HR (1985) Dependence of spatial heterogeneity of myocardial blood flow on mean blood flow rate in the rabbit heart. Cardiovasc Res 19:160–168
5. Domenech RJ, Hoffman JIE, Noble MIM, Saunders KB, Henson JR, Subijanto S (1969) Total and regional coronary blood flow measured by radioactive microspheres in conscious and anesthetized dogs. Circ Res 25:581–596
6. DuPont E, Jones CE, Luedecke RA, Smith EE (1979) Chronic ventricular sympathectomy: Effect on myocardial perfusion after ligation of the circumflex coronary artery in dogs. Circ Shock 6:323–331
7. Ebert PA, Allgood RJ, Sabiston DC Jr (1968) The anti-arrhythmic effects of cardiac denervation. Ann Surg 168:728–735
8. Falsetti HL, Carroll RJ, Marcus ML (1975) Temporal heterogeneity of myocardial blood flow in anesthetized dogs. Circulation 52:848–853
9. Feigl EO, Buffington CW, Nathan HJ (1987) Adrenergic coronary vasoconstriction during myocardial underperfusion. Circulation 75 [Suppl I]:I1–5
10. Fishbein MC, Meerbaum S, Rit J, Ando V, Kanmatsuse K, Mercier JC, Corday E, Ganz W (1981) Early phase acute myocardial infarct size quantification: Validation of the triphenyl tetrazolium chloride tissue enzyme staining technique. Am Heart J 101:593–600
11. Franzen D, Conway RS, Zhang H, Sonnenblick EH, Eng C (1988) Spatial heterogeneity of local blood flow and metabolite content in dog hearts. Am J Physiol 254:H344–H353
12. Harrison DG, Chilian WM, Marcus ML (1986) Absence of functioning α-adrenergic receptors in mature canine coronary collaterals. Circ Res 59:133–142
13. Heusch G, Deussen A (1983) The effects of cardiac sympathetic nerve stimulation on perfusion of stenotic coronary arteries in the dog. Circ Res 53:8–15
14. Jones CE, Beck LY, DuPont E, Barnes GE (1978) Effects of coronary ligation on the chronically sympathectomized dog ventricle. Am J Physiol 235:H429–H434
15. Jones CE, Devous MD, Thomas JX Jr, DuPont E (1978) The effect of chronic cardiac denervation on infarct size following acute coronary occlusion. Am Heart J 95:738–746

16. Jones CE, Scheel KW (1980) Reduced coronary collateral resistances after chronic ventricular sympathectomy. Am J Physiol 238:H196–H201
17. Jones CE, Hurst TW, Randall JR (1982) Reduced oxygen and blood flow demands in the chronically sympathectomized heart. Circ Shock 9:469–480
18. Jones CE, Liang IYS, Mass HJ, Gwirtz PA (1987) Response to brief coronary stenosis in conscious dogs after ventricular sympathectomy. Am J Physiol 252:H923–H932
19. Jugdutt BI (1986) Difference in the relation between infarct and occluded bed in pentobarbital-anesthetized and conscious dogs. Can J Physiol Pharmacol 64:254–262
20. Jugdutt BI, Hutchins GM, Bulkley BH, Becker LC (1979) The loss of radioactive microspheres from canine necrotic myocardium. Circ Res 45:746–756
21. Kaye MP, Priola DV, Coyle J (1970) Depletion of ventricular catecholamine levels following peripulmonary neurectomy. Proc Soc Exp Biol Med 135:844–848
22. King RB, Bassingthwaighte JB, Hales JRS, Rowell LB (1985) Stability of heterogeneity of myocardial blood flow in normal awake baboons. Circ Res 57:285–295
23. Knight DR, Thomas JX Jr, Randall WC, Vatner SF (1987) Effects of left circumflex flow transducer implantation on posterior wall innervation. Am J Physiol 252:H536–H539
24. Lange R, Nieminen MS, Kloner RA (1984) Failure of pindolol and metoprolol to reduce the size of non-reperfused infarcts in dogs using area at risk techniques. Cardiovasc Res 18:37–43
25. Lavallee M, Amano J, Vatner SF, Manders WT, Randall WC, Thomas JX Jr (1985) Adverse effects of chronic cardiac denervation in conscious dogs with myocardial ischemia. Circ Res 57:383–392
26. Martins JB, Zipes DP (1980) Epicardial phenol interrupts refractory period responses to sympathetic but not vagal stimulation in canine left ventricular epicardium and endocardium. Circ Res 47:33–40
27. Murdock RH Jr, Cobb FR (1980) Effects of infarcted myocardium on regional blood flow measurements to ischemic regions in canine heart. Circ Res 47:701–709
28. Reimer KA, Jennings RB (1979) The changing anatomic reference base of evolving myocardial infarction. Circulation 60:866–876
29. Randall WC, Ardell JL (1985) Differential innervation of the heart. Chapter 15, In: Zipes D, Jalife J (Eds) Cardiac Electrophysiology and Arrhythmias. Grune & Stratton
30. Reimer KA, Jennings RB (1979) The "Wavefront Phenomenon" of Myocardial Ischemic Cell Death. Lab Invest 40:633–644
31. Scheel KW, Jones CE (1983) Reduced resistances of septal artery collateral channels after cardiac sympathectomy. Basic Res Cardiol 78:373–383
32. Sestier FJ, Mildenberger RR, Klassen GA (1978) Role of autoregulation in spatial and temporal perfusion heterogeneity of canine myocardium. Am J Physiol 235:H63–H71
33. Shen Y-T, Knight DR, Vatner SF, Randall WC, Thomas JX Jr (1988) Responses to coronary artery occlusion in conscious dogs with regional cardiac denervation. Am J Physiol 255:H525–H533
34. Shen Y-T, Knight DR, Canfield DR, Vatner SF, Thomas JX Jr (1989) Progressive change in collateral blood flow after coronary artery occlusion. Am J Physiol 256:H478–H485
35. Thomas JX Jr, Randall WC, Jones CE (1981) Protective effect of chronic versus acute cardiac denervation on contractile force during coronary occlusion. Am Heart J 102:157–161

Authors' address:

You-Tang Shen, M.D., New England Regional Primate Research Center, One Pine Hill Drive, Southborough, MA 01772, USA.

IV. Sympathetic Activation in Myocardial Ischemia

Cardiocardiac Excitatory Reflexes during Myocardial Ischemia

A. Malliani

Istituto Ricerche Cardiovascolari, Consiglio Nazionale delle Ricerche; Ospedale "L. Sacco", Centro "Fidia", Patologia Medica, Università degli Studi, Milano, Italy

Summary

Myocardial ischemia represents a stimulus capable of exciting the receptive endings of both the vagal and sympathetic sensory fibers which innervate the heart. As a consequence, multiple and simultaneous reflexes can occur, mediated by the vagal and the sympathetic afferent limbs which, in turn, can modify both the vagal and the sympathetic outflows. Reflexes mediated by vagal afferent fibers are predominantly inhibitory, while those mediated by sympathetic afferent fibers are largely excitatory. In the experimental laboratory the various reflexes can be artificially isolated and analyzed, while in the clinical setting this is not feasible.

However, in recent years a computer analysis of cardiovascular short-term variability seems to furnish new indices of vagal and sympathetic modulatory activities. Such analysis in the frequency domain thus seems to provide a tool for both experimental and clinical evaluation of the neural mechanisms accompanying various physiological and pathophysiological events, including acute myocardial ischemia.

Introduction

It is difficult to affirm when reflexes arising from the heart began to be truly appreciated as important participants in the complexity of the hemodynamic profile that accompanies acute myocardial ischemia.

Experimental evidence that myocardial ischemia is a stimulus capable of eliciting powerful neural reflexes is relatively recent. In the initial studies in anesthetized dogs, vagally mediated depressor reflexes were consistently observed [7, 8]. This was interpreted as a sign that cardiovascular reflexes arising from the heart and mediated by vagal afferent fibers serve the teleologic purpose of reducing cardiac load whenever this might be beneficial to cardiac function, as in the case of acute ischemia.

Subsequently, it was proved that myocardial ischemia can also be accompanied by excitatory reflexes mediated by afferent cardiac sympathetic fibers and that vagotomy is a procedure which often facilitates their appearance [3, 22, 25].

In clinical terms, we now know that angina can be accompanied by: a) hypotension and bradycardia [11], b) hypotension and tachycardia [11, 12, 27], and c) hypertension and tachycardia [12, 16, 17, 27, 33].

The usual interpretation of these different possibilities assumes that hypotension is always due to some degree of cardiac failure, that tachycardia, in such a case, reflects a baroreceptive mechanism and finally, that hypertension and tachycardia are a reflection of pain and emotion. However, these mechanisms alone would not explain those cases in which a) hypotension is accompanied by bradycardia [11], and b) hypertension and tachycardia occur simultaneously with electrocardiographic changes but in the absence of pain [4, 12].

It is the aim of this report to utilize some experimental laboratory findings to advance the

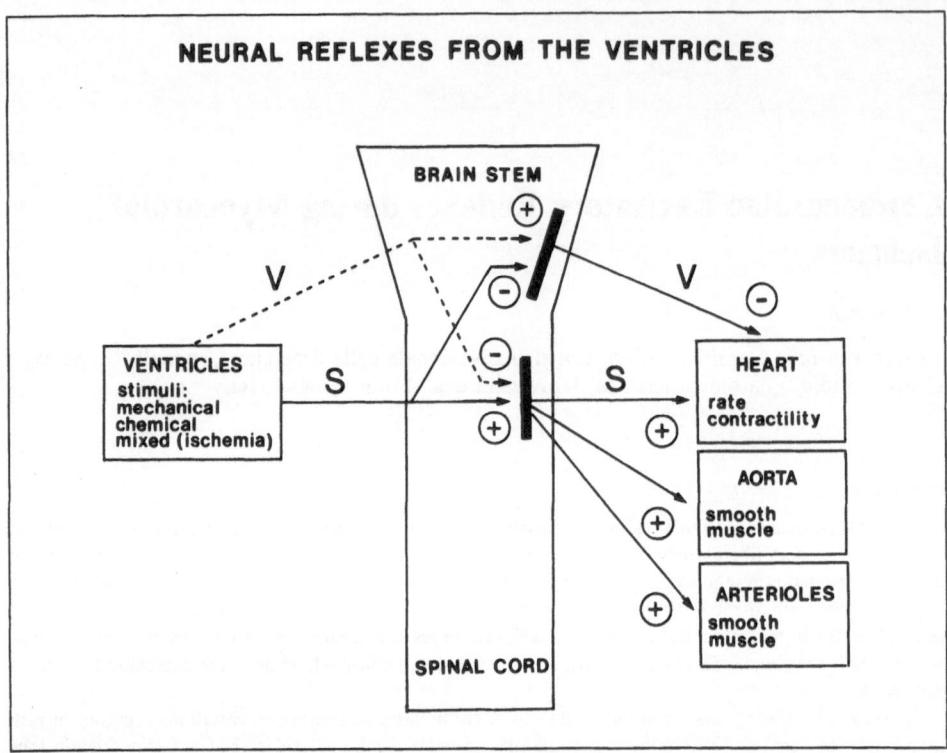

Fig. 1. Schematic representation of neural reflexes induced by activation of left ventricular vagal and sympathetic afferent fibers.

hypothesis that the hemodynamic picture which accompanies acute myocardial ischemia is likely to be the result of the complex interplay of direct depressive effects of ischemia and of multiple neural reflexes (Fig. 1). It will be proposed that the simultaneous excitation of cardiac vagal and sympathetic afferent fibers elicits reflexes from the heart that are opposite in sign and which, in addition, interact with other reflexes, such as those arising from arterial baroreceptive areas.

The role of the extent of myocardial ischemia in determining the sign of the prevailing response

We have pusblished [18] a series of experiments undertaken in anesthetized cats in order to analyze the relative contribution of reflexes mediated by vagal and sympathetic cardiac afferent fibers, respectively, to the hemodynamic and neural changes that accompany brief periods of myocardial ischemia. To simplify the analysis, the experiments were carried out on animals in which the arterial baroreceptor areas had been previously denervated. Moreover, to avoid the immediate changes in the cardiovascular regulatory mechanisms which are induced by acute baroreceptor denervation, the animals studied had undergone the denervation procedure at least one week prior to the experimental session.

In this experimental model, in addition to common hemodynamic variables, the impulse activity from a preganglionic sympathetic output largely directed to the heart was recorded while myocardial ischemia could be obtained by occluding either the left main ("global" ischemia) or the distal left anterior descending ("regional" ischemia) coronary artery.

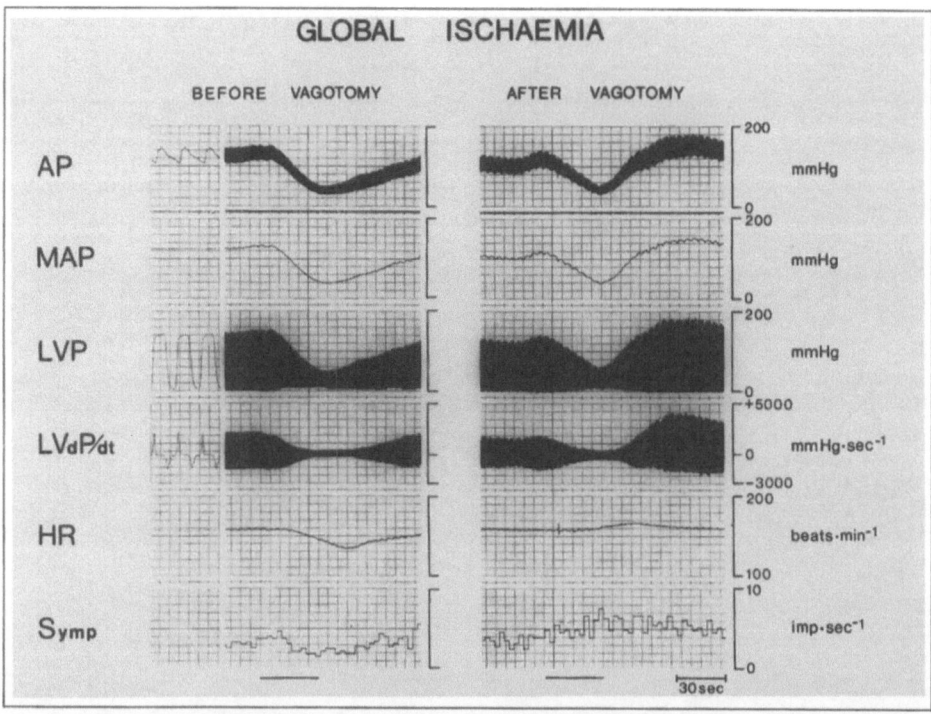

Fig. 2. Effects of "global" ischemia in a cat with chronic sinaortic denervation, before and after sectioning the vagi. Before vagotomy a significant activation of sympathetic discharge was detectable during the early phase of occlusion. Thereafter, the neural discharge was markedly inhibited. After vagotomy "global" ischemia resulted in a significant activation of sympathetic discharge throughout the occlusion period. In both experimental conditions there was a marked hemodynamic depressor response. The tracings represent, from top to bottom: AP, arterial pressure; MAP, mean arterial pressure; LVP, left ventricular pressure; LVdP/dt, left ventricular dP/dt; HR, heart rate; Symp, cardiac efferent sympathetic nerve activity. Bars indicate the periods of the left main coronary artery occlusions (from [18]).

"Global" ischemia (Fig. 2) before vagotomy, resulted in a significant reduction of mean arterial pressure (MAP), left ventricular pressure (LVP), and left ventricular dP/dt_{max} while sympathetic efferent impulse activity was significantly augmented during the initial 15 ± 2 s of occlusion (early phase) and inhibited during the subsequent 20 ± 2 s of occlusion (late phase). Vagotomy did not modify the hemodynamic responses; however, a significant increase in sympathetic discharge was detectable during the whole occlusion period (early and late phases).

"Regional" ischemia (Fig. 3) before vagotomy resulted in a significant increase in sympathetic neural discharge and mean arterial pressure, with no changes in left ventricular function. After vagotomy the occlusion elicited a significant increase in mean arterial pressure, left ventricular pressure, left ventricular dP/dt_{max}, and efferent sympathetic neural activity. These excitatory responses were abolished after the interruption of a large part of the cardiac sympathetic afferents. Thus, coronary artery occlusion induced hemodynamic and sympathetic reflex responses that were dependent upon the interaction of opposite influences mediated by the simultaneous activation of cardiac vagal and sympathetic afferents. In this experimental model the extent of "ischemic myocardium" represented a crucial factor for the prevailing type of neural response.

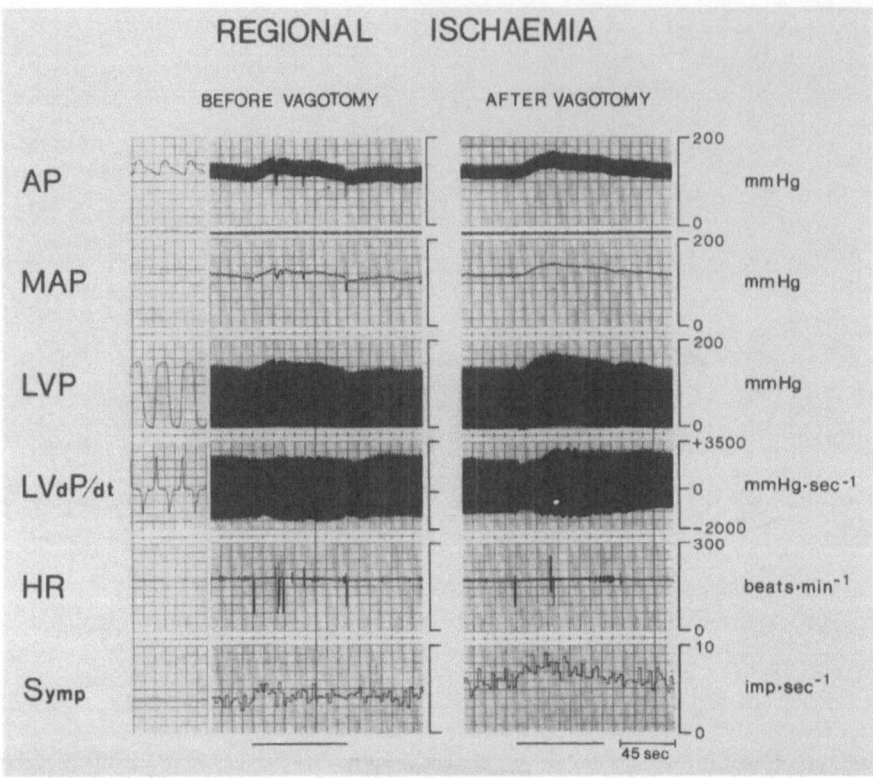

Fig. 3. Effects of "regional" ischemia in a cat with chronic sinoaortic baroreceptor denervation, before and after sectioning the vagi. A pressor response together with an increase in sympathetic efferent impulse activity was observed in both experimental conditions. Tracings as in Fig. 2. Bars indicate the periods of the left anterior descending artery occlusions (from [18]).

The site of myocardial injury as another determinant of the prevailing reflex response

The clinical report by Webb et al. [36] introduced the notion that during the hyperacute phases of human myocardial infarction the cardiovascular changes vary importantly in relation to the site of the infarction. Inferior wall infarcts were more often associated with increased parasympathetic activity (sinus bradycardia, transient hypotension or atrioventricular block), whereas anterior wall infarcts were more often associated with sympathetic overactivity (sinus tachycardia or transient hypertension). In the cat, Corr et al. [6] found that such differences are mirrored by ligation, respectively, of the right or left anterior descending coronary artery. Following that report, other studies have confirmed that the depressive responses were more easily elicited from the infero-posterior part of the left ventricle, both in humans [31] and in experimental animals [35]. On the contrary, the occlusion of the left anterior descending coronary artery provided the clearest excitatory responses [18, 25]. Obviously, great caution should accompany any comparison between myocardial infarction and transient myocardial ischemia in terms of their capabilities of affecting directly or indirectly nerve terminals, both afferent and efferent, and, hence, in producing local release of mediators or in initiating reflexes.

However, since the observation by Webb et al. [36] it was evident that some other factors were involved, because the type of "autonomic disturbance" associated with a particular localization of the infarction was only a matter of frequency and not an absolute rule. In the

246

clinical setting of transient myocardial ischemia, our hypothesis, obviously, is that an additional factor is likely to be represented by the extent of myocardial mass made ischemic. This would imply that transient ischemic episodes associated with hypotension and bradycardia or hypotension without the tachycardia which could be expected from a baroreceptive mechanism, i.e., events likely to reflect a vagally mediated depressor reflex would characterize more severe episodes of ischemia. Indeed Guazzi et al. [11] found them to be associated with acute ventricular failure. Furthermore, it is important to note the similarities between the "supernormal" phase described by Guazzi et al. [11] at the end of the ischemic episodes in humans and that observed in our experiments during the recovery from "global" ischemia (Fig. 2).

However, pressor reflexes from the heart would be the most frequent accompaniments of less severe ischemic episodes, whether or not signalled by anginal pain [20–22]. It is quite clear that these neural mechanisms should not be interpreted in a teleologic manner and, indeed, the loss of finality is likely to be a fundamental difference between physiology and pathophysiology.

A recent study by Chierchia et al. [4], carried out in patients free to move and presenting frequent episodes of transient myocardial ischemia, has shown that hypertension and tachycardia seem to accompany, as a rule, ischemic episodes characterized by ST-segment depression, independently from the presence or absence of pain.

Neural and cardiovascular rhythms as markers of functional states

Rhythmicity is an intrinsic property of the nervous system. Various rhythms can be markers of normal events such as wakefulness or sleep and of abnormal conditions such as epilepsy. However, a rhythm is rarely univocally linked with a function: thus an atropinized cat can walk around with an electroencephalogram simulating placid sleep. However, circulation and respiration – strictly related transport functions – are both based on discontinuous events, and oscillations of various orders characterize them, in particular cardiovascular variables. It has been a traditional endeavour of experimental physiology to describe such oscillations, to investigate their causes, and, more recently, the links existing between neural and cardiovascular rhythms in view of a possible functional significance. Surely the oldest are the observations on rhythmic fluctuations of systemic arterial pressure begun in 1733 with Hales' experiment [13], during which oscillations of first and second order were observed, i.e., those related to cardiac cycle and respiration. As to the third order oscillations, i.e., those with a period apparently longer than the respiratory cycle, Koepchen [15] recently reviewed the history and the various interpretations which flourished around them: according to his conclusions we shall assume the waves described by Mayer [28] and having a period of about 10 s as the prototype of the third order oscillations of arterial pressure. However, the period of this type of oscillation is highly variable.

The possibility recently offered by computer techniques to quantify the small spontaneous beat-by-beat oscillations in cardiovascular variables and, in particular, in the electrocardiographic R-R interval, aroused a growing interest in view of the hypothesis that these rhythmical oscillations could provide some insight into the neural regulatory mechanisms operating in the intact organism under real life conditions.

Indeed, even somewhat crude analyses of the variability phenomena such as those offered by the use of standard deviation, frequency histograms or simple filtering techniques provided important information in the course of pathophysiological processes such as diabetes [9] and myocardial infarction [14].

However, the application of computationally efficient spectral techniques [5] offered the opportunity of assessing specifically the non random components of heart rate variability thus quantifying the possible rhythmicity hidden in the signal. Sayers [34], for instance, employing the Fast Fourier Transform technique, reported the existence in humans of three major components in R-R variability that he observed in specific bands of predetermined frequencies around 0.25, 0.10, and 0.03 Hz, i.e., a component related to respiration (0.25 Hz), and two others at lower frequencies. Following this pioneering work several other investigators applied this technique and, in spite of the consideration that the heart rate variability signal is not strictly periodical, as requested by the deterministic nature of the algorithm, it became clear that it could be used as a quantitative probe to assess heart rate fluctuations [1].

A similar computer analysis can be applied to other cardiovascular variables, such as systolic or diastolic arterial pressure [29, 30], left ventricular pressure, and its derivative [32]. The LF component of systolic arterial pressure variability, corresponding to the Mayer waves, appears to furnish a good index of the sympathetic modulation of peripheral arterial vasomotion.

A recent development of the methodology refers to the possibility of analyzing long periods of analog recordings by using a recursive version of the program of spectral analysis. For instance, Holter tapes digitized at appropriate speed can furnish a quantitative assessment of R-R variability throughout the 24 h period of the recording [10].

Further computations can be performed on the data, such as cross-spectral analysis, which can provide quantitative information on the coherence function, i.e., a measure of the statistical link between the variability signals at any given frequency, and on the phase relationship for heart period and systolic arterial pressure [30]. On the basis of this more complex analysis the continuous "closed-loop" relationship linking heart period and the arterial pressure variabilities can be studied [30]: in our last attempt the closed-loop model also includes respiration [2].

However, in the present context we shall only report data on heart rate variability extracted from the electrocardiographic signal.

A simplifying physiological hypothesis and its relation to cardiovascular rhythms

The neural regulation of circulatory functions is mainly effected through the interplay of the sympathetic and vagal outflows which are tonically and phasically modulated by means of the interaction of at least three major black boxes: the "central command", the reflex negative feedback, and the reflex positive feedback mechanisms [20, 23]. To study this whole interplay only through the action of single reflexes appears as an unsound illusion since the fragmented pieces of knowledge are difficult if not impossible to reassemble into a unitary regulatory system. Simplified but general hypotheses might be necessary. In this respect the sympathovagal interaction might be viewed as a push-pull or reciprocal relationship: that is to say that in most conditions, as far as we know, the activation of either outflow is accompanied by the inhibition of the other.

It is the core hypothesis of our recent work in this field that this interaction can be broadly explored in the frequency domain. In fact, we have collected numerous data in support of the assumption that: i) the second-order rhythm, i.e., the high-frequency (HF) spectral component is a marker of vagal activity (a fact well documented in current literature); ii) the third-order rhythm, i.e., the low-frequency (LF) spectral component is a marker of sympathetic activity (a hypothesis that we introduced into the current research). A push-pull relationship has been found to exist between the second- and the third-order rhythm, which, in our opinion, parallels a similar functional relationship existing between the two neurovegetative outflows. In short, a new tool seems to be available for both the experimental laboratory and for clinical investigation, allowing a quantification of the state of the sympatho-vagal balance [24, 29, 32].

Experimental alterations of the sympatho-vagal balance

In our studies in man, various manoeuvres are used to shift the sympatho-vagal balance towards sympathetic predominance such as passive tilt to 90° [29], light physical exercise [30], mental stress. Since the earliest attempts it appeared that under normal conditions these manoeuvres are always accompanied by a relative increase in the power of the LF component, and by a decrease of the HF component (Fig. 4, lower right panel). Conversely, vagal predominance is mainly obtained with metronome breathing at about 15–20 breaths per min, with water immersion of the face, and with head down tilt. In all cases the HF component increased and the LF component decreased.

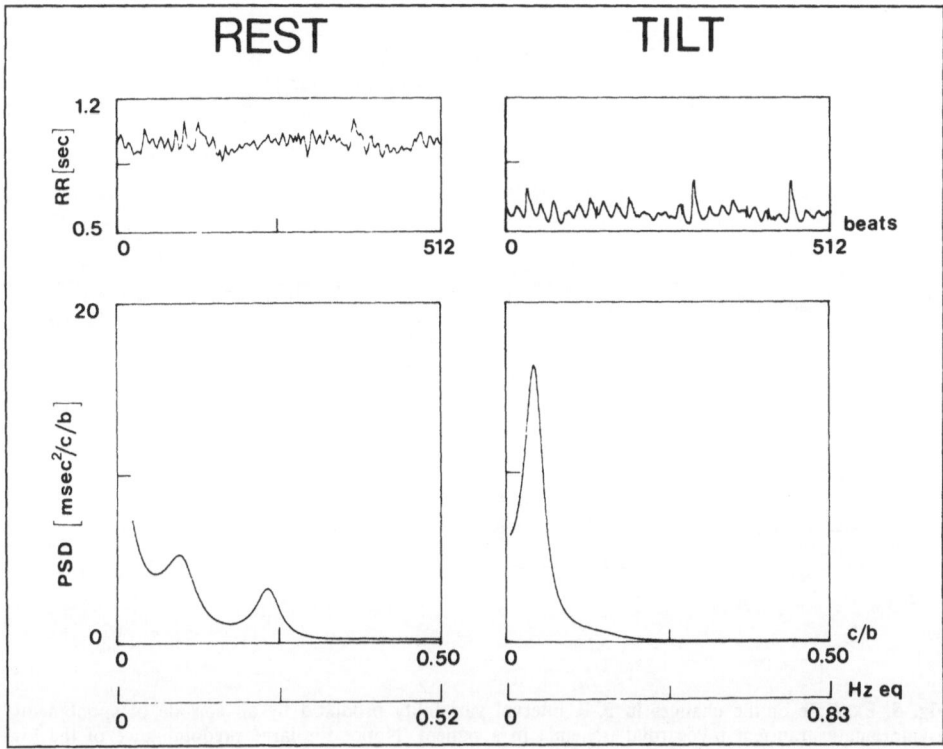

Fig. 4. R-R interval series, i.e. tachogram at rest and during passive upright 90° tilt. On the auto-spectra (bottom panels), two clearly separated low- and high-frequency components are present at rest. During tilt, the low-frequency component becomes preponderant. (from [29]).

In the conscious dog [32], sympathetic activation is obtained by moderate hypotension (produced by i.v. nitroglycerin infusion), by physical exercise, by acute baroreceptor deactivation; conversely, the normal predominance of vagal tone is further increased by baroreceptor stimulation (produced by phenylephrine infusion).

Thus the methodological soundness of this simple attributive stage has been checked throughout quite numerous and different models: the collected data always fit the basic hypothesis that the HF rhythm of R-R variability is a marker of vagal tone, while the LF spectral component of R-R and arterial pressure variability is a marker of sympathetic tone.

Experimental coronary artery occlusion and clinical observations on transient myocardial ischemia and myocardial infarction

In a recent study [32] carried out on conscious unanesthetized dogs, the occlusion of a distal branch of either the left circumflex or anterior descending coronary artery maintained for ~ 120 s did not induce behavioral signs attributable to pain. Mean arterial pressure was not significantly modified, whereas heart rate increased by 51 ± 9% and R-R and its variance were reduced significantly. The power spectral analysis of both R-R and systolic arterial pressure variabilities revealed that during the occlusion there was a marked increase in the normalized power of the LF component. Taking into consideration some obvious differences in the baseline conditions, this is quite similar to what can be observed in an ambulatory patient in the course of an asymptomatic episode of transient myocardial ischemia (Fig. 5). As to the baseline differences they consist of a strong background predominance of vagal tone in the dog, while

249

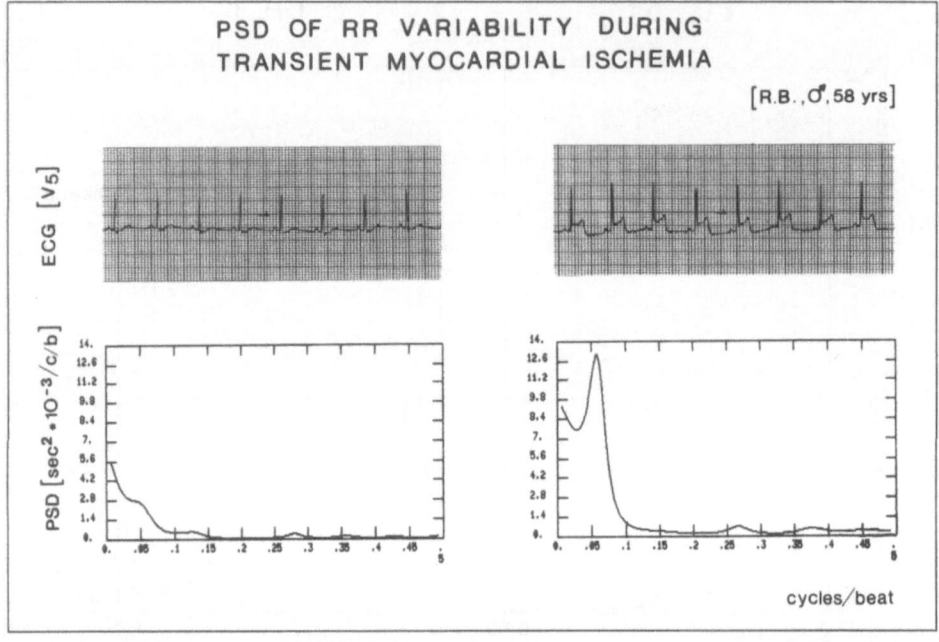

Fig. 5. Example of the changes in R-R interval variability produced by an episode of spontaneous asymptomatic transient myocardial ischemia in a patient. Notice the large predominance of the low frequency component during the ischemic episode. PSD = power spectral density. (From Lombardi et al. work in progress.)

in the case of the patient illustrated in Fig. 5 a predominance of the LF component is already present before the ischemic episode.

We also studied [19] a population of patients 2 weeks after acute myocardial infarction and after 6 and 12 months. At 2 weeks the LF component was greater (LF = 69 ± 2 vs 53 ± 3, normalized units, n.u.) and the HF component was smaller (HF = 17 ± 1 vs 35 ± 3 n.u.) than in age-matched control subjects: thus, there were clear signs of a sympathetic predominance. At 6 and 12 months after myocardial infarction there was a progressive decrease in the LF (62 ± 2 and 54 ± 3 n.u.) and a progressive increase (23 ± 2 and 30 ± 2 n.u.) in the HF components. Moreover, in a subgroup of patients 2 weeks after myocardial infarction, tilt did not further modify the LF component (from 74 ± 3 to 78 ± 3 n.u.). Conversely, 1 year after infarction, this maneouvre was accompanied by an increase in the LF component of a magnitude similar to that observed in control subjects (from 53 ± 3 to 77 ± 3 n.u.).

Thus, with this approach it was possible to conclude that the sympathetic predominance that is detectable 2 weeks after myocardial infarction is subsequently followed by a recovery of vagal tone and a normalization of the sympatho-vagal interaction, not only during resting conditions but also in response to a sympathetic stimulus.

Conclusions

Multifarious experimental and clinical evidence indicates that acute myocardial ischemia is a condition capable of eliciting powerful neural reflexes. These reflexes, in their complexity, cannot be interpreted with a finalistic conception advocating homeostasis above all. In fact, the excitatory sympathetic reflexes appear to subserve not only heterostasis [19] but, in concrete clinical terms, to facilitate life-threatening arrhythmias [26].

Cardiocardiac excitatory reflexes during myocardial ischemia

The new methodology centred on analysis in the frequency domain seems to represent a highly useful tool with which to investigate both experimental and clinical problems; this might facilitate a better integration of pathophysiological and clinical interpretations.

References

1. Akselrod S, Gordon D, Ubel FA, Shannon DC, Barger AC, Cohen RJ (1981) Power spectrum analysis of heart rate fluctuation: A quantitative probe of beat-to-beat cardiovascular control. Science 213:220–222
2. Baselli G, Cerutti S, Civardi S, Liberati D, Lombardi F, Malliani A, Pagani M (1986) Spectral and cross-spectral analysis of heart rate and arterial blood pressure variability signals. Comput Biomed Res 19:520–534
3. Brown AM, Malliani A (1971) Spinal sympathetic reflexes initiated by coronary receptors. J Physiol 212:685–705
4. Chierchia S, Muiesan L, Davies A, Balasubramian V, Gerosa S, Raftery EB (1990) Role of the sympathetic nervous system in the pathogenesis of chronic stable angina. Circulation (in press)
5. Cooley JW, Tukey JW (1965) An algorithm for the machine calculation of complex Fourier Series. Math Comput 19:297–301
6. Corr PB, Pearle DL, Hinton JR, Roberts WC, Gillis RA (1976) Site of myocardial infarction. A determinant of the cardiovascular changes induced in the cat by coronary occlusion. Circ Res 39:840–847
7. Costantin LR (1963) Extracardiac factors contributing to hypotension during coronary occlusion. Am J Cardiol 11:205–217
8. Dokukin AV (1964) Role of the stretch receptors of the left ventricle in reflex hemodynamic changes in myocardial ischemia. Fed Proc 23:T296–T297
9. Ewing DJ, Neilson JMM and Travis P (1984) New method for assessing cardiac parasympathetic activity using 24 hour electrocardiograms. Br Heart J 52:396–402
10. Furlan R, Guzzetti S, Crivellaro W, Dassi S, Tinelli M, Baselli G, Cerutti S, Lombardi F, Pagani M, Malliani A (1990) Continuous 24-hour assessment of the neural regulation of systemic arterial pressure and R-R variabilities in ambulant subjects. Circulation 81:537–547
11. Guazzi M, Polese A, Fiorentini C, Magrini F, Bartorelli C (1971) Left ventricular performance and related haemodynamic changes in Prinzmetal's variant angina pectoris. Br Heart J 33:84–94
12. Guazzi M, Polese A, Fiorentini C, Magrini F, Olivari MT, Bartorelli C (1975) Left and right heart haemodynamics during spontaneous angina pectoris. Comparisons between angina with ST segment depression and angina with ST segment elevation. Br Heart J 37:401–413
13. Hales S (1733) Statical Essays: Containing Haemastaticks. Innys, Manby and Woodward, London (vol 2)
14. Kleiger RE, Miller JP, Bigger JT, Moss AJ and the multicenter post-infaction research group (1987) Decreased heart rate variability and its association with increased mortality after acute myocardial infarction. Am J Cardiol 59:256–262
15. Koepchen HP (1984) History of studies and concepts of blood pressure waves. K. Miyakawa, H.P. Koepchen and C. Polosa (Eds) Japan Sci Soc Press, Tokyo, Springer-Verlag, Berlin, pp 3–23
16. Lewis T (1931) Angina pectoris associated with high blood pressure and its relief with amyl nitrite; with a note on Nothnagel's syndrome. Heart 15:305–327
17. Littler WA, Honour J, Sleight P, Stott FD (1973) Direct arterial pressure and electrocardiogram in unrestricted patients with angina pectoris. Circulation 48:125–134
18. Lombardi F, Casalone C, Della Bella P, Malfatto G, Pagani M, Malliani A (1984) Global versus regional myocardial ischemia: differences in cardiovascular and sympathetic responses in cats. Cardiovasc Res 18:14–23
19. Lombardi F, Sandrone G, Pernpruner S, Sala R, Garimoldi M, Cerutti S, Baselli G, Pagani M, Malliani A (1987) Heart rate variability as an index of sympathovagal interaction after acute myocardial infarction. Am J Cardiol 60:1239–1245
20. Malliani A (1982) Cardiovascular sympathetic afferent fibers. Rev Physiol Biochem Pharmacol 94:11–74
21. Malliani A (1986) The elusive link between transient myocardial ischemia and pain. Circulation 73:201–204
22. Malliani A, Lombardi F (1978) Neural reflexes associated with myocardial ischemia. In: Schwartz PJ, Brown AM, Malliani A, Zanchetti A (eds) Neural Mechanisms in Cardiac Arrhythmias. New York: Raven Press, pp 209–219

23. Malliani A, Pagani M, Lombardi F (1986) Positive feedback reflexes. In: Zanchetti A, Tarazi RC (eds) Handbook of Hypertension vol 8: Pathophysiology of Hypertension – Regulatory Mechanisms. Elsevier Science Publishers B.V.
24. Malliani A, Pagani M, Lombardi F, Cerutti S (1990) Clinical and experimental evaluation of sympatho-vagal interaction: power spectral analysis of heart rate and arterial pressure variabilities. In: Zucker IH, Gilmore JP (eds) Reflex Control of the Circulation. The Telford Press, New Jersey
25. Malliani A, Schwartz PJ, Zanchetti A (1969) A sympathetic reflex elicited by experimental coronary occlusion. Am J Physiol 217:703–709
26. Malliani A, Schwartz PJ, Zanchetti A (1980) Neural mechanisms in life-threatening arrhythmias. Am Heart J 100:705–715
27. Maseri A, Severi S, De Nes M, L'Abbate A, Chierchia S, Marzilli M, Ballestra AM, Parodi O, Biagini A, Distante A (1978) "Variant" angina: one aspect of a continuous spectrum of vasospastic myocardial ischemia. Am J Cardiol 42:1019–1035
28. Mayer S (1876) Studien zur Physiologie des Herzens und der Blutgefässe: 5. Abhandlung: Über spontane Blutdruckschwankungen. Sber Akad Wiss Wien, 3. Abt 74:281–307
29. Pagani M, Lombardi F, Guzzetti S, Rimoldi O, Furlan R, Pizzinelli P, Sandrone G, Malfatto G, Dell'Orto S, Piccaluga E, Turiel M, Baselli G, Cerutti S, Malliani A (1986) Power spectral analysis of heart rate and arterial pressure variabilities as a marker of sympatho-vagal interaction in man and conscious dog. Circ Res 59:178–193
30. Pagani M, Somers V, Furlan R, Dell'Orto S, Conway J, Baselli G, Cerutti S, Sleight P, Malliani A (1988) Changes in autonomic regulation induced by physical training in mild hypertension. Hypertension 12:600–610
31. Perez-Gomez F, Garcia-Aguado A (1977) Origin of ventricular reflexes caused by coronary arteriography. Br Heart J 39:967–973
32. Rimoldi O, Pierini S, Ferrari A, Cerutti S, Pagani M, Malliani A (1990) Analysis of the short term oscillations of RR and arterial pressure in conscious dogs. Am J Physiol 258:967–976
33. Roughgarden JW (1966) Circulatory changes associated with spontaneous angina pectoris. Am J Med 41:947–961
34. Sayers B, McA (1973) Analysis of heart rate variability. Ergonomics 16:17–32
35. Thames MD, Klopfenstein HS, Abboud FM, Mark AL, Walker JL (1978) Preferential distribution of inhibitory cardiac receptors with vagal afferents to the inferoposterior wall of the left ventricle activated during coronary occlusion in the dog. Circ Res 43:512–519
36. Webb SW, Adgey AAJ, Pantridge JF (1972) Autonomic disturbance at onset of acute myocardial infarction. Br Med J 3:89–92

Author's address:

Alberto Malliani, Istituto Ricerche Cardiovascolari, via Bonfadini 214, 20138 Milano, Italy.

Pain and Myocardial Ischemia: the Role of Sympathetic Activation

V. Thämer, A. Deussen, J. D. Schipke[†], T. Tölle and G. Heusch[*]

Inst. of Physiology, Univ. Düsseldorf, Düsseldorf, FRG; [†]Dept. of Experimental Surgery, Univ. Düsseldorf, FRG; [*]Dept. of Pathophysiology, Univ. Essen, Essen, FRG

Summary

In a first series, we tested whether the relative ischemia distal to a severe stenosis on the left circumflex coronary (CX) artery increases the activity of cardiac sympathetic (CS) nerves which, in turn, may result in a poststenotic vasoconstriction and an aggravation of ischemia. In 23 anesthetized, vagotomized dogs, an acute stenosis that reduced CX blood flow to 50% of control was produced and maintained for 20 min. The activity of postganglionic CS nerves increased by $23 \pm 4\%$ within 20 min. In parallel, poststenotic coronary resistance increased from 0.48 ± 0.03 (SEM) to 0.61 ± 0.03 mm Hg·min·100 g/ml, resulting in a net lactate production after 15 min. The selective α_2-adrenoceptor antagonist rauwolscine (0.2 mg/kg i.v.; $n = 6$) and the calcium antagonist nifedipine (10 µg/kg i.v.; $n = 6$) prevented the progressive increase in poststenotic resistance and the net lactate production, but still permitted an increase in CS activity. Segmental anesthesia of CS nerves with epidural infiltration of procaine at segments C7-T6 ($n = 6$) prevented the sympathetic activation, the progressive increase in poststenotic resistance and the net lactate production. In six additional dogs with intact vagus nerves, CS activation and a concomitant increase in poststenotic resistance resulting in myocardial ischemia were also found. These data suggest a vicious cycle between poststenotic coronary vasoconstriction and CS activation, resulting in severe myocardial ischemia. In a second series, stimulation of high-threshold somatic afferents (= nociceptive stimulation: NCS) was used to cause reflex CS activation. The superficial peroneal nerve was electrically stimulated in 14 anesthetized, vagotomized dogs. With intact CX arteries, a 1 min stimulation resulted in a pronounced increase in CX blood flow and perfusion pressure. In contrast, NCS in the presence of a severe stenosis on the CX artery increased end-diastolic poststenotic coronary resistance by $96 \pm 15\%$ due to a reflex activation of CS nerve fibers. This activation was markedly reduced after injection of fentanyl (27 µg/kg i.v.; $n = 6$). Injection of naloxone (60 µg/kg) restored the original effect. Systolic wall thickening (WT; sonomicrometry) in the CX artery-perfused myocardium was increased during NCS (10.9 ± 3.9 (SD) vs. $13.6 \pm 5.0\%$) in additional five dogs with intact coronary arteries. In the presence of a stenosis on the CX artery, systolic WT was reduced to $7.0 \pm 2.5\%$ and was further decreased to $4.6 \pm 2.3\%$ during NCS. The additional deterioration of systolic regional function during NCS was prevented after i.v.-injection of fentanyl, as was the increase in poststenotic coronary resistance. The present data suggest that CS nerve activation, both by a spinal cardiocardiac reflex and by somatic afferents, increases poststenotic coronary resistance and induces or aggravates myocardial ischemia distal to coronary stenoses.

Pain evoked by noxious stimulation can be elicited from cardiac, i.e., visceral, as well as from somatic receptors. By means of a reflex mechanism, nociceptive stimulation impacts on the activity of the autonomic nervous system and, subsequently, the cardiovascular system [18]. In studies designed to examine mechanisms involved in the sensation of cardiac pain, noxious stimuli such as coronary artery occlusion or the injection of bradykinin are usually applied.

These studies demonstrate that information regarding noxious cardiac events reaches the spinal cord via sympathetic afferent Aδ- and C-fibers [22]. The information is processed in cells of the spinoreticular and spinothalamic tract and reaches the medial medulla (for review see [3]). The aim of this report is not to describe the pathway of nociceptive information, but to demonstrate the effect of noxious stimuli on cardiac sympathetic nerve activity and myocardial perfusion.

It is generally accepted that myocardial ischemia following a coronary artery occlusion leads to an increase in sympathetic nerve activity by a cardiocardiac sympathetic reflex [17, 22]. This is not self-evident, because two opposing reflexes are evoked at the same time. The one is the pressor reflex, which increases both sympathetic nerve activity and heart rate. The Bezold-Jarisch reflex, on the other hand, inhibits nerve activity and heart rate through activation of vagal afferents [2, 23]. Studies by Lombardi et al. [15] demonstrate that the prevailing reflex pattern is determined by the extent of the ischemic region. Large ischemic areas result in cardiac dilatation and elicit a depressor reflex. Conversely, small ischemic areas elicit a pressor reflex. Also, the prevailing reflex pattern is determined by the site of myocardial ischemia, as posterior-inferior ischemia preferentially elicits depressor reflexes and anterior ischemia preferentially elicits pressor reflexes [19]. Finally, the Bezold-Jarisch reflex is only a shortlasting reflex and can, therefore, be overriden by an increase in sympathetic nervous activity. Still, this reflex could possibly explain the bradycardia often seen in patients shortly after the onset of a coronary artery occlusion [24].

We investigated the reflex effect of myocardial ischemia in open-chest dogs. In these experiments, not only the changes in sympathetic nerve activity were measured, but also their effect on coronary resistance, contractile function, and metabolism in the poststenotic area.

Reflex aggravation of myocardial ischemia

A stenosis of a large coronary artery is compensated as long as the poststenotic vascular bed can still dilate [5]. After exhaustion of the coronary reserve, further decreases in the stenotic cross-sectional area will impair myocardial perfusion and, in consequence, myocardial performance. Such a situation occurs as soon as cardiac sympathetic nerve activity is enhanced by either electrical stimulation or bilateral carotid occlusion [1, 7, 8, 12]. The poststenotic vasoconstriction is mediated by activation of postsynaptic vascular α_2-adrenoceptors and can be prevented by the selective α_2-adrenoceptor antagonist rauwolscine and by the calcium antagonist nifedipine [8, 9].

Several studies demonstrate that the occlusion of a large coronary artery can stimulate sympathetic nerve activity by a cardio-cardiac reflex [17, 22]. Thus, ischemia itself could possibly become the basis for a deleterious vicious cycle, resulting in a progressive aggravation of myocardial ischemia. For such a vicious cycle to become operative, two prerequisites are necessary: 1) The stenosis is sufficient to provoke sympathetic activation, and 2) this sympathetic activation is strong enough to induce coronary vasoconstriction. We tested this hypothesis in experiments using anesthetized dogs.

Methods

The methods have been described in detail elsewhere [10]. In brief: 29 dogs weighing 23–29 kg were anesthetized with α-chloralose (50 mg/kg i.v.) and urethane (250 mg/kg i.v.). They were ventilated using a respirator (874502 Braun-Melsungen, FRG). Blood gases were kept in the normal range. Left ventricular pressure was measured with a cathetertip-manometer (PC 350, Millar, Houston, Texas, USA) advanced from the right femoral artery. Left ventricular dP/dt and heart rate were derived from the left ventricular pressure signal. A left thoracotomy was performed, and the heart was suspended in a pericardial cradle. The left circumflex coronary artery was isolated distal to its origin for placement of an electromagnetic flowprobe (Statham SP 7515, Gould, Oxnard, California, USA). Blood flow was measured with an electromagnetic flowmeter (SP 2202, Statham). A small side branch of the left circumflex coronary artery distal to the flowprobe was cannulated with a polyethylene catheter for measurement of peripheral coronary pressure (P23ID, Statham).

A stenosis of the left circumflex coronary artery could be produced by a circumferential constrictor device distal to the flowprobe and proximal to the cannulated side branch.

To sample venous circumflex coronary blood, a 5F Cournand catheter was advanced via the coronary sinus into a vein draining the myocardium supplied by the left circumflex coronary artery. Oxygen content and plasma lactate concentration were determined both in these samples and in samples from the left femoral artery (Lex-O_2-Con, Lexington Instruments, Waltham, USA and Monotest Lactat, Boehringer-Mannheim, FRG).

The electrical activity of some postganglionic sympathetic fibers of the left ventrolateral cervical cardiac nerve projecting on the circumflex-perfused myocardium was recorded in a multifiber preparation. The fibers were cut distally and placed on a bipolar silver electrode. The signals were displayed on an oscilloscope and passed through a rate meter, which counted all impulses with amplitudes between two predetermined levels. The number of spikes per second, as well as the integrated sympathetic activity were continuously recorded on a Brush-recorder (Mark 260, Gould, Cleveland, Ohio, USA) together with the hemodynamic parameters.

Experimental protocol

In 23 dogs, both vagus nerves were cut in the midcervical region to prevent vagal reflex effects secondary to changes in blood pressure and cardiac dimensions. The vagal nerves were left intact in the remaining six dogs. After recording control conditions, a stenosis of the left circumflex coronary artery was produced by the constrictor device. The severity was adjusted such that coronary blood flow was initially reduced by about 50% of control. The stenosis was maintained for 20 min. Blood samples were taken at 1, 5, and 20 min after the onset of stenosis, as well as 1 and 5 min after its release.

At the end of each experiment, the myocardium perfused by the left circumflex coronary artery was visualized by injection of barium-sulfate, cut out, and weighed. Coronary blood flow was normalized by the weight of the perfused myocardium. End-diastolic coronary resistance was calculated from peripheral coronary pressure, circumflex coronary venous pressure and circumflex coronary blood flow.

Data analysis

Data were analyzed by a one-way analysis of variance with five levels (control, 1, 5, 10, and 20 min of stenosis). If time significantly contributed to over-all variance, the values after 20 min of stenosis were compared to control values (sympathetic activity) or to values after 1 min stenosis (hemodynamics and metabolism) using F-tests. A p value < 0.05 was considered to exhibit statistical significance. All values are given as mean $\pm$ SEM.

Results

Figure 1 demonstrates a representative experiment. The mechanical stenosis of the left circumflex coronary artery was used to reduce mean coronary blood flow by about 50%. This reduction was associated with a considerable decrease in distal perfusion pressure. In contrast, peak left ventricular pressure, used as an index of global myocardial function, was unaffected. During the 20 min stenosis, blood flow continued to decrease, whereas perfusion pressure increased during this time. This increase in coronary resistance corresponded with an increase in sympathetic nerve activity. Progressive ischemia in the poststenotic area was evidenced by decreasing myocardial lactate consumption which even turned into a net lactate production.

Figure 2 and Table 1 summarize the results of experiments in six dogs with intact vagus nerves. Ischemia increased the activity of cardiac sympathetic nerves to about 20% above

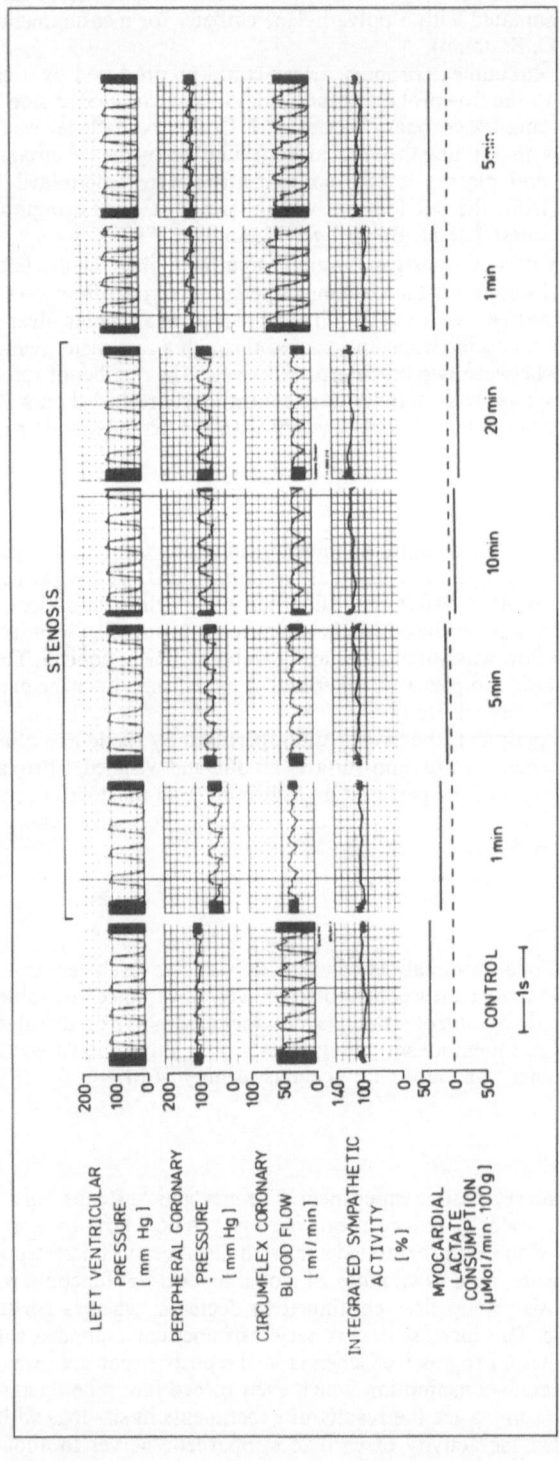

Fig. 1. Original recording of the effect of 20 min acute left circumflex coronary stenosis on left ventricular pressure, peripheral coronary pressure, coronary blood flow, cardiac sympathetic nerve activity, and calculated myocardial lactate consumption. During 20 min coronary stenosis, cardiac sympathetic nerve activity increases, and coronary blood flow progressively decreases, resulting finally in net lactate production (from [10] by permission).

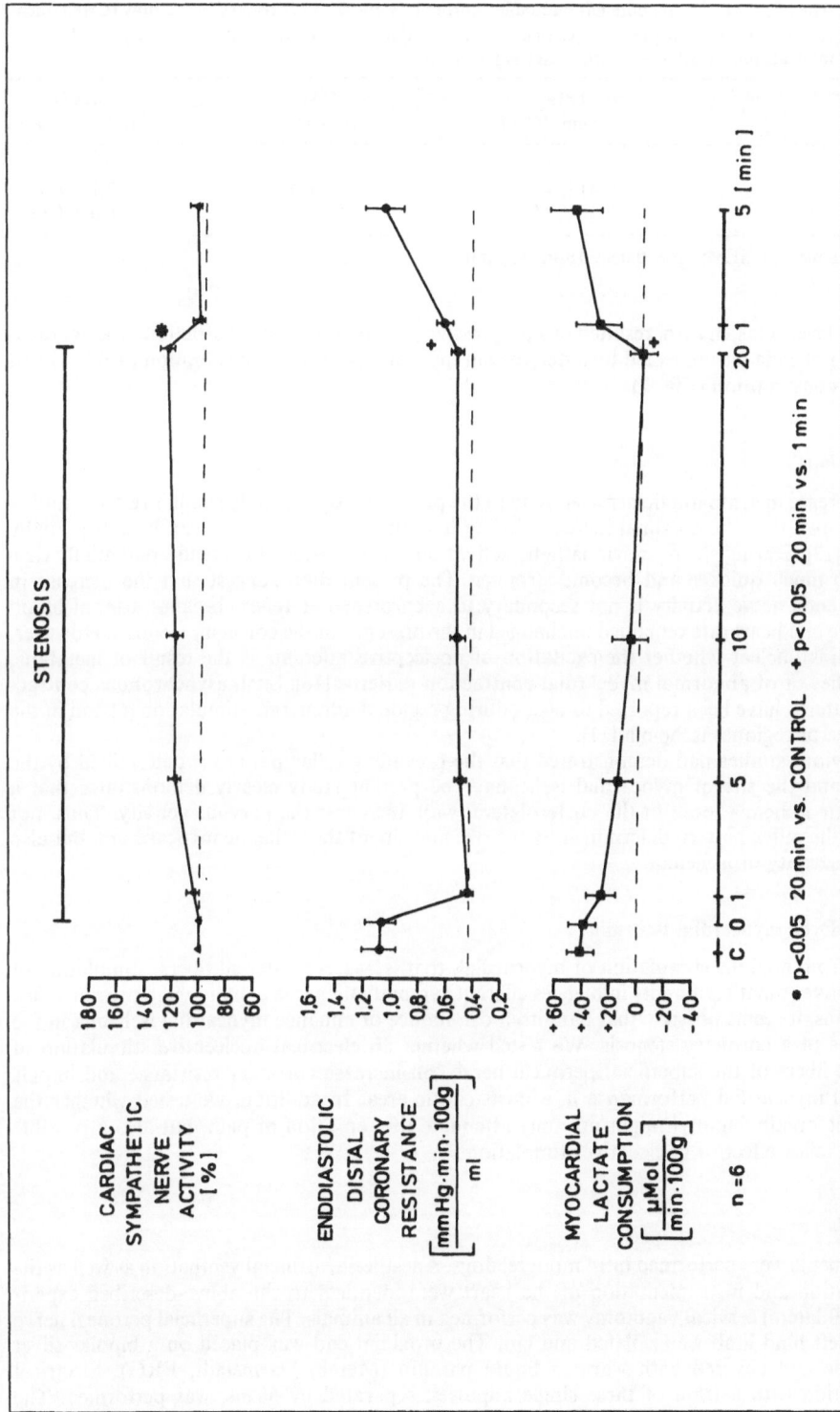

Fig. 2. The effects of 20 min acute left circumflex coronary stenosis on cardiac sympathetic nerve activity, coronary resistance, and myocardial lactate consumption in six dogs with intact vagus nerves. An increase in cardiac sympathetic nerve activity is associated with a progressive increase in poststenotic resistance and a decrease in lactate consumption (from [10] by permission).

Table 1. The effect of 20 min acute left circumflex coronary stenosis on coronary blood flow (CBF), mean peripheral coronary arterial pressure (CAP), and myocardial oxygen consumption (MV̇O₂) of the poststenotic myocardium in six dogs with intact vagus nerves.

		CBF (ml/min·100 g)	CAP (mm Hg)	MV̇O₂ (ml/min·100 g)
Control		78 ± 5	104 ± 4	9.0 ± 1.3
Stenosis	1 min	41 ± 4	40 ± 4	7.1 ± 0.9
	20 min	36 ± 4*	44 ± 4*	6.0 ± 0.8*

Data are mean ± SEM; *p < 0.05: 20 min vs. 1 min.

control level. This, in turn, resulted in a progressive poststenotic vasoconstriction. The aggravation of ischemia was reflected by a decrease in myocardial oxygen consumption (Table 1) and lactate consumption (Fig. 2).

Discussion

The increase in sympathetic nerve activity in the present study was only small. Previous studies have shown that the maximal increase during hypothalamic stimulation can be substantially higher [13]. During stronger sympathetic activation, the feedback effect could potentially then develop much quicker and become stronger. The present data suggest that the increase in sympathetic nerve activity is not secondary to a baroreceptor reflex, because arterial blood pressure and heart rate remained unchanged in the presence of the coronary stenosis. However, it remains unclear whether the excitation of nociceptive afferents is the result of metabolic properties or of abnormal myocardial contraction patterns [16]. Such asynchronous contraction patterns have been reported to occur during regional adrenergic stimulation [6] and in the presence of regional ischemia [11].

Previous studies had demonstrated that the prevailing reflex pattern is determined by the extent and the site of myocardial ischemia. The present study clearly demonstrates that a moderate ischemia, even in the posterolateral wall, increases the nervous activity. Thus, not only is the reflex pattern determined by the size and site of the ischemic myocardium, but also by the severity of ischemia.

Pain-induced myocardial ischemia

Apart from noxious stimulation of myocardial, that is, visceral afferent fibers, stimulation of nociceptive somatic afferents influences efferent sympathetic nerve activity via a pressor reflex [21]. Thus, it seems possible that pain itself can induce or enhance myocardial ischemia in the presence of a coronary stenosis. We tested whether an electrical nociceptive stimulation of afferent fibers of the superficial peroneal nerve can increase coronary resistance and impair regional myocardial performance in a poststenotic area. In addition, we tested whether the synthetic opiate fentanyl might not only attenuate the sensation of pain, but also a possible pressor reflex effect of nociceptive stimulation.

Methods

Experiments were performed in 14 mongrel dogs. Anesthesia, artificial ventilation as well as the preparation and instrumentation of the heart were identical to the above-described experiments. Bilateral cervical vagotomy was performed in all animals. The superficial peroneal nerve of the left hind limb was isolated and cut. The proximal end was placed on a bipolar silver electrode and covered with warmed liquid paraffin (Merck, Darmstadt, FRG). Electrical stimulation with a train of three single impulses, separated by 50 ms, was performed. The repetition rate was 2 s, the impulse amplitude 20 to 25 V, and the stimulation period lasted 1 min.

Sympathetic activity of a multifiber filament of the left ventrolateral cervical cardiac nerve was displayed on an oscilloscope and simultaneously recorded on a tape-recorder (Hewlett Packard). After the end of the experiments, the spontaneous activity and the activity during the stimulation period were rectified and averaged (Instrument Computer 1072, Nicolet, Madison, Wisconsin, USA). The sympathetic nerve activity is expressed in arbitrary units. In five additional dogs regional myocardial function was measured using sonomicrometry (for details see [14]). Function was assessed in the area perfused by the left anterior descending coronary artery (control) and in the area perfused by the left circumflex coronary artery (to be rendered ischemic).

Experimental protocol

Nociceptive stimulation was first performed under control conditions and then in the presence of a severe stenosis of the left circumflex coronary artery. The stimulation was repeated after intravenous injection of fentanyl (27 μg/kg) and, finally, after injection of naloxone (60 μg/kg).

Data analysis

For evaluation of regional function, end-diastole was defined as the point when left ventricular dP/dt started its rapid upstroke after crossing the zero line. End-systole was defined as the point of maximal wall excursion within 20 ms before peak negative dP/dt [20]. Regional systolic wall excursion was calculated as percent of end-diastolic wall thickness. All data are given as mean $\pm$ SD. The effects of nociceptive stimulation were analyzed using a paired t-test. A p-value < 0.05 was considered to exhibit statistical significance.

Results

Figure 3 shows the effects of an electrical nociceptive stimulation of the superficial peronaeus nerve on left ventricular hemodynamics. With intact coronary arteries, stimulation for 1 min resulted in a pronounced increase in left ventricular pressure, dP/dt_{max}, coronary blood flow and perfusion pressure. In contrast, nociceptive stimulation in the presence of a severe stenosis on the left circumflex coronary artery still increased left ventricular pressure and dP/dt_{max}, but left blood flow almost unchanged, whereas perfusion pressure increased significantly. Thus, poststenotic end-diastolic coronary resistance increased (Fig. 6).

From Figs. 4 and 5 it can be seen that these changes in coronary resistance were due to a reflex activation of cardiac sympathetic nerve fibers. The upper panel of Fig. 4 shows the result of a train of three single impulses. The two separate peaks of the sympathetic reflex activation can be attributed, based on the conduction velocities, to the stimulation of afferent Aδ-fibers and of C-fibers, respectively. During a period of 2 s stimulation, the activity of cardiac sympathetic fibers was markedly enhanced (96 $\pm$ 15%) above their spontaneous activity. A very similar activation pattern was observed with a severe stenosis on the left circumflex coronary artery.

Reflex activation of sympathetic fibers was markedly reduced after the injection of 27 μg/kg fentanyl, this reduction being apparent both in the first and the second peak of sympathetic reflex activation. The spontaneous sympathetic activity was unaffected. Injection of naloxone (60 μg/kg) restored the original reflex effect of nociceptive stimulation. These reactions of cardiac sympathetic nerve activity were paralleled by changes in end-diastolic poststenotic resistance (Fig. 6).

Figure 7 demonstrates the effect of nociceptive stimulation on myocardial wall thickening in the circumflex-perfused area. Systolic wall thickening was increased during nociceptive stimulation in dogs with intact coronary arteries. In the presence of a stenosis on the left

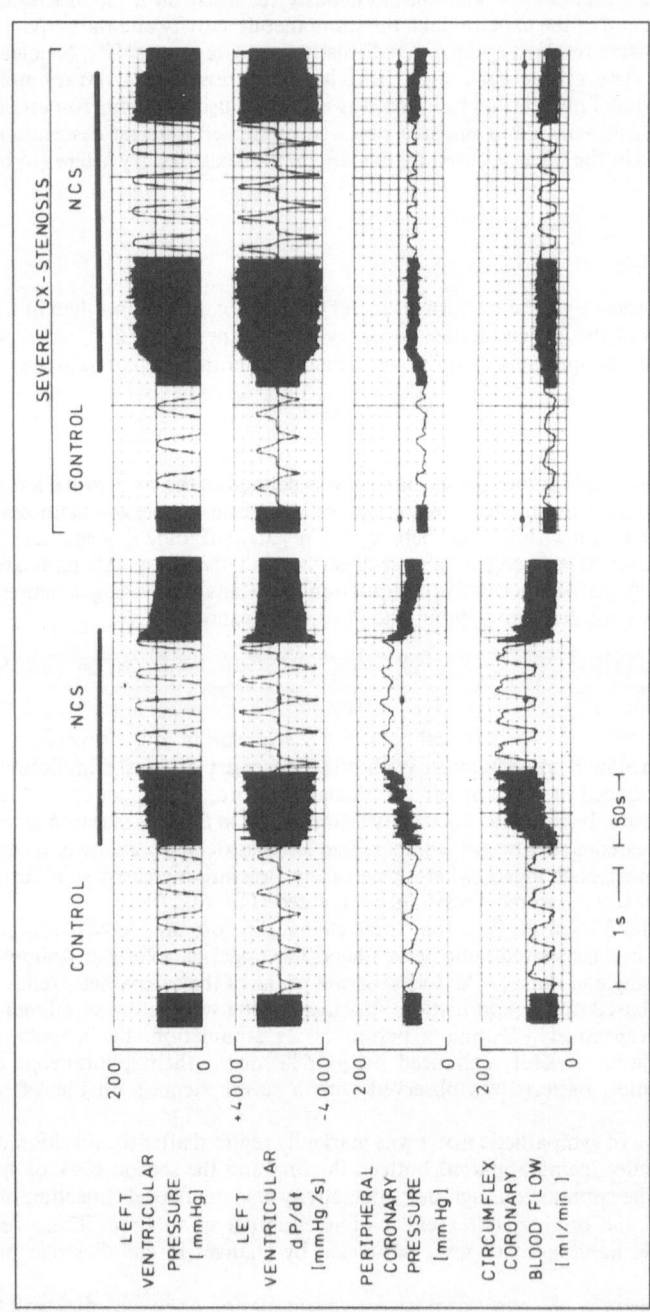

Fig. 3. Original recording of the effect of 1 min electrical nociceptive stimulation (NCS) of the superficial peronaeus nerve. NCS increases left ventricular pressure, dP/dt, peripheral coronary pressure and coronary blood flow. In the presence of a severe stenosis on the left circumflex coronary artery, NCS still increases left ventricular pressure and dP/dt whereas the increases in peripheral coronary pressure and even more so in coronary blood flow are blunted.

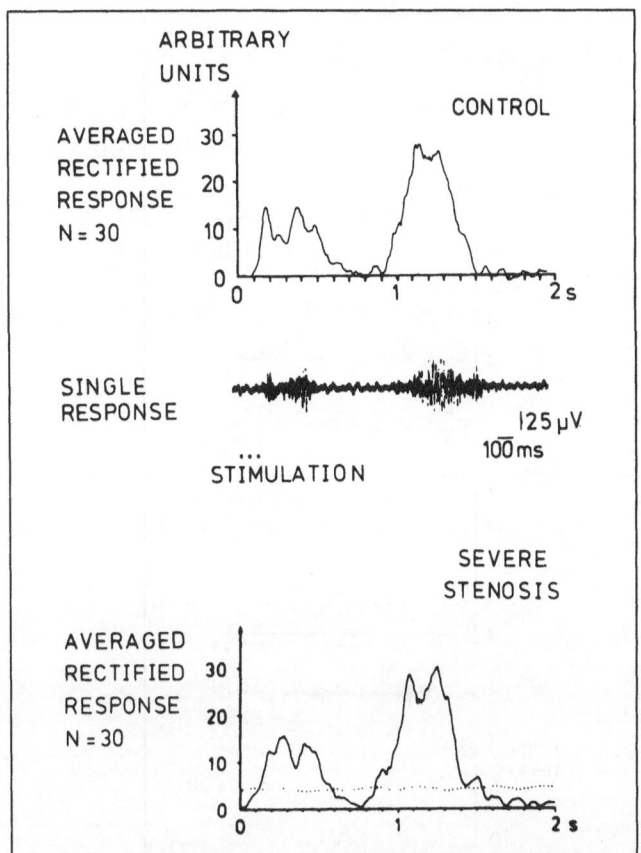

Fig. 4. Averaged effect of 30 single electrical stimulations of the superficial peronaeus nerve. Every stimulation consists of a train of three separated short stimulations. Recording of the sympathetic activity of a multifiber filament dissected out of the left ventrolateral cervical cardiac nerve. The dotted line represents the spontaneous nervous activity before stimulation. Nociceptive stimulation increases sympathetic nerve activity. The first peak corresponds to activation of $A\delta$-, the second to that of C-fibers. Sympathetic activation is similar in the presence of a severe coronary stenosis.

circumflex coronary artery, systolic wall thickening was already reduced at rest. Systolic wall thickening was then further impaired during nociceptive stimulation. The additional deterioration of regional function during nociceptive stimulation could be prevented by fentanyl. Systemic hemodynamic and regional myocardial data for all five animals of this group are summarized in Table 2.

Discussion

Our experiments confirm that noxious stimuli can induce sympathetic activation. Myocardial ischemia as such can also operate as such a noxious stimulus and, like nociceptive stimulation of a skin nerve, can activate cardiac sympathetic nerves by a reflex. This activation is apparently strong enough to impact on myocardial perfusion. In the presence of a severe coronary stenosis, systolic wall thickening in the poststenotic area is reduced. In the present study, myocardial blood flow was not measured. However, myocardial blood flow strongly correlates with regional wall thickening during myocardial ischemia [4]. Thus, the deterioration of regional contractile function can be regarded as evidence for an aggravation of myocardial ischemia.

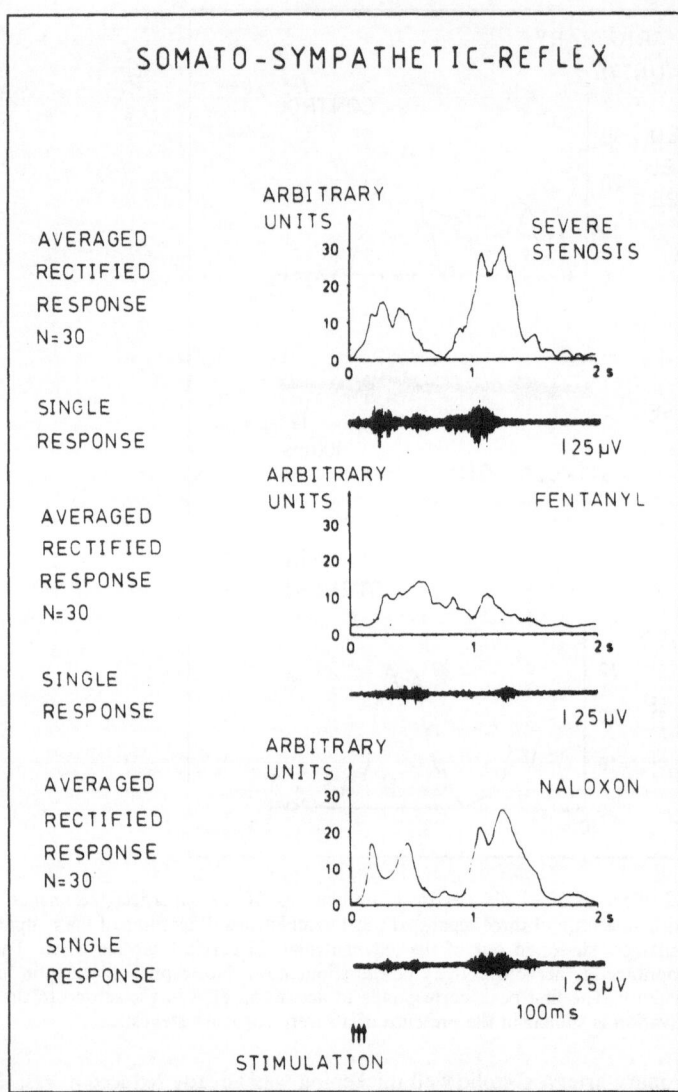

Fig. 5. Averaged effect of 30 single electrical stimulations of the superficial peronaeus nerve. Recording of the sympathetic activity of a multifiber filament dissected out of the left ventrolateral cervical cardiac nerve. The dotted line represents the spontaneous nervous activity before nociceptive stimulation. Fentanyl depresses the reflex sympathetic activation during nociceptive stimulation. The original effect is restored by naloxon.

The ischemia resulting from the reflex sympathetic activation triggers a feedback mechanism which operates in a vicious cycle. Treatment with an analgesic substance like fentanyl can, therefore, probably not only prevent the sensation of pain, but also prevent the reflex impairment of coronary blood flow and myocardial performance.

It seems possible that pain evoked by activation of somatic afferent fibers of different origin can precipitate myocardial ischemia in patients suffering from coronary artery disease. The relation between pain and myocardial ischemia is complex. Pain and psychological stress

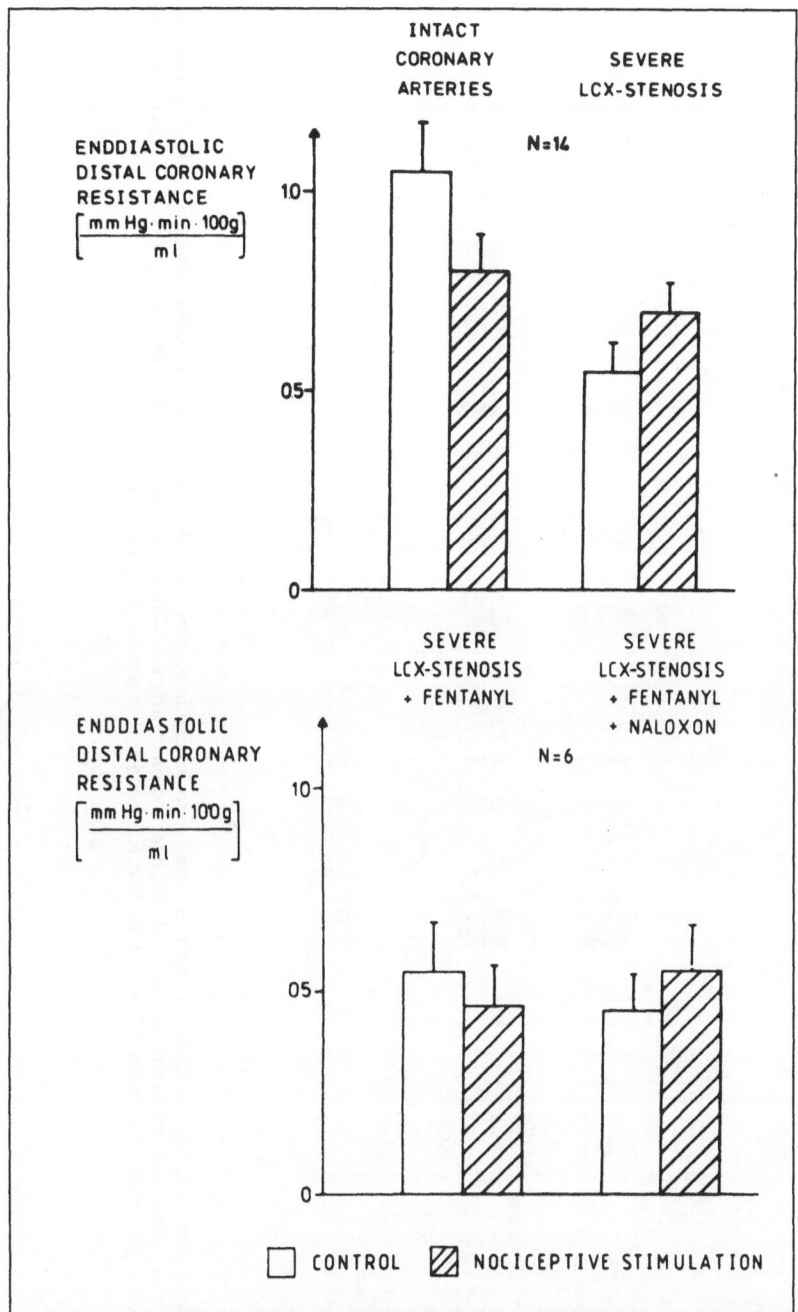

Fig. 6. End-diastolic coronary resistance before and during nociceptive stimulation of the superficial peronaeus nerve. Nociceptive stimulation significantly decreased the end-diastolic distal coronary resistance of intact coronary arteries. In contrast, with a severe stenosis on the left circumflex coronary artery, this resistance increased significantly during nociceptive stimulation. Under this condition, fentanyl prevented the increase in end-diastolic resistance. This latter effect could be prevented by naloxon.

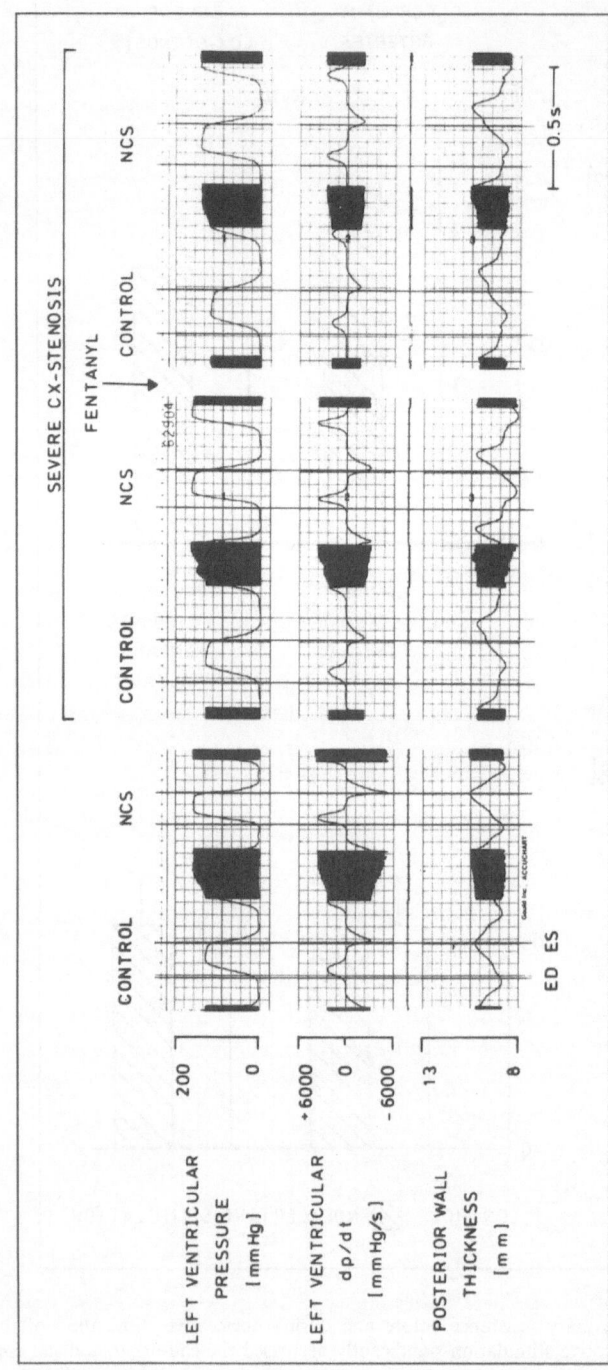

Fig. 7. Original tracing demonstrating the changes in systemic hemodynamics and in posterior wall thickness during nociceptive stimulation (NCS). There is an increase in posterior systolic wall thickening during NCS with intact coronary arteries. With a severe stenosis, NCS decreases systolic wall thickening. The ischemic dysfunction is prevented by fentanyl. ED = end-diastole; ES = end-systole (from [14] by permission).

Reflex sympathetic activation and myocardial ischemia

Table 2. Effects of nociceptive stimulation (NCS) on global and regional ventricular function. Whereas NCS improves both global and regional systolic function with intact left circumflex coronary artery (LCX), regional function in the poststenotic posterior myocardium (WTP) is depressed during NCS. With fentanyl (27 μg/kg i.v.) the decrease in regional poststenotic function during NCS can be prevented.

n = 5	Intact LCX		LCX-Stenosis			
			Without Fentanyl		With Fentanyl	
	Rest	NCS	Rest	NCS	Rest	NCS
LVP (mm Hg)	135 ± 27	151 ± 17*	131 ± 15	155 ± 16*	121 ± 11	142 ± 19*
dP/dt$_{max}$ (mm Hg/s)	2440 ± 554	3540 ± 536*	2576 ± 393	3640 ± 618*	2440 ± 219	3360 ± 577*
WTP (%)	10.9 ± 3.9	13.6 ± 5.0*	7.0 ± 2.5	4.6 ± 2.3*	6.6 ± 2.9	7.3 ± 1.9
WTA (%)	18.6 ± 3.6	19.0 ± 0.8	24.4 ± 1.8	28.1 ± 4.3*	25.5 ± 3.2	27.6 ± 3.7

LCX = left circumflex coronary artery, LVP = peak left ventricular pressure, WT = systolic wall thickening of the posterior (P) and anterior (A) myocardium. Data are mean ± SD, *p < 0.05 Rest vs. NCS.

reactions occur almost simultaneously at the onset of myocardial ischemia and may be potentiating one another in the clinical setting.

References

1. Deussen A, Heusch G (1983) Reflektorische Vasokonstriktion hochgradig stenosierter Koronararterien bei Karotisokklusion. Z Kardiol 72 Suppl 2:28 (abstr.)
2. Felder RB, Thames MD (1981) The cardiocardiac sympathetic reflex during coronary occlusion in anesthetized dogs. Circ Res 48:685–692
3. Foreman RD, Blair RW (1988) Central organization of sympathetic cardiovascular response to pain. Ann Rev Physiol 50:607–622
4. Gallagher KP, Matsuzaki M, Osakada G, Kemper WS, Ross Jr J (1983) Effect of exercise on the relationship between myocardial blood flow and systolic wall thickening in dogs with acute coronary stenosis. Circ Res 52:716–729
5. Gould KL, Lipscomb K, Calvert C (1975) Compensatory changes of the distal coronary vascular bed during progressive coronary constriction. Circulation 51:1085–1094
6. Gwirtz PA, Franklin D, Mass HJ (1986) Modulation of synchrony of left ventricular contraction by regional adrenergic stimulation in conscious dogs. Am J Physiol 251:490–495
7. Heusch G (1985) Sympathische Herznerven und Myokardischämie. Tierexperimentelle Untersuchungen, Thieme, Stuttgart
8. Heusch G, Deussen A (1983) The effects of cardiac sympathetic nerve stimulation on the perfusion of stenotic coronary arteries in the dog. Circ Res 53:8–15
9. Heusch G, Deussen A (1984) Nifedipine prevents sympathetic vasoconstriction distal to severe coronary stenoses. J Cardiovasc Pharmacol 6:378–383
10. Heusch G, Deussen A, Thämer V (1985) Cardiac sympathetic nerve activity and progressive vasoconstriction distal to coronary stenoses: feedback aggravation of myocardial ischemia. J Auton Nerv Syst 13:311–326
11. Heusch G, Guth BD, Widmann T, Peterson KL, Ross J Jr (1987) Ischemic myocardial dysfunction assessed by temporal Fourier transform of regional myocardial wall thickening. Am Heart J 113:116–124
12. Heusch G, Thämer V (1984) Die Bedeutung des sympathischen Nervensystems für die Koronardurchblutung. Z Kardiol 73:543–551
13. Horeyseck G, Jänig W, Kirchner F, Thämer V (1976) Activation and inhibition of muscle and cutaneous postganglionic neurons to hind limbs during hypothalamically induced vasoconstriction and atropine-sensitive vasodilation. Pflügers Arch 361:231–240
14. Kröger K, Schipke J, Thämer V, Heusch G (1989) Poststenotic ischemic myocardial dysfunction induced by peripheral nociceptive stimulation. Eur Heart J 10 Suppl F:179–182

15. Lombardi F, Casalone C, Della Bella P, Malfatto G, Pagani M, Malliani A (1984) Global versus regional myocardial ischemia: differences in cardiovascular and sympathetic responses in cats. Cardiovasc Res 18:14–23
16. Malliani A (1986) The elusive link between transient myocardial ischemia and pain. Circulation 73:201–204
17. Malliani A, Schwartz PJ, Zanchetti A (1969) A sympathetic reflex elicited by experimental coronary occlusion. Am J Physiol 217:703–709
18. Randich A, Maixner W (1984) Interactions between cardiovascular and pain regulatory systems. Neurosc Biobehav Rev 8:343–367
19. Thames MD, Klopfenstein HS, Abboud FM, Mark AL, Walker JL (1978) Preferential distribution of inhibitory cardiac receptors with vagal afferents to the inferoposterior wall of the left ventricle activated during coronary occlusion in the dog. Circ Res 43:512–519
20. Theroux P, Ross J Jr, Franklin D, Kemper WS, Sasayama S (1976) Regional myocardial function in the conscious dog during acute coronary occlusion and response to morphine, propranolol, nitroglycerin and lidocaine. Circulation 53:302–314
21. Tölle T, Schipke JD, Schulz R, Thämer V, Haase J (1986) Nociceptive activation of cardiac sympathetic nerves induces poststenotic coronary vasoconstriction. Pflügers Arch 406 suppl:R43 (abstr.)
22. Uchida Y, Murao S (1974) Excitation of afferent cardiac sympathetic nerve fibers during coronary occlusion. Am J Physiol 226:1094–1099
23. Vogt A, Thämer V (1980) Vagal and sympathetic reflexes of left ventricular origin on the efferent activity of cardiac and renal nerves on anaesthetized cats. Basic Res Cardiol 75:635–645
24. Webb SW, Adgey AAJ, Pantridge JF (1972) Autonomic disturbance at onset of acute myocardial infarction. Br Med J 3:89–92

Authors' address:

Prof. Dr. Gerd Heusch, Abteilung für Pathophysiologie, Zentrum für Innere Medizin, Universitätsklinikum Essen, Hufelandstr. 55, 4300 Essen, FRG

No Impairment of Sympathetic Neurotransmission in Stunned Myocardium

R. Schulz, D. Frehen and G. Heusch*

Abteilung für Pathophysiologie, Universität Essen, FRG

Summary

Reversibly injured myocardium after short periods of ischemia is characterized by a prolonged depression of contractile function which can, however, be enhanced by inotropic interventions. Thus, a lack of inotropic stimulation due to ischemic damage of cardiac sympathetic nerves has been suggested as a mechanism underlying postischemic myocardial dysfunction. We tested this hypothesis in nine anesthetized, vagotomized dogs with left cardiac sympathetic nerve stimulation (CSNS) at 1, 2, 5, 10, and 20 Hz and compared this response to that of intravenous norepinephrine infusion (NE, 0.5–1 μg/kg·min). Regional myocardial wall thickness was measured using sonomicrometry, and mean systolic wall thickening velocity (MSTV) was determined. CSNS was performed before and at 0, 1, 2, 3, 4, 8, 12, 16, 20, and 24 h after release of a 15 min occlusion of a left circumflex coronary artery branch. Before coronary artery occlusion MSTV was increased in a frequency-dependent way from 7.5 ± 2.7 (S.D.) (rest) to 8.1 ± 3.1 (1 Hz), 9.4 ± 3.2 (2 Hz), 11.4 ± 2.7 (5 Hz), 13.4 ± 2.4 (10 Hz), and 16.8 ± 2.1 (20 Hz) by CSNS, and to 12.6 ± 3.4 mm/s by NE. Immediately upon reperfusion CSNS increased MSTV from 2.9 ± 2.0 to 2.9 ± 2.8, 4.1 ± 3.0, 5.4 ± 4.6, 6.9 ± 4.5 and 9.4 ± 5.9, and NE increased MSTV to 7.8 ± 1.9 mm/s. Baseline function recovered over 24 h, as did the response to CSNS and NE. Since the recovery of baseline function paralleled the increases in regional contractile function achieved by CSNS or NE, we conclude that there is no impairment of sympathetic neurotransmission in the stunned myocardium.

Introduction

Reversibly injured myocardium after short periods of myocardial ischemia is characterized by a prolonged depression of contractile function even after full restoration of myocardial blood flow [3, 13, 14, 37, 39]. Baseline contractile function after 15 min of severe ischemia gradually returns to preischemic values only over several hours [13, 14]. Decreases in myocardial energy-rich phosphates during ischemia with only slow repletion of ATP during reperfusion have been demonstrated and related to the prolonged myocardial dysfunction [6, 9, 36]. However, enhanced repletion of myocardial ATP by infusion of nucleotide precursors does not result in enhanced recovery of myocardial function during reperfusion [15]. Despite the depressed contractile function, it can be enhanced by inotropic stimulation with dopamine [8, 28], epinephrine [1], norepinephrine [4], isoproterenol [2], adrenoceptor-independent stimulation by the cardiotonic agent ARL-57 [12], postextrasystolic potentiation [1, 21, 33] and intracoronary Ca-infusion [21], thus excluding a persistent perturbation of energy supply as the important underlying mechanism. Alternatively, a lack of endogenous inotropic stimulation could be involved in the depression of contractile function in reperfused myocardium.

Supported by grants He 1320/2-1 and 3-1 of the German Research Foundation.

Such lack of endogenous inotropic stimulation could be due to an ischemic damage to cardiac sympathetic nerves, which has been demonstrated to result in impaired sympathetic neurotransmission *during* 20 min ischemia in isolated rat hearts [5]. Confirming an impaired sympathetic neurotransmission *during* ischemia, Inoue and coworkers [18] have shown that after 13 min occlusion of the left anterior descending (LAD) coronary artery in dogs the vasopressor response to epicardial application of bradykinin was attenuated. However, this ischemia-induced damage of afferent sympathetic nerve fibers was reversible within 10 min of reperfusion.

During the reperfusion phase following a 25 min occlusion of the LAD coronary artery in dogs, Ciuffo et al. [4] could not demonstrate any response to cardiac sympathetic nerve stimulation and found only an attenuated response to intravenous norepinephrine infusion. Therefore, they suggested that in the stunned myocardium sympathetic neurotransmission may still be impaired by the preceding ischemia. This hypothesis was supported by a recent study by Schwaiger et al. [35], demonstrating a sustained catecholamine depletion over 24 h following a 20 min occlusion of the LAD coronary artery in dogs.

While this is a very attractive and potentially important hypothesis for the underlying mechanism of the stunned myocardium, there are several methodological concerns about the studies of Ciuffo and Schwaiger. First, Ciuffo [4] and Schwaiger [35] used more than 20 min of ischemia which may already induce irreversible damage, both in the myocardium [32] and the cardiac sympathetic nerves. Second, in both studies the LAD coronary artery was occluded by use of an intracoronary balloon. However, the LAD-perfused myocardium has been demonstrated to be very susceptible to sympathetic denervation after LAD-instrumentation, whereas such denervation does not occur after LCX-instrumentation [7, 11, 24]. Third, Ciuffo et al. stimulated cardiac sympathetic nerves using varying voltages and frequencies and did not establish a restoration of responses to sympathetic stimulation over time. Thus, stimulation of varying nerve fiber populations and an experimental damage to the stimulated nerves cannot be excluded. Finally, a sustained catecholamine depletion over 24 hours following a 20 min occlusion of the LAD coronary artery in dogs, as demonstrated in Schwaiger's study [35], does not necessarily indicate ischemia-impaired neurotransmission, but may also reflect a surgical denervation. Furthermore, a reduction of the catecholamine content of the sympathetic nerve terminals does not necessarily inhibit sympathetic neurotransmission, because only a small percentage of the stored catecholamines is released during each stimulation to induce a full postsynaptic response.

To reexamine the hypothesis of an impaired sympathetic neurotransmission in reperfused myocardium and address these critical concerns, we performed a 15 min LCX-occlusion in anesthetized dogs and established dose-response curves of the responses of regional contractile function to cardiac sympathetic nerve stimulation at a constant supramaximal voltage and increasing frequencies during 24 h of reperfusion.

Methods

Experimental preparation

Nine mongrel dogs of either sex weighing 23 to 35 kg were pretreated with ampicillin (3 g i.m.) on the day before the experiment was conducted. Dogs were anesthetized with an initial bolus of sodium-thiamylal (15 mg/kg i.v.). Anesthesia was then maintained with α-chloralose (50 mg/kg i.v.) and urethane (250 mg/kg i.v.); supplements were given as needed. After endotracheal intubation the dogs were artificially ventilated. Arterial blood gases and hematocrit were checked regularly (BMS 2 Radiometer, Copenhagen). pO_2 was held above 100 mm Hg by enriching the inspired air with oxygen when necessary. pCO_2 was adjusted to 35 mm Hg by the rate of ventilation. pH was corrected when needed by infusion of isotonic sodium bicarbonate. For the maintenance of normal arterial pressure and a hematocrit above 30% intravenous infusions of 40% dextran and anticoagulated donor blood were used. Rectal temperature was recorded and kept above 37°C with a heating pad.

Sympathetic neurotransmission in stunned myocardium

After a left thoracotomy and pericardiotomy the heart was suspended in a pericardial cradle. Left ventricular pressure was measured with a cathetertip-manometer (PC 350 Millar, Houston). Heart rate and left ventricular dP/dt were derived from the left ventricular pressure signal. Wall thickness in the LAD-perfused (control) and in the LCX-perfused (area of interest) myocardium was assessed using sonomicrometry [34]. The control area wall thickness served as a reference for the stability of the experimental preparation of both the myocardium and the repetitively stimulated nerve over time. The left ventrolateral cervical cardiac sympathetic nerve was isolated, then cut, and its distal part stimulated at supramaximal voltage (7–10 V) with 2 ms pulses at 1, 2, 5, 10 and 20 Hz. Stimulation at each frequency was performed until a steady-state response of systemic hemodynamic and regional myocardial parameters was established (approximately after 30–60 s). Between stimulations at different frequencies, full recovery of the recorded parameters was assured (30–120 s).

Experimental protocol

A bilateral cervical vagotomy was performed to prevent reflex vagal effects secondary to cardiac sympathetic nerve stimulation and to sympathetically induced changes in blood pressure and cardiac dimensions. The heart rate response to cardiac sympathetic nerve stimulation at 20 Hz was initially determined and throughout the subsequent experimental protocol the heart rate was kept slightly above this value by left atrial pacing. After complete instrumentation a first cardiac sympathetic nerve stimulation series was performed at increasing stimulation frequencies. To assess the dependence of the recorded inotropic responses on the sequence of the stimulation, a second cardiac sympathetic nerve stimulation series was performed at decreasing stimulation frequencies. To compare the effects of neuronally released and humoral norepinephrine, norepinephrine was infused intravenously in a dose of 0.5–1.0 μg/kg·min to match the left ventricular dP/dt value observed during the preceding sympathetic nerve stimulation at 10 Hz. Once established, this dosage was given throughout the subsequent experimental protocol. After these control responses to cardiac sympathetic nerve stimulation and intravenous norepinephrine were obtained, the one or two side branches of the left circumflex coronary artery supplying the myocardium that contained the sonomicrometers were carefully dissected and sutures were loosely passed around the vessels. To account for a possible damage to cardiac sympathetic nerves during this instrumentation, a cardiac sympathetic nerve stimulation series and an intravenous infusion of norepinephrine at the previously determined dose were repeated. The previously dissected LCX-branches were occluded completely for 15 min and then reperfused for 24 h. During the whole occlusion and reperfusion period no antiarrhythmic drugs were used, due to possible interactions with cardiac sympathetic nerve traffic [29] or neuronal norepinephrine release [4]. One dog with ventricular fibrillation occurring during coronary occlusion and two dogs with ventricular fibrillation upon reperfusion were excluded from further analyses. A series of cardiac sympathetic nerve stimulation and intravenous norepinephrine infusion were performed immediately upon reperfusion (starting 1–3 min after release of coronary occlusion when a reasonable steady state was achieved) and then repeated at 1, 2, 3, 4, 8, 12, 16, 20, and 24 h of reperfusion.

Data analysis

The recorded systemic and regional hemodynamic parameters were continuously monitored on a Brush-forced ink recorder (Mark 260) and stored on videotape (Sony SL-C30PS) using PCM-modulation (Heim KG) for later playback and computation. Data from 5–15 consecutive cardiac cycles were digitized. End-diastole was defined as the zero crossing point of the left ventricular dP/dt before its maximum value. End-systole was defined as the time of maximum wall thickness within 20 ms before peak negative dP/dt [38]. End-diastolic and end-systolic wall thickness were determined, and regional myocardial function was assessed by calculating systolic wall thickening expressed as a percentage of the end-diastolic wall thickness and by calculating the mean systolic wall thickening velocity. Velocity values were normalized for an end-diastolic wall thickness of 10 mm [31]. Data are given as mean values $\pm$ SD. The stimulation series at increasing and decreasing frequencies were compared using a two-tailed paired t-test. Similarly, the cardiac sympathetic nerve stimulation series and the intravenous norepinephrine administration before and after coronary artery instrumentation were compared with a two-tailed paired t-test. The responses to cardiac sympathetic nerve stimulation and intravenous norepinephrine before coronary occlusion and during 24 h of reperfusion were analyzed by a two-way analysis of variance for repeated measures. When a significant overall effect was detected, single mean values were compared by F-tests. A p value less than 0.05 was considered significant.

Histological analysis

After 24 h of reperfusion, dogs were killed by intravenous KCL infusion and the hearts excised. The myocardial regions containing the control zone and area of interest zone crystals were cut out and frozen with liquid nitrogen. 10–30 transmural slices of 10 μm thickness from each myocardial region were cut on a cryostat and then stained with hemalaun-eosin. Necroses were characterized by enhanced acidophilia, nucleolysis, and leukocyte infiltration. The area of necrosis was expressed as percent of the respective specimen. The areas of necrosis in the control and in the reperfused region were compared by a two-tailed t-test.

Results

Morphological findings

The area of necrosis was 7.5 $\pm$ 3.4% in the control and 5.5 $\pm$ 2.9% in the reperfused region, this difference being not significant. The necroses were centered along the channel for implantation of the subendocardial crystal and, thus, represent traumatic rather than ischemic damage.

Evaluation of sympathetic stimulation characteristics

There was no significant difference in systemic hemodynamic and regional myocardial responses to cardiac sympathetic nerve stimulation at increasing vs decreasing frequencies. Also, the systemic hemodynamic and regional myocardial responses to cardiac sympathetic nerve stimulation and intravenous norepinephrine were not different before and after circumflex coronary artery instrumentation (Fig. 1).

Systemic hemodynamics

With heart rate maintained constant at 199 $\pm$ 15 beats/min by atrial pacing, there were stepwise increases in left ventricular pressure and maximum left ventricular dP/dt (Table 1) during cardiac sympathetic nerve stimulation under control conditions. The left ventricular end-diastolic pressure (7.2 $\pm$ 4.7 mm Hg) was not significantly changed throughout this protocol. During the time course of the experiment, peak left ventricular pressure at rest decreased from 131 $\pm$ 17 mm Hg before occlusion to 127 $\pm$ 21, 118 $\pm$ 30 and 101 $\pm$ 42 mm Hg at 10 min, 12 h, and 24 h reperfusion. This decrease, however, was not significant in the analysis of variance. Maximum left ventricular dP/dt at rest was 1981 $\pm$ 575 mm Hg/s before occlusion and 2036 $\pm$ 647, 2266 $\pm$ 725 and 2458 $\pm$ 967 mm Hg/s at 10 min, 12 h and 24 h of reperfusion. Intravenous norepinephrine also induced an increase in left ventricular pressure and maximum left ventricular dP/dt (Table 1).

Regional myocardial function in the control area

Percent wall thickening was 24.7 $\pm$ 8.8% immediately after crystal implantation at a heart rate of 158 $\pm$ 21 beats/min; it then decreased to 20.3 $\pm$ 7.2% after bilateral vagotomy and atrial

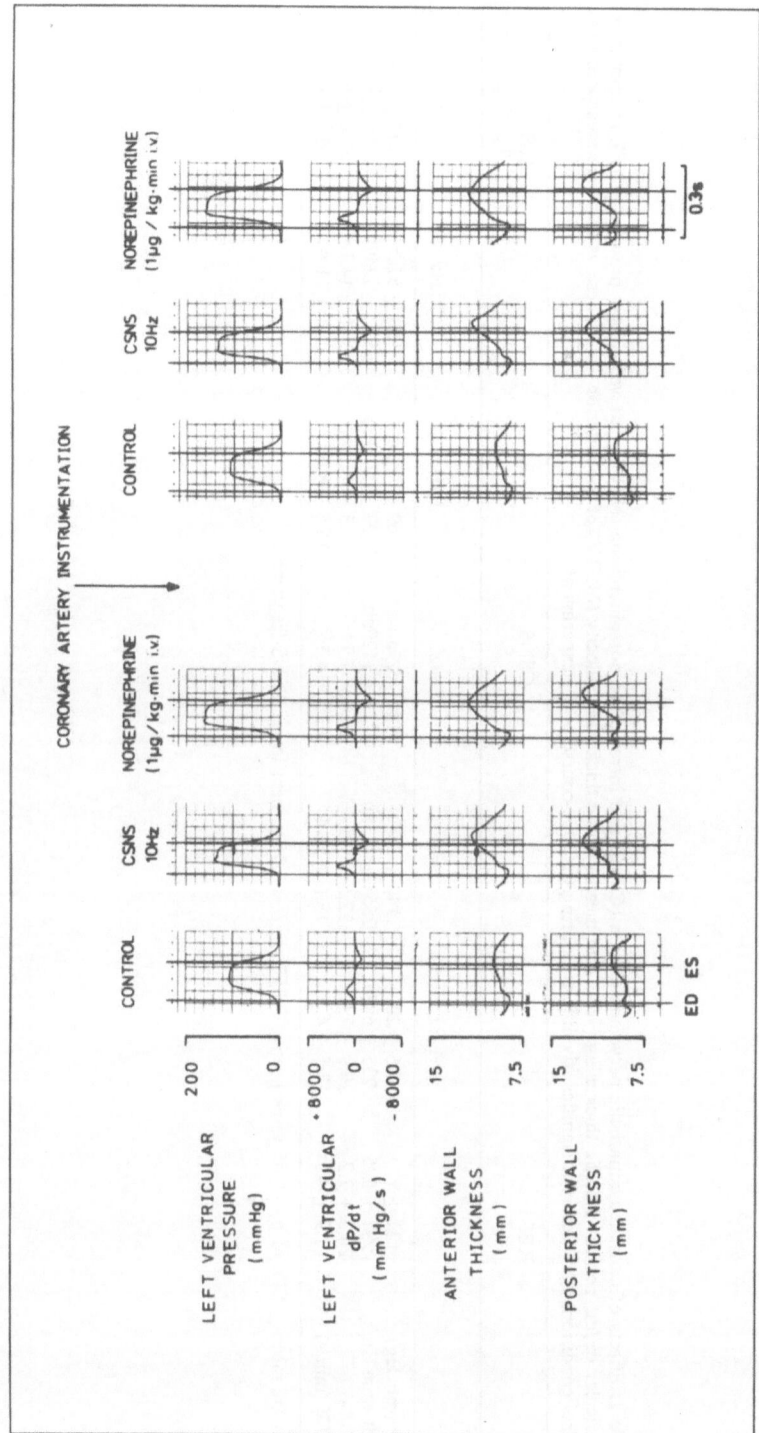

Fig. 1. The effects of cardiac sympathetic nerve stimulation (CSNS) at 10 Hz and intravenous norepinephrine on systemic hemodynamics and regional myocardial dimensions. There is no difference in inotropic responses before and after circumflex coronary artery instrumentation. Solid vertical bars indicate end-diastole (ED) and end-systole (ES).

Table 1. Effects of cardiac sympathetic nerve stimulation (CSNS) and intravenous norepinephrine (NE) on left ventricular pressure (LVP), maximum left ventricular dP/dt, percent systolic thickening (WT), and mean systolic thickening velocity (MSTV) of the anterior wall. These values were measured under control conditions and were not significantly different during the time-course of the experiment.

	REST	CSNS					NE (0.5–1 µg/kg·min)
		1 Hz	2 Hz	5 Hz	10 Hz	20 Hz	
LVP (mm Hg)	131 ± 17	136 ± 20	140 ± 24a	148 ± 29ab	169 ± 45ab	180 ± 57ab	172 ± 29a
dP/dt (mm Hg/s)	1987 ± 575	2235 ± 688a	2557 ± 823ab	3233 ± 1129ab	3747 ± 1143ab	4512 ± 1249ab	3955 ± 1405a
WT (%)	20.3 ± 7.2	20.9 ± 7.9	22.5 ± 9.5	20.4 ± 7.7	23.3 ± 12.8	25.1 ± 14.1a	27.8 ± 9.1a
MSTV (mm/s)	11.9 ± 4.3	12.5 ± 4.7	13.6 ± 5.4a	13.7 ± 4.3a	16.5 ± 8.0a	19.4 ± 11.4ab	20.3 ± 6.4a

Data are mean ± SD. a: $p < 0.05$ vs respective REST. b: $p < 0.05$ vs next lower CSNS-frequency.

pacing to 199 ± 15 beats/min. At increasing frequencies of 1, 2, 5, 10, 20 Hz of cardiac sympathetic nerve stimulation, there were only minor increases in percent wall thickening and mean systolic wall thickening velocity (Table 1). During cardiac sympathetic nerve stimulation a significant post-ejection thickening in the control zone developed that was 0.01 ± 0.02 mm at rest, and 0.04 ± 0.09, 0.09 ± 0.16, 0.17 ± 0.24, 0.22 ± 0.3, 0.24 ± 0.29 mm at 1, 2, 5, 10 and 20 Hz stimulation frequency, respectively. Intravenous norepinephrine increased percent wall thickening from 20.3 ± 7.2 (REST) to $27.8 \pm 9.1\%$ ($p < .05$ vs REST) and mean systolic wall thickening velocity from 11.9 ± 4.3 (REST) to 20.3 ± 6.4 mm/s ($p < .05$ vs REST). The post-ejection thickening measured during intravenous norepinephrine infusion was 0.01 ± 0.01 mm. During the time-course of the experiment there was a continuous decrease in percent wall thickening to 14.8% (73% of control) and in mean systolic wall thickening velocity to 9.5 mm/s (80% of control) after 24 h of reperfusion, commensurate with a continuous decrease in systemic hemodynamics. However, these decreases in percent wall thickening and mean systolic wall thickening velocity were not significant in the analysis of variance.

Regional myocardial function in the area of interest

Percent wall thickening was $16.3 \pm 4.9\%$ immediately after crystal implantation at a heart rate of 158 ± 21 beats/min; it then decreased to $12.5 \pm 3.8\%$ after bilateral vagotomy and atrial pacing to 199 ± 15 beats/min. With increasing frequencies of cardiac sympathetic nerve stimulation there were stepwise increases in percent wall thickening (Fig. 2, Table 2) and mean systolic wall thickening velocity (Fig. 2, Fig. 3, Table 3) during control conditions, as well as during 24 h of reperfusion. Baseline contractile function was severely depressed during coronary occlusion (percent wall thickening: $-5.3 \pm 4.3\%$, and mean systolic wall thickening velocity: -3.6 ± 1.5 mm/s after 15 min coronary occlusion) and only gradually returned to preischemic values over 24 h of reperfusion (Tables 2 and 3; Fig. 3). At any given time during reperfusion, the absolute increase in mean systolic wall thickening velocity during CSNS was not different from that during the respective stimulation frequency before ischemia. The increase in regional contractile function achieved by cardiac sympathetic nerve stimulation and intravenous norepinephrine *paralleled* the recovery of baseline function (Tables 2 and 3; Fig. 3).

Discussion

This study demonstrates that the contractile function of postischemic reperfused myocardium can be enhanced by cardiac sympathetic nerve stimulation. The inotropic response to intravenous norepinephrine infusion and the frequency-response curves between regional myocardial function and the frequency of cardiac sympathetic nerve stimulation exhibited a parallel downward displacement immediately upon reperfusion with a gradual recovery towards the control relationship over 24 h of reperfusion (Fig. 3).

In the present study mean systolic wall thickening velocity was used to assess the regional myocardial inotropic state. We have previously shown that at increasing left ventricular pressure during sympathetic activation changes in mean systolic excursion velocity are more sensitive than changes in systolic excursion related to the end-diastolic dimension. This superiority of velocity measures was particularly evident in an anesthetized open-chest preparation [11].

We realize that velocity of wall thickening is an afterload-sensitive measure of regional myocardial function, and that afterload, as indicated by peak left ventricular pressure, decreased over 24 h of reperfusion. Thus, a recovery of thickening velocity over reperfusion time will not necessarily reflect an improvement in regional contractile state. However, the decrease in baseline anterior myocardial wall thickening velocity over 24 h reperfusion, due to the natural deterioration of an acute open-chest preparation, indicates that the recovery of baseline posterior wall thickening velocity is not simply a consequence of decreased afterload. The

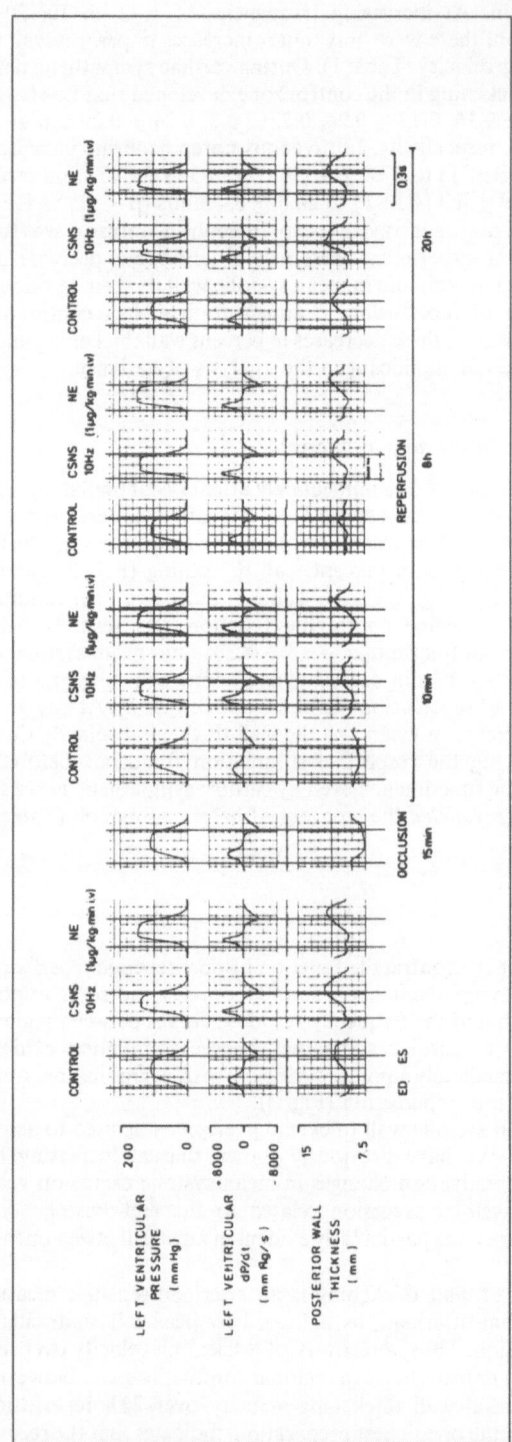

Fig. 2. The effects of cardiac sympathetic nerve stimulation (CSNS) at 10 Hz and intravenous norepinephrine (NE) on systemic hemodynamics and posterior myocardial wall thickness. Baseline regional myocardial function after 15 min coronary occlusion is severely depressed and only gradually returns to control values over 20 h of reperfusion. Regional myocardial function can, however, be increased by sympathetic stimulation and intravenous norepinephrine after only 10 min of reperfusion. The recovery of contractile responses to sympathetic stimulation and intravenous norepinephrine parallels the recovery of baseline function over 20 h of reperfusion. Solid vertical bars indicate end-diastole (ED) and end-systole (ES).

Table 2. Effects of cardiac sympathetic nerve stimulation (CSNS) and intravenous norepinephrine (NE) on percent systolic thickening of the posterior wall under control conditions and during 24 h reperfusion following a 15 min left circumflex coronary occlusion.

| | REST | CSNS | | | | | NE (0.5–1 μg/kg·min) |
		1 Hz	2 Hz	5 Hz	10 Hz	20 Hz	
CONTROL	12.5 ± 3.8	13.3 ± 4.5a	14.5 ± 4.3ab	16.0 ± 3.3a	17.3 ± 2.3a	19.7 ± 2.6ab	17.1 ± 4.3a
REPERFUSION							
10 min	5.3 ± 3.5c	5.1 ± 5.0c	7.2 ± 5.4c	8.3 ± 6.8c	9.5 ± 6.1ac	11.9 ± 7.5ac	11.0 ± 3.3ac
1 h	4.6 ± 4.3c	5.0 ± 4.8c	6.8 ± 5.3abc	9.7 ± 5.5abc	11.9 ± 5.2abc	12.6 ± 4.4ac	10.8 ± 4.4ac
2 h	5.4 ± 4.5c	6.2 ± 5.5c	8.1 ± 5.3abc	10.5 ± 5.1abc	12.8 ± 6.4ac	13.9 ± 5.7ac	12.6 ± 3.6ac
3 h	5.5 ± 4.7c	5.6 ± 4.8c	7.3 ± 5.5ac	10.6 ± 5.5abc	12.0 ± 5.7abc	13.7 ± 5.0abc	12.4 ± 4.3ac
4 h	6.3 ± 3.9c	6.7 ± 4.5c	7.8 ± 4.2abc	10.9 ± 4.8abc	13.2 ± 5.1abc	13.6 ± 5.1ac	11.8 ± 5.1ac
8 h	8.5 ± 2.8	8.1 ± 3.6	9.7 ± 3.8b	12.8 ± 3.8ab	14.0 ± 3.8a	14.6 ± 4.1a	13.7 ± 4.8a
12 h	9.2 ± 2.7	9.5 ± 3.3	9.9 ± 3.2a	10.8 ± 3.1abc	13.2 ± 3.9abcd	14.1 ± 3.8acd	13.4 ± 5.0a
16 h	12.0 ± 5.3	11.6 ± 5.4	12.4 ± 5.2b	13.9 ± 4.8ab	16.1 ± 4.5ab	17.5 ± 4.2ab	15.1 ± 5.0a
20 h	10.9 ± 4.4	10.9 ± 4.8	12.2 ± 4.4ab	14.4 ± 4.5ab	15.8 ± 5.2ab	17.8 ± 5.8a	15.6 ± 4.9a
24 h	10.1 ± 5.5	8.7 ± 3.3	10.8 ± 4.9	13.1 ± 6.3a	14.5 ± 5.9a	16.7 ± 5.5a	13.3 ± 4.3a

Data are mean ± SD. a: $p < 0.05$ vs respective REST. b: $p < 0.05$ vs next lower CSNS-frequency. c: $p < 0.05$ vs respective CONTROL. d: $p < 0.05$ vs preceding REPERFUSION time.

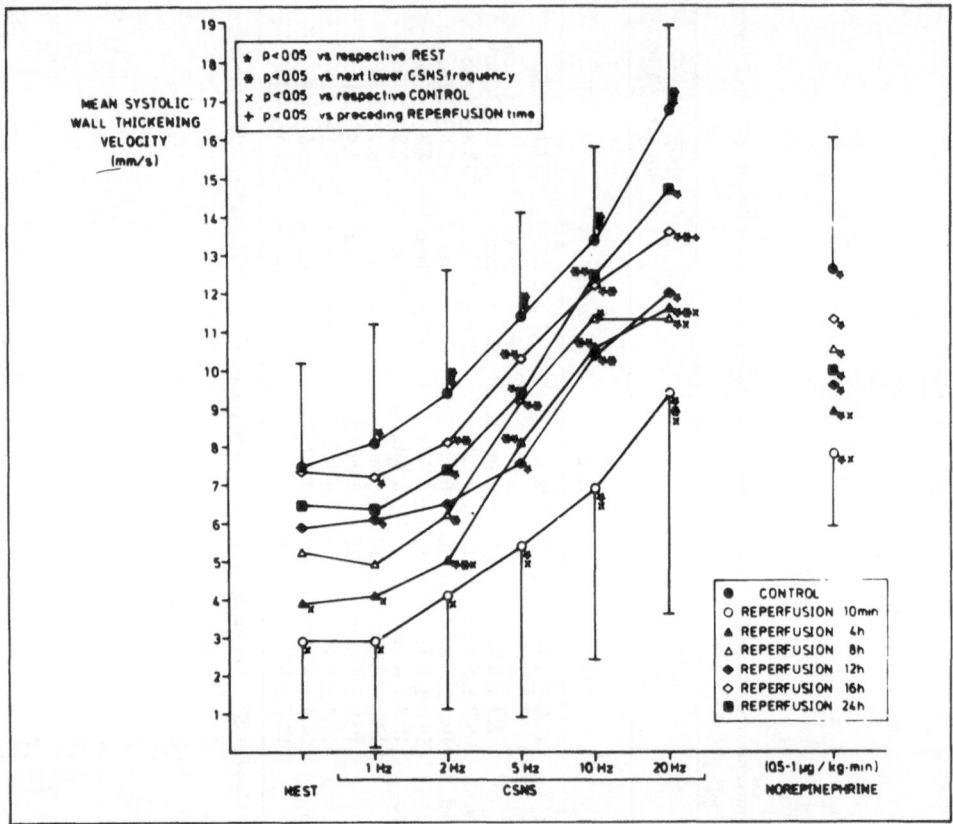

Fig. 3. Dose-response curves of mean systolic wall thickening velocity at different frequencies of cardiac sympathetic nerve stimulation and intravenous norepinephrine infusion. The control relationship is displaced in a parallel fashion to the bottom after 10 min of reperfusion following 15 min coronary occlusion. Over 24 h of reperfusion this parallel displacement is gradually reversible.

frequency-dependent increases in posterior wall thickening velocity against frequency-dependent increases in peak left ventricular pressure were consistent throughout the reperfusion period. Thus, the increase in the wall thickening velocity during reperfusion reflects a recovery of baseline regional myocardial inotropic state, and the increase in thickening velocity during cardiac sympathetic nerve stimulation reflects a regional myocardial inotropic effect.

Due to the anatomic distribution of the perfusion areas, the posterolateral wall thickness crystals (area of interest; within the perfusion bed of the left circumflex coronary artery) were placed near the base of the heart, while the anterior wall thickness crystals (within the perfusion bed of the left anterior descending coronary artery) were implanted closer to the apex of the heart. Baseline function in the posterolateral myocardium was considerably lower than in the anterior myocardium, reflecting a well-established base- to apex-gradient of regional myocardial function [27]. Surprisingly, cardiac sympathetic nerve stimulation induced frequency-dependent stepwise increases in regional inotropic state of the circumflex-perfused posterolateral myocardium, but only minor increases in that of the anterior myocardium, whereas intravenous norepinephrine increased contractile function in both walls to a similar degree. Also, a significant post-ejection thickening developed in the anterior myocardium with increasing frequencies of cardiac sympathetic nerve stimulation, but not with intravenous norepinephrine. We interpret these findings as the result of a preferential projection of the stimulated cardiac sympathetic nerves to the posterolateral myocardium. Cardiac sympathetic nerve

Table 3. Effects of cardiac sympathetic nerve stimulation (CSNS) and intravenous norepinephrine (NE) on mean systolic thickening velocity of the posterior wall under control conditions and during 24 h reperfusion following a 15 min left circumflex coronary occlusion.

	REST	CSNS					NE (0.5–1 µg/kg·min)
		1 Hz	2 Hz	5 Hz	10 Hz	20 Hz	
CONTROL	7.5 ± 2.7	8.1 ± 3.1a	9.4 ± 3.2ab	11.4 ± 2.7ab	13.4 ± 2.4ab	16.8 ± 2.1ab	12.6 ± 3.4a
REPERFUSION							
10 min	2.9 ± 2.0c	2.9 ± 2.8c	4.1 ± 3.0c	5.4 ± 4.6ac	6.9 ± 4.5ac	9.4 ± 5.9abc	7.8 ± 1.9ac
1 h	2.8 ± 2.7c	3.1 ± 2.9c	4.4 ± 3.2abc	5.7 ± 3.6abc	9.5 ± 4.1abcd	11.0 ± 3.9ac	8.2 ± 3.2ac
2 h	3.5 ± 3.3c	4.0 ± 3.7c	5.4 ± 3.6abc	7.5 ± 3.6abc	11.2 ± 4.1abcd	11.9 ± 4.1ac	9.7 ± 2.9a
3 h	3.3 ± 3.1c	3.4 ± 3.0c	4.5 ± 3.4abc	7.6 ± 3.6abc	9.5 ± 4.3abc	11.4 ± 3.2abc	9.1 ± 2.4a
4 h	3.9 ± 2.3c	4.1 ± 2.5c	5.0 ± 2.4abc	8.1 ± 2.9ab	10.6 ± 3.3ab	11.6 ± 3.5abc	8.9 ± 3.1ac
8 h	5.2 ± 1.5	4.9 ± 2.1	6.2 ± 2.3b	9.2 ± 2.5ab	11.3 ± 3.2a	11.3 ± 3.4ac	10.5 ± 3.9a
12 h	5.9 ± 1.6	6.2 ± 2.3d	6.5 ± 2.0	7.6 ± 2.0a	10.9 ± 4.0ab	12.0 ± 3.3a	9.6 ± 3.5a
16 h	7.6 ± 3.1	7.2 ± 3.2a	8.1 ± 3.4ab	10.0 ± 3.1abd	12.2 ± 2.9ab	13.6 ± 2.9abd	11.3 ± 3.1a
20 h	6.8 ± 2.9	6.9 ± 3.4	7.6 ± 3.1ab	9.8 ± 3.3ab	11.7 ± 4.1ab	13.4 ± 4.6ab	10.8 ± 4.0
24 h	6.5 ± 3.1	5.8 ± 2.3	7.4 ± 3.3a	9.4 ± 4.5a	12.4 ± 4.6ab	14.7 ± 2.7a	10.1 ± 2.7a

Data are mean ± SD. a: $p < 0.05$ vs respective REST. b: $p < 0.05$ vs respective CONTROL. c: $p < 0.05$ vs next lower CSNS-frequency. d: $p < 0.05$ vs preceding REPERFUSION time.

277

stimulation increased the inotropic state of the posterolateral wall, thereby imposing an increased regional load on the anterior myocardium without a concomitant increase in its inotropic state. Similar left ventricular asynergy resulting from only regional inotropic stimulation has been previously demonstrated during intracoronary catecholamine infusion [10, 17] and reflex sympathetic stimulation of a heart with regional denervation [11, 23].

Conceivably, the asynergy observed in our study during only left ventrolateral cervical cardiac sympathetic nerve stimulation would be less pronounced or even missing during global left or simultaneous left and right cardiac sympathetic nerve stimulation. With the selective stimulation of the left ventrolateral cervical cardiac sympathetic nerve we found an increase in posterior myocardial function without a concomitant increase in anterior wall function. Conversely, Ciuffo et al. [4] using stimulation of the left ansa subclavia argued that dysfunction in the postischemic anterior area that was potentially denervated in their study was augmented by the increase in function of the nonischemic areas. Although selective stimulation of only the left ventrolateral cervical cardiac sympathetic nerve is certainly unphysiological, the integrity of sympathetic neurotransmission can only be adequately assessed using a strong inotropic stimulation of the area of interest, as done in the present study.

Our experiments confirm that a prolonged depression of regional contractile function occurs after only 15 min of ischemia [13, 14], and that this dysfunction can be reversed by inotropic stimulation [1, 2, 4, 8, 12, 21, 28, 33]. Previous studies showing inotropic responses to dopamine [8, 28], epinephrine [1], norepinephrine [4], and isoproterenol [2] have already excluded the possibility that a persistent damage to myocardial adrenoceptors depresses contractile function of reperfused myocardium. However, an impairment of sympathetic neurotransmission during 20 min ischemia has been demonstrated in a Langendorff-perfused rat heart preparation [5]. A persistent impairment of sympathetic neurotransmission has also been suggested as a possible mechanism of postischemic contractile dysfunction by Ciuffo et al. [4] and Schwaiger et al. [35]. However, our experiments, while avoiding experimental damage to the cardiac sympathetic nerves during coronary instrumentation and repeated electrical stimulation and establishing a full recovery of responses over 24 h of reperfusion do not confirm these conclusions. In contrast, regional myocardial function was substantially increased by cardiac sympathetic nerve stimulation after only 10 min of reperfusion. During 24 h of reperfusion, the increase in regional contractile function achieved by cardiac sympathetic nerve stimulation and intravenous norepinephrine *paralleled* the recovery of baseline function. If the regional inotropic responses to cardiac sympathetic nerve stimulation and norepinephrine were calculated as a percentage of their respective resting values (Tables 2 and 3), the responses to adrenergic stimulation would even appear to be enhanced. Such enhanced response to adrenergic stimulation could be due to an increased density of β-adrenoceptors during the recovery process from myocardial stunning, which has been demonstrated by Hori and coworkers [16].

If the regional myocardial function values achieved by a given cardiac sympathetic nerve stimulation frequency during reperfusion would be compared to the same frequency values under control conditions (Tables 2 and 3), a greater discrepancy at the low frequencies than at the high frequencies would appear. Therefore, impaired sympathetic neurotransmission may be responsible for the postischemic depression in baseline function, especially given the high resting sympathetic tone of an anesthetized, vagotomized preparation. However, complete elimination of sympathetic efferent innervation of the posterolateral myocardium by epicardial phenol did not reduce regional myocardial function in an almost identical preparation [11], thus demonstrating a minimal role of sympathetic tone for resting regional myocardial function.

Our results of unimpaired sympathetic neurotransmission in stunned myocardium are in agreement with prior studies [19, 20, 22, 30] that did not find denervation of efferent sympathetic fibers with myocardial ischemia or even nontransmural myocardial infarction [20, 22, 30]. Only after 3–12 h of transmural myocardial infarction caused by a complete coronary occlusion could sympathetic denervation be observed. We conclude that there is no impairment of sympathetic neurotransmission in the stunned myocardium. The mechanisms underlying postischemic dysfunction, therefore, do not involve impaired sympathetic neurotransmission, but may rather relate to intracellular Ca-kinetics [26] which can be increased by intracoronary

Sympathetic neurotransmission in stunned myocardium

Ca-infusion [21], postextrasystolic potentiation [1, 21, 33], and adrenergic stimulation [1, 2, 4, 8, 28]. This hypothesis was further strengthened by a recent study of Krause et al. [25], who showed a reduced Ca-uptake and a reduced Ca-ATPase activity of the sarcoplasmic reticulum after 8–12 repetitive 5 min occlusions with a 10 min reperfusion period.

References

1. Becker LC, Levine JH, DiPaula AF, Guarnieri T, Aversano T (1986) Reversal of dysfunction in postischemic stunned myocardium by epinephrine and postextrasystolic potentiation. J Am Coll Cardiol 7:580–589
2. Bolli R, Zhu W-X, Myers ML, Hartley CJ, Roberts R (1985) Beta-adrenergic stimulation reverses postischemic myocardial dysfunction without producing subsequent deterioration. Am J Cardiol 56:964–968
3. Braunwald E, Kloner RA (1982) The stunned myocardium: prolonged, postischemic ventricular dysfunction. Circulation 66:1146–1149
4. Ciuffo AA, Ouyang P, Becker LC, Levin L, Weisfeldt ML (1985) Reduction of sympathetic inotropic response after ischemia in dogs. Contributor to stunned myocardium. J Clin Invest 75:1504–1509
5. Dart AM, Schömig A, Dietz R, Mayer E, Kübler W (1984) Release of endogenous catecholamines in the ischemic myocardium of the rat. Part B: Effect of sympathetic nerve stimulation. Circ Res 55:702–706
6. DeBoer LWV, Ingwall JS, Kloner RA, Braunwald E (1980) Prolonged derangements of canine myocardial metabolism after a brief coronary artery occlusion not associated with anatomic evidence of necrosis. Proc Natl Acad Sci USA 77:5471–5475
7. Dolezel S, Gerova M, Hartmannova B, Dostal M, Janeckova H, Vasku J (1984) Cardiac adrenergic innervation after instrumentation of the coronary artery in dog. Am J Physiol 246:H459–H465
8. Ellis SG, Wynne J, Braunwald E, Henschke CI, Sandor T, Kloner RA (1984) Response of reperfusion-salvaged, stunned myocardium to inotropic stimulation. Am Heart J 107:13–19
9. Guth BD, Martin JF, Heusch G, Ross Jr J (1987) Regional myocardial blood flow, function and metabolism using phosphorus-31 nuclear magnetic resonance spectroscopy during ischemia and reperfusion. J Am Coll Cardiol 10:673–681
10. Gwirtz PA, Franklin D, Mass HJ (1986) Modulation of synchrony of left ventricular contraction by regional adrenergic stimulation in conscious dogs. Am J Physiol 251:H490–H495
11. Heusch G, Guth BD, Roth DM, Seitelberger R, Ross Jr J (1987) Contractile responses to sympathetic activation after coronary artery instrumentation. Am J Physiol 252:H1059–H1069
12. Heusch G, Schäfer S, Kröger K (1988) Recruitment of inotropic reserve in "stunned" myocardium by the cardiotonic agent AR-L 57. Basic Res Cardiol 83:602–610
13. Heyndrickx GR, Baig H, Nellens P, Leusen I, Fishbein MC, Vatner SF (1978) Depression of regional blood flow and wall thickening after brief coronary occlusions. Am J Physiol 234:H653–H659
14. Heyndrickx GR, Millard RW, McRitchie RJ, Maroko PR, Vatner SF (1975) Regional myocardial functional and electrophysiological alterations after brief coronary artery occlusion in conscious dogs. J Clin Invest 56:978–985
15. Hoffmeister HM, Mauser M, Schaper W (1985) Effect of adenosine and AICAR on ATP content and regional contractile function in reperfused canine myocardium. Basic Res Cardiol 80:445–458
16. Hori M, Koretsune Y, Kagiya T, Watanabe Y, Iwahura K, Iwai K, Kitabatake A, Yoshida H, Inoue M, Kamada T (1989) An increase in myocardial β-adrenoceptors to compensate for postischaemic dysfunction following coronary micro-embolisation in dogs. Cardiovasc Res 23:424–431
17. Ilebekk A, Lekven J, Kiil F (1980) Left ventricular asynergy during intracoronary isoproterenol infusion in dogs. Am J Physiol 239:H594–H600
18. Inoue H, Skale BT, Zipes DP (1988) Effects of ischemia on cardiac afferent sympathetic and vagal reflexes in dogs. Am J Physiol 255:H26–H35
19. Inoue H, Zipes DP (1987) Results of sympathetic denervation in the canine heart: supersensitivity that may be arrhythmogenic. Circulation 75:877–887
20. Inoue H, Zipes DP (1988) Time course of denervation of efferent sympathetic and vagal nerves after occlusion of the coronary artery in the canine heart. Circ Res 62:1111–1120
21. Ito BR, Tate H, Kobayashi M, Schaper W (1987) Reversibly injured, postischemic canine myocardium retains normal contractile reserve. Circ Res 61:834–846
22. Janes RD, Johnstone DE, Armour JA (1987) Functional integrity of intrinsic cardiac nerves located over an acute transmural myocardial infarction. Can J Physiol Pharmacol 65:64–69

23. Knight DR, Shen Y-T, Thomas Jr JX, Randall WC, Vatner SF (1988) Sympathetic activation induces asynchronous contraction in awake dogs with regional denervation. Am J Physiol 255:H358–H365
24. Knight DR, Thomas Jr JX, Randall WC, Vatner SF (1987) Effects of left circumflex coronary flow transducer implantation on posterior wall innervation. Am J Physiol 252:H536–H539
25. Krause SM, Jacobus WE, Becker LC (1989) Alterations in cardiac sarcoplasmic reticulum calcium transport in the postischemic "stunned" myocardium. Circ Res 65:526–530
26. Kusuoka H, Porterfield JK, Weisman HF, Weisfeldt ML, Marban E (1987) Pathophysiology and pathogenesis of stunned myocardium: Depressed Ca-activation of contraction as a consequence of reperfusion-induced cellular calcium overload in ferret hearts. J Clin Invest 79:950–961
27. LeWinter MM, Kent RS, Kroener JM, Carew TE, Covell JW (1975) Regional differences in myocardial performance in the left ventricle of the dog. Circ Res 37:191–199
28. Mercier JC, Lando U, Kanmatsuse K, Ninomiya K, Meerbaum S, Fishbein MC, Swan HJC, Ganz W (1982) Divergent effects of inotropic stimulation on the ischemic and severely depressed reperfused myocardium. Circulation 66:397–400
29. Miller BD, Thames MD, Mark AL (1983) Inhibition of cardiac sympathetic nerve activity during intravenous administration of lidocaine. J Clin Invest 71:1247–1253
30. Miyazaki T, Zipes DP (1989) Protection against autonomic denervation following acute myocardial infarction by preconditioning ischemia. Circ Res 64:437–448
31. Osakada G, Hess OM, Gallagher KP, Kemper WS, Ross Jr J (1983) End-systolic dimension-wall thickness relations during myocardial ischemia in conscious dogs. A new approach for defining regional function. Am J Cardiol 51:1750–1758
32. Reimer KA, Jennings RB, Tatum AH (1983) Pathobiology of acute myocardial ischemia: metabolic, functional and ultrastructural studies. Am J Cardiol 52:72A–81A
33. Schäfer S, Heusch G (1990) Recruitment of a time-dependent inotropic reserve by postextrasystolic potentiation in normal and reperfused myocardium. Basic Res Cardiol 85:257–269
34. Schipke J, Heusch G, Schulz R, Thämer V (1987) An easy and quick implantation procedure for the measurement of myocardial wall thickness using sonomicrometry. Basic Res Cardiol 82:411–414
35. Schwaiger M, Araujo L, Luxen A, Buxton D, Krivokapich J, Phelps ME, Schelbert HR (1987) Sustained catecholamine depletion in stunned canine myocardium. Circulation 76 suppl IV:378 (abstr)
36. Swain JL, Sabina RL, McHale PA, Greenfield JC, Holmes EW (1982) Prolonged myocardial nucleotide depletion after brief ischemia in the open-chest dog. Am J Physiol 242:H818–H826
37. Theroux P, Ross Jr J, Franklin D, Kemper WS, Sasayama S (1976) Coronary arterial reperfusion. III. Early and late effects on regional myocardial function and dimensions in conscious dogs. Am J Cardiol 38:599–606
38. Theroux P, Ross Jr J, Franklin D, Kemper WS, Sasayama S (1976) Regional myocardial function in the conscious dog during acute coronary occlusion and responses to morphine, propranolol, nitroglycerin, and lidocaine. Circulation 53:302–314
39. Weiner JM, Apstein CS, Arthur JH, Pirzada FA, Hood Jr WB (1976) Persistence of myocardial injury following brief periods of coronary occlusion. Cardiovasc Res 10:678–686

Authors' address:

Prof. Dr. Gerd Heusch, Abteilung für Pathophysiologie, Zentrum für Innere Medizin, Universität Essen, Hufelandstr. 55, 4300 Essen, FRG.

V. Adrenergic Mechanisms Triggering Arrhythmias During Myocardial Ischemia

Does Noradrenaline Influence the Extracellular Accumulation of Potassium, Sodium, Calcium, and Hydrogen Ions ($[K^+]_e$, $[Na^+]_e$, $[Ca^{2+}]_e$, $[H^+]_e$) during Global Ischemia in Isolated Rat Hearts?

Hj. Hirche, H. Knopf, H. Homburg and R. Walser

Institute of Vegetative Physiology, University of Cologne, Cologne, West Germany

Summary

The influence of noradrenaline (NA) on net cation fluxes during global ischemia (gI) was investigated in isolated rat hearts. The hearts were perfused according to the Langendorff technique and left ventricular pressure (LVP), the first derivative of the LVP (dP/dt_{max}, dP/dt_{min}), coronary perfusion pressure (CPP), and heart rate (HF) were measured. In the control group, the perfusion medium was either Krebs-Henseleit's solution (KHS), or KHS and tetramethylammonium-chloride (TMA; 100 μM). In this study TMA was used as a marker to determine changes in the extracellular space (ECS) size during gI. The hearts were subjected to 40 min of gI. Changes in the size of the ECS, and net cation movements were calculated during the first 20 min of gI. In treated hearts, NA (50 nM) was added to the perfusate 15 min before the onset of gI. Extracellular concentrations of K^+, Na^+, Ca^{2+}, H^+, and TMA were measured using double-barreled polyvinyl-chloride (PVC) mini-electrodes. Relative changes in the ECS size and net cation movements were calculated from the extracellular TMA and cation concentrations.

In separate experiments, the hemodynamics and lactate overflow of treated hearts were compared with control hearts prior to and following a brief period (1 min) of gI.

Addition of NA to the perfusate significantly:

1) increased CPP, LVP, dP/dt_{max}, dP/dt_{min}, and HF prior to the onset of gI, and increased cell swelling during gI;
2) diminished K^+-release from the cells, but significantly increased the influx of sodium and calcium into the intracellular space (ICS);
3) increased lactate overflow prior to and following 1 min of gI.

We assume that catecholamines increase ECS shrinkage before and during gI, probably by increased lactate production. NA stimulates the Na^+/K^+ pump, thereby reducing $[K^+]_e$ accumulation. The increased $[Ca^{2+}]_i$ and intracellular acidification promote sodium entry into the cells during gI.

Introduction

In the ischemic myocardium, an increased efflux of K^+ and H^+ has been observed [16]. Recently, it has been suggested that the early release of K^+ may occur via ATP-dependent K^+ channels [12]. After 3 to 4 min of ischemia, $[K^+]_e$ accumulation stops for 5 to 10 min. During this plateau phase, $[K^+]_e$ is maintained between 6 and 10 mM/l, probably by increased activity of the Na^+/K^+ pump.

During this period, ischemia-induced arrhythmias are rare [22] and reperfusion rapidly restores normal myocardial function. After 15 to 20 min of ischemia, $[K^+]_e$ again starts to rise [3] and ischemic damage becomes irreversible.

Conflicting results have been described for Na^+ redistribution in ischemic myocardium. Although Kléber [20] did not find an increased sodium concentration in the intracellular space (ICS) of ischemic myocardium, Tani and Neely [47] estimated a Na^+ influx into ischemic cells. Furthermore, a gain in intracellular Ca^{2+} has been observed in the ischemic myocardium [27,46].

Anaerobic metabolism during ischemia increases the intracellular osmolarity [48], leading to a water shift from the ECS into the ICS. In globally ischemic hearts with constant tissue water, any change in the extracellular concentration of an ion reflects a shift of the ion and/or water between the ECS and the ICS [21]. The time course and the extent of the ischemia-induced water shift are the subject of investigations.

Acute myocardial ischemia causes an increased local release of noradrenaline (NA) in hearts in situ [29] and in isolated hearts [23,43]. High NA concentrations have deleterious effects on ischemic myocardium [for review see: 4]. This is mainly due to increased energy requirements and the increased Ca^{2+} uptake caused by NA. Catecholamines directly stimulate the Na^+/K^+ pump [6] and may, therefore, influence the redistribution of K^+ and Na^+ in the myocardium.

The aim of this study was to investigate in globally ischemic rat hearts whether NA influences: 1) the net transmembrane ion flux rates of K^+, Na^+, and Ca^{2+} and 2) the water shift from ECS to ICS.

Materials/Methods

Double-barreled PCV-mini-electrodes were prepared according to the technique of Hill et al. [14], and were inserted into the left mid-ventricular myocardium. Ion-selective membranes were prepared by embedding the ion exchanger in PVC. The electrodes exhibited a linear response when calibrated in solutions with known ion concentrations. Tetramethylammonium-chloride (TMA) was detected using the Corning 477317 resin which, although designated as a K^+-exchanger, is a tetraalkyl-ammonium sensor [34]. Details of the method have been described recently [24]. The electrode potentials measured were corrected with respect to the concentrations of the interfering ions. Only electrodes with a Nernstian slope ($\pm$ 5%) were used for the experiments. The electrode recordings were amplified by high-input-impedance amplifiers and displayed on a chart recorder. After the experiments, the hearts were reperfused with the pre-ischemic perfusion medium, in order to re-equilibrate the ECS with the perfusate. A difference between the electrode signals before ischemia and after reperfusion was regarded as drift of the electrodes, and the readings of the electrodes obtained during ischemia were corrected for this drift.

Hearts from male Wistar rats (250–400 g) were excised and perfused retrogradely at constant flow using the Langendorff technique. The perfusion medium (Krebs-Henseleit solution; KHS) was heated (37°C), charged with carbogen (95% 0_2, 5% CO_2) to maintain the pH at approximately 7.4 and pO_2 above 500 mmHg, and was not recirculated. A latex balloon was inserted into the left ventricle via the mitral valve and the hearts were connected to a systemic circulation according to the technique of Bardenheuer and Schrader [1]. The preload was adjusted to 10 mmHg (the normal end-diastolic pressure of rats [8]). In order to enable pressure-volume work in our experimental setup, the afterload was maintained at 40 mmHg. The coronary perfusion pressure (CPP), the left ventricular pressure (LVP), and the first derivatives of the LVP (dP/dt_{max}, dP/dt_{min}) were recorded continously, and the heart rate (HF) was determined from the LVP recording. The hearts were immersed in an oil bath, to prevent cooling and desiccation during ischemia. Global ischemia (gI) was induced by stopping the perfusion. The pre-ischemic ECS volume was determined using the method provided by Scheufler and Peters [42].

Protocol

Control hearts were perfused with KHS containing TMA (100 μM) for about 30 min; treated hearts were perfused with KHS + TMA (100 μM) for about 15 min and with

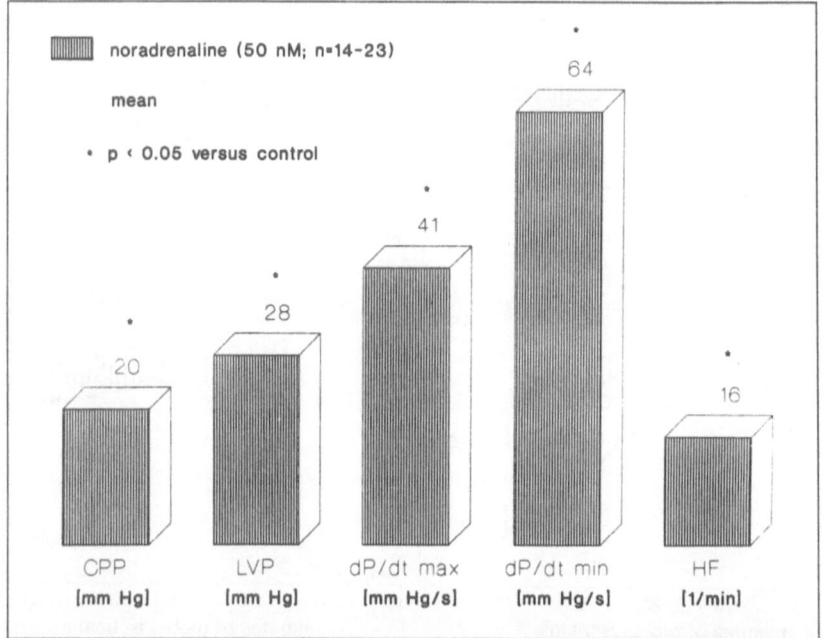

Fig. 1. Switching the perfusion from KHS to KHS containing noradrenaline (50 nM) significantly increased the hemodynamic parameters prior to the onset of global ischemia. Percent changes in parameters are given. (coronary perfusion pressure = CPP; left ventricular pressure = LVP; first derivative of the LVP = dP/dt; heart rate = HF).

KHS + TMA + noradrenaline (NA; 50 nM) for another 15 min. The pre-ischemic hemodynamics were compared in treated and non-treated hearts (KHS + TMA + NA vs KHS + TMA). The hearts were subjected to 40 min of gI and the changes in extracellular ion concentrations were measured using ion-selective mini-electrodes. During the first 20 min of gI net ion movements were calculated from changes in extracellular TMA and cation concentrations.

In separate experiments, the pre-ischemic perfusion period of control and treated hearts was followed by 1 min of gI. The hearts were reperfused with the pre-ischemic perfusion medium. The hemodynamics during reperfusion were compared with the hemodynamics prior to the onset of gI. Lactate overflow was measured during the pre-ischemic perfusion period and during reperfusion.

Statistics

All data are expressed as means ± SEM. Significance was tested using the two-tailed Student's t-test for paired or unpaired data. A P-value of ≤ 0.05 was considered to be significant.

Results

Switching the perfusion from KHS to KHS + NA (50 nM) significantly increased CPP (52 vs 62 mmHg; + 20%), LVP (47 vs 61 mmHg; + 28%), dP/dt_{max} (1550 vs 2180 mmHg/s; + 41%), dP/dt_{min} (1026 vs 1680 mmHg/s; + 64%), and HF (225 vs 261 l/min; + 16%) (Fig. 1). In control hearts, the pre-ischemic ECS, measured with cobalt-III-chloride, was about 30% of the total tissue water [24]. ECS declined to about 52% of its initial size within the first 20 min of gI. Treatment with NA (50 nM) led to only a slight reduction of ECS (26% of total tissue water) before the onset of gI. However, during the first 20 min of gI, ECS shrinkage was significantly

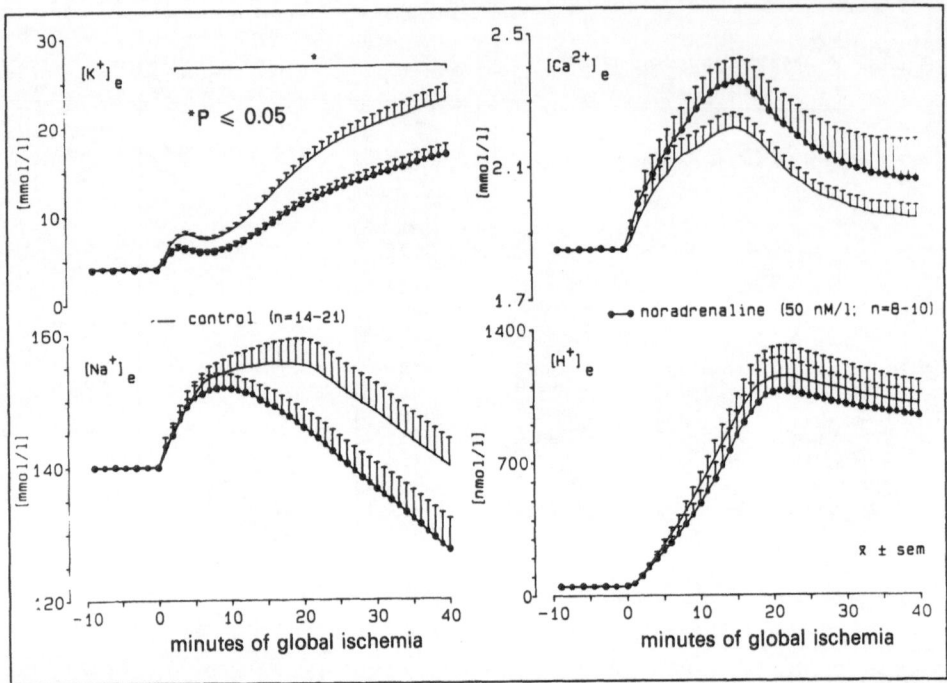

Fig. 2. Changes of extracellular cation concentrations in the left mid-ventricular myocardium during 40 min of global ischemia in the isolated rat heart.

higher compared with the control hearts (38% vs 52% of the initial size). In separate experiments, perfusion with KHS + NA increased lactate overflow during the pre-ischemic perfusion period (34 vs 10 μM/min), and during reperfusion following 1 min of gI (149 vs 58 μM/min).

The changes in extracellular cation concentrations are shown in Fig. 2. In the control hearts, $[K^+]_e$ started to increase about 10 s after the onset of gI and showed the typical triphasic pattern. There was a rapid increase from 4.0 to 7.9 ± 0.3 mM within the first 5 min of gI, followed by a slight decrease (plateau phase) to about 7.4 ± 0.2 mM after 7 min of ischemia. The plateau phase was followed by a second but slower increase in $[K^+]_e$, reaching a concentration of 23 ± 1.6 mM after 40 min of gI. NA (50 nM) significantly reduced the $[K^+]_e$ increase 3 to 40 min after the onset of gI. The changes in $[Na^+]_e$ and $[Ca^{2+}]_e$ were biphasic, showing an initial increase 20–30 s after the onset of gI followed by a decrease. In control hearts, $[Na^+]_e$ increased from 140 to 155 ± 3.7 mM during the first 18 min of gI and then declined to nearly the pre-ischemic value (140 ± 4.2 mM). $[Ca^{2+}]_e$ increased from 1.85 mM to about 2.16 ± 0.04 mM during the first 19 min of ischemia and then slowly decreased to about 1.95 ± 0.04 mM after 40 min of ischemia. NA (50 nM) tended to reduce the $[Na^+]_e$ increase and tended to augment the $[Ca^{2+}]_e$ increase, but no significance could be achieved. In control and treated hearts, $[H^+]_e$ increased from 40 ± 3 nM (pH 7.4) to 1150 ± 160 nM (pH 5.9) during the first 20 min of ischemia, and then slowly decreased to about 1000 ± 125 nM (pH 6.0) after 40 min of gI.

Ischemia-induced net cation movements of $[K^+]_e$, $[Na^+]_e$, and $[Ca^{2+}]_e$ were calculated from changes in $[TMA]_e$ and the extracellular ion concentrations [for calculation see: 24], and are shown in Fig. 3. Addition of NA (50 nM) to the perfusate significantly increased net sodium and calcium influx and reduced potassium efflux during the first 20 min of gI.

Relative changes in the hemodynamics during reperfusion following a 1 min period of gI are shown in Fig. 4. As can be seen, addition of NA (50 nM) to the perfusate significantly diminished functional recovery during reperfusion.

286

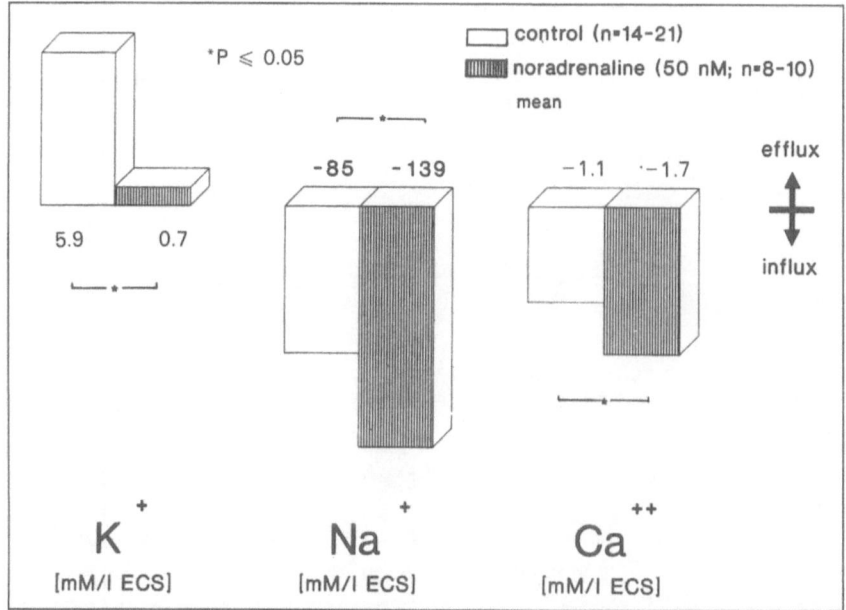

Fig. 3. Net cation movements from/into the ECS during the first 20 min of global ischemia. The appearance and disappearance rates were calculated from changes of the extracellular cation concentrations and of the ECS size.

Discussion

ECS

Quarternary ammonium ions have been used to measure the extracellular volume in brain [7]. In this study tetramethylammonium-chloride (TMA) was used to determine changes in the size of the ECS during gI. Although TMA may interfere with the transport of potassium ions, the concentration of TMA used in this study did not influence the extracellular potassium accumulation during gI [24]. An extensive review on this topic has been provided by Stanfield [45].

Isolated saline-perfused hearts take up water which expands the ECS [19]. Polimeni and Buraczewski [38] have shown that there is a rapid uptake of water by the myocardium within the first 5 min of perfusion and that this water is distributed proportionally between the intracellular and extracellular spaces.

The pre-ischemic ECS measured in control hearts (about 30% of total tissue water) is in good agreement with the data published by Scheufler and Peters [42]. The reduction of the ECS during control perfusion in NA treated hearts (26% of total tissue water) may result from the enhanced production of osmotically active compounds (e.g., lactate). In early ischemia the majority of metabolic compounds, such as lactate [41], remain intracellular, and consequently water enters the cell. This may explain our finding that ECS shrinkage during ischemia is significantly increased following perfusion with NA.

Potassium

It is well documented that during regional and global myocardial ischemia extracellular potassium accumulates in a triphasic pattern [15,16,23]. This increase in $[K^+]_e$ may be attributed to a decreased K^+ influx and/or an increased K^+ conductance. Although an impaired

287

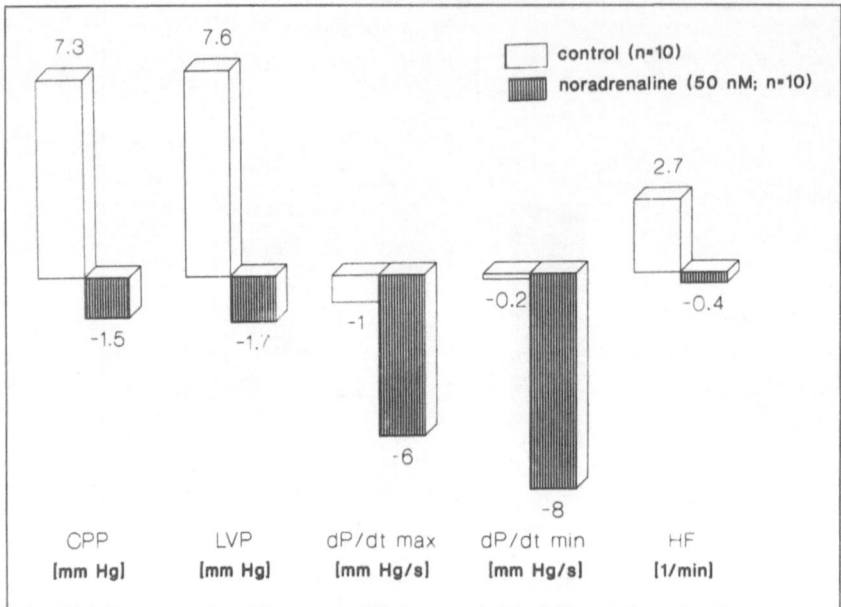

Fig. 4. The influence of a brief period of global ischemia (1 min) on the hemodynamics of isolated working rat hearts. The hemodynamics during reperfusion were compared with those prior to the onset of global ischemia. Percent changes in parameters are given. (Abbreviations as in Fig. 1).

Na$^+$/K$^+$ exchange mechanism has been suggested [10], there is evidence that the pumping is only partially inhibited, and therefore cannot account for the first rapid rise of [K$^+$]$_e$ [20]. A loss of membrane integrity seems unlikely, as the changes in [K$^+$]$_e$ are reversible if reperfusion is started within 15 to 20 min of ischemia [15].

There is experimental evidence suggesting that energy-rich phosphate compounds decline rapidly during the early phase of ischemia [18]; however, the intracellular ATP levels necessary to activate ATP-sensitive K$^+$ channels (< 1 mM; [35]) are only reached during later stages of ischemia [10].

An earlier study with radiolabeled K$^+$ revealed that hypoxia-induced K$^+$ loss is caused by an increased K$^+$ efflux, whereas the energy-dependent influx was unaffected [39]. In that study, the application of catecholamines reduced the accumulation of [K$^+$]$_e$ during gI. This may be explained by the fact that catecholamines exert a stimulatory effect on the cardiac Na$^+$/K$^+$ pump [26].

Sodium

During ischemia, catecholamines are released from the neurogenic sites [23,43] and may directly or indirectly stimulate (via the cAMP pathway) the Na$^+$ pump, thereby reducing the intracellular sodium level [37]. However, in isolated perfused guinea pig hearts, internal Na$^+$ levels are unchanged during the first 15 min of ischemia [20] which was explained by a decrease in the resting membrane potential [49]. This decrease in resting membrane potential stimulates the Na$^+$/K$^+$ pump [13], thereby promoting Na$^+$ efflux. However, consistent with our results, Tani and Neely [47] found that [Na$^+$]$_i$ increased by about 50% when rat hearts were exposed to 15 min of zero-flow ischemia.

As can be seen from the results, NA increased net sodium influx during the first 20 min of gI. When the internal pH is lowered, the Na$^+$/H$^+$ antiporter serves as a pH$_i$ regulatory system [25] leading to an increase in sodium influx. The Na$^+$/H$^+$ exchange system is activated by treatments which decrease pH_i (e.g., application of catecholamines) or increase pH_e, and it is a major pathway for Na^+ uptake by cardiac cells [11].

Calcium

The application of catecholamines in isolated heart muscle preparations produces an increase in the force of contraction [31] due to an increased calcium influx. This is supported by the results in this study. The increase in $[Ca^{2+}]_i$ is probably further maintained by the release of Ca^{2+} from intracellular stores due to a Ca^{2+}-induced Ca^{2+} release [9]. The increase in $[Ca^{2+}]_i$ could activate a calcium-dependent potassium conductance [17], which may contribute to the K^+ increase early in ischemia. The extrusion of Ca^{2+} could be accomplished by the Na^+/Ca^{2+} exchanger in the forward mode with the Na^+ gradient as the energy source, thereby increasing $[Na^+]_i$. The increase in $[Ca^{2+}]_i$ may also contribute to the entry of sodium into the cells via the activation of nonspecific cation channels [36].

The accumulation of $[Na^+]_i$ is known to trigger Ca^{2+} entry in cardiac cells through the Na^+/Ca^{2+} exchange system [2]. Sodium pump inhibition could increase $[Ca^{2+}]_i$ fivefold via the sodium calcium exchanger [30] and may play an important role in regulating intracellular calcium concentration. In contrast, Deitmer and Ellis [5] suggested that the Na^+/Ca^{2+} exchanger cannot increase intracellular Na^+ during ischemia because it is inhibited by a decrease in intracellular pH. The increase in free cytosolic calcium depletes the energy supply to the cells by enhanced use of ATP due to an increased Ca^{2+}-ATPase activity and by uncoupling oxidative phosphorylation [28].

Hydrogen ions

Our data indicate that in ischemic rat hearts, anaerobic metabolism (generation of hydrogen ions) is active during the first 20 to 25 min of ischemia as the capacity of the myocardium to generate ATP is limited and since the metabolic pathway is pH-sensitive [32]. Frelin et al. [11] have demonstrated the presence of a coupled Na^+/H^+ exchange system. When Na^+ influx via the Na^+/H^+ exchange system is blocked with amiloride, ouabain is unable to provoke an accumulation of Na^+ and the subsequent entry of Ca^{2+} into the cells. Treatment with catecholamines enhances H^+ generation, which may in turn release calcium from binding-proteins and thereby increase $[Ca^{2+}]_i$. Extrusion of this excess calcium can be accomplished by both the Na^+/Ca^{2+} exchanger in the forward mode with the electrochemical Na^+ gradient as the energy source, and the ATP-dependent Ca^{2+} pump. The lack of an increase in $[H^+]_e$ following treatment with NA may indicate an increased intra- or extracellular buffering capacity.

Recently, Neely and Grotoyohann [33] have shown that recovery of ventricular function following 30 min of gI in rat hearts is inversely related to the accumulated levels of lactate. If lactate accumulation was prevented by a previous period of high-flow anoxia (to deplete myocardial glycogen), functional recovery following ischemia was improved. This is supported by our observation that the application of NA increases lactate overflow and decreases functional recovery following a brief period of gI. Application of catecholamines enhances the lactate overflow during the pre-ischemic perfusion period (this study), suggesting that the internal pH of the cells is lowered. The internal acidification may favor sodium entry into the cells via the Na^+/H^+ exchange system. However, from our results it cannot be excluded that the internal sodium level is higher in treated hearts during the pre-ischemic perfusion period. During reperfusion with a medium at normal pH, intracellular H^+ ions are removed in exchange with sodium ions. This gain in intracellular sodium may favor calcium entry via the Na^+/Ca^{2+} exchange system leading to an increase in intracellular calcium. When $[Ca^{2+}]_i$ is high, the availability of ATP and ATP synthesis are reduced [44]. Thus, our results are consistent with the idea that the elevated $[Ca^{2+}]_i$ and $[H^+]_i$ impair mechanical recovery during reperfusion. Reibel and Rovetto [40] demonstrated a positive correlation between ATP levels at the end of the ischemic period and the capacity to recover mechanically upon reperfusion.

This indicates, that in treated hearts the pre-ischemic ATP level may be reduced due to enhanced energy demand during the pre-ischemic perfusion period. The results from our study suggest that norepinephrine:

1) increases the production of lactate before and during gI, leading to a decreased functional recovery during reperfusion;
2) increases calcium influx into the cells and promotes sodium entry via the Na^+/Ca^{2+} exchange system; and
3) stimulates the Na^+/K^+ pump and decreases K^+ efflux during gI.

References

1. Bardenheuer H, Schrader J (1983) Relationship between myocardial oxygen consumption, coronary flow, and adenosine release in an improved isolated working heart preparation of guinea pig. Circ Res 51: 263–271
2. Bersohn MM, Philipson KD, Fukushima JY (1982) Sodium-calcium exchange and sarcolemmal enzymes in ischemic rabbit hearts. Am J Physiol 242: C288–C295
3. Bös L, Franz Chr, Hirche Hj (1978) Cardiac arrhythmia and increase of local myocardial extracellular K^+ activity in pigs. J Physiol 284: 88 P (abstr.)
4. Corr PB, Yamada KA, Witkowski FX (1986) Mechanisms controlling cardiac autonomic function and their relation to arrhythmogenesis. In: Fozzard HA, Haber E, Jennings RB, Katz AM, Morgan HE (Eds.) The Heart and Cardiovascular System. Raven Press, New York, pp 1343–1403
5. Deitmer JW, Ellis DW (1980) Interactions between the regulation of the intracellular pH and sodium activity of sheep cardiac Purkinje fibers. J Physiol 304: 471–488
6. Désilets M, Baumgarten CM (1986) K^+, Na^+, and Cl^- activities in ventricular myocytes from rabbit heart. Am J Physiol 251: C197–C208
7. Dietzel I, Heinemann U, Hofmeier G, Lux HD (1980) Transient changes in the size of the extracellular space in the sensorimotor cortex of cats in relation to stimulus induced changes in potassium concentration. Exp Brain Res 40: 432–439
8. Döhring HJ, Dehnert H (1985) Das isolierte perfundierte Warmblüter-Herz nach Langendorff. In: Methoden der experimentellen Physiologie und Pharmakologie. Biomesstechnik (5), Biomesstechnik Verlag March GmbH, March
9. Fabiato A, Fabiato F (1977) Calcium release from the sarcoplasmic reticulum. Circ Res 40: 119–129
10. Fiolet JWT, Baartscheer A, Schumacher CA, Coronel R, ter Welle HF (1984) The change of the free energy of ATP hydrolysis during global ischemia and anoxia in the rat heart. Its possible role in the regulation of the transsarcolemmal sodium and potassium gradients. J Mol Cell Cardiol 16: 1023–1026
11. Frelin C, Vigne P, Lazdunski M (1984) The role of the Na^+/H^+ exchange system in cardiac cells in relation to the control of the internal Na^+ concentration. J Biol Chem 259: 8880–8885
12. Friedrich M, Benndorf K, Schwalb M, Hirche Hj (1990) Effects of anoxia on K^+ and Ca^{2+} currents in isolated guinea pig cardiocytes. Pflügers Arch 416: 207–209
13. Gadsby DC, Kimura J, Noma A (1985) Voltage dependence of Na/K pump current in isolated heart cells. Nature 315: 63–65
14. Hill JL, Gettes LS, Lynch MR, Hebert NC (1978) Flexible valinomycin electrodes for on-line determination of intravascular and myocardial K^+. Am J Physiol 235: H455–H459
15. Hill JL, Gettes LS (1980) Effect of acute coronary artery occlusion on local myocardial extracellular K^+ activity in swine. Circulation 61: 768–778
16. Hirche Hj, Franz Chr, Bös L, Bissig R, Lang R, Schramm M (1980) Myocardial extracellular K^+ and H^+ increase and noradrenaline increase as possible cause of early arrhythmias following acute coronary artery occlusion in pigs. J Mol Cell Cardiol 12: 579–593
17. Isenberg G (1977) Cardiac Purkinje fibers $[Ca^{2+}]_i$ controls the potassium permeability via the conductance components g_{k1} and g_{k2}. Pflügers Arch 371: 77–85
18. Jennings RB, Reimer KA, Hill ML, Mayer SE (1981) Total ischemia in dog hearts in vitro. 1. Comparison of high energy phosphate production, utilization and depletion and of adenosine nucleotide catabolism in total ischemia in vitro vs. severe ischemia in vivo. Circ Res 49: 892–900
19. Jennings RB, Reimer KA, Steenbergen C (1986) Myocardial ischemia revisited. The osmolar load, membrane damage, and reperfusion. J Mol Cell Cardiol 18: 769–780
20. Kléber AG (1983) Resting membrane potential, extracellular potassium activity and intracellular sodium activity during acute global ischemia in the isolated guinea pig heart. Circ Res 52: 442–450
21. Kléber AG (1984) Extracellular potassium accumulation in acute myocardial ischemia. J Mol Cell Cardiol 16: 389–394

22. Knopf H, McDonald FM, Bischoff A, Hirche Hj, Addicks K (1988a) Effect of propranolol on early postischemic arrhythmias and noradrenaline and potassium release of ischemic myocardium in anaesthetized pigs. J Cardiovasc Pharmacol 12 (suppl. 1): S41–S47

23. Knopf H, Theising R, Hirche Hj (1988b) The effect of desipramine on ischemia-induced changes in extracellular K^+, Na^+, and H^+ concentrations and noradrenaline release in the isolated rat heart during global ischemia. J Cardiovasc Pharmacol 12: 8–14

24. Knopf H, Theising R, Moon CH, Hirche Hj (1990) Continuous determination of extracellular space (ECS) and net flux rates of K^+, Na^+, Ca^{2+}, and H^+ during global ischemia (GI) in isolated rat hearts. J Mol Cell Cardiol (in press)

25. Lazdunski M, Frelin C, Vigne P (1985) The sodium/hydrogen exchange system in cardiac cells: its biochemical and pharmacological properties and its role in regulating internal concentrations of sodium and internal pH. J Mol Cell Cardiol 17: 1029–1042

26. Lee CO, Vasalle M (1983) Modulation of intracellular Na^+ activity and cardiac force by norepinephrine and Ca^{2+}. Am J Physiol 244: C110–C114

27. Lee HC, Mohabir R, Smith N, Franz MR, Clusin WT (1988) Effect of ischemia on Ca-dependent fluorescence transients in rabbit hearts containing Indo 1. Circulation 78: 1047–1059

28. Lehninger AL (1970) Mitochondria and calcium ion transport. Biochem J 119: 129–138

29. McDonald FM, Knopf H, Hartono S, Polwin W, Bischoff A, Hirche Hj, Addicks K (1986) Acute myocardial ischaemia in the anaesthetised pig: local catecholamine release and its relation to ventricular fibrillation. Basic Res Cardiol 81: 636–645

30. Murphy E, Jacob R, Lieberman M (1985) Cytosolic free calcium in chick heart cells. J Mol Cell Cardiol 17: 221–231

31. Nawrath H (1989) Adrenoceptor mediated changes of excitation and contraction in isolated heart muscle preparations. J Cardiovasc Pharmacol 14: 1–10

32. Neely JR, Whitmer JT, Rovetto MJ (1975) Effect of coronary flow on glycolytic flux and intracellular pH of isolated rat hearts. Circ Res 37: 733–741

33. Neely JR, Grotoyohann LW (1984) Role of glycolytic products in damage to ischemic myocardium. Dissociation of adenosine triphosphate levels and recovery of function of reperfused ischemic hearts. Circ Res 55: 816–824

34. Neher E, Lux HD (1973) Rapid changes of potassium concentration at the outer surface of exposed single neurons during membrane current flow. J Gen Physiol 61: 385–399

35. Noma A (1983) ATP-regulated K^+ channels in cardiac muscle. Nature 305: 147–148

36. Partridge LD, Swandulla D (1988) Calcium activated non-specific cation channels. TINS 11: 69–72

37. Pecker MS, Im WB, Sonn JK, Lee CO (1986) Effect of norepinephrine and cyclic AMP on intracellular sodium ion activity and contractile force in canine cardiac Purkinje fibers. Circ Res 59: 390–397

38. Polimeni PI, Buraczewski SI (1988) Expansion of extracellular tracer spaces in the isolated heart perfused with crystalloid solutions: expansion of extracellular space, trans-sarcolemmal leakage, or both? J Mol Cell Cardiol 20: 15–22

39. Rau EE, Shine KI, Langer GA (1977) Potassium exchange and mechanical performance in anoxic mammalian myocardium. Am J Physiol 232: H85–H94

40. Reibel DK, Rovetto MJ (1978) Myocardial ATP synthesis and mechanical function following oxygen deficiency. Am J Physiol 234: H620–H624

41. Rovetto MJ, Whitmer JT, Neely JR (1973) Comparison of the effects of anoxia and whole heart ischemia on carbohydrate utilization in isolated working rat hearts. Circ Res 22: 699–711

42. Scheufler E, Peters T (1987) Determination of the extracellular space with nonradioactive Co^{3+}EDTA and simultaneous estimation of Na, K, Ca, and Mg contents in isolated guinea-pig heart preparations by atomic absorptions spectrometry. Basic Res Cardiol 82: 341–347

43. Schömig A, Fischer S, Kurz T, Richardt G, Schömig E (1987) Nonexocytotic release of noradrenaline in the ischemic and anoxic rat heart: mechanism and metabolic requirements. Circ Res 60: 194–205

44. Smith GL, Allen DG (1988) Effects of metabolic blockade on intracellular calcium concentration in isolated ferret ventricular muscle. Circ Res 62: 1223–1236

45. Stanfield PR (1983) Tetraethylammonium ions and the potassium permeability of excitable cells. Rev Physiol Biochem Pharmacol 97: 1–67

46. Steenbergen C, Murphy E, Levy L, London RE (1987) Elevation in cytosolic free calcium concentration early in myocardial ischemia in perfused rat hearts. Circ Res 60: 700–707

47. Tani M, Neely JR (1989) Role of intracellular Na^+ in Ca^{2+} overload and depressed recovery of ventricular function of reperfused ischemic rat hearts. Circ Res 65: 1045–1056

48. Tranum-Jensen J, Janse MJ, Fiolet JWT, Krieger WJG (1981) Tissue osmolality, cell swelling, and reperfusion in acute regional myocardial ischemia in the isolated porcine heart. Circ Res 49: 364–381

49. Wilde AAM, Kléber AG (1986) The combined effects of hypoxia, high K$^+$, and acidosis on the intracellular sodium activity and resting potential in guinea pig papillary muscle. Circ Res 58: 249–256

Authors' address:

Prof. Dr. Hj. Hirche, Institut für Vegetative Physiologie, Robert-Koch-Straße 39, 5000 Köln 41, West Germany

Effects of β-Blockade on the Incidence of Ventricular Tachyarrhythmias during Acute Myocardial Ischemia: Experimental Findings and Clinical Implications

W. Haverkamp*, H. Gülker**, G. Hindricks* and G. Breithardt*

Medical Hospital of the Westfälische Wilhelms Universität of Münster, Department of Cardiology and Angiology*, Münster, FRG and the Medical Clinic of the Klinikum Barmen**, Wuppertal, FRG

Summary

Myocardial ischemia and infarction are the most common substrates for life-threatening ventricular tachyarrhythmias. Experimental and clinical evidence suggests that enhanced activity of the sympathetic nervous system plays an important role in the genesis of ischemia-related arrhythmias. In animal experiments, β-blockers display significant antifibrillatory effects during the acute phase of myocardial ischemia. Preconditions for their antifibrillatory effects are high serum- and tissue-concentrations, and absence of a significant partial agonist activity. During the delayed phase of ischemic arrhythmias which starts 6–8 h after coronary occlusion, β-blockers gain significance as antiarrhythmic and potentially antifibrillatory drugs, if sympathetic activity is enhanced.

The presently available evidence suggests that the potentially antifibrillatory effects of β-blockers are at least one of the major mechanisms by which these drugs may decrease mortality when given prophylactically in patients after myocardial infarction. However, it remains to be explained why β-blockers, in a great number of prospective randomized trials, have reduced the incidence of sudden death only by an average of about 30%. This may be the result of their "specific" mechanisms acting in the setting of acute myocardial ischemia with enhanced adrenergic tone, whereas in the remaining patients other mechanisms such as a chronic arrhythmogenic substrate may be operative. A clearer separation of these various mechanisms seems mandatory in order to allow a more specific "targeted" administration of β-blockers. This is the more important since none of the available prospective studies that used antiarrhythmic agents has shown an improvement of prognosis, but, instead showed a worsening of the mortality rate.

Introduction

Myocardial ischemia and infarction are the most common substrates for life-threatening ventricular tachyarrhythmias. In experimental models of infarction, as well as in patients, arrhythmias related to ischemia occur in several phases. In the ischemic dog heart at least two phases can be separated: the acute and the delayed phase [14, 18, 19, 21, 22].

The acute phase (phase I) begins within minutes of coronary occlusion and lasts for approximately 30 min, while the delayed phase (phase II) starts 6–8 h after occlusion and lasts approximately 72 h. The electrophysiologic substrates and arrhythmic mechanisms underlying spontaneously occurring ventricular arrhythmias show distinct differences during these two phases. Acute phase arrhythmias are mainly due to reentrant activation resulting from a deterioration of conduction in the ischemic area [14, 19, 30], whereas arrhythmias occurring during the delayed phase have been attributed to abnormal automaticity and triggered activity [11, 12, 14, 15, 19].

Acute phase of ischemic arrhythmias: autonomic neural influences and effects of β-blockade

During the acute phase of ischemic arrhythmias, ventricular premature beats occur and often lead to ventricular tachycardia and ventricular fibrillation. The occurrence of the arrhythmia depends on several factors. Major determinants are the mode of occlusion (e.g., one-stage vs two-stage occlusion) [39], the amount of myocardium at risk (the size of the infarct) [19, 39], oxygen consumption and general metabolic changes [42], and autonomic neural influences [8, 17]. When occlusion of a coronary artery is performed slowly, the amount of myocardium at risk is small and oxygen consumption is relatively low; myocardial infarction may develop without the occurrence of severe ventricular tachyarrhythmias [19, 39].

Several investigators have provided links between the activation of the sympathetic nervous system and the occurrence of ventricular fibrillation after acute coronary occlusion. Experimental evidence suggests that acute ischemia results in stimulation of sympathetic afferent and efferent fibers [36, 37, 54]. Lombardi and coworkers [36] were able to demonstrate that activation of sympathetic preganglionic fibers during coronary occlusion corresponds to the period of maximal vulnerability to ventricular fibrillation. Likewise, stellate ganglion stimulation exerts pronounced arrhythmogenic effects during acute myocardial ischemia [31, 47].

A significant increase in the incidence of ventricular fibrillation after acute coronary artery ligation can also be observed when the serum-concentration of circulating catecholamines is increased [23]. Figure 1 presents the effects of combined infusion of epinephrine and norepinephrine on the occurrence of ischemic ventricular fibrillation in anesthetized dogs. To gain further insight into the mechanisms of arrhythmogenesis, epicardial conduction times in the ischemic area were assessed by a 42 bipolar mapping electrode array. The increased incidence of ventricular fibrillation during intravenous application of catecholamines was associated with a significant prolongation of epicardial conduction times in the ischemic area, while the number of electrodes displaying complete conduction block was decreased compared to control animals (Fig. 2). An increase in epicardial conduction times may be interpreted as slowing of conduction, but the opposite effect on intramyocardial conduction characteristics (i.e., increase in conduction velocity with a longer pathway of conduction through intramural layers) cannot be excluded. One possible interpretation of the results may be that the arrhythmogenic effect of catecholamines may be due to an increased likelihood for reentry in the ischemic myocardium. The finding of a reduced incidence of complete conduction block in the epicardial layer suggests that overall slowing of conduction may be achieved by recruitment of previously non-conducting tissue due to a catecholamine-induced enhanced activation of the slow inward current [9]. However, to date the occurrence of slow response potentials during ischemia has not been documented in vivo and calcium antagonists decrease conduction delay in the ischemic myocardium [24]. A reduction in refractory periods induced by enhanced adrenergic stimulation may also contribute to arrhythmogenesis [55].

In contrast, chronic left and bilateral stellectomy exert antifibrillatory effects, while acute denervation does not reduce the incidence of ventricular fibrillation after coronary occlusion [10, 48]. The lack of effect of acute stellectomy has been attributed to local release of norepinephrine from intramyocardial catecholamine stores, whereas chronic denervation results in almost complete depletion of myocardial catecholamines [10]. However, right stellectomy may be associated with an increased incidence of ventricular fibrillation [48]. The mechanisms underlying this apparently paradoxic effect have not been fully elucidated. Martin and Meesmann [38] demonstrated that regional chemical sympathectomy also exerts antifibrillatory effects after coronary occlusion. In their study, selective infusion of 6-hydroxydopamine into the left coronary artery was used to reduce the norepinephrine content of the myocardium below 1.5% of normal. After acute ligation of the left circumflex coronary artery, the incidence of ventricular fibrillation was significantly reduced compared to control animals.

In the animal experiment, β-adrenoceptor blockade significantly reduces the incidence of

Ischemic tachyarrhythmias and β-blockade

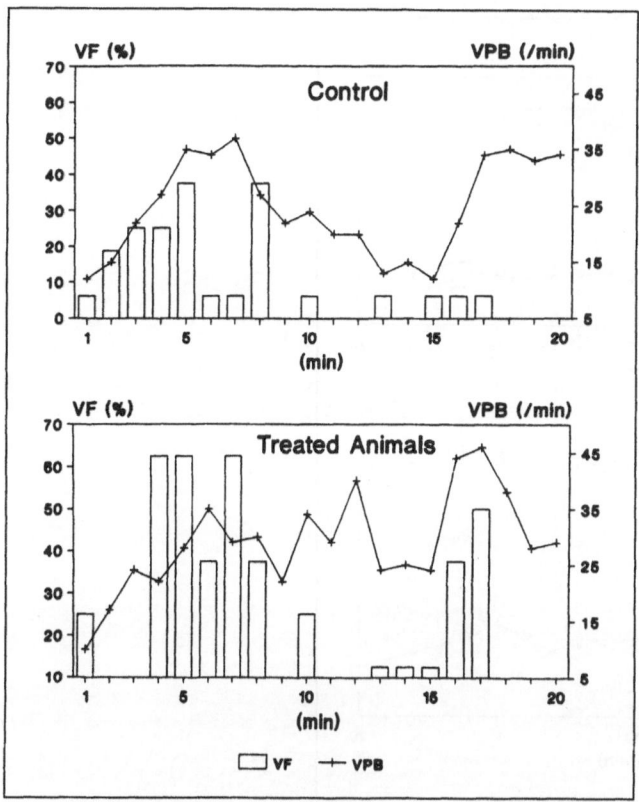

Fig. 1. Effect of combined infusion of epinephrine (1.5 μg/kg × min) and norepinephrine (0.2 μg/kg × min) on the incidence of ventricular premature beats (VPB) and ventricular fibrillation (VF) after acute coronary artery occlusion compared to untreated animals. (From [23].)

ventricular fibrillation at the very onset of acute myocardial ischemia [1, 13, 25, 32, 40, 44, 49], at least by inhibiting adrenergic influences on the heart. In the non-ischemic dog heart, the administration of β-blockers results in a dose-related increase in the ventricular fibrillation threshold [2, 35, 45]. However, this only holds true when drugs without partial agonist activity are applied. When β-blockers with intrinsic sympathomimetic activity are used, a fall in the ventricular fibrillation threshold occurs at higher doses, which has been attributed to the partial agonist activity which itself is dose-dependent [35]. The effects of β-blockade on the ventricular fibrillation threshold persist after coronary occlusion, although the relative fall in the ventricular fibrillation threshold is not prevented (Fig. 3) [19]. The effects of β-blockade on fibrillation threshold during acute ischemia are associated with a less pronounced deterioration of epicardial conduction in the ischemic zone compared to control animals (Fig. 4). As a result, a significant decrease in the incidence of ventricular fibrillation can be observed although ventricular premature beats are only partially suppressed (Fig. 5). Since measurements of epicardial conduction times were performed during constant atrial pacing, the beneficial effects of β-blockade cannot be considered to solely result from a reduction in heart rate that may reduce oxygen consumption and tissue damage. To date, despite the many experimental observations, no detailed information is available concerning the precise mechanisms responsible for the antifibrillatory potency of β-adrenergic blockade during ischemia.

The effects of β-blockade on the incidence of ischemia-induced ventricular fibrillation are

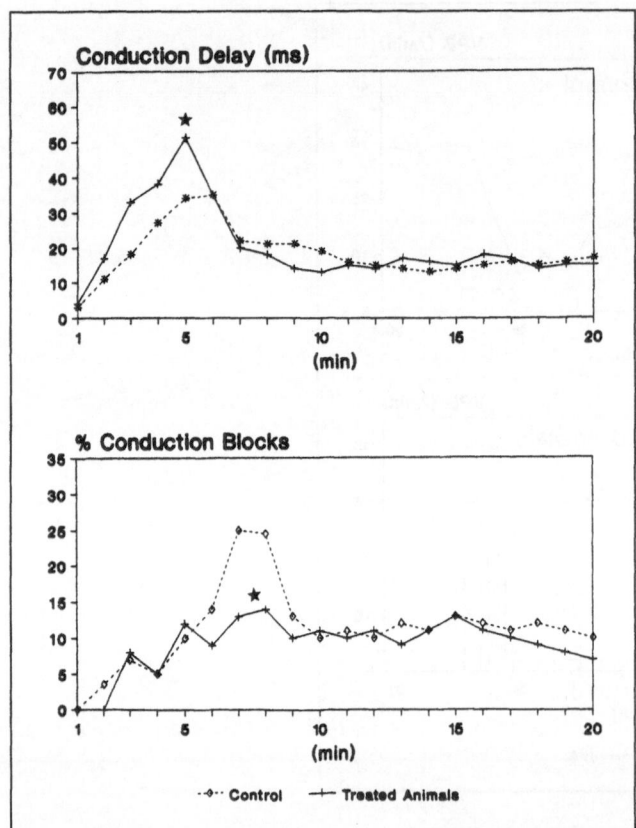

Fig. 2. Effect of combined infusion of epinephrine (1.5 µg/kg × min) and norepinephrine (0.2 µg/kg × min) on epicardial conduction delay and the incidence of complete conduction block in the ischemic zone after acute coronary artery occlusion as assessed by a 42 bipolar mapping electrode array during constant atrial pacing at a rate of 160 bpm. (*p < 0.05, compared to untreated animals). (From [23].)

strongly dose-dependent, as clearly demonstrated by Menken and coworkers [40]. In their study, animals treated orally with atenolol (10 mg for 5 days) displayed a concentration-dependent decrease in the number of episodes of ventricular fibrillation after coronary ligation. In the control group, all animals died of ventricular fibrillation. Animals given the last dose of atenolol 6 h before coronary occlusion showed a significant decrease in the incidence of ventricular fibrillation. However, the drug proved ineffective when the last dose was given 18 h before occlusion. Plasma concentrations were 0.82 and 0.15 µg/ml, respectively. The administration of inadequate doses, the application of β-blockers with significant partial agonist activity, and the use of high doses of β-blockers with additional electrophysiologic effects (for example, local anesthethic effects as in the case of propranolol [50]) may explain the lack of effect of β-blockade on the incidence of ischemia-related ventricular fibrillation reported in some studies [19, 32, 33, 43].

There is ample experimental evidence that, in contrast to β-blockers, class I agents do exert pronounced proarrhythmic effects during acute myocardial ischemia. In contrast to β-blockers, prophylactic administration of class I agents results in a significant increase in epicardial conduction delay in the ischemic area [19]. Additional β-blocking activity may significantly modify the effects of class I agents on ischemic ventricular fibrillation. Evidence for this is given

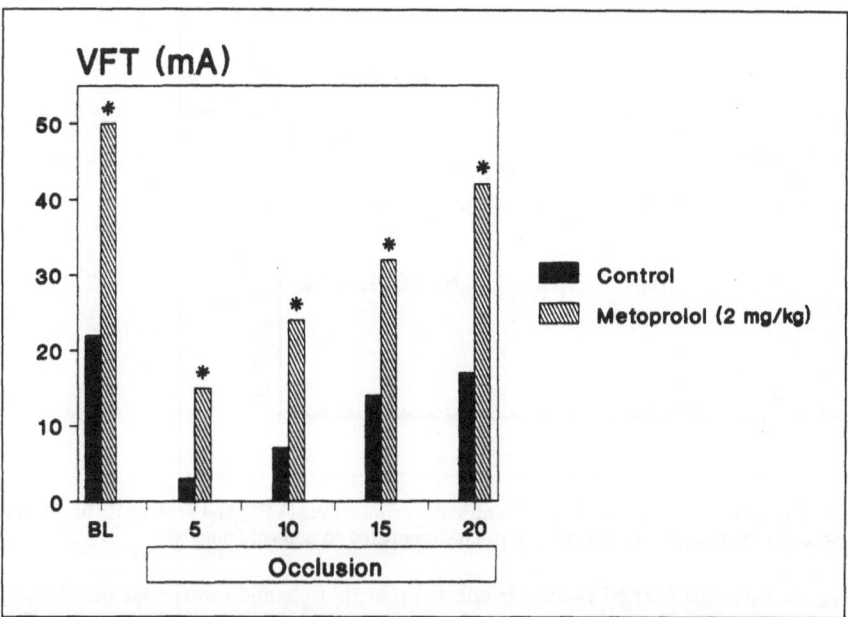

Fig. 3. Effects of metoprolol (2 mg/kg) on ventricular fibrillation threshold (VFT) before (baseline, BL) and after acute coronary artery occlusion. (*p < 0.05, compared to controls).

by a series of experiments previously performed in our laboratory in anesthetized dogs. Figure 6 shows the effects of diprafenone, a propafenone derivative, on the incidence of ventricular fibrillation after coronary occlusion [53]. The β-blocking activity of diprafenone can be considered as clinically relevant and is more marked compared to propafenone [20, 51]. At a dosage

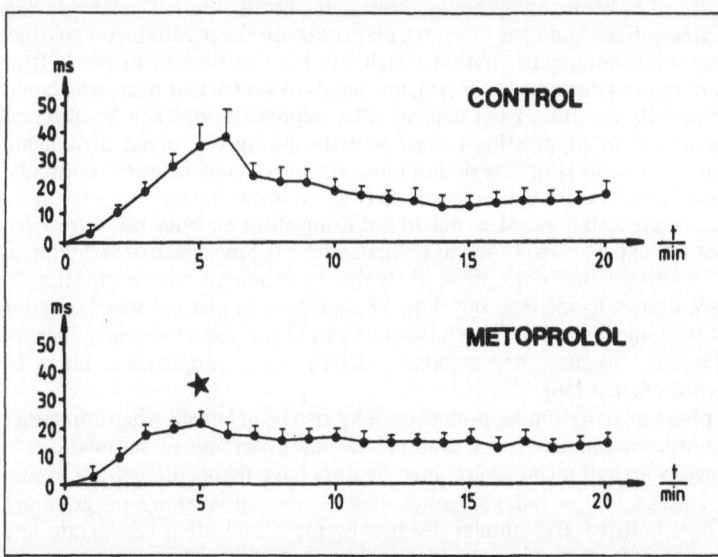

Fig. 4. Effects of metoprolol (2 mg/kg) on epicardial conduction delay in the ischemic area after acute coronary artery occlusion. Measurements of epicardial conduction times were performed at constant atrial pacing at a rate of 160 bpm. (*p < 0.05, compared to control animals).

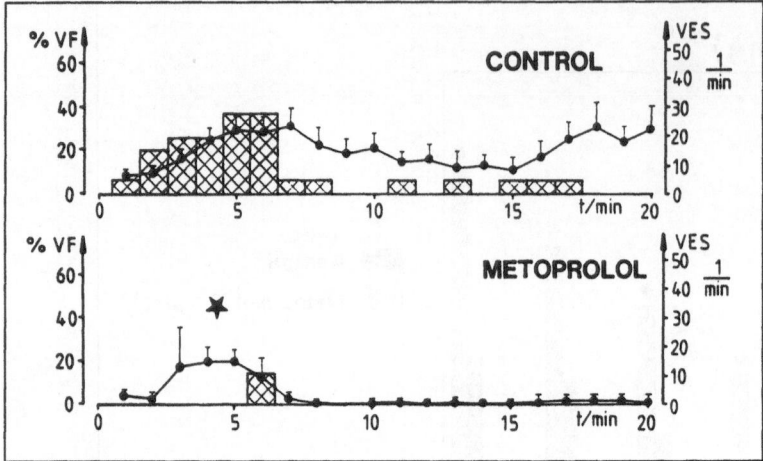

Fig. 5. Effects of metoprolol on the incidence of ventricular extrasystoles (VES) and ventricular fibrillation (VF) after acute coronary artery occlusion. (*p < 0.05, compared to control animals).

of 2.4 mg/kg, diprafenone exerted favorable effects on acute ischemic ventricular fibrillation, whereas the incidence of ventricular extrasystoles and ventricular tachycardia was not influenced. In contrast, when the dose was increased to 3.5 mg/kg, all treated animals fibrillated. It can be assumed that the local anesthetic action of the drug prevails at higher dosages, whereas at the lower dose the degeneration of ischemic ventricular tachycardias into ventricular fibrillation is partially prevented by virtue of its additional sympatholytic activity.

Delayed phase of ischemic arrhythmias: autonomic neural influences and effects of β-blockade

During the delayed phase of ischemic arrhythmias, ventricular fibrillation is a relatively rare event and ventricular extrasystoles and runs of ventricular beats are the predominant rhythms [19, 22]. Besides this slower spontaneous rhythm which has been attributed to result from abnormal automaticity of injured Purkinje fibers [16], fast bursts of ventricular beats which may be caused by triggered activity resulting from delayed after depolarizations can be observed [12]. β-Blockers are ineffective in suppressing ectopic activity during this phase of ischemic arrhythmias in anesthetized animals [19]; they do not affect the overall incidence of ventricular extrasystoles, salvoes and runs of ventricular tachycardia (Fig. 7). In contrast, in some instances the number of premature beats rather increases due to a slowing effect on sinus rate. However, some beneficial effects can be expected in conscious animals where the incidence of arrhythmias is increased compared to anesthetized dogs, most likely due to enhanced adrenergic stimulation. Catecholamines are known to increase the slope of spontaneous phase 4 depolarization in normal and diseased Purkinje fibers [9]. Catecholamines can also cause or augment delayed afterdepolarizations, possibly because they enhance calcium entry into cardiac fibers by increasing the slow inward current [56].

During the delayed phase of arrhythmias, β-blockers may also be of benefit when adrenergic activity is enhanced by other manoeuvers, for example, during programmed stimulation. In addition to the spontaneous arrhythmias, several investigators have mapped reentrant excitation in the epicardial surface of the ischemic zone that can be initiated by programmed stimulation [11, 16]. These induced arrhythmias are usually rapid and often degenerate into ventricular fibrillation. The induction of ventricular tachyarrhythmias during this phase of arrhythmias is facilitated by application of class I antiarrhythmic agents [26, 34]. Their proarrhythmic effects are associated with a significant rate-dependent increase in conduction times which is more pronounced in the ischemic than in the non-ischemic myocardium. A close

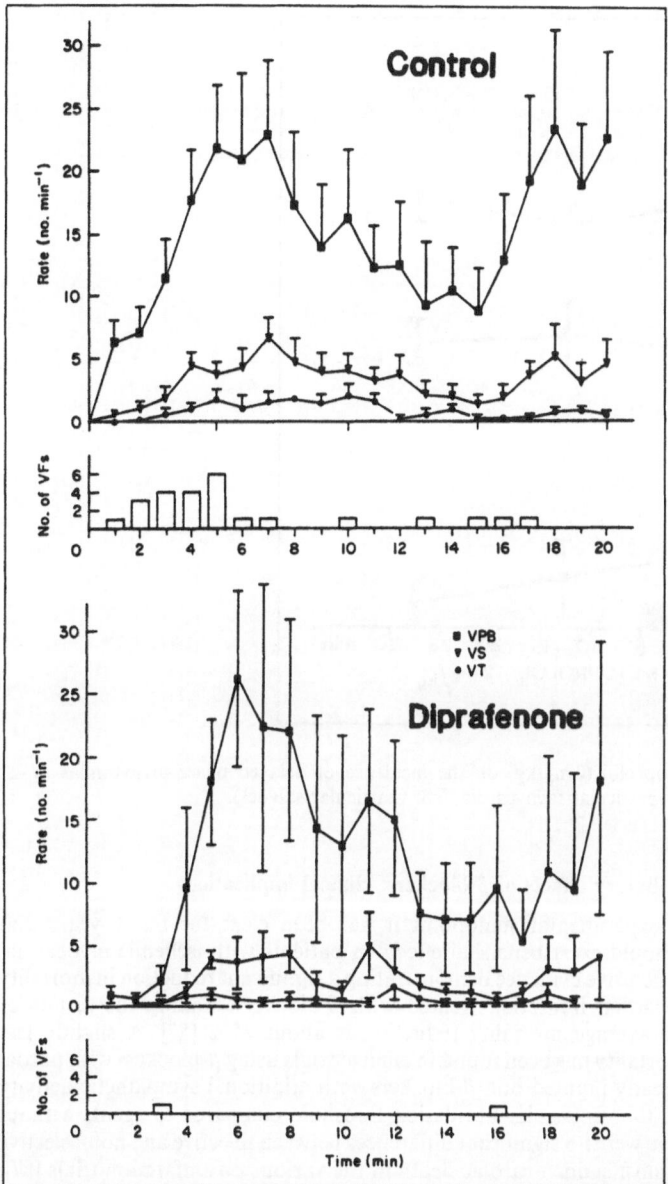

Fig. 6. Effects of diprafenone (2.4 mg/kg) on the incidence of ventricular premature beats (VPB), ventricular salvoes (VS), ventricular tachycardia (VT), and ventricular fibrillation (VF) after acute coronary artery occlusion. Diprafenone at a dosage of 3.5 mg/kg failed to exert antifibrillatory effects (not shown). (From [53]).

correlation between the extent of delay in activation in the epicardium induced by class I agents and the inducibility of ventricular tachyarrhythmias has been demonstrated by several investigators [26, 34]. β-Blockade prevents the induction of ventricular tachycardia in almost 50% of dogs [29]. In contrast to class I agents, β-blockers do not affect conduction in the ischemic zone [29].

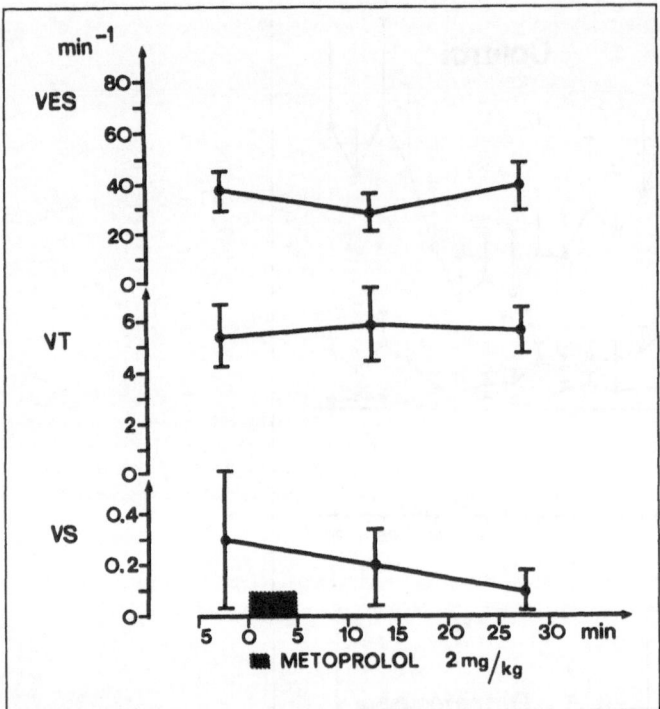

Fig. 7. Lack of effect of metoprolol (2 mg/kg) on the incidence of delayed phase arrhythmias (VES: ventricular extrasystoles, VT: ventricular tachycardia, VS: ventricular salvoes).

Antiarrhythmic and antifibrillatory effects of β-blockers: clinical implications

From the theoretical and experimental standpoint it has been clear for many years that β-adrenoceptor blockade should exert beneficial effects in patients with ischemic myocardial syndromes. There is now conclusive evidence demonstrating a significant reduction in mortality in survivors of acute myocardial infarction treated with β-blockers. Taking the results of various trials together, the average mortality reduction is about 25% [57]. A slightly less pronounced reduction in mortality has been found in clinical trials using β-blockers with partial agonist activity [57]. As already pointed out, β-blockers with additional sympathetic activity have less marked effects on the ventricular fibrillation threshold compared to agents without partial agonist activity. There were no significant differences between selective and non-selective β-blockers in protection against sudden cardiac death in the various postinfarction trials [57].

Although several studies have shown that β-blockade decreases the incidence of ventricular premature beats in patients with myocardial infarction [28, 46], others have demonstrated that patients without ventricular premature beats benefit as much from β-blockade as those with arrhythmias. In the Beta Blocker Heart Attack Trial (BHAT) the effect of propranolol on mortality was independent of the degree and type of ventricular ectopic activity [15]. Presently available data suggest that antifibrillatory effects, rather than direct antiarrhythmic effects, are at least one of the mechanisms by which β-blockers (when given prophylactically) may reduce mortality in patients with acute or recent myocardial infarction.

Recently, the role of β-blockers as antiarrhythmic or antifibrillatory agents in patients after myocardial infarction has gained new interest in the light of the results of the Cardiac Arrhythmia Suppression Trial (CAST) [7]. The major result of this trial was a significant increase in total and cardiac mortality in patients after myocardial infarction who presented with asymptomatic ventricular arrhythmias when treated with either flecainide or encainide, two class Ic antiarrhythmic agents. Several mechanisms may have been operative leading to this unfavorable outcome, as recently discussed in a Task Force Committee Report [52]. This may include selection factors leading to exclusion of ischemic mechanisms of sudden death, since a high percentage of patients underwent revascularisation or PTCA before inclusion in the study. This may have favored the inclusion of patients with a chronic arrhythmogenic substrate [4] that may be especially vulnerable to class I antiarrhythmic drugs. In contrast, the early β-blocker trials did not select patients this way, and therefore included more patients with (undetected) propensity to ischemia which would better respond to β-blockers. As already mentioned above, additional β-sympatholytic activity significantly modifies the effects of local anesthetic agents during experimental myocardial ischemia. To date, whether combined administration of β-blockers and class I agents in patients who have had myocardial infarction may improve prognosis, has not yet been shown by prospective clinical studies. However, although the exact mechanisms of the salutary effects of β-blockade have not been fully established, currently available evidence suggests that β-blockers play a role in the treatment of patients after myocardial infarction though the subgroups of patients who benefit cannot yet be identified. However, high-risk patients with documented malignant ventricular arrhythmias need a more rigid treatment; β-blockers are not the drugs of first choice [27, 41]. Treatment should be guided by electrophysiologic testing and individualized pharmacological or non-pharmacological therapeutic modalities [3–5].

References

1. Abendroth RR, Meesmann W, Stephan K, Schley G, Hübner H (1977) Wirkung des β-Sympatholytikums Atenolol auf die Arrhythmien, speziell Kammerflimmern und die Flimmerschwelle des Herzens beim akuten experimentellen Koronarverschluß. Z Kardiol 66:341–350
2. Anderson JL, Rodier HE, Green LS (1983) Comparative effects of beta blocking drugs on experimental ventricular fibrillation threshold. Am J Cardiol 51:1196–1202
3. Anderson JL (1990) Clinical implications of new studies in the treatment of benign, potentially malignant and malignant ventricular arrhythmias. Am J Cardiol 65 (suppl B):36B–42B
4. Breithardt G, Borggrefe M, Martinez-Rubio A, Budde T (1989) Pathophysiological mechanisms of ventricular tachyarrhythmias. Eur Heart J 10 (suppl E):9–18
5. Breithardt G, Borggrefe M, Haerten K (1985) Role of programmed ventricular stimulation and non-invasive recording of ventricular late potentials for the identification of patients at risk of ventricular tacharrhythmias after acute myocardial infarction. In: Zipes DP, Jalife J (eds) Cardiac Electrophysiology and Arrhythmias; Grune and Stratton, Orlando, 553–561
6. Breithardt G, Borggrefe M, Zipes DP (eds) (1988) Nonpharmacological Therapy of Tachyarrhythmias. Futura Publishing Company, Mount Kisco, New York
7. The Cardiac Arrhythmia Suppression Trial (CAST) Investigators (1989) Preliminary report: effect of encainide and flecainide on mortality in a randomized trial of arrhythmia suppression after myocardial infarction. N Engl J Med 321:406–412
8. Corr PB, Gillis RA (1975) Effect of autonomic neural influences on the cardiovascular changes induced by coronary occlusion. Am Heart J 89:766–774
9. Cranefield PF (1975) The Conduction of the Cardiac Impulse. Futura Publishing Company, Mount Kisco, New York
10. Ebert PA, Allgood RJ, Sabiston DC (1968) The antiarrhythmic effects of cardiac denervation. Ann Surg 168:728–735
11. El-Sherif N, Mehra R, Gough WB, Zeiler RH (1982) Ventricular activation of spontaneous ventricular rhythms in the canine one-day-old myocardial infarction. Circ Res 51:152–166
12. El-Sherif N, Gough WB, Zeiler RH, Mehra R (1983) Triggered ventricular rhythms in one-day-old myocardial infarction in dogs. Circ Res 52:566–573
13. Fearon RE (1967) Propranolol in the prevention of ventricular fibrillation due to experimental coronary artery occlusion. Am J Cardiol 20:222–228
14. Frame LH (1988) Ischemia and infarction: the evolving substrate for arrhythmias. Prog Cardiol 1:87–107
15. Friedman LM, Byington RP, Capone RJ, Furberg CD, Goldstein S, Lichstein E (1986) Effect of

propranolol in patients with myocardial infarction and ventricular arrhythmias. J Am Coll Cardiol 7:1–8

16. Friedman PL, Stewart JR, Wit AL (1973) Spontaneous and induced cardiac arrhythmias in subendocardial Purkinje fibers surviving extensive myocardial infarction in dogs. Circ Res 33:612–621
17. Gillis RA (1971) Role of the nervous system in the arrhythmias produced by coronary occlusion in the cat. Am Heart J 81:677–684
18. Gülker H, Thale J, Heuer H, Zurstegge KM, Bender F (1983) Divergierende Zeitverläufe der ventrikulären Vulnerabilität des Herzens für Kammertachykardien und Kammerflimmern im frühen Nekrosestadium des akuten experimentellen Herzinfarktes. Z Kardiol 72:711–717
19. Gülker H (1989) Herzrhythmusstörungen bei Myokardischämie und Myokardnekrose. Pathophysiologische Grundlagen, therapeutische Beeinflußbarkeit und Aspekte der Prävention. Springer, Berlin Heidelberg, New York
20. Gülker H, Thale J, Olbing B, Heuer H, Frenking B, Bender F (1985) Assessment of the antiarrhythmic profile of the new class-I agent diprafenone. Drug Res 35:1387–1393
21. Harris AS, Estandia A, Tillotson RF (1943) Ventricular ectopic rhythms and ventricular fibrillation following cardiac sympathectomy and coronary occlusion. Exp Med Surg 1:105–122
22. Harris AS (1950) Delayed development of ventricular ectopic rhythms following experimental coronary occlusion. Am Heart J 1:1318–1328
23. Haverkamp W, Thale J, Gülker H, Hindricks G, Mann B, Bender F (1986) Einfluß erhöhter Katecholamin-Stimulation auf ventrikuläre Arrhythmien und epikardiale Leitungsverzögerungen bei akuter Myokardischämie und Reperfusion. Z Kardiol 75 (Suppl 1):29 (abstr.)
24. Haverkamp W, Thale J, Gülker H, Hindricks G, Bender F (1987) Comparative investigations on the antiarrhythmic and electrophysiologic effects of various calcium antagonists (diltiazem, verapamil, gallopamil, nifedipine) following acute transient coronary artery occlusion and reperfusion. Eur Heart J 8 (suppl D):117–128
25. Haverkamp W, Thale J, Gülker H, Hindricks G, Bender F (1986) Antiarrhythmic and electrophysiologic effects of class I–IV antiarrhythmic agents in the ischemic canine heart. In: Santini M, Pistolese M, Alliegro A (eds) Progress in Clinical Pacing. TF Begliomini, Rome, pp 759–764
26. Haverkamp W, Hindricks G, Gülker H, Kottkamp H, Borggrefe M, Breithardt G (1990) Proarrhythmic effects of flecainide: possible mechanisms. New Trends Arrhythm IV:1065–1069
27. Haverkamp W, Hindricks G, Gülker H (1990) Antiarrhythmic properties of β-blockers. J Cardiovasc Pharmacol, in press
28. Herlitz J, Edvardsson N, Holmberg S, Ryden L, Waagstein F, Waldenström A, Hjalmarson A (1984) Göteborg metoprolol trial: effects on arrhythmias. Am J Cardiol 53 (suppl D):27D–31D
29. Hindricks G, Haverkamp W, Wiesmann H, Wiedhold D, Teutemacher H, Gülker H (1988) Effects of bisoprolol on ventricular tachyarrhythmias induced by acute myocardial infarction. Naunyn-Schmiedeberg's Arch Pharmacol 338 (suppl):R41 (abstr.)
30. Janse MJ, Kleber AG (1981) Electrophysiological changes and ventricular arrhythmias in the early phase of regional myocardial ischemia. Circ Res 49:1069–1079
31. Janse MJ, Schwartz PJ, Wilms-Schopman F, Peters JG, Durrer D (1985) Effects of unilateral stellate ganglion stimulation and ablation on electrophysiologic changes induced by acute myocardial ischemia in dogs. Circulation 72:585–595
32. Khan MI, Hamilton JT, Manning GW (1972) Protective effect of beta adrenoceptor blockade in experimental coronary occlusion in conscious dogs. Am J Cardiol 30:832–851
33. Knopf H, McDonald FM, Bischoff A, Hirche H, Addicks K (1988) Effect of propranolol on early postischemic arrhythmias and noradrenaline and potassium release of ischemic myocardium in anesthetized pigs. J Cardiovasc Pharmacol 12:41–47
34. Kou WH, Nelson SD, Lynch JJ, Montgomery DG, Lucchesi BR (1987) Effect of flecainide acetate on prevention of electrical induction of ventricular tachycardia and occurrence of ischemic ventricular fibrillation during the early postmyocardial infarction period: evaluation in a conscious canine model of sudden death. J Am Coll Cardiol 9:359–366
35. Krämer B, Gülker H, Stephan K, Meesmann W (1977) Influence of increasing doses of β-sympatholytic activity (ISA) on the ventricular fibrillation threshold (VFT) of the heart. Naunyn-Schmiedeberg's Arch Pharmacol 297 (suppl):290 (abstr.)
36. Lombardi F, Verrier RL, Lown B (1983) Relationship between sympathetic neural activity, coronary dynamics, and vulnerability to ventricular fibrillation during myocardial ischemia and reperfusion. Am Heart J 105:958–965
37. Malliani A, Schwartz PJ, Zanchetti A (1969) A sympathetic reflex elicited by experimental coronary occlusion. Am J Physiol 217:703–709

Ischemic tachyarrhythmias and β-blockade

38. Martin C, Meesmann W (1985) Antiarrhythmic effect of regional myocardial chemical sympathectomy in the early phase of coronary artery occlusion in dogs. J Cardiovasc Pharmacol 7 (suppl 5):S76–S80
39. Meesmann W, Gülker H, Krämer B, Stephan K (1976) Time course of changes in ventricular fibrillation threshold in myocardial infarction: characteristics of acute and slow occlusion with respect to the collateral vessels of the heart. Cardiovasc Res 10:466–473
40. Menken U, Wiegand V, Bucher P, Meesmann W (1979) Prophylaxis of ventricular fibrillation after acute experimental coronary occlusion by chronic beta-adrenoceptor blockade with atenolol. Cardiovasc Res 13:588–594
41. Morganroth J (1987) Antiarrhythmic effects of beta-adrenergic blocking agents in benign or potentially lethal ventricular arrhythmias. Am J Cardiol 60 (suppl D):10D–14D
42. Opie LH, Nathan D, Lubbe WF (1979) Biochemical aspects of arrhythmogenesis and ventricular fibrillation. Am J Cardiol 43:131–148
43. Pearle DL, Williford D, Gillis RA (1978) Superiority of practolol versus propranolol in protection against ventricular fibrillation induced by coronary occlusion. Am J Cardiol 42:960–964
44. Pentecost BL, Austen WG (1966) Beta adrenergic blockade in experimental myocardial infarction. Am Heart J 72:790–796
45. Raeder EA, Verrier RL, Lown B (1983) Intrinsic sympathomimetic activity and the effects of beta-adrenergic blocking drugs on vulnerability to ventricular fibrillation. J Am Coll Cardiol 1:1442–1446
46. Rehnquist N, Olsson G, Erhardt L, Ekman AM (1987) Metoprolol in acute myocardial infarction reduces ventricular arrhythmias both in the early stage and after the acute event. Int J Cardiol 15:301–308
47. Schwartz PJ, Vanoli E (1981) Cardiac arrhythmias elicited by interaction between acute myocardial ischemia and sympathetic hyperactivity: a new model for the study of antiarrhythmic drugs. J Cardiovasc Pharmacol 3:1251–1259
48. Schwartz PJ, Stone HL, Brown AM (1976) Effects of unilateral stellate ganglion blockade on the arrhythmias associated with coronary occlusion. Am Heart J 92:589–599
49. Sethi V, Haider B, Ahmed S, Oldewurtel HA, Regan RJ (1973) Influence of beta blockade and chemical sympathectomy on myocardial function and arrhythmias in acute ischemia. Cardiovasc Res 7:740–748
50. Shand DG (1976) The pharmacokinetics of propranolol: a review. Postgrad Med J 52:22–25
51. Sullivan ME, Reiser HJ (1989) Preclinical pharmacology of diprafenone. In: Bender F, Gülker H (eds) Neue Aspekte in der Therapie mit Antiarrhythmika. Präklinische und klinische Ergebnisse mit Diprafenon. Steinkopff Verlag, Darmstadt, pp 29–51
52. The Task Force of the Working Groups on Arrhythmias of the European Society of Cardiology (1990) CAST and beyond. Implications of the Cardiac Arrhythmia Suppression Trial. Circulation 81:1123–1127
53. Thale J, Gülker H, Hindricks G, Haverkamp W, Bender F (1987) Use of diprafenone, a new potent propafenone-derivative, in acute experimental myocardial ischaemia and infarction. Eur Heart J 8:107–115
54. Uchida Y, Murao S (1973) Excitation of afferent cardiac sympathetic nerve fibers during coronary occlusion. Am J Physiol 226:1094–1099
55. Vaughan Williams EN (1989) Adrenergic arrhythmogenicity. In: Vaughan Williams EN (ed) Antiarrhythmic Drugs. Springer, Berlin Heidelberg New York, pp 303–308
56. Wit AL, Cranefield PF (1977) Triggered and automatic activity in the canine coronary sinus. Circ Res 44:435–445
57. Yusuf S, Peto R, Lewis J, Collins R, Sleight P (1985) Beta blockade during and after myocardial infarction: an overview of the randomized trials. Prog Cardiovasc Dis 27:335–371

Authors' address:

Wilhelm Haverkamp, M.D., Medical Hospital of the Westfälische Wilhelms Universität, Dept. of Cardiology – Angiology, Albert-Schweitzer-Straße 33, 4400 Münster, FRG.

Sympathetic – Parasympathetic Interaction and Sudden Death

E. Vanoli and P. J. Schwartz

Unita' di Studio delle Aritmie, Centro di Fisiologia Clinica e Ipertensione, Istituto di Clinica Medica II, Universita' degli Studi di Milano, Milan, Italy

Summary

The effects of the autonomic nervous system on malignant arrhythmias, particularly in the setting of ischemic heart disease, have been widely investigated and described. Specifically, it has been shown that while sympathetic hyperactivity is arrhythmogenic, an increased vagal activity often exerts a beneficial effect. New insights on the relationship between autonomic activity and sudden cardiac death have been obtained in conscious dogs in which a healed myocardial infarction, acute myocardial ischemia, and exercise are combined. In this chronic animal model it was shown that myocardial infarction reduces baroreflex sensitivity and heart rate variability (markers of vagal reflex and tonic activity to the heart) and that a depressed baroreflex sensitivity or a reduced heart rate variability after myocardial infarction indicate an increased risk for ventricular fibrillation. The validity of these experimental observations was confirmed in clinical studies in patients with a myocardial infarction. The protective effect of vagal activity was further confirmed in two experimental studies in which muscarinic stimulation, both electrically and pharmacologically induced, was able to prevent ventricular fibrillation during acute myocardial ischemia. These observations have led to new research directions. At the experimental level, the effect of Gi proteins activity blockade by pertussis toxin on the cardiac response to vagal activation is currently evaluated in conscious dogs. At the clinical level, the prognostic value after myocardial infarction of baroreflex sensitivity and of heart rate variability will be tested in a large, multicenter, prospective study.

Introduction

The last 20 years have provided clear evidence of the tight relationship existing between the autonomic nervous system and sudden cardiac death [10, 28, 39, 45, 48]. Specifically, sympathetic hyperactivity favors the occurrence of ventricular tachyarrhythmias while vagal activation exerts a protective and antifibrillatory effect. On this basis it is logical to examine the possibility that the analysis of some aspects of the autonomic control of the heart might contribute to the identification of subjects at higher risk of sudden death.

In this chapter we will discuss some of the more recent experimental and clinical observations that opened new perspectives in the understanding of the pathophysiologic mechanisms underlying the onset of malignant arrhythmias and in the development of new approaches in risk stratification of patients with coronary artery disease.

Baroreceptive reflexes – experimental studies

The availability of an experimental model in conscious dogs for the study of sudden cardiac death allowed several observations relevant to the present discussion. This animal model, which has been already described in detail [6, 41], combines three elements that are highly relevant to the genesis of malignant arrhythmias: a healed myocardial infarction (MI), acute myocardial ischemia, and elevated sympathetic activity. In brief, 30 days after an anterior myocardial infarction, conscious dogs are exposed to a submaximal exercise stress test. When heart rate is approximately 210–220 beats/min, a 2 min occlusion of the circumflex coronary artery is performed by means of a hydraulic occluder, previously positioned around the vessel. After 1 min, the exercise ends but the occlusion is continued for an additional 1 min. The outcome of this test is highly reproducible over time in the same animal and allows to identify dogs *susceptible* to sudden death, i.e., those that develop ventricular fibrillation during acute myocardial ischemia, and dogs that are *resistant*, who survive the test.

A critical observation made in this model was that, in susceptible dogs, acute myocardial ischemia produced a marked reflex increase in heart rate while resistant dogs had a surprising reduction in heart rate despite the ongoing exercise [42]. It is important to note that in susceptible dogs, the reflex tachycardia was not due to a greater hemodynamic impairment during acute myocardial ischemia since mean blood pressure just before the occurrence of ventricular fibrillation was not different from that of resistant dogs at the same moment. However, bradycardia in resistant dogs was clearly dependent on a powerful vagal reflex as it could be prevented by atropine [52].

The observation that the two groups of dogs had overall opposite autonomic reflex responses to acute myocardial ischemia represented the rationale for further investigations in this model, which ultimately led to clinically relevant findings. Indeed, it became conceivable that analysis of cardiac reflexes at rest could help in the recognition of animals susceptible or resistant to sudden death. Were this hypothesis confirmed, it would have represented a new approach for risk stratification of patients with coronary artery disease.

Accordingly, before undergoing the exercise and ischemia protocol, the animals were tested by a method described in 1969 by the Oxford group [50] and already used in the clinical setting. The capability of vagal activity to reflexly increase following an adequate stimulus was evaluated by the analysis of baroreflex sensitivity (BRS), expressed by the regression line correlating R-R interval lengthening with blood pressure increases, induced by phenylephrine.

Baroreflex sensitivity was significantly lower in 192 dogs studied 30 days after a myocardial infarction than in a group of 86 dogs studied in control conditions (12.9 ± 7.6 vs 19.6 ± 7.9 ms/mm Hg, $p < 0.001$) [46]. The hypothesis that myocardial infarction could impair vagal reflexes was further validated by the internal control analysis in a subgroup of 55 dogs in which BRS was measured before and 30 days after MI. Baroreflex sensitivity was indeed markedly reduced by the MI (from 17.8 ± 6.6 to 13.5 ± 6.7 ms/mm Hg, $p < 0.001$). It is important to underline that a reduction of BRS occurred in 75% of the dogs while in 18% BRS did not change and in 7% increased. This observation, together with the wide distribution of BRS values, might explain the variety of results obtained in different clinical studies in which a relatively small number of patients was used.

Relevant to the problem of early recognition of individuals at high risk for sudden death was the finding that BRS was significantly lower in susceptible dogs than in resistant dogs. This observation was originally made in a small group of dogs [4] and subsequently confirmed in a much larger population [46]. In this latter study made in 192 dogs, BRS was 9.1 ± 6 ms/mm Hg in susceptible dogs and 17.7 ± 6.5 ms/mm Hg in resistant dogs, ($p < 0.001$) (Fig. 1), thus indicating that the capability of reflexly increasing vagal activity was significantly lower in dogs with an increased risk for sudden death.

A further analysis in the group of 55 dogs in which BRS was assessed before and after MI, indicated that the difference between susceptible and resistant dogs was not due to a diverse effect of the MI in the two groups of dogs, as the reductions in BRS after MI were similar. Indeed, already *before* MI, dogs that would have died in the recovery phase after the anterior MI or during the exercise and ischemia test had a lower BRS when compared to dogs that would have survived (Fig. 2).

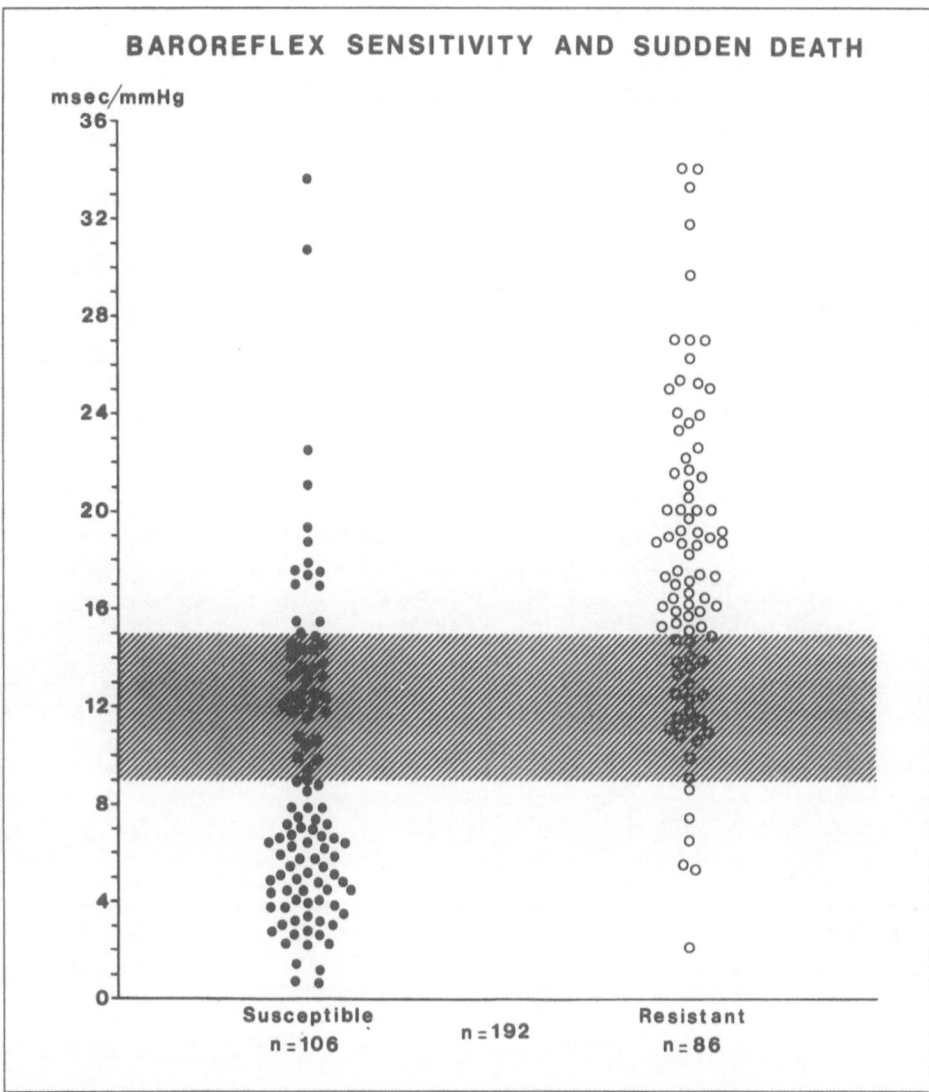

Fig. 1. Plot of baroreflex sensitivity in 192 dogs after infarction and its relation with susceptibility to sudden death. Dashed area is an arbitrary gray zone. At a baroreflex sensitivity of less than 9 ms/mm Hg, 91% of the dogs were susceptible to sudden death, whereas at greater than 15 ms/mm Hg, 80% of the dogs survived during the exercise and ischemia test. Note the large number of animals with baroreflex sensitivity of less than 9 ms/mm Hg. (From [46].)

Baroreceptive reflexes – clinical studies

The potential clinical relevance of the results obtained in the experimental studies described above represented a strong rationale to investigate whether these findings could effectively contribute to the identification of individuals at higher risk of sudden death after a myocardial infarction.

As a first step, it was evaluated if BRS was different in patients with a two- to three-week-old

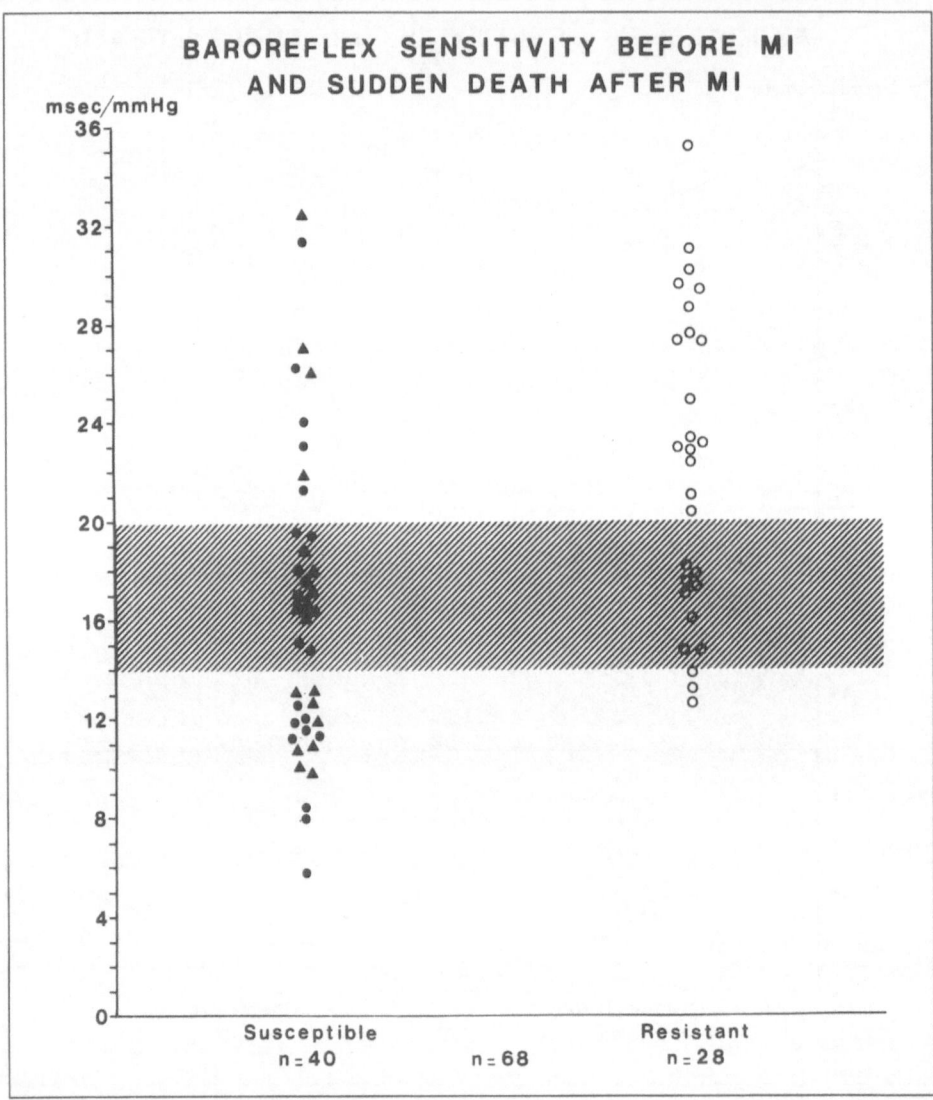

Fig. 2. Plot of baroreflex sensitivity before myocardial infarction (MI) in 68 animals and its relation with susceptibility to sudden death after MI. In this case, the arbitrary gray zone extends from 14 to 20 ms/mm Hg. ● = Animals susceptible to sudden death during the exercise and ischemia test. ▲ = Animals that died suddenly during the first 4 weeks after MI. ○ = Animals that survived during the exercise and ischemia test. Note how few animals have baroreflex sensitivity of less than 9 ms/mm Hg. (From [46].)

myocardial infarction when compared to age- and sex-matched normal subjects [47]. As in experimental studies, BRS was significantly lower in patients with a prior myocardial infarction than in normal subjects (8.2 ± 3.7 vs 12.3 ± 2.9 ms/mm Hg, $p = 0.0001$) (Fig. 3). The same study showed also that this alteration in cardiac reflexes was transient, as the depressed BRS in post-MI patients significantly increased within the first 6–12 months after the MI. The finding that cardiac vagal activity was often reduced early after a myocardial infarction was

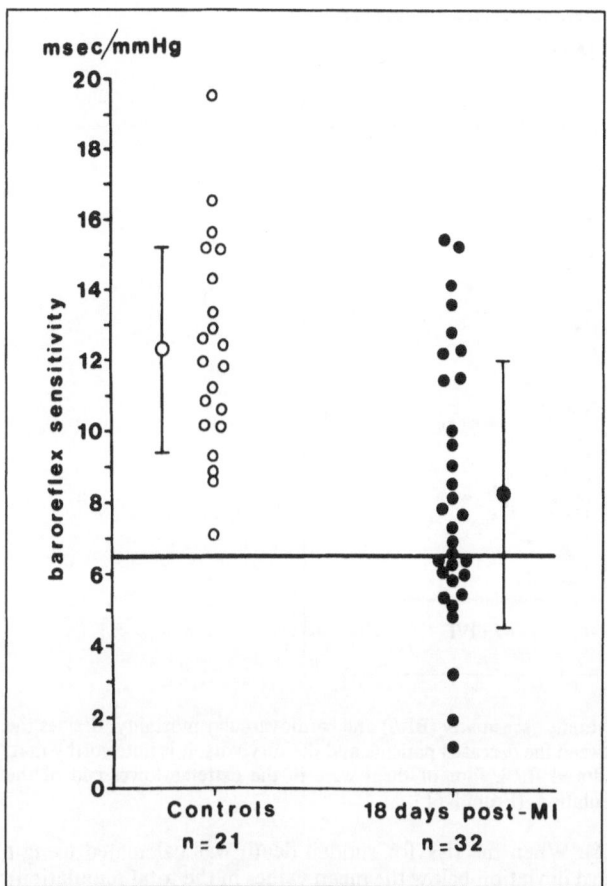

Fig. 3. Individual values of baroreflex sensitivity among 21 control subjects and 32 patients, 18 days after myocardial infarction (post-MI). The *horizontal line* at 6.5 ms/mm Hg represents 2 SD below the mean value for the control subjects. The mean values ± standard deviation are also indicated. (From [47].)

also confirmed by consecutive analyses of heart rate variability and of its spectral components during the first year after a myocardial infarction [27, 35]. Indeed, similarly to what happens with BRS, the high-frequency component of the power spectrum (an index of vagal activity directed to the sinus node) 2 weeks after a myocardial infarction was significantly smaller than in control subjects. It is important to note that in this latter study, the three patients who died within 3 months of follow-up all had a very low heart rate variability. Furthermore, even in this study, 6 to 12 months after MI a progressive recovery of cardiac vagal activity became evident as the high-frequency component showed a significant increase.

Thus, the two studies [27, 47] may indicate a correlation between the well known progressive reduction of risk for sudden death within 6 to 12 months after MI and the progressive increase in vagal activity to the heart.

As a consequence of the experimental observations just described, a prospective study was designed with the aim of evaluating if analysis of BRS could contribute to risk stratification of post-MI patients. Baroreflex sensitivity was evaluated in 78 patients with a 1 month-old myocardial infarction [26], who were then followed for a mean period of 2 years. During this time there were seven cardiovascular deaths, four of which were sudden. The analysis of individual values clearly indicated that BRS of patients who later died was significantly lower

309

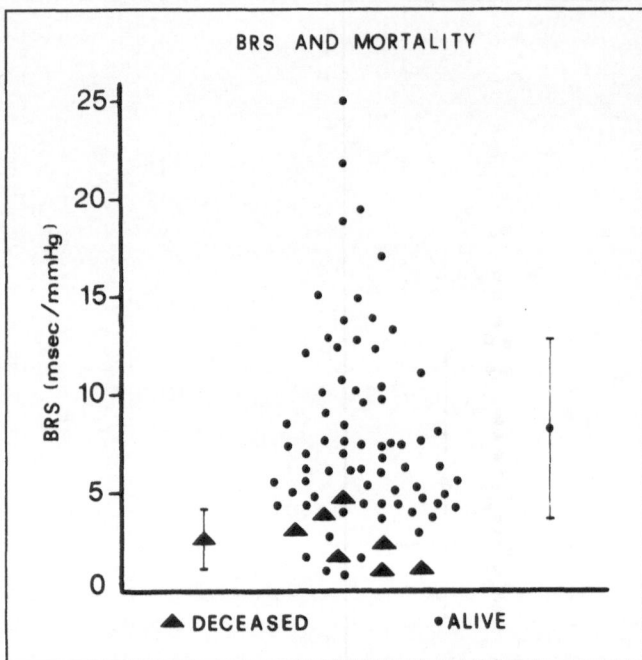

Fig. 4. Plot of relationship between baroreflex sensitivity (BRS) and cardiovascular mortality. Besides the clear difference in BRS (p = 0.004) between the deceased patients and the survivors, it is noteworthy that, while all patients who died had a reduced BRS, four of them were in the extreme lower end of the distribution of BRS for the entire population. (From [26].)

than the BRS of survivors (Fig. 4). When the risk for sudden death was calculated using a cut-off at 3 ms/mm Hg (one standard deviation below the mean values in the total population), it was evident that this risk increased from 2.9% for subjects with a BRS higher than 3 ms/mm Hg to 50% for subjects with a BRS lower than 3 ms/mm Hg. The relatively small number of patients enrolled in this study calls for caution in the interpretation of these results, but the striking difference observed in the two subgroups confirms the validity of the experimental observation that a depressed BRS after MI identifies a subgroup of individuals at higher risk for sudden death.

Hemodynamic evaluation of the patients enrolled in the study ruled out the possibility that a more depressed BRS in high risk patients simply reflected a poorer cardiac function. Indeed, linear regression analysis indicated a lack of correlation between BRS and ejection fraction (Fig. 5). However, as in other studies, patients with an ejection fraction lower than 40% had a higher risk of sudden death, but such a risk was significantly enhanced if depressed cardiac function was associated with a depressed BRS (Fig. 6). Therefore, it is possible that a depressed cardiac function represents a substrate on which a reduced vagal activity, probably combined with an increased sympathetic activity, can favor malignant arrhythmias.

Heart rate variability

Clinical and experimental studies

Beat to beat variability of heart rate strictly depends upon instantaneous variations in the balance of the two limbs of the autonomic nervous system and it is well recognized that measurements of heart rate variability represent a good marker of cardiac vagal tone [13, 21,

Sympathetic — parasympathetic interaction and sudden death

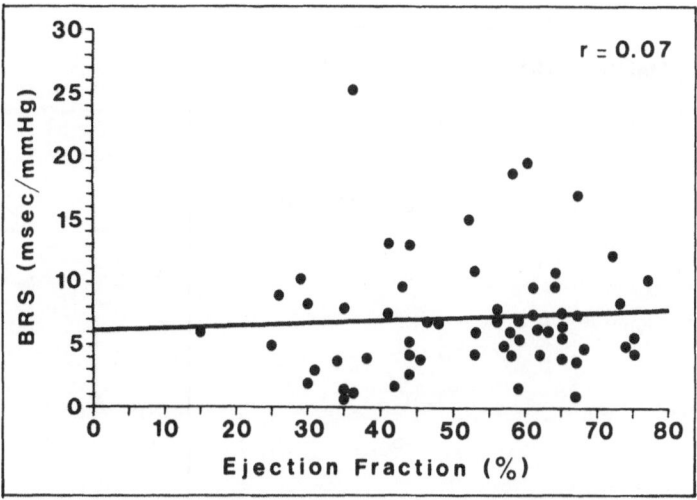

Fig. 5. Plot of relation between baroreflex sensitivity (BRS) and left ventricular ejection fraction at rest. (From [26].)

32]. The widespread availability of equipment for Holter monitoring and the simplicity of the test has stimulated, in the last few years, the development of new techniques of analysis of the 24 h ECG recordings [1, 9, 14, 22, 31]. Specifically, new criteria have been described to evaluate risk in patients with a prior myocardial infarction [2].

Among the several clinical studies in which measurements of heart rate variability were used in patients with coronary artery disease [24, 30, 34, 49], one is particularly important. In that study [34], which involved a large number of patients, heart rate variability (HRV, measured as the standard deviation of R-R intervals over 24 h) significantly correlated with mortality in patients with a prior myocardial infarction. Indeed, the relative risk of mortality was 5.3 times higher in patients with a HRV lower than 50 ms when compared to patients with a HRV higher than 100 ms.

Thus, the combined information from the two studies in which BRS or HRV were used strongly supports the hypothesis of an inverse relationship between cardiac vagal activity and cardiovascular mortality after a myocardial infarction.

The limitations inherent to clinical investigations still left open the question of whether the differences observed in low- and high-risk patients after MI were due to a different effect of the MI on the autonomic control of the heart, or if they reflected, at least in part, an intrinsic characteristic of individuals potentially *already* at higher risk *before* the occurrence of an ischemic event. This latter possibility was raised by the observation, presented above, that dogs at higher risk of dying suddenly after MI had, in control conditions, a BRS at the lower end of the distribution of the normal population.

Thirty days after a myocardial infarction, HRV was significantly lower in 25 dogs susceptible to sudden death than in a group of 25 resistant dogs (118 ± 9 ms vs 209 ± 13 ms, p < 0.001) [20]. The different HRV in susceptible and resistant dogs was independent from the difference observed in resting heart rate in the two groups of dogs; the analysis of the coefficient of variance (the ratio between mean of R-R intervals and its standard deviation) also provided a similar result.

The possibility of studying the same animals before and after MI allowed to examine, by an internal control analysis, the effect of the MI on the HRV in the two groups of dogs. This analysis showed that, at variance from BRS, in control conditions susceptible and resistant dogs had similar HRV, and that the MI produced a significant reduction of HRV in susceptible

311

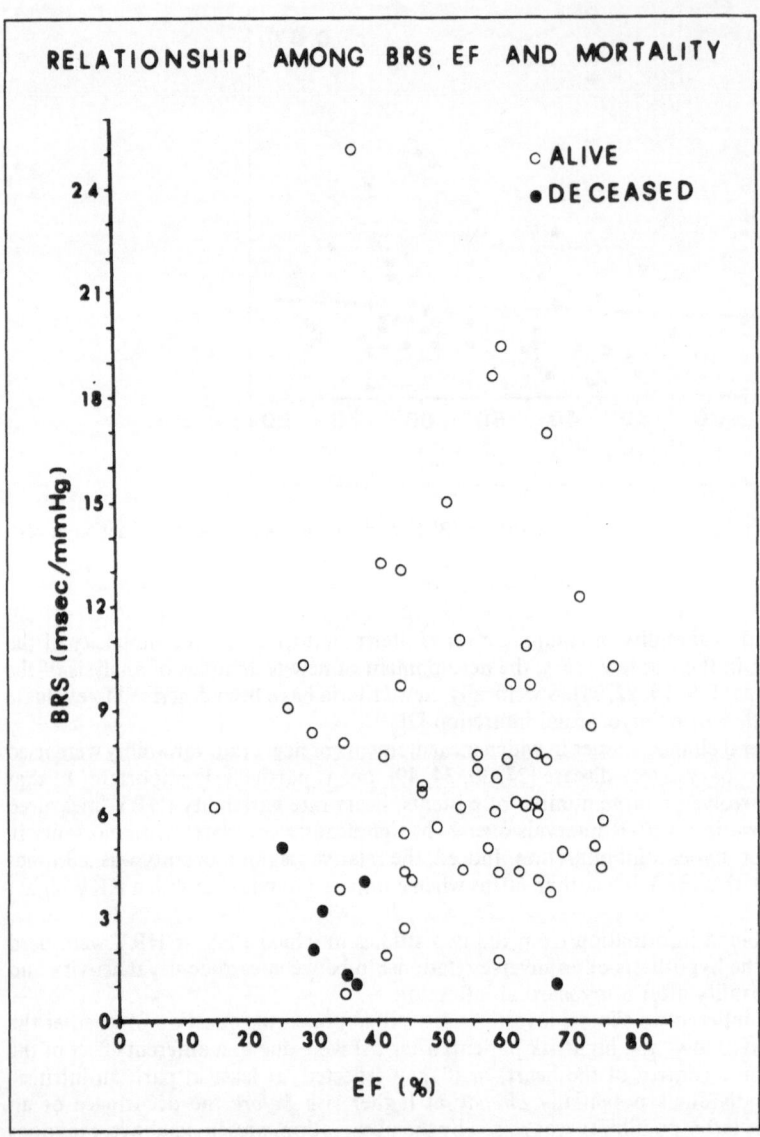

Fig. 6. Plot of relation between baroreflex sensitivity (BRS), left ventricular ejection fraction, and cardiovascular mortality. It is evident that, among patients with depressed left ventricular function, the prediction of mortality is enhanced by the analysis of BRS.

dogs (from 226 ± 30 ms before MI to 118 ± 9 ms, p < 0.001) but not in resistant dogs (from 233 ± 30 ms to 209 ± 13 ms, NS) (Fig. 7). Therefore, the analysis of HRV before MI could not predict the consequences of the subsequent MI. This finding represents an important difference in respect to what was observed with the analysis of BRS.

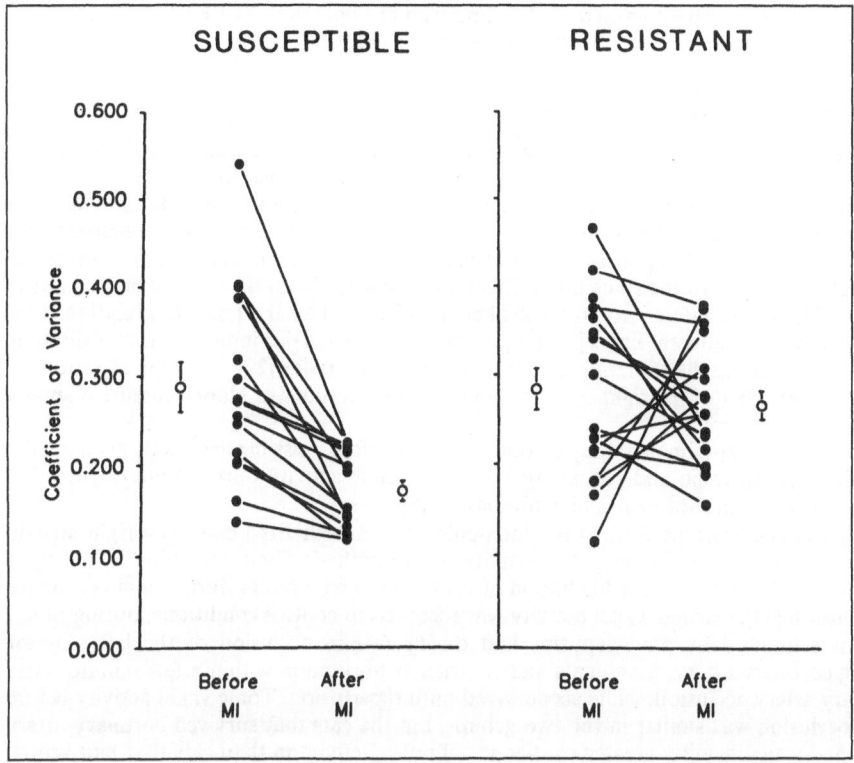

Fig. 7. Effect of myocardial infarction on heart rate variability in susceptible and resistant dogs. Heart rate variability is expressed by the coefficient of variance (ratio between standard deviation of the mean of R-R intervals and mean of R-R intervals). (From [20].)

Relationship between baroreflex sensitivity and heart rate variability

The evidence that the analysis of BRS and of HRV has a prognostic value after myocardial infarction and that it may provide different informations made it important to test whether or not HRV would predict BRS. Indeed, if BRS strictly correlated with HRV, analysis of reflex vagal activity would add little, if anything, to the analysis of tonic vagal activity.

To test the two opposite hypotheses that HRV does or does not predict BRS, the two variables were evaluated and compared in 32 patients with a prior myocardial infarction [3]. BRS was assessed by the phenylephrine method while four different measurements of HRV were performed, three in the time domain and one in the frequency domain. The variables in the time domain were the standard deviation of R-R intervals, the only parameter documented to predict risk for sudden death after MI; pRR50, the percentage of R-R intervals exceeding by at least 50 ms the preceding R-R intervals; and MSSD, the mean square root of the difference of successive R-R intervals. In the frequency domain the spectral analysis of the components of HRV was used.

The first evidence provided by this study was that three different methods of measuring HRV (pRR50, MSSD, and high-frequency peak of power spectrum analysis) highly correlated among each other with a correlation coefficient varying between 0.94 and 0.97, but they did not correlate as highly with the standard deviation of R-R intervals. Second, and probably more important was the finding that, although the different measures of HRV did significantly correlate with BRS, the degree of this correlation was rather weak, considering that correlation

313

coefficients varied between 0.57 and 0.63. It is important to note that the BRS correlated better with HRV during the night, when vagal tone is higher (Fig. 8).

Vagal activity to the heart

Baroreflex sensitivity and heart rate variability are only "markers" of cardiac vagal activity. As a consequence, the neural mechanisms involved in the increased risk for sudden death after myocardial infarction can only be indirectly speculated, but cannot be directly studied. Experimental preparations in which neural activity to the heart can be directly recorded and measured allow to properly approach this problem. In these experimental preparations vagal activity mainly directed to the sinus node [25] can be recorded from the cardiac branch of the right vagus. Efferent vagal activity in these fibers is influenced by the reflex modulation from vagal and sympathetic afferent fibers [37] (Fig. 9) and receives a major input by the carotid sinus nerve, thus representing the efferent part of the baroreceptive reflex [29]. The typical pattern of activity of a single fiber of the right cardiac vagus during variations of blood pressure is shown in Fig. 10.

By means of this experimental preparation it was, therefore, possible to directly evaluate the relationship between tonic and reflex vagal activity, reflex activity during acute myocardial ischemia, and occurrence of ventricular fibrillation [8].

The experiments were performed in alpha-chloralose-anesthetized cats. The right stellate ganglion was removed to eliminate the activity of sympathetic fibers travelling in the right cardiac vagus and to increase the likelihood of ventricular arrhythmias during acute coronary artery occlusion [38]. Cardiac vagal activity was recorded in control conditions, during blood pressure rises induced by phenylephrine and during 60 min occlusion of the left anterior descending coronary artery. Seven cats had ventricular fibrillation within a few minutes after the coronary artery occlusion, while six survived until reperfusion. Tonic vagal activity before coronary occlusion was similar in the two groups, but the cats that survived coronary artery occlusion had a significantly greater cardiac vagal reflex activation than cats that had ventricular fibrillation, both during blood pressure rises and in the first few minutes of acute myocardial ischemia.

These data are in agreement with the observation made in conscious animals by means of "markers" of vagal activity. Indeed, animals with a higher reflex vagal activity survived an episode of acute myocardial ischemia; such a response during coronary artery occlusion could be predicted by the analysis, in control conditions, of cardiac reflex vagal activity, but not of tonic vagal activity. "Tonic" vagal activity, as defined here, is the resting activity present in anesthetized cats, which can, only with caution, be extrapolated to the conscious state.

In these experiments the effect of sympathetic afferent traffic on vagal efferent activity was also investigated. In four cats tonic and reflex vagal activity during a blood pressure rise was significantly higher when sympathetic afferent traffic from the heart was abolished by the addition of left stellectomy. This observation correlates with the decrease in vagal efferent activity following cardiac sympathetic afferent stimulation [37] (Fig. 11) and with the increase in heart rate after dorsal root section [18]. It also supports the hypothesis that a reduction in BRS after MI may depend upon an increased afferent sympathetic activity, possibly originating from strech receptors located in the dyskinetic areas surrounding the scar [46].

Electrical and pharmacological vagal stimulation

If a reduced vagal activity is associated with an increased risk for ventricular fibrillation during acute myocardial ischemia, it is rational to test if interventions able to increase this activity in those individuals in which it is depressed may affect their susceptibility to sudden death. Exercise training is a physiological means to increase vagal activity [36, 51] and, in the animal model for sudden death described above, is able to produce an increase in BRS, when depressed, and to favorably affect the electrical instability of the ischemic heart [5].

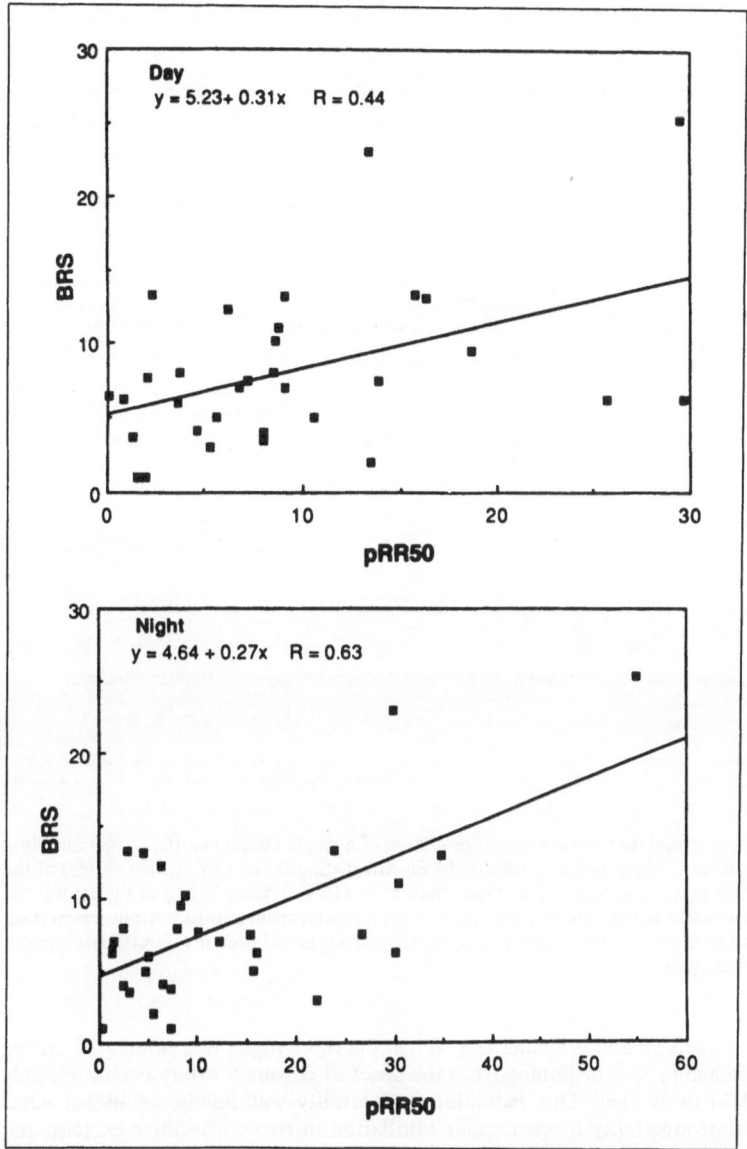

Fig. 8. Correlation between baroreflex sensitivity (BRS) in ms/mm Hg and the percent of successive normal RR-intervals differing > 50 ms (pRR50) plotted separately for day and night. (From [3].)

The antiarrhythmic effect during acute myocardial ischemia of increased vagal activity, either pharmacologically or electrically induced, was explored in two experimental preparations.

In the first study the feasibility of vagal stimulation and its potential antifibrillatory effect during coronary artery occlusion was evaluated in conscious dogs [11]. By using a chronic implantable electrode, the right cervical vagus was stimulated during acute myocardial ischemia in 26 conscious dogs. All dogs had already been identified to be susceptible to ventricular

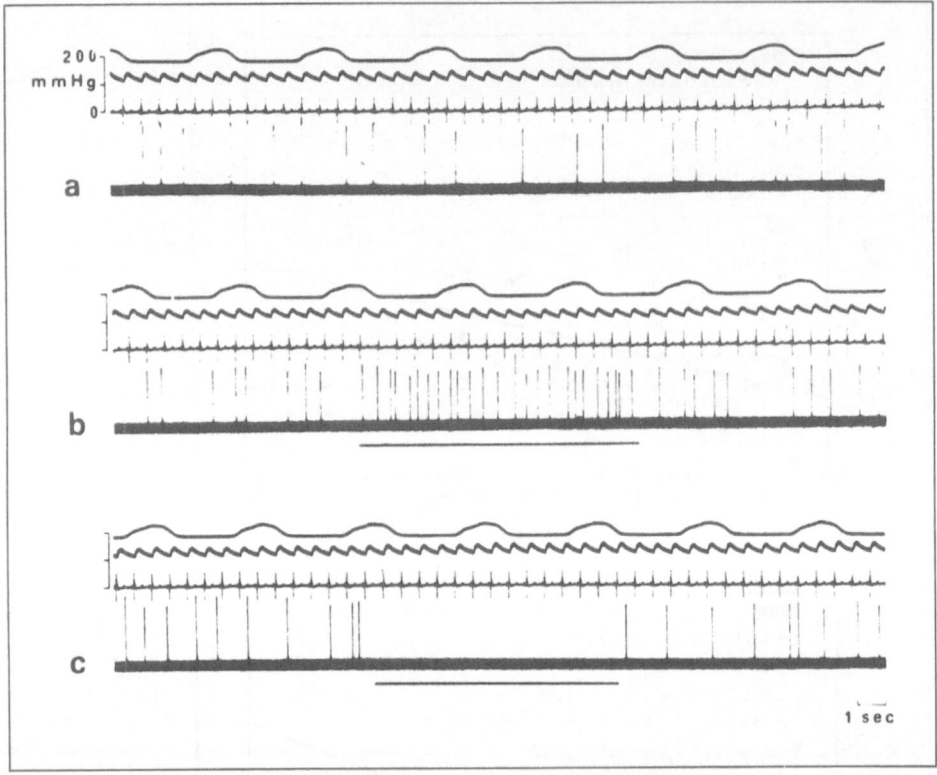

Fig. 9. Effects of electrical stimulation on the neural discharge of a single efferent cardiac vagal fiber in a decerebrate, anesthetized cat. a) Spontaneous activity; b) Electrical stimulation (5 V, 1.5 ms, 30 Hz) of the cut central end of the left cervical vagus; c) Electrical stimulation (10 V, 1.5 ms, 30 Hz) of the central cut end of the left inferior cardiac nerve. The tracings in each section are, from top to bottom: respiration (positive-pressure inflation is an upward deflection), systemic arterial blood pressure, electrocardiogram, and neural activity. (From [37].)

fibrillation during the exercise and ischemia test. When the right vagus was stimulated during a new exercise and ischemia test, beginning from the onset of coronary artery occlusion, only three of the 26 (12%) dogs died. This reduction in mortality was highly significant when compared with the reproducibility of ventricular fibrillation in two consecutive exercise and ischemia tests in a group of 24 dogs that served as control, where ventricular fibrillation occurred in all dogs in the first test and in 22/24 (92%) dogs in the second test. In these experiments the antifibrillatory effect of vagal stimulation was only in part dependent upon the attendant reduction of heart rate. Indeed, when heart rate during vagal stimulation was kept at the same level as in control conditions by atrial pacing, five of the nine (55%) dogs remained protected.

A potential antifibrillatory effect of vagal stimulation had already been proposed in previous experimental observations made in anesthetized animal models [23, 53]. Nevertheless, this information remained limited to experimental pathophysiology without any obvious prospective of future application. The demonstration that vagal stimulation is feasible in conscious individuals and has a powerful antifibrillatory effect in a clinically relevant setting may open a new area of research with a potential clinical relevance.

316

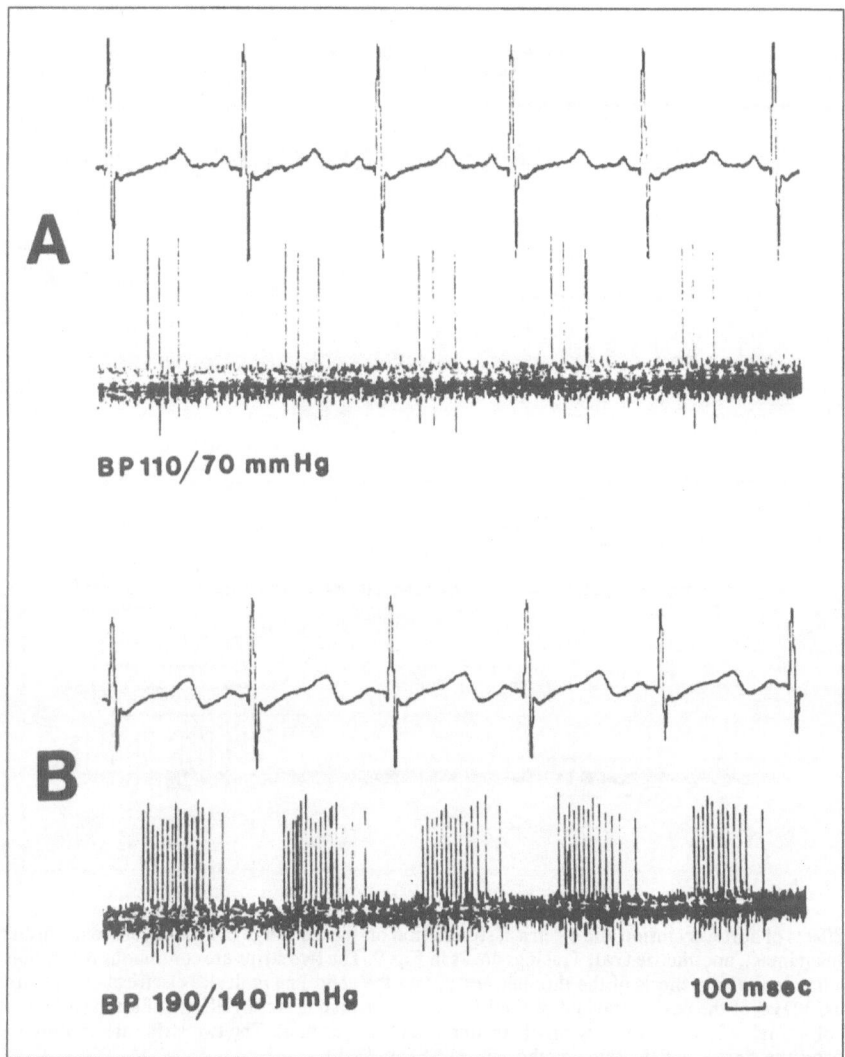

Fig. 10. Activity of a single cardiac vagal fiber in an alpha-chloralose-anesthetized cat. In each panel the upper trace shows the electrocardiogram, the lower trace shows the electroneurogram at two different blood pressure levels. A) At the resting level of blood pressure a pulse-synchronous activity with three impulses in each burst was present. The calculated delay from the beginning of the QRS was approximately 80 ms. B) When blood pressure was raised, the activity of the single fiber increased; the activity was still synchronous with the cardiac cycle, but it started earlier (60 ms) with a higher number of impulses in each burst.

Pharmacologic muscarinic stimulation is an alternative to the electrical stimulation of the vagus and was tested in anesthetized cats in which ventricular tachyarrhythmias can be reproducibly induced in seven to eight consecutive trials [40]. In this animal model ventricular arrhythmias (mainly ventricular tachycardia and fibrillation) are induced by the combination of a 2 min occlusion of the left anterior descending coronary artery and a 30 s stimulation of the left stellate ganglion. Several antiarrhythmic drugs were evaluated over the years in this

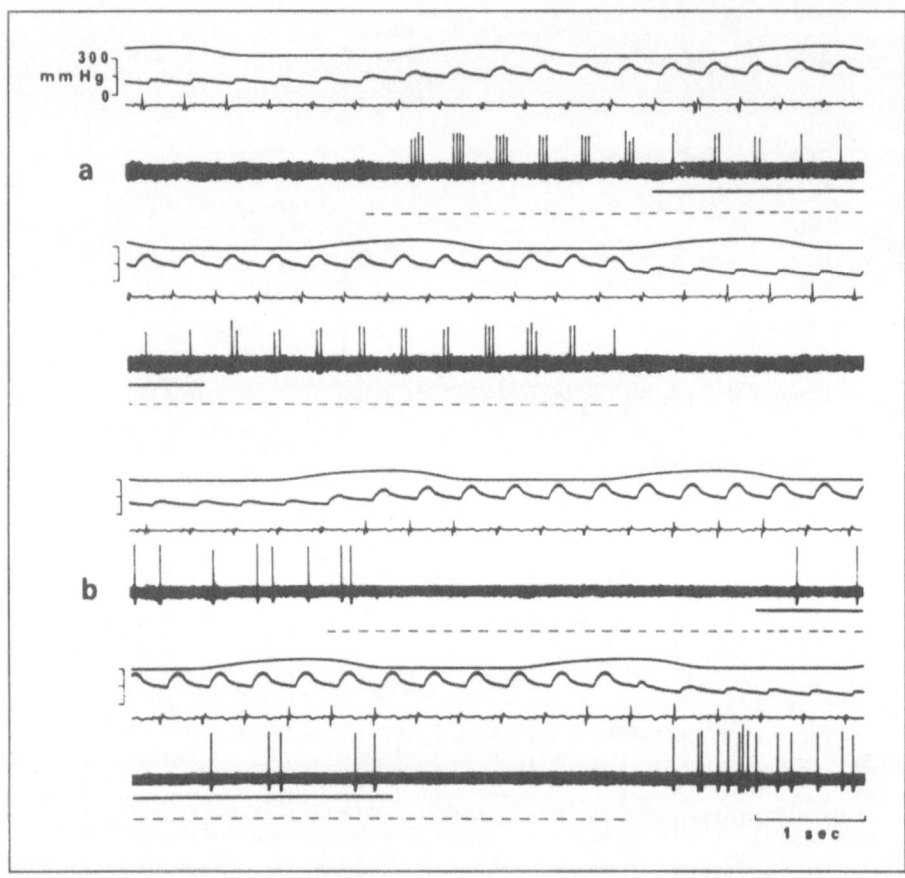

Fig. 11. a) Effects of aortic occlusion and electrical stimulation on the discharge of a single, efferent, cardiac vagal fiber in an intact, anesthetized cat. Tracings are as in Fig. 9. The two strips are continuous recordings. The broken line indicates stenosis of the thoracic aorta, and the solid line indicates electrical stimulation (10 V, 1.5 ms, 30 Hz) of the cut central end of the left inferior cardiac nerve; b) Effects of same stimuli on the activity of a single efferent cardiac sympathetic fiber. Same cat as in a). The two strips are continuous recordings, and the signals are the same as those in a). (From [37].)

model [33, 43, 44]. β-Blocking agents (propranolol), calcium entry blockers (verapamil, diltiazem, nifedipine), and amiodarone exerted a powerful protective effect, while class-I antiarrhythmic drugs (lidocaine, mexiletine, and propafenon) were ineffective. In this experimental model the muscarinic agonist oxotremorine has been evaluated [12]. In 13 cats oxotremorine provided a remarkable antiarrhythmic effect in all animals tested. Specifically, ventricular tachycardia or fibrillation occurred in 12 cats in the control trials and in only two cats in the trials performed after oxotremorine was given. The protective effect of muscarinic activation in these experiments was mostly independent from a reduction in heart rate. Indeed, when heart rate was kept at the same levels observed in the control trials by atrial pacing, eight of the 11 (73%) animals tested remained free of malignant arrhythmias.

The evidence of the powerful antiarrhythmic effect of muscarinic stimulation represents a new pharmacologic approach to antagonize the detrimental effects of sympathetic hyperactivity, particularly during acute myocardial ischemia.

New experimental perspectives

In the last few years, the possibility of modulating the autonomic control of the heart by acting on the primary intracellular mediators of membrane receptors has generated a growing interest. The activity of these mediators, the so-called G proteins [17], is mainly targeted to the modulation of adenylcyclase activity. Bacterial toxins have been used to study the mechanisms of actions of these proteins. Specifically, it has been shown that cholera toxin blocks the activity of stimulatory G proteins (Gs), while pertussis toxin blocks the activity of inhibitory G proteins (Gi) [7, 15].

The knowledge about G proteins is, so far, limited by the fact that the effects of these proteins have always been studied in vitro and the potential physiological effects of G proteins have been suggested, but never proven.

Recently, it has been shown that pertussis toxin administered in vivo in conscious dogs is able to induce a modification of the intracellular activity of the Gi proteins, as it does in in vitro experiments [16]. On the basis of this observation it was tested whether these intracellular modifications would have a physiologic effect in the conscious animal.

To accomplish this goal, BRS and HRV were measured in eight conscious dogs in control conditions and 72 h after the intravenous administration of $30 \mu g/kg$ of pertussis toxin [19]. Pertussis toxin produced a significant reduction of the two parameters tested in all animals studied. BRS decreased from 18.7 ± 2.6 to 6.8 ± 1.4 ms/mm Hg and HRV from 176 ± 17 to 61 ± 7 ms. These changes in the vagal control of the heart occurred together with changes in Gi activity, as proven by in vitro assay of tissue samples from the ventricles of the dogs treated with pertussis toxin.

The possibility to modify the end organ response to neural activity by acting at the subreceptorial level indicates that G proteins may represent the site of action of new pharmacologic interventions able to selectively modulate autonomic control of the heart.

New clinical strategies

The hypothesis that BRS and HRV may be independent markers of risk for sudden death, and that they may contribute in an additive manner to an early detection of high-risk patients, without being redundant, needs to be tested in a large follow-up study.

For these reasons the multicenter study ATRAMI (Autonomic Tone and Reflexes After Myocardial Infarction) has been designed. In this study, 1200 patients with a prior myocardial infarction will be enrolled from 30 centers in Europe and in USA. In all patients BRS, HRV, and other parameters commonly used for risk stratification after MI will be measured. BRS will be evaluated by the phenylephrine method, while HRV will be measured from 24 h Holter recording by analysis in the time and in the frequency domain, as already described.

The realization of ATRAMI, independently from the final results, represents a significant clinical evolution of new concepts developed in experimental research.

References

1. Berger RD, Akselrod S, Gordon D, Cohen RJ (1986) An efficient algorithm for spectral analysis of heart rate variability. IEEE Trans Biomed Eng 9:900–904
2. Bigger JT Jr, Kleiger RE, Fleiss JL, Rolnitzky LM, Steinman RC, Miller JP and The Multicenter Post-Infarction Research Group (1988) Components of heart rate variability measured during healing of acute myocardial infarction. Am J Cardiol 61:208–215
3. Bigger JT Jr., La Rovere MT, Steinman RC, Fleiss JL, Rottman JN, Rolnitzky LM, Schwartz PJ (1989) Comparison of baroreflex sensitivity and heart period variability after myocardial infarction. J Am Coll Cardiol 14:1511–1518
4. Billman GE, Schwartz PJ, Stone HL (1982) Baroreceptor reflex control of heart rate: A predictor of sudden cardiac death. Circulation 66:874–880
5. Billman GE, Schwartz PJ, Stone HL (1984) The effects of daily exercise on susceptibility to sudden cardiac death. Circulation 69:1182–1189

6. Billman GE, Schwartz PJ, Gagnol JP, Stone HL (1985) The cardiac response to submaximal exercise in dogs susceptible to sudden cardiac death. J Appl Physiol 59:890–897
7. Brown AM, Birnbaumer L (1988) Direct G protein gating of ion channels. Am J Physiol 254:H401–H410
8. Cerati D, Schwartz PJ (1989) Vagal reflexes and survival during acute myocardial ischemia in cats. Circulation 80 (Suppl II):522 (abstr.)
9. Cohen J (1969) Statistical power analysis for behavioral sciences. New York, Academic Press 89
10. Corr PB, Yamada KA, Witkowski FX (1986) Mechanisms controlling cardiac autonomic function and their relation to arrhythmogenesis. In: Fozzard HA, Haber E, Jennings RB, Katz AM, Morgan HE (Eds.) The Heart and Cardiovascular System. Vol. II, Raven Press, New York, pp 1343–1403
11. De Ferrari GM, Vanoli E, Stramba-Badiale M, Foreman RD, Schwartz PJ (1987) Vagal stimulation and sudden death in conscious dogs with a healed myocardial infarction. Circulation 76(suppl IV):107 (abstr.)
12. De Ferrari GM, Vanoli E, Tommasini G, Grossoni M, Uckmar G, Patrono C, Schwartz PJ (1989) Antiarrhythmic effect of muscarinic agonists during acute myocardial ischemia. Circulation 80 (Suppl IV):202 (abstr.)
13. Eckberg DL (1983) Human sinus arrhythmia as an index of vagal cardiac outflow. J Appl Physiol 54:961–966
14. Ewing DJ, Neilson JMM, Travis P (1984) New method for assessing cardiac parasympathetic activity using 24-hour electrocardiograms. Br Heart J 52:396–402
15. Fleming JW, Watanabe AM (1986) Biochemical mechanisms of parasympathetic regulation of cardiac function. In: Fozzard HA, Haber E, Jennings RB, Katz AM, Morgan HE (Eds.) The Heart and Cardiovascular System, Raven Press, New York, pp 1679–1688
16. Fleming JW, Hodges TD, Watanabe AM (1988) Pertussis toxin-treated dog: A whole animal model of impaired inhibitory regulation of adenylate cyclase. Circ Res 62:992–1000
17. Gilman AG (1984) G proteins and dual control of adenylate cyclase. Cell 36:577–579
18. Gnecchi Ruscone T, Lombardi F, Malfatto G, Malliani A (1987) Attenuation of baroreceptive mechanisms by cardiovascular sympathetic afferent fibers. Am J Physiol 253:H787–H791
19. Hull SS, Vanoli E, Adamson PB, Green FJ, De Ferrari GM, Foreman RD, Schwartz PJ, Watanabe AM (1989) Pertussis toxin alters vagal control of heart rate in conscious dogs. Circulation 80 (Suppl II):12 (abstr.)
20. Hull SS, Evans A, Vanoli E, Adamson PB, Albert DE, Foreman RD, Stramba-Badiale M, Schwartz PJ (1990) Heart rate variability before and after myocardial infarction in conscious dogs at high and low risk of sudden death. J Am Coll Cardiol 16: 978–985
21. Katona PG, Jih F (1975) Respiratory sinus arrhythmia: Noninvasive measure of parasympathetic cardiac control. J Appl Physiol 39:801–805
22. Kay SM, Marple SL (1981) Spectrum analysis: A modern perspective. Proc IEEE 69:1380–1419
23. Kent KM, Smith ER, Redwood DR, Epstein SE (1973) Electrical stability of acutely ischemic myocardium: Influence of heart rate and vagal stimulation. Circulation 47:291–298
24. Kleiger RE, Miller JP, Bigger JT Jr, Moss AJ, and The Multicenter Post-Infarction Research Group (1987) Decreased heart rate variability and its association with increased mortality after acute myocardial infarction. Am J Cardiol 59:256–262
25. Kunze DL (1972) Reflex discharge patterns of cardiac vagal efferent fibres. J Physiol 222:1–15
26. La Rovere MT, Specchia G, Mortara A, Schwartz PJ (1988) Baroreflex sensitivity, clinical correlates and cardiovascular mortality among patients with a first myocardial infarction: A prospective study. Circulation 78:816–824
27. Lombardi F, Sandrone G, Pernpruner S, Sala R, Garimoldi M, Cerutti S, Baselli G, Pagani M, Malliani A (1987) Heart rate variability as an index of sympathovagal interaction after acute myocardial infarction. Am J Cardiol 60:1239–1245
28. Lown B (1979) Sudden cardiac death: The major challenge confronting contemporary cardiology. Am J Cardiol 43:313–320
29. Mancia G, Mark AL (1983) Arterial baroreflexes in humans. In: Shephard JL, Abboud FM (Eds.) Handbook of Physiology: The Cardiovascular System, Vol. III. Bethesda, Md, Am Physiol Soc, pp 755–793
30. Myers GA, Martin GJ, Magid NM, Barnett PS, Schaad JW, Weiss JS, Lesch M, Singer DH (1986) Power spectral analysis of heart rate variability in sudden cardiac death: Comparison to other methods. IEEE Trans Biomed Eng 33:1149–1156
31. Pagani M, Lombardi F, Guzzetti S, Rimoldi O, Furlan R, Pizzinelli P, Sandrone G, Malfatto G, Dell'Orto S, Piccaluga E, Turiel M, Baselli G, Cerutti S, Malliani A (1986) Power spectral analysis of heart rate and arterial pressure variabilities as a marker of sympatho-vagal interaction in man and conscious dog. Circ Res 59:178–193

32. Pomeranz B, Macaulay RJB, Caudill MA, Kutz I, Adam D, Gordon D, Kilborn KM, Barger AC, Shannon DC, Cohen RJ, Benson H (1985) Assessment of autonomic function in humans by heart rate spectral analysis. Am J Physiol 248:H151–H153

33. Priori SG, Zuanetti G, Schwartz PJ (1988) Ventricular fibrillation induced by the interaction between acute myocardial ischemia and sympathetic hyperactivity: Effect of nifedipine. Am Heart J 116:37–43

34. Rich MW, Saini JS, Kleiger RE, Carney RM, teVelde A, Freedland KE (1988) Correlation of heart rate variability with clinical and angiographic variables and late mortality after coronary angiography. Am J Cardiol 62:714–717

35. Rothschild M, Rothschild A, Pfeifer M (1988) Temporary decrease in cardiac parasympathetic tone after acute myocardial infarction. Am J Cardiol 62:637–639

36. Scheuer J, Tipton CM (1979) Cardiovascular adaptation to physical training. Ann Rev Physiol 39:221–251

37. Schwartz PJ, Pagani M, Lombardi F, Malliani A, Brown AM (1973) A cardiocardiac sympatho-vagal reflex in the cat. Circ Res 32:215–220

38. Schwartz PJ, Stone HL, Brown AM (1976) Effects of unilateral stellate ganglion blockade on the arrhythmias associated with coronary occlusion. Am Heart J 92:589–599

39. Schwartz PJ, Brown AM, Malliani A, Zanchetti A (Eds.) (1978) Neural Mechanisms in Cardiac Arrhythmias. Raven Press, New York

40. Schwartz PJ, Vanoli E (1981) Cardiac arrhythmias elicited by interaction between acute myocardial ischemia and sympathetic hyperactivity: a new experimental model for the study of antiarrhythmic drugs. J Cardiovasc Pharmacol 3:1251–1259

41. Schwartz PJ, Billman GE, Stone HL (1984) Autonomic mechanisms in ventricular fibrillation induced by myocardial ischemia during exercise in dogs with a healed myocardial infarction. An experimental preparation for sudden cardiac death. Circulation 69:780–790

42. Schwartz PJ, Stone HL (1985) The analysis and modulation of autonomic reflexes in the prediction and prevention of sudden death. In: Zipes DP, Jalife J (Eds.) Cardiac Arrhythmias: Mechanisms and Management, Grune and Stratton, New York, pp 165–176

43. Schwartz PJ, Vanoli E, Zaza A, Zuanetti G (1985) The effect of antiarrhythmic drugs on life-threatening arrhythmias induced by the interaction between acute myocardial ischemia and sympathetic hyperactivity. Am Heart J 109:937–948

44. Schwartz PJ, Priori SG, Vanoli E, Zaza A, Zuanetti G (1986) Efficacy of diltiazem in two experimental feline models of sudden cardiac death. J Am Coll Cardiol 8:661–668

45. Schwartz PJ, Stramba-Badiale M (1988) Parasympathetic nervous system and cardiac arrhythmias. In: Kulbertus HE, Frank G (Eds.), Neurocardiology, Futura Publishing Co., Mount Kisco, NY, pp 179–200

46. Schwartz PJ, Vanoli E, Stramba-Badiale M, De Ferrari GM, Billman GE, Foreman RD (1988) Autonomic mechanisms and sudden death. New insight from the analysis of baroreceptor reflexes in conscious dogs with and without a myocardial infarction. Circulation 78:969–973

47. Schwartz PJ, Zaza A, Pala M, Locati E, Beria G, Zanchetti A (1988) Baroreflex sensitivity and its evolution during the first year after a myocardial infarction. J Am Coll Cardiol 12:629–636

48. Schwartz PJ, Priori SG (1990) Sympathetic nervous system and sudden death. In: Zipes DP, Jalife J (Eds.) Cardiac Electrophysiology. From Cell to Bedside, WB Saunders Co, Philadelphia, (In press)

49. Singer DH, Martin GJ, Magid N, Weiss JS, Schaad JW, Kehoe R, Zheutlin T, Fintel DJ, Hsieh AM, Lesch M (1988) Low heart rate variability and sudden cardiac death. J Electrocardiol Suppl:S46–S65

50. Smyth HS, Sleight P, Pickering GW (1969) Reflex regulation of arterial pressure during sleep in man. Circ Res 24:109–121

51. Stone HL (1977) Cardiac function and exercise training in conscious dogs. J Appl Physiol 42:824–832

52. Stramba-Badiale M, De Ferrari GM, Vanoli E, Billman GE, Hull SS, Foreman RD, Schwartz PJ (1988) Myocardial ischemia and ventricular fibrillation in conscious dogs. Atropine and the role of vagal reflex. Fed Proc 47:A327 (abstr.)

53. Verrier RL (1980) Neural factors and ventricular electrical instability. In: Kulbertus HE, Wellens HJJ (Eds.) Sudden Death, Martinus Nijhoff, The Hague, pp 137–155

Authors' address:

Prof. Dr. Peter J. Schwartz, Unita' di Studio delle Aritmie, Centro di Fisiologia Clinica e Ipertensione, Istituto di Clinica Medica II, Universita' degli Studi di Milano, Via Francesco Sforza 35, I-20122 Milano, Italia.

VI. Adrenergic Mechanisms in Clinical Coronary Heart Disease

Heart Rate and β-Adrenergic Mechanisms in Acute Myocardial Infarction

Å. Hjalmarson

Division of Cardiology, Department of Medicine I and Wallenberg Laboratory, University of Göteborg, Sahlgren's Hospital, Göteborg, Sweden

Summary

Increased heart rate is an independent predictor of mortality in patients with acute myocardial infarction. Elevated heart rate is due to increased sympathetic activity and/or decreased parasympathetic activity. In placebo-controlled trials β-blockers are known to reduce mortality as well as morbidity and these effects are most evident among patients with elevated heart rate. Studies on circadian variation have demonstrated that there is an increased sympathetic activity in the morning as well as a more frequent onset of ischemic attacked including acute myocardial infarction and sudden death. There seems to be a close relationship between increased sympathetic activity and the onset of ischemic events which can be prevented by prophylactic institution of a β-blocker.

It is well known that there is an increase in sympathetic activity and a decrease in parasympathetic activity in stress situations. Such changes in autonomic tone have been described after acute myocardial infarction [25,39], heart failure [32], cardiac arrhythmias [47], and in patients prone to sudden death [19,21,40]. The changes in sympathetic and parasympathetic tone have lately been indirectly evaluated as a decrease in heart rate variability and found to be related to poor prognosis.

The aim of this clinical review is to describe the role of heart rate in acute myocardial infarction and the β-adrenergic mechanisms. Firstly, heart rate as a predictor of the development of an acute myocardial infarction and its complications will be discussed; secondly, pathophysiological aspects related to heart rate as a predictor; thirdly, effects of β-adrenergic receptor blockade on ischemic events and mortality.

Heart rate as a predictor in acute myocardial infarction

Several studies on primary and secondary prevention have suggested that elevated heart rate is a predictor, both of risk of developing acute myocardial infarction, as well as a predictor for later complications, including death. The elevated heart rate has generally been considered to be a marker of depressed left ventricular heart function, and especially in studies in patients who have suffered acute myocardial infarction, heart rate has been thought to reflect the severity of myocardial damage. Several reports have, however, reported that increased heart rate is an independent predictor of mortality [3,4,7,8,22,23,31]. Even if heart rate was identified as an independent predictor, heart rate per se has received little attention as a risk factor in the practical treatment of patients with acute myocardial infarction. Also, in a number of trials on short- and long-term treatment with β-blockers in patients with acute myocardial infarction, mortality tended to be higher with higher heart rate at entry to the trial [2,5,10,18,44]. In one of the postinfarction trials, the Norwegian Timolol Multicenter Study [5], a logistic regression

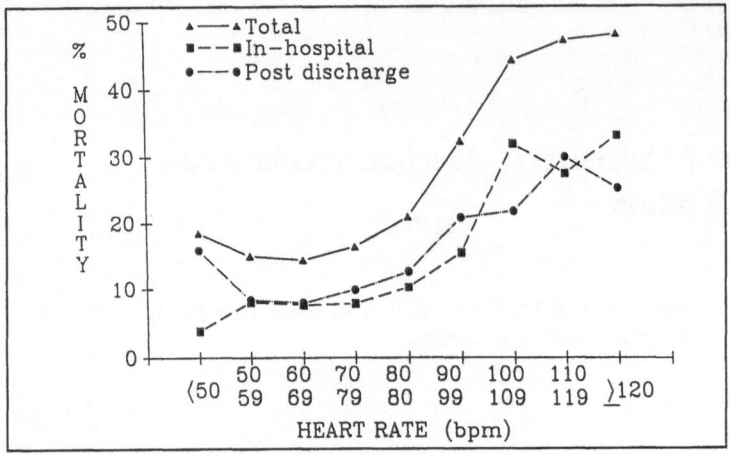

Fig. 1. Mortality with different heart rate upon hospital admission. Solid line shows total mortality from day 2 through 1 year, dashed line in-hospital mortality, and dash-dot line mortality from discharge to 1 year [13].

analysis was performed, showing that heart rate was a significant predictor of long-term mortality among patients treated with timolol, as well as with placebo. A pooling of randomized long-term placebo-controlled trials on β-blockers in postinfarction patients also showed that mortality after 1 to 2 years of myocardial infarction was higher among those with higher heart rate shortly after their infarction [42].

In one recent trial [13], a careful analysis was done to characterize in a large population the relationship between heart rate among patients with acute myocardial infarction and mortality. In this study including 1807 patients of all ages with acute myocardial infarction admitted to hospital within 24 h of chest pain, the level of heart rate at hospital admission, maximal heart rate during the coronary care unit stay and at discharge were related to in-hospital and post-discharge mortality. In Fig. 1 is shown the relationship between heart rate on hospital admission related to in-hospital, post-discharge, and total 1-year mortality. In this study maximum heart rate during the stay in the coronary care unit and heart rate at discharge also predicted mortality from hospital discharge to 1 year. Patients with higher heart rate on admission, during hospital stay, and at discharge were found to more often have a history of cardiovascular disease and more complications during hospital stay, including left ventricular heart failure and arrhythmias. A higher proportion of the patients also had anterior-lateral location of infarction, while patients with lower heart rates more often had inferior-posterior location. However, the trend for increased mortality with increasing heart rate appeared most pronounced in patients with absent to mild, or moderate heart failure (Fig. 2). The total 1-year mortality for patients with admission heart rate ⩾ 90 bpm and absent to mild heart failure was 25%, compared to 10% (p < 0.001) for patients with admission heart rate < 90 bpm. As can be seen from the figure, also in patients with moderate heart failure during hospital stay, total mortality increased with higher admission heart rates (p < 0.01). In contrast, in patients with severe heart failure with x-ray signs of localized or diffuse alveolar edema or with signs of cardiogenic shock, there was a very high mortality from admission to 1 year (> 60%) among all patients with no significant influence of heart rate. In this study, cumulative mortality from day 2 to 1 year was related to variables from the history and clincial course, using Cox regression analysis to identify independent predictors of 1-year mortality. Admission heart rate was, among all known prognostic factors, fourth in rank according to its adjusted chi-square value (p < 0.0004) behind degree of heart failure, age, and maximal blood urea nitrogen.

In several studies on early β-blocker treatment in patients with suspected acute myocardial

Heart rate and β-adrenergic mechanisms in acute myocardial infarction

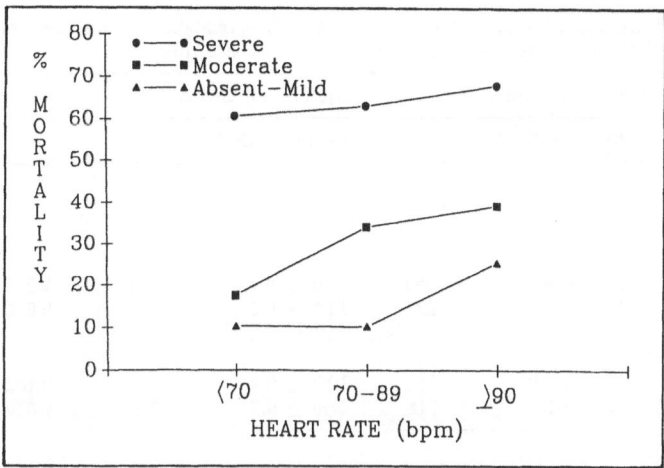

Fig. 2. Total mortality from day 2 to 1 year, related to admission heart rate in patients with absent to mild x-ray evidence of heart failure (▲), moderate heart failure (■), and severe failure including shock (●) [13].

infarction, heart rate on arrival to hospital was not only a predictor of later mortality, but these patients also had more complications [9–12]. Patients with higher heart rate tended to have more severe chest pain with a longer duration of pain, as well as a higher use of morphine analgesics. The patients with higher heart rate at entry also had signs of larger infarctions estimated from enzymes and Q- and R-wave changes, and also more often developed a definite acute myocardial infarction compared to those with lower entry heart rates. Arrhythmias during hospital stay, including supraventricular and ventricular arrhythmias, were more frequently seen among patients with higher heart rates on arrival to hospital.

Pathophysiological aspects

It is well known that heart rate is a major determinant of oxygen consumption and metabolic demand. This means that among patients with restricted coronary blood flow, an increase in heart rate causes more severe myocardial ischemia and thereby more ischemic complications, including pain, larger infarction, arrhythmias, and dysfunction. Experimental studies have demonstrated that myocardial infarct size is increased with elevated heart rate [24,37,41]. It is well documented that in studies on experimental acute ischemia, a lowering of heart rate with β-blockers and other agents reduces severity of myocardial ischemia and improves myocardial function [6,15]. The elevated heart rate can be due to increased sympathetic activity and/or reduced parasympathetic activity. Experimental studies have demonstrated that there is a relationship between high sympathetic and low parasympathetic activity and the ventricular fibrillation threshold [17,40]. In clinical studies on patients with myocardial infarction, reduced heart rate variability or signs of depressed baroreflexes have been associated with increased mortality [19,21].

A number of studies have reported circadian variations in the onset of symptoms in acute myocardial infarction, showing that this was most frequent in the late morning from 6 a.m. to noon [1,14,28,29,49]. It has also been reported that the morning peak of incidence is related to other daily rhythms, especially related to an increase in physical and mental stress and an increase in sympathetic activity [28,29]. There is an increase in sympathetic activity after waking with a rise in plasma levels of catecholamines and cortisol [46], heart rate, blood pressure, cardiovascular tone [20,26,50], platelet aggregability [35,45], and fibrinolytic activity [38]. The circadian variation is not only reported for the onset of acute myocardial infarction, but also for sudden cardiac death and symptomatic and silent ischemia. In a recent study [14], it was found that the morning peak was most marked in younger, previously healthy individuals who

Table 1. Heat-stable lactic dehydrogenase (LD) maximum activity (μkat·1^{-1}) in relation to clinical findings at entry in patients randomized 12 h or less after the onset of symptoms (see also [10]).

Characteristics	Placebo Group			Metoprolol Group			
	Mean ± SEM	n		Mean ± SEM	n		p
Clinical findings immediately before blind injection							
Heart rate (beats/min; median)							
≤ 70	10.8 ± 0.7	250		10.0 ± 0.7	246		> 0.2
> 70	16.1 ± 1.0	225		12.5 ± 0.8	215		0.006
Systolic blood pressure (mm Hg; median)							
≤ 140	12.6 ± 0.9	256		10.7 ± 0.8	235		0.102
> 140	14.1 ± 0.9	218		11.6 ± 0.7	226		0.029

are non-smokers and not receiving β-blockers. The fact that β-blocker treatment blunts the morning peak suggests that increased mental and physical stress with increased sympathetic activity may play a role for the incidence of infarct development and other ischemic manifestations in the late morning.

Effect of β-blockers on ischemic events and mortality

During the early 1970s studies were performed in man with symptoms of acute myocardial infarction showing favorable effects of β-blockers on the severity of myocardial ischemia. Early intravenous administration of a β-blocker after onset of symptoms promptly reduced ST-segment elevation and severity of chest pain, as judged from a pain score analysis [33,36,48]. Acute β-blocker administration was found to improve glucose utilization and reduce lactate production in patients with acute myocardial infarction [27]. Two large placebo-controlled trials, the Göteborg Metoprolol and the Norwegian Study on Timolol showed that intravenous injection of a β-blocker resulted in significant reduction of ST-segment elevation compared to placebo [10,43]. In both studies the β-blocker administration caused a marked reduction in a pain score. The effects on ST-segment elevation and on severity of pain were most marked among patients with elevated heart rate prior to start of treatment.

A number of studies have demonstrated that intravenous administration of β-blockers to patients with suspected acute myocardial infarction can limit later development of infarct size as estimated from serum enzyme analysis and from Q- and R-wave changes [9,12,34,51]. In the Göteborg Metoprolol Trial [10] and the MIAMI Trial [11], a more favorable effect of the β-blocker was seen among patients with higher initial heart rate and systolic blood pressure (Table 1). In these studies, early institution of metoprolol also prevented the development of a definite acute myocardial infarction, and this effect was also more marked among patients with initial higher heart rates. Pooled data from 24 trials on β-blockers compared to placebo during postinfarction follow-up have shown that β-blocker treatment significantly prevented reinfarctions [52]; this effect was also related to entry heart rate [5].

Three studies published in 1981, the Göteborg Metoprolol Trial [10], the Norwegian Timolol Trial [30], and the Beta-blocker Heart Attack Trial (BHAT) [2], demonstrated that the β-blocker therapy could reduce long-term mortality among patients with myocardial infarction. Later studies also demonstrated that early intravenous administration followed by oral treatment with atenolol and metoprolol could reduce mortality already during the first week of myocardial infarction [11,16,44]. When pooling all published early intervention trials, 28 trials including more than 27 000 patients showed a significant reduction in 1 week mortality by β-blocker treatment [12]. In all randomized β-blocker trials, early entry heart rate prior to start of treatment is related to mortality. The higher the entry heart rate, the higher the

Heart rate and β-adrenergic mechanisms in acute myocardial infarction

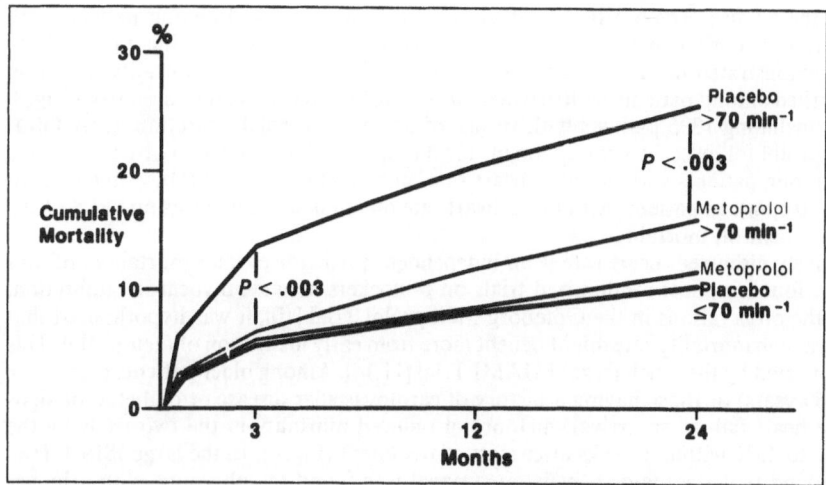

Fig. 3. Cumulative mortality in all patients during 12 months of the Göteborg Metoprolol Trial [10]. Blind medication (placebo or metoprolol) was given for 3 months, and thereafter all treated patients were given the same treatment, most on open metoprolol.

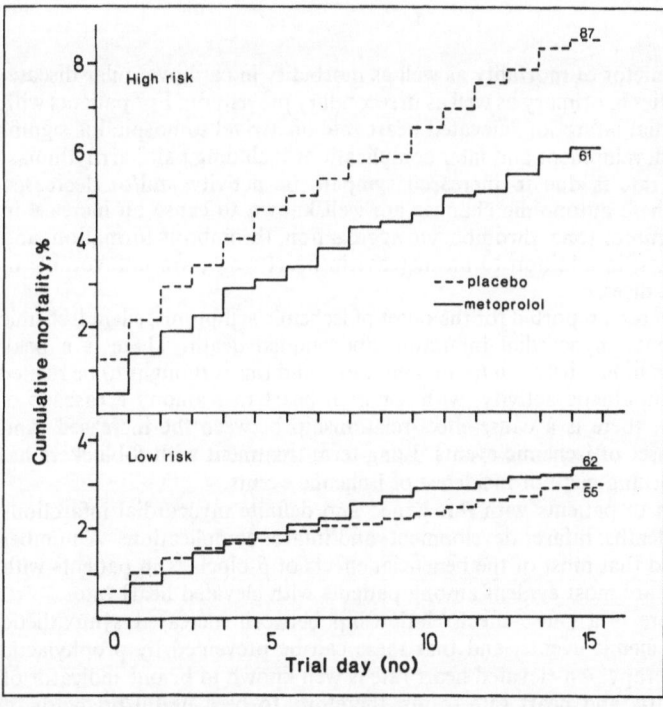

Fig. 4. Cumulative mortality rate and number of deaths in low- and high-risk subgroups of the MIAMI Trial. Using a simple model including eight well known risk predictors, a high-risk group of 2038 patients was identified in which mortality was reduced by 29% [11].

329

mortality in the studies. The mortality reduction is also more marked among patients with higher heart rates and more marked reduction of heart rate by β-blocker administration [5,18]. It was first demonstrated in the Göteborg Metoprolol Trial [10] that the mortality reduction was most marked among patients with elevated heart rate at entry. As can be seen from Fig. 3, in this study including 1395 patients with suspected acute myocardial infarction, early intravenous metoprolol followed by oral treatment for 3 months reduced mortality by about 50% (p < 0.003) among patients with elevated heart rate on arrival to hospital (above the median heart rate of 70 bpm). In patients with lower heart rate on admission there was no effect of the β-blocker treatment on mortality.

As previously discussed, heart rate is an independent predictor of later mortality and this has also been found in most randomized trials on β-blockers in acute myocardial infarction. Based upon the observations in the Göteborg Metoprolol Trial [10], it was hypothesized that patients with a high mortality rate might benefit more from early institution of metoprolol. This was later supported by the much larger MIAMI Trial [11,14]. Among older patients (above the median age 60 years) or those having a history of cardiovascular disease or diabetes, or signs of congestive heart failure on arrival, metoprolol reduced mortality in the two trials by the order of 30% to 45% within 2 weeks after start of treatment (Fig. 4). In the large ISIS-1 Trial it was also found that the major benefit on survival was found in subgroups with a higher mortality, such as patients older than 65 years, in women, in patients with extended anterior plus inferior infarction, and those with bundle branch block, short delay time, and diabetes [16]. It is well known that among most of these subgroups, sicker patients' heart rate is higher than among subgroups with a lower mortality risk.

Conclusion

Increased heart rate is a predictor of mortality as well as morbidity in cardiovascular disease, based upon results from studies in primary as well as in secondary prevention. For patients with a threatening acute myocardial infarction, elevated heart rate on arrival to hospital is significantly correlated to infarct development and later complications including pain, arrhythmias, and death. Elevated heart rate is due to increased sympathetic activity and/or decreased parasympathetic activity. These autonomic changes are well known to cause an increase in heart work, but also in vasomotor tone, thrombocyte aggregation, thrombosis formation, and fibrinolysis. All these changes, in addition to an increase in heart work, are unfavorable in patients with ischemic heart disease.

A circadian variation has been reported for the onset of ischemic symptoms, silent ischemic attacks, development of acute myocardial infarction and sudden death. There is a peak incidence in the late morning hours, between 6 a.m. and noon, and this is thought to be related to a general increase in sympathetic activity, with a rise in heart rate among a cascade of reactions. It is thought that there is a cause-effect-relationship between the increased sympathetic activity and the onset of ischemic events. Long-term treatment with β-blockers has been found to blunt the morning peak of incidence of ischemic events.

β-blocker administration to patients with threatened and definite myocardial infarctions has been found to prevent deaths, infarct development, and infarct complications. A number of studies have demonstrated that most of the beneficial effects of β-blockers in patients with acute myocardial infarction are most evident among patients with elevated heart rate.

Thus, it seems that there is a cause-effect-relationship between increased sympathetic activity and the onset of ischemic events, and that these can be prevented by prophylactic institution of a β-blocker therapy. An elevated heart rate is well known to be one indicator of increased sympathetic activity, and heart rate seems, therefore, to be a useful predictor of outcome, as well as a guide to our choice of therapy among patients during and following acute myocardial infarction. There are new bradycardiac agents which, in addition to β-blockers, may be of value in the prophylactic treatment of patients with threatened myocardial infarction, but this approach requires further study [15].

Heart rate and β-adrenergic mechanisms in acute myocardial infarction

References

1. Beamer AD, Lee H, Cook EF, Brand DA, Rouan GW, Weisberg MC, Goldman L and the Chest Pain Study Group (1987) Diagnostic implications for myocardial ischemia of the circadian variation in the onset of chest pain. Am J Cardiol 60: 998–1002
2. Beta-blocker Heart Attack Trial Research Group (1982) A randomized trial of propranolol in patients with acute myocardial infarction. I. Mortality results. JAMA 247: 1707–1714
3. Coronary Drug Project Research Group (1974) Factors influencing long-term prognosis after recovery from myocardial infarction--three-year findings of the Coronary Drug project. J Chron Dist 27: 267–285
4. Gilpin E, Olshen R, Henning H, Ross J Jr (1983) Risk prediction after myocardial infarction, comparison of three multivariate methodologies. Cardiology 70: 73–84
5. Gundersen T, Gottum P, Pedersen T, Kjekshus KJ for the Norwegian Multicenter Study Group (1986) Effect of timolol on mortality and reinfarction after acute myocardial infarction: Prognostic importance of heart rate at rest. Am J Cardiol 58: 20–25
6. Guth BD, Heusch G, Seitelberger R, Ross J Jr (1987) Mechanism of beneficial effect of beta-adrenergic blockade on exercise-induced ischemia in conscious dogs. Circ Res 60: 738–746
7. Henning H, Gilpin E, Covell JW, Swan EA, O'Rourke RH, Ross J Jr (1979) Prognosis after acute myocardial infarction: a multivariate analysis of mortality and survival. Circulation 59: 1124–1136
8. Henning R, Wedel H (1981) The long-term prognosis after myocardial infarction: a five year follow-up study. Eur Heart J 2: 65–74
9. Herlitz J, Hjalmarson Å (1986) The role of beta-blockade in the limitation of infarct development. Br Heart J 7: 916–924
10. Hjalmarson Å, Herlitz J (1984) The Göteborg Metoprolol Trial in acute myocardial infarction. Am J Cardiol 53: 1D–50D
11. Hjalmarson Å for the MIAMI Trial Research Project (1985) MIAMI: Metoprolol in Acute Myocardial Infarction. Am J Cardiol 56: 1G–57G
12. Hjalmarson Å, Herlitz J (1989) The use of beta-blockers in the treatment of suspected myocardial infarction. A clinical review. JEUR 2: 190–201
13. Hjalmarson Å, Gilpin EA, Kjekshus J, Schieman G, Nicod P, Henning H, Ross J Jr (1990) Influence of heart rate on mortality after acute myocardial infarction. Am J Cardiol. In press.
14. Hjalmarson Å, Gilpin EA, Nicod P, Dittrich H, Henning H, Engler R, Blacky AR, Smith Jr SC, Ricou F, Ross J Jr (1989) Differing circadian patterns of symptom onset in subgroups of patients with acute myocardial infarction. Circulation 80: 267–275
15. Indolfi C, Guth BD, Miura T, Miyazaki S, Schulz R, Ross J Jr (1989) Mechanisms of improved ischemic regional dysfunction by bradycardia. Studies on UL-FS49 in swine. Circulation 80: 983–993
16. ISIS-1 (First International Study of Infarct Survival) Collaborative Group. Randomized trial of intravenous atenolol among 16027 cases of suspected acute myocardial infarction: ISIS-1. Lancet 2: 57–66
17. James RGG, Arnold JMO, Allen JD, Pantridge JF, Shanks RG (1977) The effects of heart rate, myocardial ischemia and vagal stimulation on the threshold for ventricular fibrillation. Circulation 55: 311–317
18. Kjekshus JK (1986) Importance of heart rate in determining beta-blocker efficacy in acute and long-term acute myocardial infarction intervention trials. Am J Cardiol 57: 43F–49F
19. Kleiger RE, Miller JP, Bigger JT, Moss AJ and the Multicenter Post-Infarction Research Group (1987) Decreased heart rate variability and its association with increased mortality after acute myocardial infarction. Am J Cardiol 59: 256–262
20. Kostis JB, Moreyra AE, Amendo MT, Di Pietro J, Cosgrove N, Kuo PT (1982) The effect of age on heart rate in subjects free of heart disease: studies by ambulatory electrocardiography and maximal exercise stress test. Circulation 65: 141–145
21. LaRovere MT, Specchia G, Montara A, Schwartz P (1988) Baroreflex sensitivity, clinical correlates, and cardiovascular mortality among patients with a first myocardial infarction: A prospective study. Circulation 78: 800–815
22. Madsen EB, Gilpin E, Henning H, Ahnve S, LeWinter M, Ceretto W, Joswig W, Collins D, Pitt W, Ross J Jr (1984) Prediction of late mortality after myocardial infarction from variables measured at different times during hospitalization. Am J Cardiol 53: 47–54
23. Madsen EB, Gilpin E, Henning H, Ahnve S, LeWinter M, Mazur J, Shabetai R, Collins D, Ross J Jr (1984) Prognostic importance of digitalis after acute myocardial infarction. J Am Coll Cardiol 3: 681–689.

331

24. Maroko PR, Kjekshus JK, Sobel BE, Watanabe T, Covell J, Ross J Jr, Braunwald E (1971) Factors influencing infarct size following experimental coronary artery occlusion. Circulation 43: 67–82

25. McAlpine HM, Cobbe SM (1988) Neuroendocrine changes in acute myocardial infarction. Am J Med 84 (suppl 3A): 61–66

26. Millar-Craig MW, Bishop CN, Raftery EB (1978) Circadian variation of blood pressure. Lancet 1: 795–797

27. Mueller HS, Ayres SM, Religa A, Evans RG (1974) Propranolol in the treatment of acute myocardial infarction. Effect on myocardial oxygenation and hemodynamics. Circulation 49: 1078–1087

28. Muller JE, Stone PH, Turi ZG, Rutherford JD, Czeisler CA, Parker C, Poole WK, Passamani E, Roberts R, Robertson T, Sobel BE, Willerson JT, Braunwald E and the MILIS Study Group (1985) Circadian variation in the frequency of onset of acute myocardial infarction. N Engl J Med 313: 1315–1322

29. Muller JE, Ludmer PL, Willich SN, Tofler GH, Aylmer G, Klangos I, Stone PH (1987) Circadian variation in the frequency of sudden cardiac death. Circulation 75: 131–138

30. Norwegian Multicenter Study Group (1981) Timolol-induced reduction in mortality and reinfarction in patients surviving acute myocardial infarction. N Engl J Med 304: 801–807

31. Olshen RA, Gilpin E, Henning H, LeWinter M, Collins D, Ross J Jr (1985) Twelve-month prognosis following myocardial infarction: classification trees, logistic regression and stepwise linear discrimination. In: LeCam LM, Olshen RA (eds) Proceedings of the Conference in Honor of Jerzy Meyman and Jack Kiefer: Monterey, CA. Wadsworth Advanced Books and Software, pp. 245–267

32. Packer M (1988) Neurohormonal interactions and adaptations in congestive heart failure. Circulation 77: 721–730

33. Pelides LJ, Reid DS, Thomas M, Shillingford JP (1972) Inhibition by betablockade of the ST segment elevation after acute myocardial infarction in man. Cardiovasc Res 6: 295–301

34. Peter T, Norris RM, Clarke ED, Heng MK, Singh BN, Williams B, Howell DR (1978) Reduction of enzyme levels by propranolol after acute myocardial infarction. Circulation 57: 1091–1095

35. Petralito A, Mangiafico RA, Gibiino S, Cuffari MA, Miano MF, Fiore CE (1982) Daily modifications of plasma fibrinogen, platelets aggregation, Howell's time, PTT, TT and antithrombin III in normal subjects and in patients with vascular disease. Chronobiologica 9: 195–201

36. Ramsdale DR, Faragher EB, Bennett DH, Bray CL, Ward C, Cruickshank JM, Yusuf S, Sleight P (1982) Ischemic pain relief in patients with acute myocardial infarction by intravenous atenolol. Am Heart J 103: 459–467

37. Redwood DR, Smith ER, Epstein S (1972) Coronary artery occlusion in the conscious dog. Effects of alterations in heart rate and arterial pressure on the degree of myocardial ischemia. Circulation 46: 323–332

38. Rosing DR, Brakman P, Redwood DR, Goldstein RE, Beiser GD, Astrupt T, Epstein SE (1970) Blood fibrinolytic activity in man: diurnal variation and the response to varying intensities of exercise. Circ Res 27: 171–184

39. Ryan C, Hollenberg M, Harvey DB, Gwynn R (1976) Impaired parasympathetic responses in patients after myocardial infarction. Am J Cardiol 37: 1013–1018

40. Schwartz PJ, Vanoli E, Stramba-Badiale M, DeFerrari GM, Billman GE, Foreman RD (1988) Autonomic mechanisms and sudden death. New insights from analysis of baroreceptor reflexes in conscious dogs with and without a myocardial infarction. Circulation 78: 969–979

41. Strangeland L, Grong K, Vik-Mo H, Andersen KS, Lekven J (1986) Is reduced cardiac performance the only mechanism for myocardial infarct size reduction during beta-adrenergic blockade? Circ Res 20: 322–330

42. The Beta Blocker Pooling Project (BBPP) (1988) Subgroup findings from randomized trials in post-infarction patients. Eur Heart J 9: 8–16

43. The International Collaborative Study Group (1984) Reduction of infarct size with the early use of timolol in acute myocardial infarction. N Engl J Med 310: 9–15

44. The MIAMI Trial Research Group (1985) Metoprolol in acute myocardial infarction (MIAMI). A randomised placebo-controlled international trial. Eur Heart J 6: 199–226

45. Tofler GH, Brezinski D, Schaffer AI, Czeisler CA, Rutherford JD, Willich SN, Gleason RE, Williams GH, Muller JE (1987) Concurrent morning increase in platelet aggregability and the risk of myocardial infarction and sudden cardiac death. N Engl J Med 316: 1514–1518

46. Turton MB, Deegan T (1974) Circadian variations of plasma catecholamine, cortisol and immuno-reactive insulin concentrations in supine subjects. Clin Chim Acta 55: 389–397

47. Verrier RL, Hohnloser SH (1987) How is the nervous system implicated in the genesis of cardiac arrhythmias In: Hearse A, Janse M (eds) Life-Threatening Arrhythmias During Ischemia and Infarction. New York, Raven Press, pp. 153–169

48. Waagstein F, Hjalmarson Å (1975) Double-blind study of the effect of cardioselective beta-blockade on chest pain in acute myocardial infarction. Acta Med Scand 587: 201–208
49. Willich SN, Levy D, Rocco MB, Tofler GH, Stone PH, Muller JE (1987) Circadian variation in the incidence of sudden cardiac death in the Framingham heart study population. Am J Cardiol 60: 801–806
50. Yasue H, Omote S, Takizawa A, Nagao M, Miwa K, Tanaka S (1979 Circadian variation of exercise capacity in patients with Prinzmetal's variant angina: role of exercise-induced coronary arterial spasm. Circulation 59: 938–948
51. Yusuf S, Peto R, Bennet D, Ramsdale D, Furze L, Bary C, Sleight P (1980) Early intravenous atenolol treatment in suspected acute myocardial infarction. Lancet 2: 273—276
52. Yusuf S, Peto R, Lewis J, Collins R, Sleight P (1985) Beta Blockade during and after myocardial infarction: An overview of the randomized trials. Prog Cardiovasc Dis 27: 335–371

Author's address:

Åke Hjalmarson, Division of Cardiology, Department of Medicine I and Wallenberg Laboratory, University of Göteborg, Sahlgren's Hospital, Göteborg, Sweden

Coronary Vasomotor Tone in Large Epicardial Coronary Arteries with Special Emphasis on β-Adrenergic Vasomotion, Effects of β-Blockade

P. R. Lichtlen, W. Rafflenbeul, St. Jost and C. Berger

Division of Cardiology, Department of Medicine, Hannover Medical School

Summary

Changes in coronary vasomotor tone of large epicardial coronary arteries today can be assessed quite accurately by exact measurements of coronary diameters applying computer assisted systems. The effect of various vasodilators (nitrates, calcium antagonists, EDRF-dependent compounds) was tested in this way. It appears that normal coronary artery segments reach a maximum of dilator reserve with an increase of luminal diameter of approximately 30–40%; different patterns of kinetics were, however, encountered. β-Blocking agents, both non-selective (propranolol) and selective (atenolol), were found to lead to a gradual vasoconstriction, i.e., a decrease in diameter by approximately 20–25% over 20 min, an effect which is overcome by nitrates. New β-blocking compounds with vasodilator properties, such as celiprolol, show no constriction. The vasoconstrictor effect of propranolol and atenolol may not only be due to the decrease of flow following the drop in myocardial oxygen consumption, but could also reflect an unopposed α-adrenergic tone. The clinical aspects of this observation are discussed.

Introduction

For a long time, coronary vasomotor tone in man could only be assessed as total resistance, combining proximal, extramural and peripheral arteriolar resistance. Calculations were based on measurements of mean aortic pressure and transmural coronary blood flow, usually derived from several regions [31]. Only during recent years, the evaluation of proximal resistance became possible, even under abnormal conditions provoked by the formation of atherosclerotic plaques or due to functional narrowings, i.e., an increased vasomotor tone. Coronary angiograms with optimal visualization of proximal coronary artery segments allow exact measurements of coronary diameters, both normal and narrowed following coronary artery disease. By applying computer-assisted systems [25,43,44] accurate data are obtained, resulting in absolute values when calibration with the catheter is performed. This, for the first time enables the study of proximal coronary vasomotor tone under vasodilating and vasoconstricting drugs.

Methods

Angiographic assessment of luminal changes

In patients undergoing analysis of proximal coronary artery vasomotor tone during diagnostic coronary angiography, all vasoactive drugs had been withheld for at least 24 h prior to angiography. Premedication consisted of 5 mg diazepam (valium) administered 30 min before angiography. The study was always started with a left ventricular angiogram. Both left

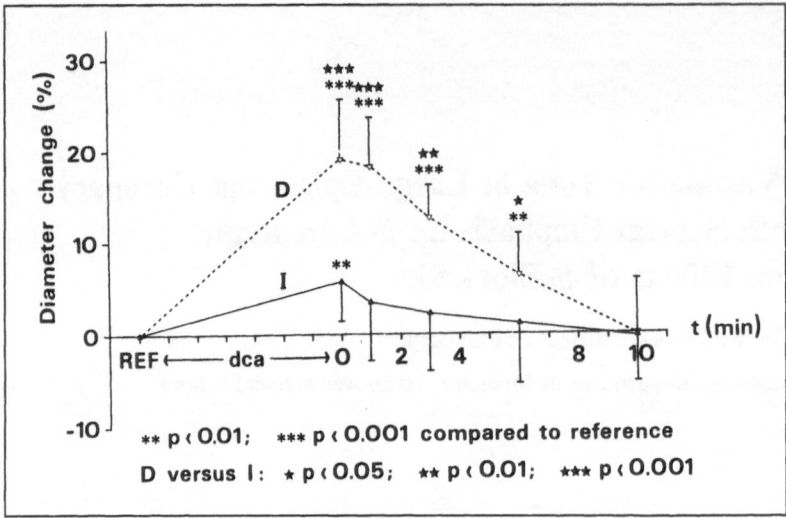

Fig. 1. Diameter changes of large epicardial coronary arteries under an ionic (diatrizoate, D) and a non-ionic (iopromide, I) contrast medium. Ref = reference, dca = diagnostic coronary angiography. The maximum change was observed after several contrast medium injections, beginning at REF and ending at time 0. The subsequent measurements of diameters at 2 min intervals demonstrate that with ionic contrast media one has to wait at least 8 min to be back to reference [26].

ventricular and coronary angiograms were performed by administering non-ionic contrast media, as they proved to influence coronary vasomotor tone only minimally, if compared to the ionic ones (Fig. 1) [26]. Quantitative evaluation of luminal diameter changes of large epicardial coronary arteries (proximal and middle portion of the LAD, RCA and LCX, in the latter including proximal regions of the first and second posterolateral branch) was performed by the aid of the CAAS-system, which operates with an automatic contour-recognition algorithm (for details of the algorithm and performance with CAAS, see [25,43,44]). After a cardiologist identified the coronary segment to be analyzed (preferably running parallel to the cineplane, stretched in end-diastole) on a Tagarno projector, the entire cineframe was digitized by a videoconverter; for this purpose the segment was optically enlarged and a few points in the middle of the lumen were marked by the operator, using a mouse and a writing table. After the contours of the coronary segment were identified by the computer algorithm (Fig. 2), mean segment diameters were computed by averaging individual diameter measurements performed in intervals of 0.1 mm along the longitudinal axis of the segment. After the analysis procedure, cineframes were corrected for pincushion distortion of the image amplifier using special grid films (1 cm distances) and a special algorithm. For calibration of cineframes the diameter of the automatically determined tip of the catheter on the cineframe was compared with the actual diameter measured immediately after angiography, using a precision caliper [25].

Two computer programs were applied, one for assessment of average diameters of angiographically normal coronary segments, usually not longer than 3–4 cm, and a second program to assess minimal stenosis diameters and percent stenosis. The latter algorithm extrapolated the original "normal" diameter at the site of the stenosis by a special contour-finding program which allowed to calculate percent stenosis as well as "length" and "area of the plaque".

The variability of the system in repeated measurements in identical projections is small. For repeated measurements of identical cineframes, standard deviation of the mean values of segment diameters amounts to ± 0.11 mm, for the degree of stenosis to ± 3.9% [25,43,44].

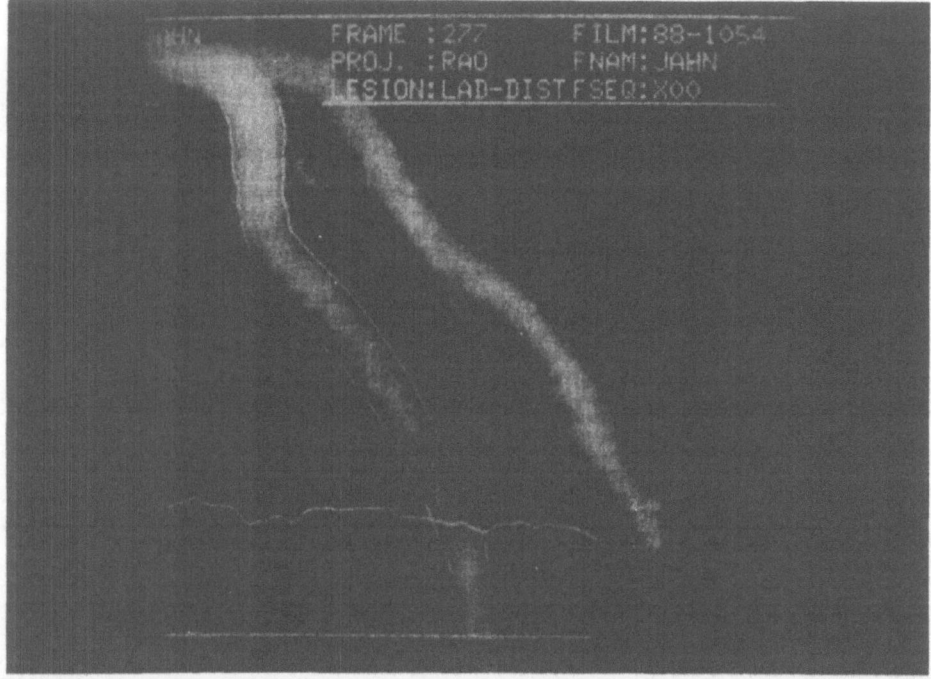

Fig. 2. Computer measurement of the diameter of a large epicardial coronary artery (LAD) by the CAAS system. At the bottom, the diameter function curve indicating the changes of the diameter at 0.1 mm intervals over the entire length of the segment, decreasing from 2.4 mm proximally to 1.7 mm distally; this is typical for the normal segment of a coronary branch.

Statistics

The differences of mean values within one group were tested by the paired Student's t-test, a P-value of 0.05 or less being accepted as significant.

Drugs tested

In recent years, a number of drugs known to dilate the large epicardial coronary arteries, as well as contrast media influencing vasomotion were tested in a still ongoing program. Repeat injections of ionic contrast media (diatrizoate 75%) induced a maximal dilation averaging $18.9 \pm 6.7\%$ ($P < 0.001$) of large epicardial coronary arteries, which correlated positively with the number of contrast injections performed per minute (Fig. 1); coronary dilation was still present after a 6 min interval ($6.2\% \pm 4.6\%$) ($P < 0.01$). In contrast, after repeat injections of a non-ionic contrast medium (iopromide 370) coronary dilation was mild (maximum $5.4 \pm 4.3\%$; $P < 0.05$) and insignificant 2 min after injection [26] (Fig. 1). For this reason, all studies on changes in coronary vasomotor tone induced by drugs were performed by applying a non-ionic contrast medium.

The influence of the following drugs will be presented here:
- Isosorbide dinitrate, 10 mg sublingually (chewing capsules) [27,29];
- Nifedipine, 2 mg intravenously administered over 4 min [28];
- Nisoldipine, 1 mg intravenously administered over 4 min [24,28];
- Propranolol 0.1 mg/kg, administered intravenously over 4 min [40,41];

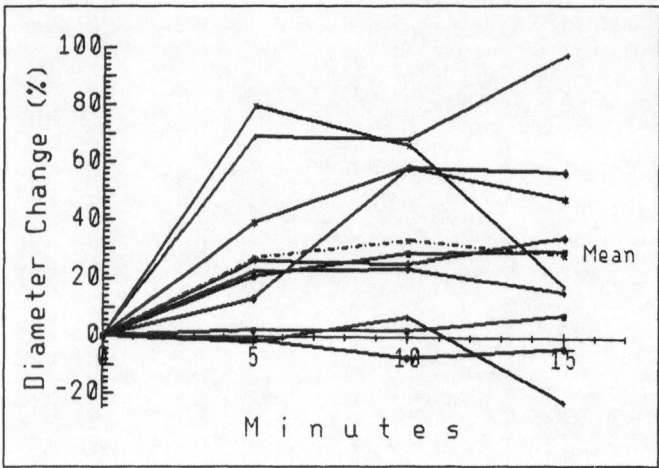

Fig. 3. Diameter changes of various stenoses over time under 10 mg isosorbide dinitrate (chewing capsule). There was a mean change of 28%, reached at the 10th min; three concentric stenoses showed no change; the other stenoses were eccentric in nature and showed a maximum dilatation of 60–80%.

- Atenolol 0.12 mg/kg applied over 4 min, and
- Celiprolol 0.4 mg/kg.

Besides these drugs a number of other compounds were studied, including bradykinin [42] substance P [2,4], acetylcholine [3], and ATP [35]. For a number of these compounds (nitrates, nifedipine, nisoldipine) plasma levels were also assessed simultaneously.

Hemodynamic measurements

Before injections, blood pressure was measured through the aortic catheter and will be indicated as systolic and diastolic blood pressure; heart rate from the ECG was also recorded continously throughout the study.

Results

1. Isosorbide dinitrate, 10 mg sublingually, chewing capsules [27]

This effect was analyzed in 10 patients (nine male, one female, average age 57 years; range 48–68 years) with coronary artery disease, six with a previous myocardial infarction.

Plasma levels showed a maximum increase to 138 ± 73 ng/ml at 5 min after intake; after 15 min plasma levels had dropped to 62 ± 34 ng/ml. Blood pressure dropped from $147 \pm 19/76 \pm 8$ to $115 \pm 15/78 \pm 9$ mmHg; heart rate increased from 66 ± 4 to 80 ± 11 beats/min (all values $P < 0.05$–0.001).

Maximal coronary dilatation of the 64 segments analyzed was $26.8\% \pm 13.2\%$ on the average, and in two patients was even 52% and 43%, respectively ($P < 0.001$). This maximum was reached at 15 min after intake.

In addition, the minimal diameter of 10 stenoses (Fig. 3) was analyzed; three stenoses – probably of concentric nature – showed no changes. The seven reacting to ISDN had a maximum dilation between 23% and 98%.

Hence, in a large dose of 10 mg applied sublingually as chewing capsules (Sorbidilate®), ISDN showed a rapid and consistent dilation of large epicardial coronary arteries; a considerable dilation was also found in eccentric stenoses.

338

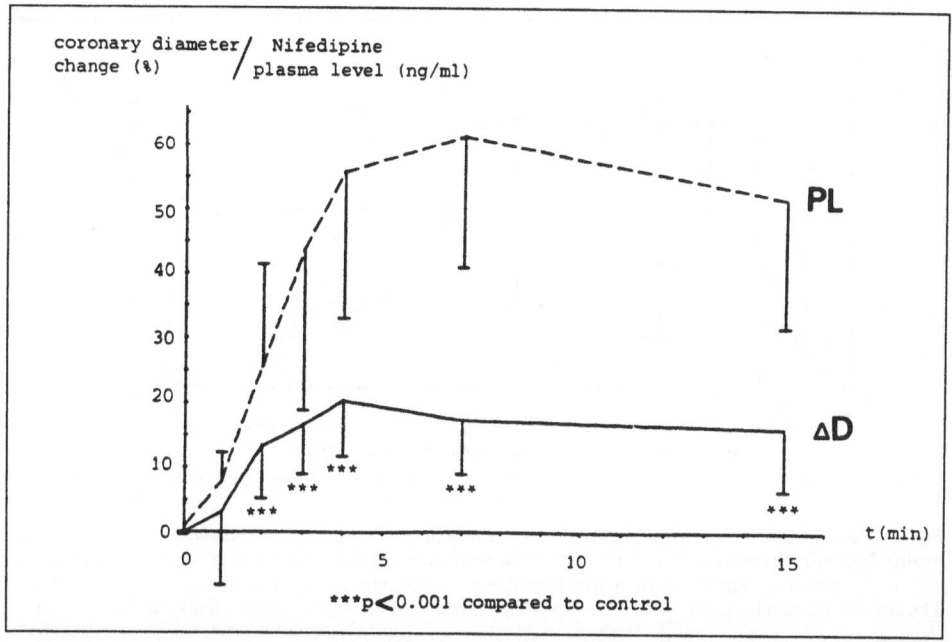

Fig. 4. Diameter changes under 2 mg nifedipine infused over the first 4 min. Plasma levels (PL) reached a maximum of approximately 60 ng/ml, diameter changes reached an average of 20% (D).

2. Calcium channel entry blocking agents; dihydropyridines, 2 mg nifedipine and 1 mg nisoldipine, short infusions over 4 min [24,28]

This effect was studied in 26 patients; 13 received nifedipine (12 male, one female; average age 58 years, range 37–64 years) and 13 received nisoldipine (12 male, one female; average age 54 years, range 34–59 years), all with coronary artery disease.

*Nifedipine:*Maximal plasma levels averaged 62 ± 21 ng/ml at the 7th min (3 min after completion of the infusion) (Fig. 4), declining to 50 ± 20 ng/ml at the 15th min. Maximum average coronary dilation was $20 \pm 9\%$ at the 4th min (P < 0.001) (range 5–37%), being sustained up to the 15th min ($16 \pm 10\%$) (Fig. 4). Blood pressure dropped from $132 \pm 21/75 \pm 8$ to $91 \pm 11/43 \pm 10$ mmHg at the 4th min and heart rate increased from 74 ± 12 to 84 ± 14 beats/min (P < 0.001).

*Nisoldipine [24]:*In contrast to nifedipine, nisoldipine plasma levels reached only an average of 16.9 ± 7.1 ng/ml (4th min) and dropped to 9.1 ± 2.6 ng/ml at the 15th min.

Maximum coronary dilation was reached at the end of the study period, at 15 min, averaging $18.1 \pm 9.0\%$ (range 7–36%) (p < 0.001); at the 4th min (end of infusion period), it averaged only $12.7 \pm 6.9\%$ (Fig. 5). Under nisoldipine, where minimal stenosis diameters were assessed, eight out of nine dilated by a maximum averaging $28 \pm 15\%$ (16–65%) (P < 0.01). Hence, during nisoldipine coronary diameters steadily increased up to the 15th minute in spite of a decline of plasma levels after the 4th min, indicating a relatively strong longlasting receptor binding. Blood pressure under nisoldipine dropped from $143 \pm 19/74 \pm 11$ to $96 \pm 12/61 \pm 9$ mmHg at the 4th min (P < 0.001) and increased to $109 \pm 13/58 \pm 9$ mmHg at the 15th min; heart rate increased from 70 ± 10 to 91 ± 15 beats/min. Hence, blood pressure dropped maximally at the 4th minute, while coronary dilation reached its maximum only at the end of the observation period at 15 min.

Finally, with both nifedipine and nisoldipine a negative correlation was found between the initial diameters of individual coronary segments ranging from 1.0 to 3.5, and 3.7 mm, respec-

339

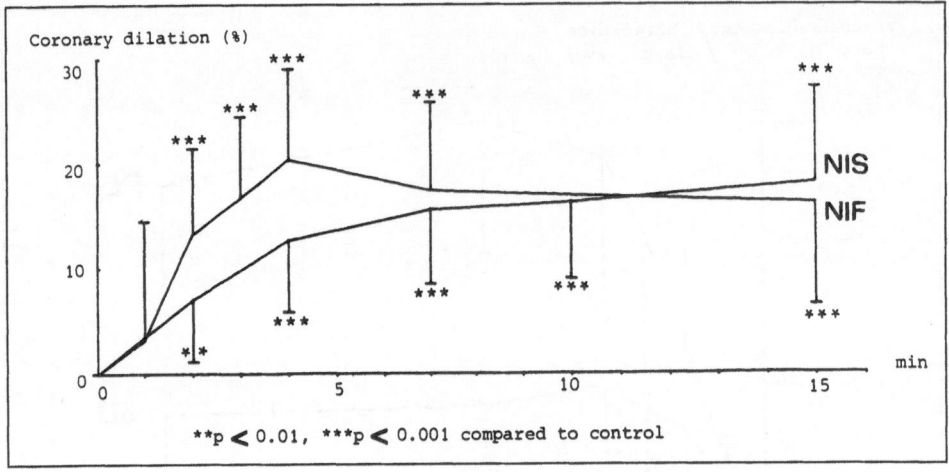

Fig. 5. Comparison of coronary dilation between nifedipine (2 mg i.v.) and nisoldipine (1 mg i.v. over 4 min). Nisoldipine shows a steady increase in diameter size although the infusion was stopped at the 4th min, maximum dilatation being reached at 15 min, while nifedipine shows a mild decrease after the 4th min. The continuous and lasting dilation under nisoldipine is due to the high receptor affinity of this compound.

tively, and the corresponding maximum dilation (0–54 and 0–61%, respectively), the correlation coefficients amounting to r = − 0.21 (P < 0.05) and r = − 0.51 (P < 0.001).

3. β-receptor blocking agents

A different behavior of the vasomotor tone in large epicardial coronary arteries was observed under β-blocking agents.

*Propranolol (Figs. 6,7):*Propranolol, 0.1 mg/kg bodyweight, was administered over 4 min and the vasoactive reaction was tested in coronary segments of three different sizes (diameter > 2.8 mm, 2.2–2.8 mm, < 2.2 mm); the entire measurement period lasted for 20 min. This differentiation between various diameters was done, because already for nitrates and calcium antagonists, i.e., for coronary dilators, a negative correlation was found between the initial diameters of individual coronary segments and the corresponding maximal dilatation.

Twelve patients, average age of 53 years, received propranolol intravenously, 0.1 mg/kg bodyweight; changes in luminal diameters were assessed in 60 segments up to 20 min in 2–4 min intervals. In all segments below 2.8 mm, most pronounced in those below 2.2 mm, a close and steady decrease in lumen size was observed, reaching a maximum of − 23% ± 6% in segments < 2.2 mm diameter, of − 14 ± 4% in segments between 2.2 and 2.8 mm, and of − 5% ± 4% in segments > 2.8 mm (P < 0.05) (Fig. 7). A mild, insignificant vasodilation, probably caused by the contrast medium was only observed in the large segments > 2.8 mm and this only during the first 6 min, followed by a continous vasoconstriction thereafter.

Systolic blood pressure rose slightly from 127 ± 15 mmHg to 135 ± 12 mmHg; heart rate did not change (average 77 beats/min).

*Atenolol (Fig. 8):*A similar study was performed with the selective β_1-blocking agent atenolol, administered in doses of 0.12 mg/kg bodyweight intravenously. The protocol was similar to the one applied for propranolol; it included 12 patients, with an average age of 56 years, all with coronary artery disease. The average maximum decrease in luminal diameter for segments with a diameter > 3 mm was − 8.6%, for segments with a diameter of 2–3 mm it was − 10 ± 6%, and for those < 2 mm it was − 15.4% (P < 0.05 vs control). Hence, atenolol also

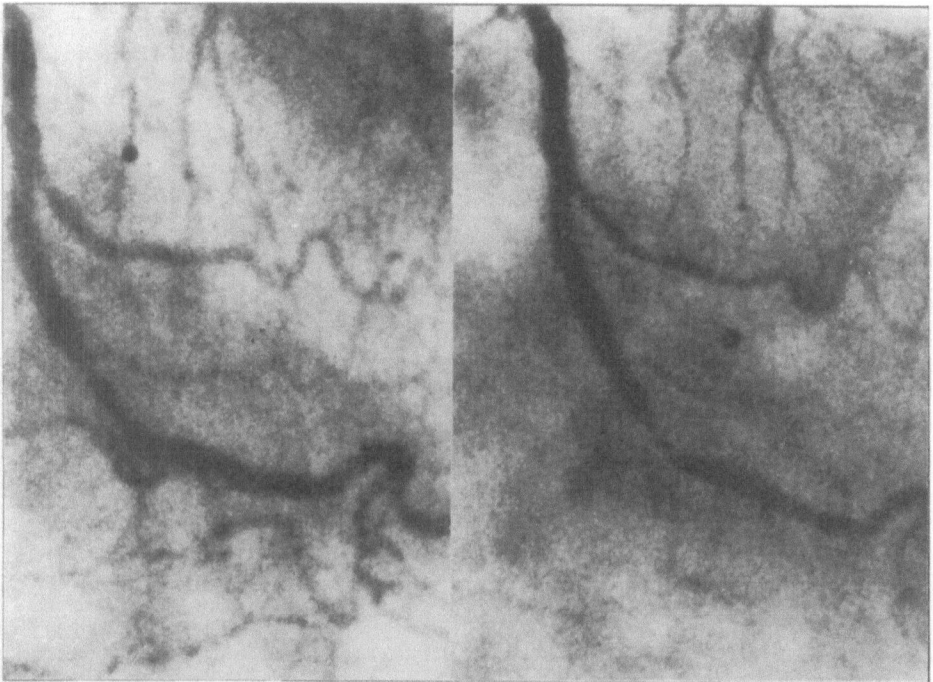

Fig. 6. Typical example of coronary vasoconstriction under propranolol. On the left, a postero-lateral branch before drug administration. On the right, 20 min after administration of 0.1 mg per kg propranolol. There was a 20% vasoconstriction, especially in the lower part of the vessel, as well as in side branches (no changes in geometry between the two cineangios performed at an interval of 20 min).

demonstrated a similar behavior as propranolol: a time-dependent decrease in luminal diameter, i.e., an increasing vasoconstriction of large epicardial coronary arteries over 20 min.

Blood pressure and heart rate remained unchanged.

Celiprolol (Fig. 9): The effect of celiprolol, 0.4 mg/kg, a β-blocking agent with vasodilating properties, was tested in 12 patients, average age 55 years. Altogether, 42 segments were studied using the identical protocol as applied to propranolol and atenolol. In contrast to the behavior observed with propranolol and atenolol, only a mild and insignificant vasoconstrictive effect was seen with celiprolol; this effect was again overcome by nitroglycerin (Fig. 9).

The blood pressure and heart rate remained unchanged during the study period.

Discussion

The vasomotor tone of large epicardial coronary arteries is only of clinical relevance during abnormal situations, especially in coronary atherosclerosis [32]. In the normal coronary system in man, proximal resistance is minimal and even a mild vasoconstriction during exercise, i.e., during increased α-tone [10,14,18,20,21,22] is negligible or rather helpful as it guarantees an equal distribution of transmural and by this preservation of endocardial flow. Under pathological conditions, especially in the presence of coronary artery disease, proximal resistance increases considerably. This is due either to anatomical changes, the formation of atherosclerotic plaques, narrowing the coronary lumen by 70% or more, and/or to functional phenomena,

341

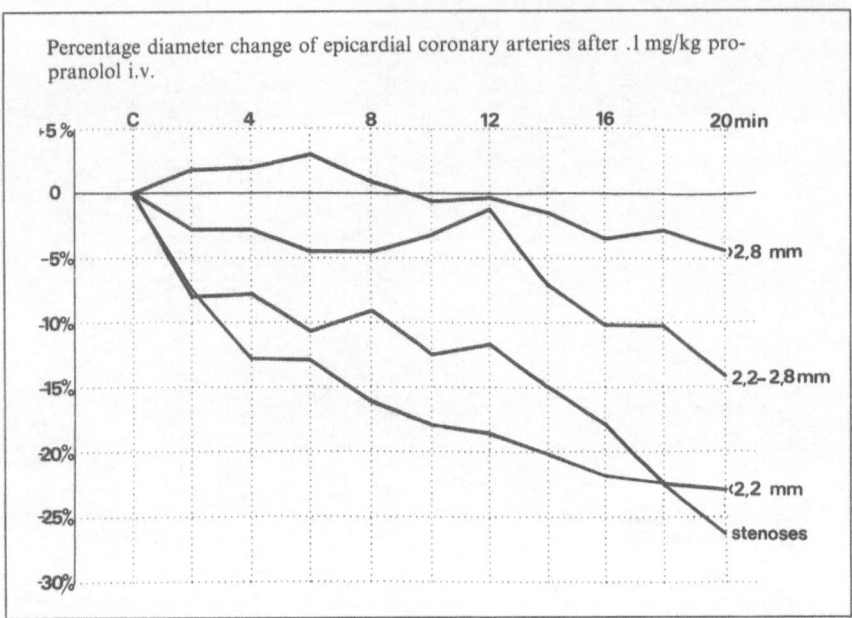

Percentage diameter change of epicardial coronary arteries after .1 mg/kg propranolol i.v.

Fig. 7. Coronary diameter changes under propranolol from control (c) up to the 20th min. Note the gradual decrease, especially of the smaller branches, reaching almost 25%; an equal decrease was also observed in eccentric stenoses (see text).

by an abnormal increase in vasomotor tone; this vasoconstriction is usually based on endothelial damage, i.e., the absence of endothelial-derived relaxing factor (EDRF) [4,7,12]; the resulting coronary spasm [16] occurs especially in the area of eccentric stenoses, which still possess considerable normal wall tissue [11] that is able to contract or dilate. Hence, it seems reasonable to assume that normally vasodilation prevails in the epicardial coronary segments and resistance changes solely or predominantly take place in the arteriolar bed. This changes drastically in the presence of atherosclerosis. During recent years it became evident that proximal coronary resistance, especially in eccentric stenoses, is amenable to medical treatment [38]; reduction of proximal resistance was accepted as a therapeutic way to influence, i.e. interrupt or prevent myocardial ischemia, both silent or clinically manifest as angina, by applying vasodilating drugs such as nitrates [34,38] or calcium entry blockers [8,9,32]. To which extent this way of intervention is involved in the relief of ischemia still remains a matter of debate as most of these drugs, especially nitrates and calcium entry blockers, also exhibit an anti-ischemic effect in other ways (decrease in pre- and afterload, cardioprotective effect, etc.) [32,34]. For all these reasons it seemed important to us to quantify the vasoactive effect of anti-ischemic drugs on the large epicardial coronary arteries, i.e., on the proximal resistance, for all of those drugs expected to exert a dilatory effect (nitrates or calcium entry blockers) or a vasoconstrictive effect (β-blockers). The problem became even more important when considering EDRF, which – in advanced atherosclerosis – seems to be lacking to a great degree [16] and might be responsible for the prevalence of coronary artery constriction and the further increase in proximal coronary resistance in these patients. Nitrates exert their dilatory effect by the same mechanisms as EDRF, nitrous oxide stimulating the soluble guanylate cyclase and, thus, the production of cGMP [23,34]; this explains their excellent anti-ischemic effect in atherosclerosis [34]. Calcium antagonists act predominantly by blocking the calcium transport in the voltage-dependent channel, inhibiting the calcium-mediated activation of the actin-myosin contractile system in vascular smooth muscle cells. It is interesting to note that

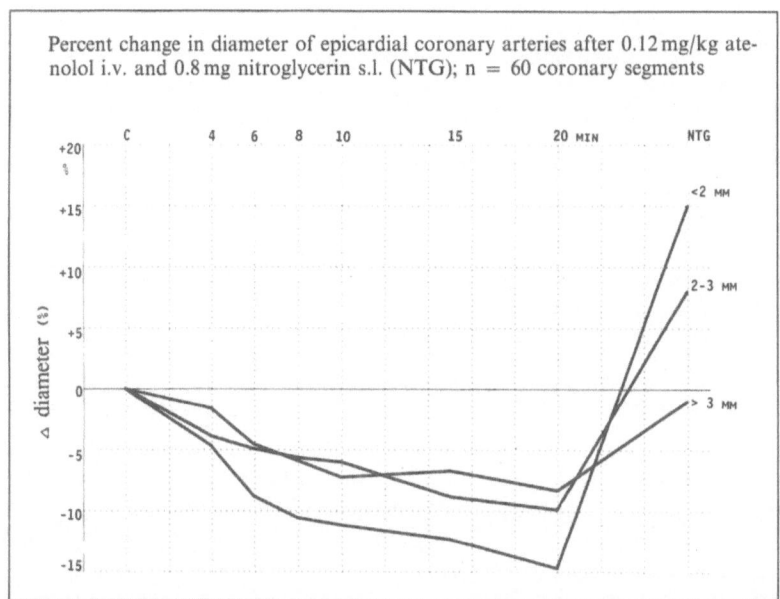

Percent change in diameter of epicardial coronary arteries after 0.12 mg/kg atenolol i.v. and 0.8 mg nitroglycerin s.l. (NTG); n = 60 coronary segments

Fig. 8. Coronary diameter changes under atenolol from control up to the 20th min; after administration of 0.8 mg nitroglycerin sublingually, a fast vasodilation is observed. Also, atenolol exhibits a significant, approximately 15% vasoconstriction.

maximum dilation of large epicardial coronary arteries, both with nitrates as well as calcium antagonists, averages approximately 30%, depending on initial diameters, i.e., vessel size. It is still debated whether a combination of both drugs could lead to still further dilation or whether each drug is able to induce maximum relaxation of the vascular smooth muscle cells, although by different mechanisms.

Finally, to which extent coronary vasodilation of high-grade, especially eccentric obstructions contributes to the inhibition of myocardial ischemia is still unclear [5,13,39]. Various clinical studies with the aim to elucidate this question showed equivocal results [13,15]. Nevertheless, recent studies confirmed that during exercise minimal diameters of eccentric stenoses further decrease [17,18], leading to a reversible, functional increase in the degree of stenosis which could further aggravate ischemia [5,18,19]. Feigl [10] and Heusch et al. [21] demonstrated that this paradoxical behavior is probably due to an increase in α-adrenergic tone during exercise.

It was, therefore, of particular interest to see whether β-blocking agents, suggested to lead to a secondary increase in α-adrenergic tone [14,46], would negatively affect coronary vasomotor tone, that is lead to vasoconstriction. From measurements of coronary blood flow at rest as well as during increased oxygen demand provoked by atrial pacing [9] or bicycle exercise [30,36] and by applying the "Precordial Xenon Clearance Technique" [31] to measure regional, transmural coronary blood flow, we know that total coronary resistance increases slightly in both situations under propranolol. This occurs in the presence of a significant decrease of transmural coronary blood flow due to a drastic drop in myocardial oxygen consumption [31]. Nevertheless, the increase in resistance was unexpected and was more than that assumed from the decrease in oxygen consumption. The primary location of the resistance changes, however, could not be detected by our methodology, i.e. large epicardial coronary arteries were not analyzed at that time; the increase in resistance was mainly attributed to the decrease in arteriolar luminal diameter, i.e., an increase in peripheral coronary resistance.

The observations from this study, however, suggest that proximal resistance is the primary location, as both the β_1 and β_2-blocker propranolol, as well as atenolol, demonstrated a

343

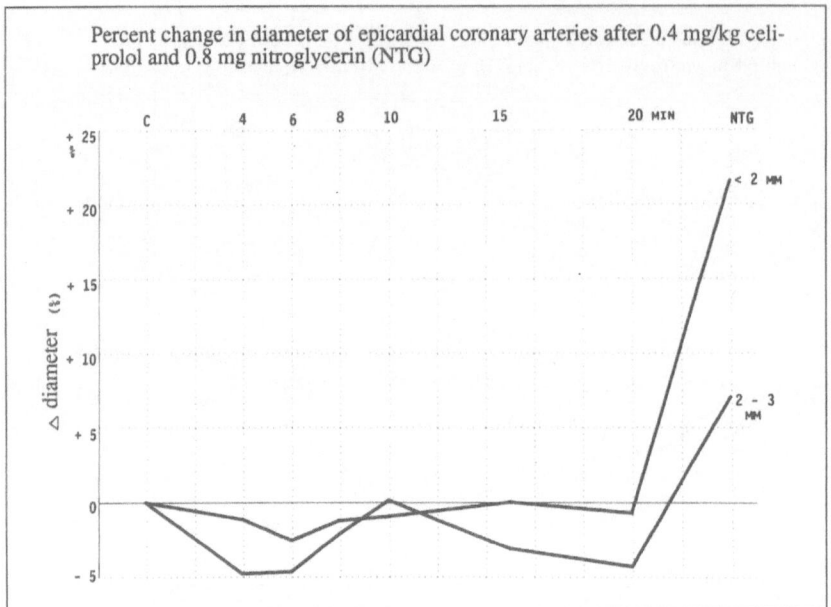

Percent change in diameter of epicardial coronary arteries after 0.4 mg/kg celiprolol and 0.8 mg nitroglycerin (NTG)

Fig. 9. Coronary diameter changes under celiprolol from control to the 20th min; in contrast to atenolol and propranolol, there was no significant decrease in the diameters, however, again a significant dilatation after nitroglycerin. (For details see text).

decrease in lumen size. This does not completely exclude that also arteriolar resistance might have increased, although this would have to be on a smaller level. It is of interest to note that there was no difference between propranolol and atenolol, but that celiprolol showed no vasoconstrictive effect, as was suggested previously [37].

The clinical significance of an increase in proximal resistance under β-blockade, especially in patients with advanced coronary artery disease, overt angina pectoris, eventually unstable angina, still needs to be further elucidated. An increased incidence of attacks of resting angina in patients with vasospastic or Prinzmetal's angina was, however, described previously [45] and also observed in some of our own patients with this type of angina [33]. In patients with stable angina pectoris, especially ischemia during exercise, caused by high-grade eccentric stenoses and a critical limitation of the increase in flow in the presence of an increased oxygen demand, the drastic drop in myocardial oxygen consumption caused by the β-blockade seems to override a further increase in resistance, even in the presence of a calcium-entry blocker [6,33]. Recent studies [1,19], however, demonstrated that the further decrease in minimal stenotic diameter under propranolol at rest was relieved during exercise, although not to the extent as observed with nitrates [1]. This was interpreted to indicate that the mild increase of flow during exercise overcomes the vasoconstricting effect of the β-receptor-blocker. This unexpected observation of Bortone et al. [1] would explain the absence of an increase of angina under β-blockade in the presence of high-grade obstructions. Nevertheless, further clinical studies, eventually combined with flow measurements, are needed to clarify this interesting, rather paradoxical behavior of β-blockade, vasoconstriction at rest, and mild dilation during exercise.

References

1. Bortone AS, Hess OM, Gaglione A, Suter T, Nonogi H, Grimm J, Krayenbühl HP (1990) Effect of intravenous propranolol on coronary vasomotion at rest and during dynamic exercise in patients with coronary artery disease. Circulation 81: 1225–1235

Coronary vasomotor tone in large epicardial coronary arteries

2. Bossaller C, Hehlert-Friedrich K, Rafflenbeul W, Jost St, Reschke V, Reil GH, Wagner TOF (1987) Endothelium-dependent vasodilation with substance P is preserved in patients with coronary artery disease. Circulation 76 (Suppl. IV): 53 (abstr.)
3. Bossaller C, Reschke V, Barthels H, Hertel R, Olbricht C, Fleck E (1988) Impaired acetylcholine – and nitroglycerin-induced vascular relaxation in cyclosporin-treated rats. Circulation 78 (Suppl. II): 452 (abstr.)
4. Bossaller C, Hehlert-Friedrich K, Jost S, Rafflenbeul W, Lichtlen P (1989) Angiographic assessment of human coronary artery endothelial function by measurement of endothelium dependent vasodilation. Eur Heart J 10 (Suppl F): 44–48
5. Brown BG, Bolson EL, Dodge HT (1984) Dynamic mechanisms in human coronary stenoses. Circulation 70: 917—922
6. Daniel W, Reil GH, Schober O, Creutzig H, Lichtlen PR (1986) Effects of the combined nifedipine and propranolol treatment on regional myocardial blood flow in coronary patients. In: Lichtlen PR (ed), 6th International Adalat Symposium, Excerpta Medica Amsterdam, pp. 414–421
7. Davies MJ, Woolf N, Rowles PM, Pepper J (1988) Morphology of the endothelium over atherosclerotic plaques in human coronary arteries. Br Heart J 60: 459–464
8. Engel HJ, Wolf R, Hundeshagen H, Lichtlen PR (1980) Different effects of nitroglycerin and nifedipine on regional myocardial blood flow during pacing-induced angina pectoris. Eur Heart J 1 (Suppl B): 53–58
9. Engel HJ, Lichtlen PR (1981) Beneficial enhancement of coronary blood flow by nifedipine; comparison with nitroglycerin and betablocking agents. Am J Med 71: 658–666
10. Feigl EO (1967) Sympathetic control of coronary circulation. Circ Res 20: 262–271
11. Freudenberg H, Lichtlen PR (1981) Das normale Wandsegment bei Koronarstenose – eine postmortale Studie. Z Kardiol 70: 863–869
12. Furchgott RF, Zawadzki JV (1980) The obligatory role of endothelial cells in the relaxation of arterial smooth muscles by acetylcholine. Nature 288: 373–376
13. Gage J, Hess O, Murakami T, Ritter M, Grimm J, Krayenbühl H (1986) Vasoconstriction of stenotic coronary arteries during dynamic exercise in patients with classic angina pectoris: Reversibility by nitroglycerin. Circulation 73: 865–876
14. Gaglione A, Hess OM, Corin WJ, Ritter M, Grimm J, Krayenbühl HP (1987) Is there coronary vasoconstriction after intra-coronary beta-adrenergic blockade in patients with coronary artery disease. J Am Coll Cardiol 10: 292–310
15. Ganz W, Marcus HS (1972) Failure of intracoronary nitroglycerin to alleviate pacing-induced angina. Circulation 46; 880–889
16. Henry PD (1984) Coronary artery spasm. In: Sperelakis N (ed). Physiology and Pathophysiology of the Heart. Boston, The Hague, Martinus Nijhoff Publishing, pp. 819–833
17. Hess OM, Gaglione A, Corin W, Grimm J, Krayenbühl HP (1986) Koronare Vasokonstriktion nach intrakoronarem Propranolol? Z Kardiol 75 (Suppl IV): 26 (abstr.)
18. Hess OM, Bortone A (1989) Coronary vasomotor tone during static and dynamic exercise. Eur Heart J 20 (Suppl F): 105–110
19. Hess OM, Bortone A, Gaglione A, Nonogi H, Grimm J, Krayenbühl HP (1989) Effect of intracoronary and intravenous propranolol on human coronary arteries. Eur Heart J 10 (Suppl F): 153–158
20. Heusch G, Deussen A (1983) The effects of cardiac sympathetic nerve stimulation on perfusion of stenotic coronary arteries in the dog. Circ Res 53: 8–15
21. Heusch G, Deussen A, Schipke JD, Thämer V (1984) Alpha-1 and alpha-2 adrenoceptor mediated vasoconstriction of large and small canine coronary arteries in vivo. J Cardiovasc Pharmacol 6: 961–968
22. Heusch G, Guth BD (1989) Neurogenic regulation of coronary vasomotor tone. Eur Heart J 10 (Suppl. F): 6–14
23. Ignarro LJ, Byrns NE, Wood KS, Chandhuri G (1988) Pharmacological evidence that endothelium-derived relaxing factor is nitric oxyde. J Pharm Exp Ther 244: 181–198
24. Jost S, Rafflenbeul W, Mogwitz B, Nellessen U, Hecker H, Lichtlen PR (1987) Effects of nisoldipine on coronary arteries – correlation to plasma levels. Eur Heart J 8 (Suppl. 2): 63 (abstr.)
25. Jost St, Deckers J, Nellessen U, Rafflenbeul W, Hecker H, Reiber JHC, Lippolt P, Hugenholtz PG, Lichtlen PR (1989) Computer-assisted contour analysis technique in coronary angiographic follow-up trials: results of the first angiograms from the INTACT-study. Z Kardiol 78: 23–32
26. Jost S, Rafflenbeul W, Gerhardt U, Hecker H, Nellessen U, Reil GH, Lichtlen PR (1989) Influence of ionic and non-ionic radiographic contrast media on the vasomotor tone of epicardial coronary arteries. Eur Heart J 10 (Suppl. F): 60–65
27. Jost S, Rafflenbeul W, Knop I, Bossaller C, Gulba D, Hecker H, Lippolt P, Lichtlen PR (1989) Drug

plasma levels and coronary vasodilation after isosorbide dinitrate chewing capsules. Eur Heart J 10 (Suppl F): 137–141

28. Jost S, Rafflenbeul W, Mogwitz B, Nellessen U, Bossaller C, Zwicky P, Hecker H, Lippolt P, Lichtlen PR (1989) Coronary vasodilation with dihydropyridines – a pharmacokinetic study. Eur Heart J 10 (Suppl F): 147–151

29. Jost St, Reil G, Knop I, Rafflenbeul W, Auricchio A, Frombach R, Gulba D, Hecker H, Lichtlen P (1989) Coronary vasodilation with nitrocompounds – is there a maximum? Z Kardiol 78 (Suppl. 2): 38–40

30. Lichtlen P, Albert H, Moccetti P (1971) Left ventricular dynamics at rest and during exercise under different betablocking agents (propranolol, LB 46) in patients with severe coronary artery disease. In: Kaltenbach M, Lichtlen P (eds) Coronary heart disease, International Symposium, Frankfurt; Thieme Stuttgart, pp. 205–221

31. Lichtlen P, Engel HJ, Hundeshagen H (1985) Assessment of regional myocardial blood flow using invasive techniques, especially the precordial Xenon Clearance Technique. In: Diethelim L, Heuck F, Olsson O, Vieten H, Zuppinger A, Hundeshagen H (eds.)Nuklearmedizin, Teil III. Springer Verlag, Berlin, Heidelberg, New York, Tokyo, pp. 65–111

32. Lichtlen P, Rafflenbeul W (1985) Effects of calcium antagonists on fixed and dynamic obstructions in patients with severe coronary artery disease. In: Fleckenstein A, van Breemen CR, Hofmeister F (eds.) Cardiovascular Effects of Dihydropyridine Type Calcium Antagonists and Agonists. Berlin, Heidelberg; Springer Verlag, pp. 381–407

33. Lichtlen P (1988) Derzeitiger Stand der Kombinationstherapie zur Behandlung der Angina pectoris bei Betablockern als Grundsubstanz. In: Kaíndl F, Kraupp O, Lehnert T, Lichtlen P, Schuster J, Siegenthaler W (eds.). Beta-Rezeptoren und Beta-Rezeptoren-Blocker, de Gruyter Verlag, Berlin, pp. 411–428

34. Lichtlen PR (1989) Wirkungsmechanismus der Nitrate, Stand 1988. Z Kardiol 78 (Suppl. II): 3–10

35. Meyer GP, Rafflenbeul W, Auricchio A, Jost S, Mügge A, Lichtlen P (1990) Einfluß von ATP auf pikardiale Koronararterien im Vergleich zu Nitroglycerin. Z Kardiol 79 (Suppl. I): 23 (abstr.)

36. Moccetti T, Halter J, Lichtlen P (1972) Koronare und linksventrikuläre Dynamik dreier Substanzen mit unterschiedlicher betablockierender Wirkung: Propranolol, Pindolol und Praktolol. Schw Med Wschr 102: 422–425

37. Raberger G (1988) Möglichkeiten zur Charakterisierung von Betarezeptoren-Blockern: Wirkung von Betarezeptoren-Blockern in Ruhe und während Belastung am wachen Hund. In: Kaíndl F, Kraupp O, Lehnert T, Lichtlen P, Schuster J, Siegenthaler W (eds.). Beta-Rezeptoren und Beta-Rezeptoren-Blocker. de Gruyter Verlag, Berlin, pp. 93–110

38. Rafflenbeul W, Urthaler F, Russell R, Lichtlen P, James TN (1980) Dilatation of coronary artery stenoses after isosorbide dinitrate in man. Br Heart J 43: 546–549

39. Rafflenbeul W, Lichtlen P (1982) Zum Konzept der dynamischen Koronarstenose. Z Kardiol 71: 439–444

40. Rafflenbeul W, Berger C, Jost S, Lichtlen PR (1987) Constriction of coronary arteries and stenoses with propranolol. Circulation 76 (Suppl IV): 276 (abstr.)

41. Rafflenbeul W, Berger C, Lichtlen P (1988) Einfluß von Betarezeptoren und Betarezeptorenblockern auf Koronargefäße. In: Kaíndl F, Kraupp O, Lehnert T, Lichtlen P, Schuster J, Siegenthaler W (eds.). Beta-Rezeptoren und Beta-Rezeptoren-Blocker. de Gruyter Verlag, Berlin, pp. 179–186

42. Rafflenbeul W, Bassenge E, Lichtlen P (1989) Competition between endothelium-dependent and nitroglycerin-induced coronary vasodilation. Z Kardiol 78 (Suppl 2):, 45–47

43. Reiber JHC, Kooijman CJ, Slager CJ, Gerbrands JJ, Schuurbiers JHC, den Boer A, Wijns W, Serruys PW, Hugenholtz PG (1984) Coronary artery dimensions with cineangiograms – methodology and validation of a computer-assisted analysis procedure. IEEE Trans Med Imag 3: 131–141

44. Reiber JHC, Serruys PW, Kooijman CJ, Wijns W, Slager CJ, Gerbrands JJ, Schuurbiers JCH, den Boer AD, Hugenholtz PG (1985) Assessment of short-, medium- and long-term variations in arterial dimensions from computer-assisted quantitation of coronary cine-angiograms. Circulation 71: 280–288

45. Robertson RM, Wood AJJ, Vaughn WK, Robertson D (1982) Exacerbation of vasotonic angina pectoris by propranolol. Circulation 65: 281–285

46. Vatner SF, Hintze TH (1983) Mechanisms of constriction of large coronary arteries by beta-adrenergic receptor blockade. Circ Res 53: 389–400

Authors' address:

Prof. Dr. P.R. Lichtlen, Division of Cardiology, Department of Medicine, Hannover Medical School, Konstanty-Gutschow-Str. 8, 3000 Hannover 61, FRG

The Poststenotic Vessel Segment during Dynamic Exercise: Effect of Oral Isosorbide-dinitrate

A. Gaglione*, O. M. Hess, C. Haemmerli, T. Suter, R. L. Kirkeeide, H. P. Osenberg, M. Muser, M. Anliker, K. L. Gould and H. P. Krayenbuehl

Department of Internal Medicine, Medical Policlinic, Cardiology, University Hospital, Institute for Biomedical Engineering, University of Zürich, Switzerland, and Division of Cardiology, University of Texas, Houston, Texas, USA

Summary

Coronary vasomotion of normal, stenotic, and poststenotic vessel segments was studied in 18 patients with coronary artery disease at rest, during submaximal bicycle exercise, and 5 min after sublingual nitroglycerin or oral isosorbide-dinitrate (ISDN) spray. Patients were divided into two groups: group 1 consisted of 10 patients with no premedication prior to exercise, and group 2 consisted of 8 patients receiving 120 mg long-acting ISDN orally 1 h before the procedure. Quantitative coronary arteriography was carried out in biplane projection using a semi-automatic computer system.

The normal vessel segment showed a trend toward a small increase in cross-sectional area during exercise in both groups (+ 3% in group 1 and + 4% in group 2, both NS). After sublingual nitroglycerin following exercise, there was a significant increase in group 1 (+ 29%, p < 0.001 vs rest) but not after ISDN spray in group 2 (+ 5%, NS vs rest). The stenotic vessel segment showed exercise-induced stenosis narrowing in group 1 (− 31%, p < 0.01 vs rest) which was prevented by oral ISDN (+ 6%, NS vs rest). After exercise, sublingual administration of nitroglycerin or ISDN spray was associated with no significant change in stenosis area in either group. The poststenotic vessel segment showed no significant vasomotion during exercise in both groups (area change + 6% in group 1 and + 7% in group 2), but poststenotic luminal area increased after sublingual nitroglycerin (group 1: + 15%, p < 0.01 vs rest) or ISDN spray (group 2: + 15%, p < 0.05 vs rest). The mean pulmonary artery pressure increased during exercise from 22 to 39 mmHg (p < 0.001) in group 1 and from 14 to 27 mmHg (p < 0.001) in group 2. At rest (p < 0.001) and during exercise (p < 0.01) mean pulmonary pressure was lower in group 2 than in group 1.

Thus, it is concluded that coronary vasomotion of the poststenotic vessel segment is only minimal during exercise and is not affected by coronary vasomotion of the stenotic vessel segment. Pretreatment with oral ISDN did not influence coronary vasomotion of the poststenotic vessel segment, but prevented exercise-induced stenosis narrowing. In the untreated patients, vasoconstriction of the stenotic vessel segment is limited to the site of the stenosis, and it appears that there is no release of vasoactive substances with vasoconstrictive influences on the poststenotic segment.

Introduction

The poststenotic vessel segment plays an important role in the pathogenesis of myocardial ischemia, since vascular tone of the poststenotic vessel segment is one of the factors regulating coronary perfusion in the poststenotic region. Most studies have focused on the anatomical and functional properties of the stenotic lesion, but only a few reports have dealt with the post-stenotic vessel segment, which might be influenced, not only by the stenosis itself, but also by the metabolism of the ischemic myocardium [10,14,19]. Exercise-induced stenosis narrowing

---

*A. Gaglione was supported by a fellowship grant of the European Society of Cardiology.

Table 1. Upright and supine bicycle exercise data

Group 1

	Upright bicycle exercise						Supine bicycle exercise					
	WL	%	HR	BP	RPP	ST	T	WL	%	HR	BP	RPP	T
Mean	133	90	119	160/68	191	0.17	5F	111	74	112	162/81	182	3A
±SD	37	23	28	22/8	52	0.12	5A	30	14	25	32/12	60	7F
p (upright vs supine)								*	*	NS	NS/*	NS	

Group 2 (ISDN)

Mean	104	72	115	153/80	177	0.11	5A	97	67	109	141/73	157	5A
±SD	23	16	20	29/16	47	0.10	3F	24	13	16	23/7	40	3F
p (upright vs supine)								NS	NS	NS	NS/NS	NS	

Legend: WL = workload at maximal exercise (W); % = workload as a percent of predicted work capacity; HR = heart rate (bpm); BP = systolic/diastolic blood presssure (mm Hg); RPP = rate-pressure product (mm Hg/min); ST = ST-segment depression (mV) during upright bicycle exercise; T = reason for termination of exercise test; A = angina pectoris; F = fatigue; *p < 0.05.

has been reported previously in patients with classical angina pectoris which was prevented by intracoronary injection of nitroglycerin prior to the exercise test. The exact mechanism of exercise-induced stenosis vasoconstriction is not clear and several factors such as a passive collapse of the disease-free vessel wall within the stenosis or active vasoconstriction either due to endothelial dysfunction or platelet aggregation with release of thromboxane A2 and serotonin have been discussed [2,4,14]. The purpose of the present study was to evaluate coronary vasomotion of the poststenotic vessel segment in patients with classical angina pectoris and to examine the effect of pretreatment with oral isosorbide-dinitrate (ISDN) on coronary vasomotion at rest and during exercise. It has to be realized, however, that only the large poststenotic arteries can be evaluated by quantitative coronary arteriography and not the small resistance vessels (i.e., arterioles) which are beyond angiographic resolution.

Patients and methods

Eighteen patients were included in the present study. A history of stable angina pectoris and/or ST-segment depression ≥ 0.1 mV in the upright bicycle exercise test was found in all patients. The patients were recruited on a consecutive basis when at arteriography a clearly visible coronary artery stenosis for quantitative evaluation was present. Patients were divided into two groups: group 1 consisted of 10 patients (mean age 53 years, range 39–61 years) with no premedication prior to exercise, and group 2 consisted of 8 patients (mean age 59 years, range 47–71 years) receiving 120 mg long-acting ISDN orally 1 h before the catheterization. There were no significant differences between the two groups in regard to clinical findings or baseline hemodynamics (Table 2). All medication was stopped at least 12–24 h before cardiac catheterization.

Study protocol

After the diagnostic coronary arteriography (Judkins technique) an interval of at least 10 min was allowed for dissipation of the effects of the contrast medium on coronary vasomotion. After the patient's feet were attached to the bicycle ergometer, biplane coronary arteriography was performed [8,13]. Aortic and pulmonary artery pressure were recorded immediately before coronary arteriography. Exercise was performed at two levels (2 min each), starting with a first

Vasomotion of the poststenotic vessel segment

level at 50–75 W, and continuing with a second level at 100–125 W. Biplane coronary arteriography was repeated at the end of each exercise level with the patient interrupting exercise
and holding his breath for 10–15 s. The exercise was terminated because of anginal pain, fatigue
or ST-segment depression of more than 0.2 mV. At the end of the exercise test, 1.6 mg
sublingual nitroglycerin in group 1 and 1 spray of ISDN (= 1.25 mg) in group 2 was administered and biplane coronary arteriography was repeated 5 min thereafter. There were no complications related to the procedure in any of the 18 patients.

Quantitative coronary arteriography

Quantitative evaluation of biplane coronary arteriograms was performed using an automatic
computer system [6,13,15]. The system is based on a 35 mm film projector (Tagarno A/S,
Horsens, Denmark), a slow-scan CCD-camera (image digitation) and a computer work-station
(Apollo DN 3000, Apollo Computer AG, Wangen, Switzerland, image storage and processing).
Calibration is performed by the isocenter technique which is based on two fixed reference points
in the center of the two image intensifiers [6]. Contour detection is carried out in biplane
projection using a geometric-densitometric edge detection algorithm (Fig. 1). Quantitative
assessment was performed in biplane projection in 11 of 18 patients. A single view was used
when difficulties were encountered in visualization of the stenosis in one of the two views
because of overlying vessels, contrast reflux into the aorta, or because the analyzed vessel was
orthogonal to one of the two projections (foreshortening). In these cases (five normal, seven
stenosed, and five poststenotic vessels), the computer constructed a circular lumen to calculate
the cross-sectional area.

Interobserver variability was found to be small for this system with a standard error of
estimate for biplane data of 0.14 mm (i.e., 4.1% of the mean vessel area). The methodology for
computerized analysis of coronary arteriograms has been described elsewhere [5,15,16]. Briefly,
a three-dimensional model of the vessel is constructed by matching center lines of the individual
biplane tracings, assuming the vessel cross-section to be ellipsoidal. The proximal and distal,
as well as the minimal cross-sectional areas of the vessel segment are calculated and printed out
by the computer [15,16]. Percent area stenosis was calculated in all patients from the minimal
luminal area and the two reference areas of the proximal and distal vessel.

Statistics

Comparisons of pressure and angiographic data in response to bicycle exercise and sublingual
nitroglycerin or ISDN spray were performed by a two-way analysis of variance for repeated
measurements. Comparisons between groups 1 and 2 were carried out by an unpaired Student's
t-test.

Results

Pressure measurements (Table 2)

Mean aortic and mean pulmonary artery pressure (Fig. 2) increased significantly in both groups
during bicycle exercise. Mean aortic pressure was lower ($p < 0.001$) at rest in group 2 than 1.
Mean pulmonary artery pressure was also lower at rest ($p < 0.001$) and during exercise
($p < 0.01$) in group 2 compared to group 1. Sublingual administration of nitroglycerin or ISDN
was associated with a fall in mean aortic pressure and mean pulmonary artery pressure in both
groups.

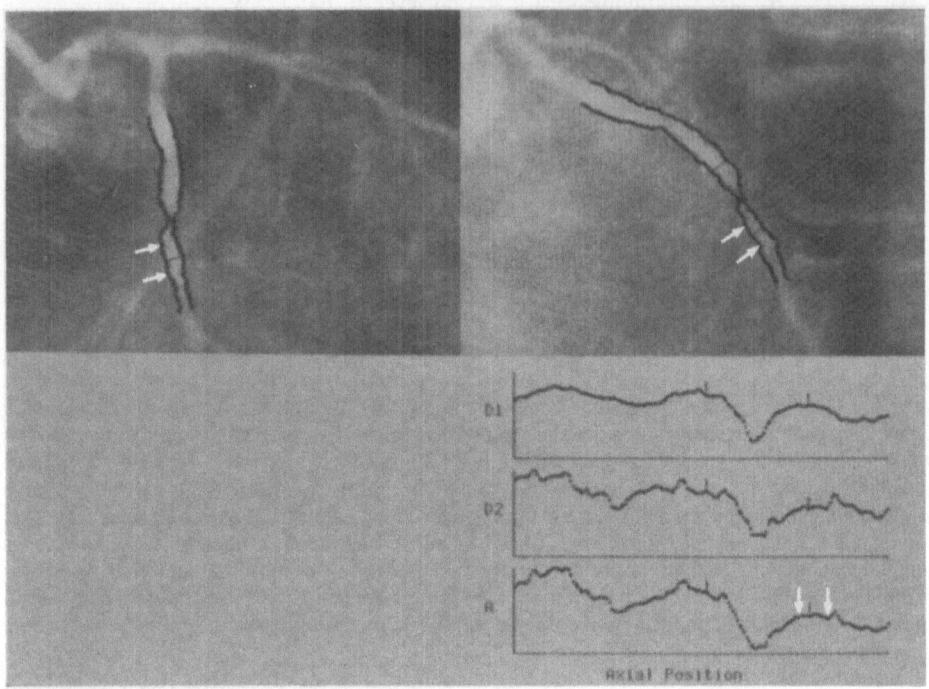

Fig. 1. Quantitative coronary arteriography is carried out in biplane projection using a semi-automatic system based on a 35 mm film projector, a slow-scan CCD-camera (image digitation) and a computer workstation (image storage and processing). Contour detection is performed in biplane projection using a geometric-densitometric edge-detection algorithm. Luminal areas of a normal, stenotic, and poststenotic (arrows) vessel segment are determined. Three to four images are analyzed for each sequence, and luminal areas of these images are averaged to reduce sampling error.

D1: Coronary diameter in the right anterior oblique projection (mm);
D2: Coronary diameter in the left anterior oblique projection (mm);
A: Coronary luminal area in biplane projection (mm^2).

Angiographic measurements (Table 3)

Changes in coronary luminal area of the normal, stenotic, and poststenotic vessel segments are shown in Fig. 3. The normal vessel segment showed only a slight, but not significant increase of cross-sectional area during exercise in both groups (+ 3% in group 1 and + 4% in group 2). After sublingual nitroglycerin there was a significant increase in group 1 (+ 29%, p < 0.001 vs rest), but not in group 2 after ISDN spray (+ 5%, NS vs rest). The stenotic vessel showed exercise-induced vasoconstriction of the luminal area in group 1 (− 31%, p < 0.01 vs rest) which was prevented by oral ISDN (+ 6%, NS vs rest) in group 2. Sublingual administration of nitroglycerin or ISDN spray was associated with a mild, but not significant increase in stenosis area. The luminal area of the poststenotic vessel showed no significant change during exercise in either group (+ 6% in group 1 and + 7% in group 2), but increased after sublingual nitroglycerin (group 1: + 15%, p < 0.01 vs rest) and ISDN spray (group 2: + 15%, p < 0.05 vs rest).

Table 2. Hemodynamic data

Group 1

	Heart rate (bpm)				Mean aortic pressure (mm Hg)				Mean pulmonary arterial pressure (mm Hg)			
	R	Ex1	Ex2	NTG	R	Ex1	Ex2	NTG	R	Ex1	Ex2	NTG
Mean	68	95	112	80	103	107	115	88	22	31	39	20
±SD	10	8	25	14	14	16	21	17	5	10	7	5
p (vs R)		*	***	NS		*	**	**		**	***	NS
p (vs EX)				***				***				***

Group 2 (ISDN)

				ISO				ISO				ISO
Mean	67	97	109	74	84	98	99	81	14	21	27	12
±SD	7	7	16	8	10	12	18	11	2	9	10	2
p (vs R)		*	***	NS		*	**	NS		*	***	NS
p (vs EX)				**				***				***
p (G1 vs G2)	NS	NS	NS	NS	**	NS	NS	NS	***	NS	**	***

Legend: R = Rest; Ex1, Ex2 = first and second level of supine bicycle exercise; NTG = 5 min after 1.6 mg sublingual nitroglycerin given at the end of exercise; ISO = 5 min after 1 spray of ISDN given at the end of exercise; G1 = Group 1; G2 = Group 2; *$p < 0.05$; **$p < 0.01$; ***$p < 0.001$.

Discussion

Previous experimental studies by Gould and coworkers [10] and Gould and Kelley [11] have shown that increasing stenosis severity is associated with a compensatory vasodilation of the poststenotic vessel segment which is greatest in the endocardium with some vasodilator reserve remaining in the epicardium. As the stenosis reaches complete occlusion, the poststenotic vessels become maximally vasodilated [10]. It has to be realized that the large poststenotic vessel segments have only a conduit function and contribute little to coronary vascular resistance in contrast to the small arteries, i.e. arterioles. Poststenotic vasodilation might have adverse effects on coronary blood flow because coronary stenosis severity increases by vasodilation of the distal coronary bed, simply due to a relative increase in percent area stenosis which leads to more ischemia and to more vasodilation downstreams to the stenosis, although minimal luminal area might not change. In contrast to these observations, Schwartz and Bache [19] demonstrated in the experimental animal that arteriolar dilation (produced by transient coronary occlusion or by intracoronary adenosine) caused a decrease in poststenotic coronary diameter which was directly related to distal coronary pressure [19]. These authors concluded that the close relationship between pressure and diameter of the poststenotic coronary arteries suggests a passive phenomenon with narrowing or collapse of the stenotic vessel segment with increasing stenosis severity.

Coronary vasomotion of the poststenotic vessel segment during exercise

In the present study a mild, insignificant increase in poststenotic vessel area of + 6% in group 1 and + 7% in group 2 was observed during exercise. Only two patients in group 1 with no pretreatment prior to exercise showed vasoconstriction of the poststenotic vessel segments. All other patients in groups 1 and 2 showed an increase in vessel area that was similar to the increase in coronary luminal area of the normal vessel segment (Fig. 3). A decrease in distal coronary luminal area during exercise, as it was suggested by Schwartz and Bache [19], was

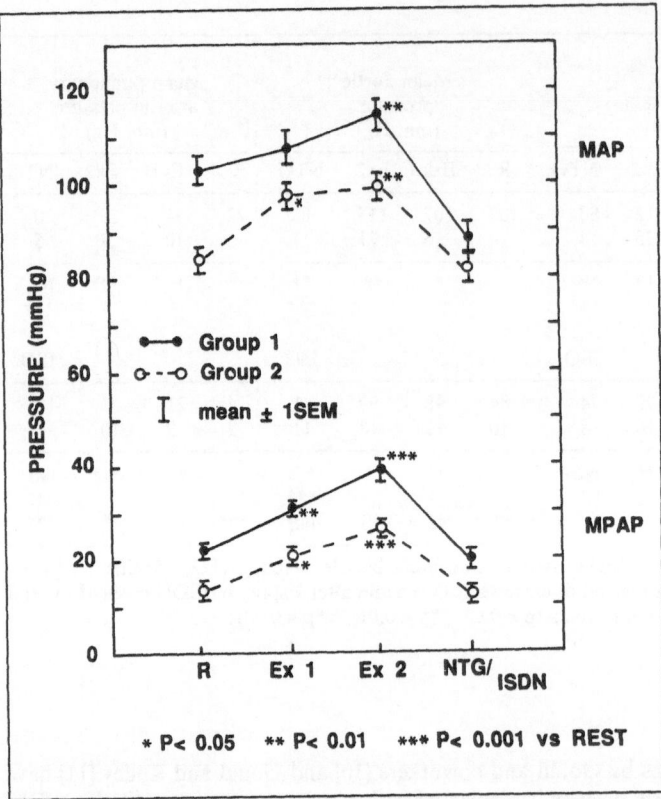

Fig. 2. Mean aortic (MAP) and mean pulmonary artery pressure (MPAP) at rest (R), during a first exercise level at 50–75 W (Ex1) and a second level of exercise at 100–125 W (Ex2), as well as 5 min after sublingual administration of 1.6 mg nitroglycerin (NTG) or 1 spray (= 1.25 mg) of ISDN (ISDN) at the end of the exercise. Group 1 (solid symbols) consisted of 10 patients with coronary artery disease who received no pretreatment with ISDN prior to exercise, and group 2 of 8 patients (open symbols) who received 120 mg ISDN orally 1 h prior to catheterization. After pretreatment with ISDN, MAP and MPAP are significantly lower in group 2 than in group 1, but MAP and MPAP increase significantly during exercise in both groups. The increase in MPAP during exercise suggests myocardial ischemia, despite pretreatment with ISDN.

observed in only two out of 18 patients; this might be due to the fact that exercise is a very potent vasodilator stimulus with a three- to five-fold increase in coronary blood flow and a significant increase in coronary driving pressure. Thus, a decrease in poststenotic pressure might be, at least in part, counterbalanced by the increase in driving pressure. However, mean aortic pressure was lower in the ISDN-treated than in the untreated patients; this might have offset the increase in driving pressure during exercise and, thus, poststenotic perfusion pressure might not have changed. If the concept of Schwartz and Bache is correct, then we have to assume that the poststenotic pressure increased slightly during exercise in all except two patients. It is likely that exercise was associated with myocardial ischemia, as indicated by the increase in mean pulmonary artery pressure and the occurrence of angina pectoris in 10 of 18 patients, and that metabolic changes with release of adenosine caused submaximal or maximal vasodilation of the large poststenotic vessel segments which might have overridden the decrease in coronary diameter due to the decrease in poststenotic distending pressure. Previous ex-

Table 3. Quantitative coronary arteriography

Group 1	Normal vessel (mm²)				Stenotic vessel (mm²)				Poststenotic vessel (mm²)			
	R	Ex1	Ex2	NTG	R	Ex1	Ex2	NTG	R	Ex1	Ex2	NTG
Mean	3.5	3.6	3.5	4.4	1.2	0.9	0.8	1.3	5.3	5.7	5.8	6.2
±SD	1.0	1.0	1.0	1.1	0.6	0.6	0.6	0.5	1.8	1.7	1.7	2.0
p (vs R)		NS	NS	***		*	**	NS		NS	NS	**
p (vs Ex)				***				***				NS
Group 2 (ISDN)				ISO				ISO				ISO
Mean	7.5	7.8	7.8	8.0	1.2	1.3	1.3	1.5	4.1	4.5	4.6	4.8
±SD	2.4	2.4	2.4	2.5	0.8	0.9	1.0	1.0	1.7	1.7	1.7	1.8
p (vs R)		NS	NS	NS		NS	NS	NS		NS	NS	*
p (vs Ex)				NS				NS				NS
p (G1 vs G2)	**	**	**	*	NS	NS	NS	NS	NS	NS	NS	NS

Legend: R = Rest; Ex1, Ex2 = first and second level of supine bicycle exercise; NTG = 5 min after 1.6 mg sublingual nitroglycerin given at the end of exercise; ISO = 5 min after 1 spray of ISDN given at the end of exercise. G1 = Group 1; G2 = Group 2; *p < 0.05; **p < 0.01; ***p < 0.001.

perimental studies have shown [1,7,18] that, even during ischemia, maximal vasodilation might not be reached in the subendocardium, since additional vasodilation can be recruited pharmacologically. Prevention of exercise-induced stenosis vasoconstriction by oral ISDN had no effect on coronary vasomotion of the large poststenotic vessel segment. This suggests that a passive phenomenon such as collapse of the stenotic vessel segment is not a major determinant of poststenotic coronary vasomotion, since otherwise, a different response of the large poststenotic vessel segment would have been expected during exercise in the two groups. Furthermore, vasoconstriction is limited to the site of the stenosis and, hence, it appears that there is no exercise-induced release of vasoactive substances with vasoconstrictive influences on the poststenotic vessel segment. Thus, coronary vasomotion of the poststenotic vessel is not only pressure-dependent, as suggested by the study of Schwartz and Bache [19], but flow-dependency might be operative as well [2] with release of the endothelium-derived relaxing factor (EDRF). This latter mechanism is likely to be initiated when flow is increased via local metabolic factors during high-demand situations such as bicycle exercise.

Effect of oral ISDN on coronary vasomotion

Administration of oral ISDN 1 h prior to coronary arteriography prevented exercise-induced stenosis vasoconstriction [8,12,13]. This effect of ISDN might be important for its anti-ischemic action during exercise, besides its effect on pre- and afterload. Previously, we have shown that intracoronary injection of nitroglycerin [8] and diltiazem [17] prevented exercise-induced stenosis narrowing, probably due to the strong vasodilating properties of the two compounds. A similar beneficial effect on coronary vasomotion was observed with oral ISDN treatment, but in contrast to intracoronary nitroglycerin and diltiazem, mean aortic (= afterload) and mean pulmonary (= preload) pressure were significantly lower at rest and during exercise with than without oral ISDN. The effect of ISDN on the normal vessel is similar to the control group, since no exercise-induced vasodilation was observed in either group. This is at variance with previous reports from our laboratory [3,8,12]; a significant increase in cross-sectional area of the normal vessel segment of approximately 20–25% was found during exercise. However, in the present study only a minimal (not significant) increase of 3% was found during exercise.

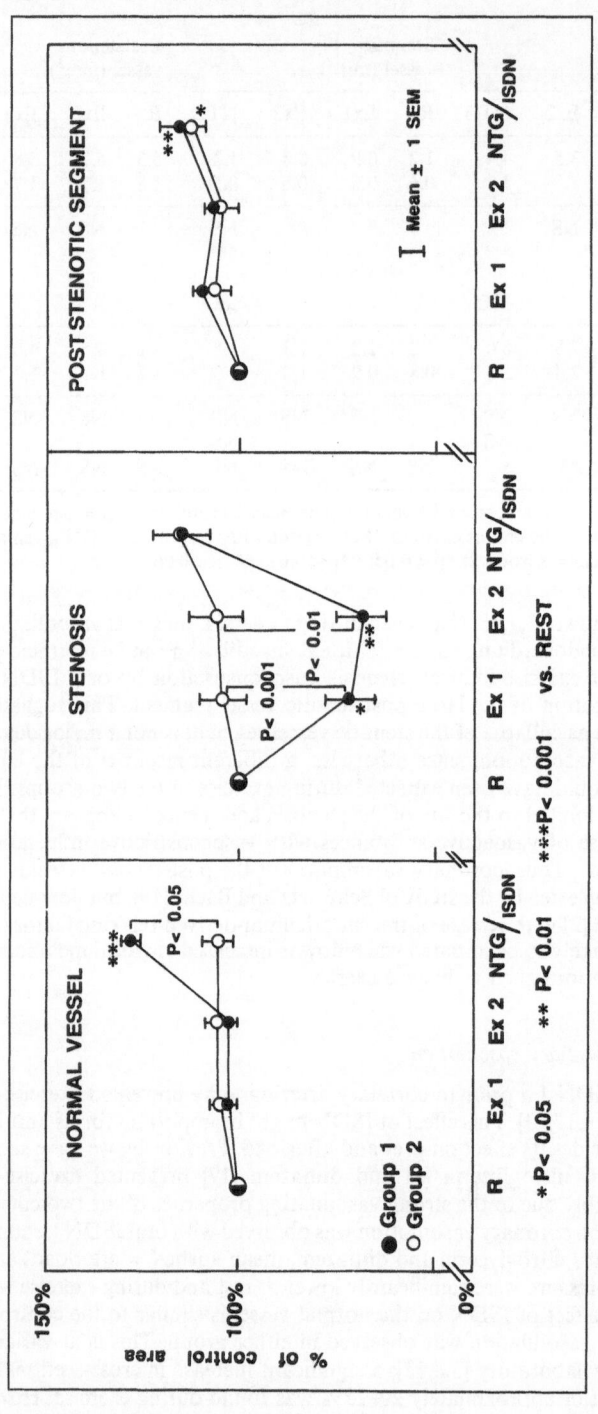

Fig. 3. Coronary vasomotion of the normal (left), stenotic (middle), and poststenotic (right) vessel segment in 18 patients with coronary artery disease. Group 1 consisted of 10 patients with coronary artery disease (solid symbols) with no pretreatment prior to exercise, and group 2 of 8 patients (open symbols) who received 120 mg ISDN orally 1 h prior to catheterization. Measurements were carried out at rest (R), during a first level (Ex1) and a second level of exercise (Ex2) as well as 5 min after 1.6 mg sublingual nitroglycerin or 1 spray (1.25 mg) of ISDN. The stenotic vessel in group 1 showed exercise-induced stenosis narrowing that was prevented by 120 mg ISDN in group 2. For further explanation see text.

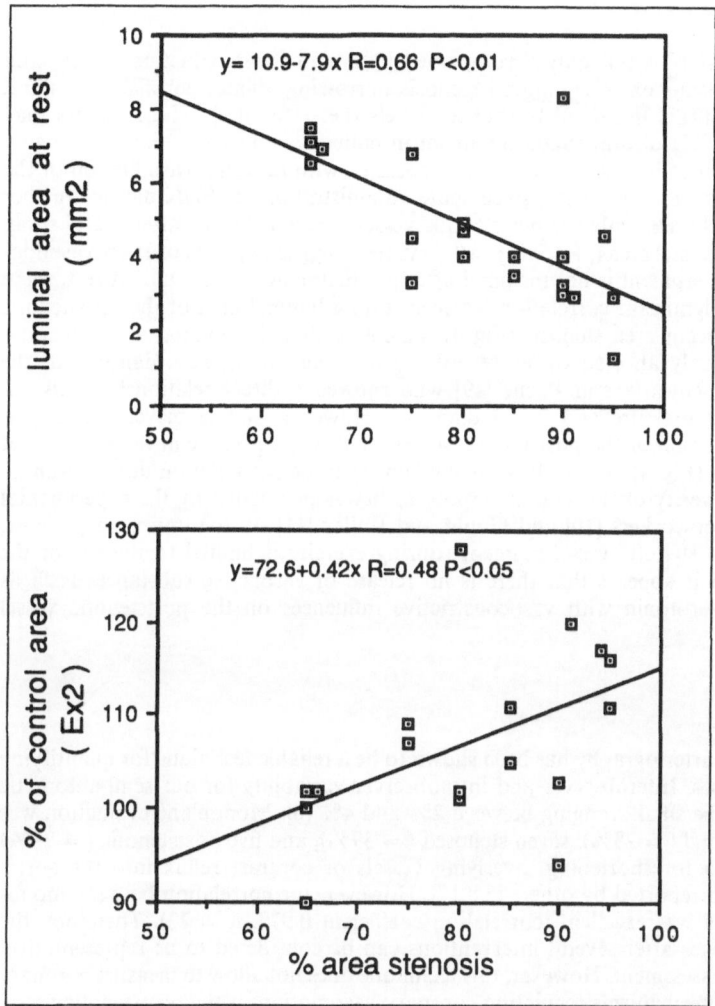

Fig. 4. Correlation between percent area stenosis and luminal area of the poststenotic vessel segment at rest (upper panel) as well as between percent area stenosis and percent change in luminal area of the poststenotic vessel segment during exercise (lower panel). There is an inverse relationship between percent area stenosis and resting luminal area of the poststenotic vessel segment, suggesting that the decrease in distending pressure is associated with a decrease in luminal area, as it was shown by Schwartz and Bache in the experimental animal [19]. However, a positive correlation is found between percent area stenosis and percent change in luminal area of the poststenotic vessel segment during exercise, suggesting that exercise-induced vasodilation of the poststenotic vessel segment is dependent on the severity of the stenotic lesion, as it was shown by Gould and Kelley [11]. There was no difference between group 1 and 2. For further explanation see text.

This has to be explained by the fact that these vessels are probably not completely normal (no luminal irregularities, but atherosclerotic changes of the wall) and react abnormally to exercise [9]. In true normal arteries an increase of 26% of the proximal and 45% of the distal epicardial cross-sectional area was reported during supine bicycle exercise [3].

Clinical implications

Oral treatment with ISDN is not only accompanied by a significant reduction in pre- and afterload, but also prevents exercise-induced stenosis narrowing, dilates coronary collateral vessels, and possibly affects the small resistance vessels (i.e. arterioles). These factors may explain the beneficial effect of oral ISDN treatment in patients with exercise-induced angina pectoris. Administration of long-acting ISDN is associated with maximal vasodilation of the normal and the stenotic vessel segment, since acute administration of ISDN did not further increase coronary vessel area. Only the poststenotic vessel segment showed after ISDN spray a significant increase in vessel area (+ 15%, $p < 0.05$ vs rest) suggesting that coronary vasodilation of the poststenotic segment is not maximal after pretreatment with ISDN. At rest, there was a weak although significant correlation between resting luminal area of the poststenotic vessel segment and percent area stenosis (Fig. 4) indicating that the luminal area beyond a stenotic lesion is inversely affected by its severity. This would be in accordance with the experimental data of Schwartz and Bache [19] who showed a direct relationship between poststenotic distending pressure and vessel diameter. However, supine bicycle exercise is associated with vasodilation of the poststenotic vessel segments, especially in the presence of a severe stenotic lesion (Fig. 4). Thus, the degree of poststenotic vasodilation during exercise is dependent on the severity of the stenotic lesion, as it was postulated by the experimental studies of Gould and coworkers [10] and Gould and Kelley [11]. In the untreated patients, vasoconstriction of the stenotic vessel segments during exercise is limited to the site of the stenosis (group 1) and it appears that there is no release of vasoactive substances such as thromboxane A2 or serotonin with vasoconstrictive influences on the poststenotic vessel segment.

Limitations

Quantitative coronary arteriography has been shown to be a reliable technique for quantifying coronary artery stenoses. Interobserver and intraobserver variability for our semi-automatic system were found to be small, ranging between 2% and 4% [6]. Monoplane evaluation was carried out in five normal (= 28%), seven stenosed (= 39%), and five poststenotic (= 28%) vessel segments due to foreshortening, overlying vessels or contrast reflux into the aorta. Similar data have been reported by others [5,8,17]. However, the correlation between mono-plane and biplane data was excellent (correlation coefficient 0.979, n = 22). Therefore, the observed relative changes after several interventions can be considered to be representative, even with monoplane assessment. However, this technique does not allow to measure coronary blood flow. Thus, the mechanisms regulating coronary vasomotion of the poststenotic vessel segment remain speculative, because we do not know poststenotic blood flow, pressure or metabolic factors. Nevertheless, the present observation of a linear inverse relationship between percent area stenosis and luminal area of the poststenotic vessel segment seems to be of particular interest and confirms that there is a close interplay between vasomotion of the stenotic and poststenotic coronary artery.

References

1. Aversano T, Becker LC (1985) Persistence of coronary vasodilator reserve despite functionally significant flow reduction. Am J Physiol 248: H403–H411
2. Bassenge E, Stewart DJ (1988) Interdependence of pharmacologically induced and endothelium-mediated coronary vasodilation in antianginal therapy. Cardiovasc Drugs Ther 1: 47–55
3. Bortone A, Hess OM, Eberli FR, Nonogi H, Marolf AP, Grimm J, Krayenbuehl HP (1989) Abnormal coronary vasomotion during exercise in patients with normal coronary arteries and reduced coronary flow reserve. Circulation 79: 516–527
4. Brown BG, Bolson EL, Dodge HT (1984) Dynamic mechanisms in human coronary stenosis. Circulation 42: 917–922

Vasomotion of the poststenotic vessel segment

5. Brown BG, Bolson EL, Frimer M, Dodge HT (1977) Quantitative coronary arteriography: estimation of dimensions, hemodynamic resistance and atheroma mass of coronary artery lesions using the arteriogram and digital computation. Circulation 55: 329–337
6. Büchi M. Hess OM, Kirkeeide R, Suter T, Muser M, Osenberg HP (1990) Niederer P, Anliker M, Gould L, Krayenbuehl HP. Validation of a new automatic system for biplane quantitative coronary arteriography. Int J Cardiovasc Imaging, in press
7. Canty JM, Klocke FJ (1985) Reduced regional myocardial perfusion in the presence of pharmacologic vasodilator reserve. Circulation 71: 370–377
8. Gage JE, Hess OM, Murakami T, Ritter M, Grimm J, Krayenbuehl HP (1986) Vasoconstriction of stenotic coronary arteries during dynamic exercise in patients with classic angina pectoris: reversibility by nitroglycerin. Circulation 73: 865–876
9. Gordon JB, Ganz P, Nabel EG, Fish RD, Zebede J, Mudge GH, Alexander RW, Selwyn AP (1989) Atherosclerosis influences the vasomotor response of epicardial coronary arteries to exercise. J Clin Invest 83: 1946–1952
10. Gould KL, Lipscomb K, Calvert C (1975) Compensatory changes of the distal coronary vascular bed during progressive coronary constriction. Circulation 51: 1085–1094
11. Gould KL, Kelley KO (1982) Physiological significance of coronary flow velocity and changing stenosis geometry during coronary vasolidation in awake dogs. Circ Res 50: 695–704
12. Hess OM, Bortone A, Eid K, Gage JE, Nonogi H, Grimm J, Krayenbuehl HP (1989) Coronary vasomotor tone during static and dynamic exercise. Eur Heart J 10 (Suppl F): 105–110
13. Hess OM, Büchi M, Kirkeeide R, Niederer P, Anliker M, Gould L, Krayenbuehl HP (1990) Potential role of coronary vasoconstriction in ischaemic heart disease: effect of exercise. Eur Heart J 11 (Suppl B): 58–64
14. Kanatsuka H, Lamping KG, Eastham CL, Marcus ML (1990) Heterogeneous changes in epimyocardial microvascular size during graded coronary stenosis. Circ Res 66: 389–396
15. Kirkeeide RL, Gould KL (1984) Cardiovascular imaging: Coronary artery stenosis. Hosp Pract 19: 160–163
16. Kirkeeide RL, Gould KL, Parsel L (1986) Assessment of coronary stenoses by myocardial perfusion imaging during pharmacologic coronary vasodilation. VII. Validation of coronary flow reserve as a single integrated functional measure of stenosis severity reflecting all its geometric dimension. J Am Coll Cardiol 7: 103–113
17. Nonogi H, Hess OM, Ritter M, Bortone AS, Corin WJ, Grimm J, Krayenbuehl HP (1988) Prevention of coronary vasoconstriction by diltiazem during dynamic exercise in patients with coronary artery disease. J Am Coll Cardiol 12: 892–899
18. Pantely GA, Bristow JD, Swenson LJ, Ladley HD, Johnson WB, Anselone CG (1985) Incomplete coronary vasodilation during myocardial ischemia in swine. Am J Physiol 249: H638–H647.
19. Schwartz JS, Bache RJ (1985) Effect of arteriolar dilation on coronary artery diameter distal to coronary stenoses. Am J Physiol 249: H981–H988

Authors' address:

Otto M. Hess, M.D., Medical Policlinic, Cardiology, University Hospital, 8091 Zürich/Switzerland

α-Adrenergic Coronary Constriction in Effort Angina

G. Berkenboom and P. Unger

Department of Cardiology, Hôpital Erasme, Bruxelles, Belgium

Summary

In order to assess the role of α-adrenergic coronary tone in exercise-induced ischemia, 23 patients with chronic stable angina underwent, after coronary angiography, a symptom-limited supine exercise test on a cyclo-ergometer.

After recovery, either phentolamine (for the first nine patients) or indoramin (for the following nine patients) was directly injected into the most diseased vessel at identical doses (2 mg over 5 min). In the remaining 5 patients, a placebo was injected. Immediately thereafter the same exercise (identical workloads and exercise duration) was repeated. During exercise 1, heart rate (HR), mean blood pressure, and cardiac index increased by 51%, 23% and 33% in the phentolamine group, and by 45%, 15%, and 33% in the indoramin group.

After intracoronary injection of phentolamine or indoramin, control values (including pulmonary artery wedge pressure (PA wedge)) at rest did not change significantly. During exercise 2, HR, mean blood pressure, and cardiac index increased in a similar way; however, the increase in PA wedge was less pronounced ($p < 0.01$ in the phentolamine group and $p < 0.05$ in the indoramin group). ST-segment depression at the end of exercise 2 was significantly smaller for identical workloads and double products in the phentolamine group: 1.5 ± 0.3 mm vs 2.5 ± 0.3 mm ($p < 0.01$). However, these changes did not reach a statistical significance in the indoramin group: 1.7 ± 0.2 mm vs 2.0 ± 0.1 mm (NS). ST/HR slope in exercise 2 decreased by 51% ($p < 0.01$) in the phentolamine group and by 34% ($p < 0.05$) in the indoramin group. In the placebo group, exercise 2 was identical to exercise 1 and the ST/HR slopes were quite reproducible. These results show a less severe ischemic response after intracoronary α-blockade. Therefore, our results argue for a role of α-adrenergic coronary tone in exertional angina. The relatively higher efficiency of phentolamine vs indoramin suggests that α_2-adrenergic mechanisms contribute to the inappropriate coronary vasoconstriction during exercise in these patients.

Introduction

Studies of patients with chronic stable angina by Holter monitoring suggest that an impairment of coronary blood flow, rather than an increase of myocardial metabolic demands, is more often responsible for the ischemia occurring during normal daily life [5]. Moreover, in these patients this impairment in coronary blood flow seems to be triggered by adrenergic stimuli such as the cold pressor test [16] or isometric exercise [4]. Congruent with this hypothesis is our finding [1] that an intracoronary injection of a non-selective α-adrenergic blocker attenuates the exercise-induced ischemia in a small group of patients with chronic stable angina.

The aims of the present study were to apply our protocol to a larger number of patients and to determine whether a selective α_1-blockade (with indoramin) in the coronary circulation could be more efficient than a non-selective one (with phentolamine).

Methods

Patient selection

For inclusion in the study each patient had to have a typical exertion angina, at least 1.5 mm of ST-segment depression during a preliminary symptom-limited exercise, and a stable symptomatology for the last 3 months.

The coronary angiography was performed for evaluation of their coronary artery disease. They were catheterized with a standard 8F pigtail, Judkins coronary and Swan-Ganz thermodilution catheters. All cardiac medications were discontinued at least 12 h before the procedure. After coronary angiography, the film was reviewed on videotape and only patients with good ejection fraction ($\geqslant 55\%$) and at least one high-grade coronary stenosis ($\geqslant 70\%$) on an artery supplying a non-infarcted area were included in the study. The last criterion of selection was the absence of significant alterations ($< 5\%$) of peripheral hemodynamic parameters after intracoronary injection of phentolamine or indoramin. Doses higher than 0.4 mg/min in 5 min induced for both drugs significant changes in heart rate and blood pressure. In addition, according to previous studies [6, 19] in case of intravenous administration 1 mg of phentolamine is equivalent to 1 mg of indoramin with regard to the fall in blood pressure. However, even with 0.4 mg/min in 5 min for both drugs, peripheral effects were observed in three patients (one in the phentolamine group and two in the indoramin group) and these patients were excluded.

Finally, 23 patients were enrolled in the study; the first nine patients were given intracoronary phentolamine, nine others were given intracoronary indoramin, and the last five patients served as control subjects: they received an intracoronary injection of 5% dextrose in water. Each patient gave informed consent to the study and the investigation protocol was approved by the Institutional Review Board of this hospital.

Study design

After completing routine coronary angiography the patients had a 25 min recovery period. Thereafter, they performed the first exercise test in supine position. The workload was increased by 30 W every 4 min and the beginning was adjusted individually from a previous exercise so that each patient reached the end point between 3 and 6 min after starting. Patients were asked to indicate the onset of angina.

The exercise was continued until the onset of moderately severe chest pain. Hemodynamic and electrocardiographic data were monitored and recorded exactly in the same manner as previously described [1]. Briefly, heart rate, blood pressure, and pulmonary artery pressure were continuously monitored and recorded on an optical strip-chart recorder. Pulmonary wedge pressure (PA wedge) was measured every minute, cardiac output was measured by thermodilution at rest and at end-exercise. On a 12-lead electrocardiogram (ink-jet Mingograph), the lead with the greatest ST-changes was continuously recorded. ST-segment depression was measured as the horizontal or downsloping depression of a point 80 ms after the QRS complex. Measurements were averaged over four consecutive complexes in which the isoelectric line was stable. R-R intervals were averaged over the same consecutive complexes and converted into beats/min. Every minute 2 means of ST-segment depression were plotted against the values of heart rate. From the onset to maximal ST changes, a regression line was calculated to determine an ST/heart rate slope.

Immediately after completion of exercise 1, the ostium of the most diseased coronary artery that did not supply an infarcted area, was catheterized; a small quantity of dye (less than 1 ml) was injected to check the position of the catheter. After all hemodynamic parameters had returned to control levels (reached in approximately 30 min) the intracoronary injection was started. Phentolamine or indoramin were injected at a rate of 0.4 mg/min for 5 min into the coronary ostium of the vessel(s) with the most severe stenosis. Intracoronary phentolamine was administered in nine patients, indoramin in nine others. The last five patients served as control subjects: they received 5% dextrose in water according to the same protocol. At the end of the coronary perfusion, the coronary catheter was removed from the ostium to prevent a wedging into the coronary artery. Immediately thereafter, and if the control hemodynamic values had not appreciably changed (as mentioned previously), patients exercised a second time under the same conditions; the same parameters were monitored. This second exercise test was discontinued as soon as the level of the first test was reached. This second exercise was, therefore, the exact reproduction of exercise 1 with regard to the workloads and the duration. Lastly, patients were asked to quantify and to compare their chest pain (on a scale of 0 to 4+) during both exercises.

Statistical analysis

The two-tailed t-test for paired data was used for statistical analysis. Values are given as mean $\pm$ SEM; $p < 0.05$ was considered significant.

Results

1. Patients characteristics

Details of clinical, electrocardiographic, and angiographic characteristics of the 23 patients studied are listed in Table 1. All these patients had at least one high-grade stenosis and a good left ventricular function (Table 1). One-vessel disease was present in one patient of the phentolamine group (No. 9), in three patients of the indoramin group (No. 14, 16 and 17), and in none of the control group. All the remaining patients had two- or three-vessel disease. Among them, four patients in the phentolamine group, one in the indoramin group, and one in the placebo group had an occluded vessel supplied by collaterals arising from an artery with a significant stenosis, into which the α-blocker was infused.

2. Phentolamine and indoramin groups

During exercise 2, anginal symptoms decreased (by 1+ on the score scale) in seven patients in the phentolamine group and in four patients in the indoramin group. In the others, these anginal scores remained unchanged. The results at rest and end-exercise in both exercises are listed in Table 2.

2.1. *Hemodynamic data*: During exercise 1, heart rate, blood pressure and cardiac index increased by 51%, 23%, 33% in the phentolamine group and by 45%, 15%, 33% in the indoramin group.

After intracoronary injection of phentolamine, the control values did not change: heart rate was 82 ± 3 vs 80 ± 3 beats/min at rest before exercise 1, mean blood pressure 101 ± 6 vs 102 ± 5 mm Hg, double product 11.186 ± 790 vs 10.950 ± 664 mm Hg·min^{-1} and cardiac index 3.0 ± 0.2 l/min/m^2 at both controls.

In the same way, after intracoronary injection of indoramin, the control values also did not change: heart rate was 75 ± 3 vs 77 ± 3 beats/min at rest before exercise 1, mean blood pressure 112 ± 4 vs 111 ± 4 mm Hg, double product 11.777 ± 758 vs 12.138 ± 750 mm Hg·min^{-1} and cardiac index was also similar (3.3 ± 0.2 l/min/m^2) at both controls.

During exercise 2, heart rate, mean blood pressure, and cardiac index rose in a similar way as during exercise 1: by 50%, 25%, 40% in the phentolamine group and by 51%, 14%, 33% in the indoramin group. However, the increase in PA wedge was smaller in the phentolamine group: at end-exercise 2, PA wedge was 19 ± 2 vs 26 ± 2 mm Hg at end-exercise 1 ($p < 0.01$). In the indoramin group, the decrease in PA wedge at the end of exercise 2 was less obvious but still significant. In two patients with three-vessel disease (No. 13 and 18), PA wedge remained unchanged at the end of exercise 2. However, for the whole group, PA wedge was 24 ± 2 mm Hg at the end of exercise 2 vs 27 ± 1 mm at the end of exercise 1 ($p < 0.05$).

361

Table 1. Physical, electrocardiographic, and angiographic characteristics of the patients

Patients	Years	Previous MI	Coronary disease Left	Right	Visible collaterals	EF %	Lead with the greatest ST-changes	Intracoronary injection
1	40	0	80% LAD / 95% CX	80% RCA		69	V 5	LCA
2	50	0	70% LAD / 70% CX			71	V 4	LCA
3	54	SMI (INF)	80% LAD / 80% CX	100% RCA	CX to RCA	59	V 6	LCA
4	57	TMI (INF)	80% LAD / 60% CX	100% RCA	LCA to RCA	63	V 4	LCA — Phento
5	52	SMI (INF)	100% CX	80% RCA	RCA to CX	65	V 4	RCA
6	57	0	100% LAD	70% RCA	RCA to LAD	61	V 5	RCA
7	54	0	90% LAD / 70% CX	–	–	73	V 5	LCA
8	61	0	80% LAD / 80% CX	–	–	69	V 4	LCA
9	69	0	50% LAD	90% RCA	–	74	II	RCA
10	52	TMI (INF)	95% CX	100% RCA	–	58	V 5	LCA
11	60	TMI (Post)	100% CX	80% RCA	–	57	V 4	RCA
12	66	TMI (INF)	90% LAD / 90% CX	100% RCA	–	55	V 5	LCA
13	47	0	70% LAD / 70% CX	100% RCA	CX to RCA	58	V 5	LCA
14	53	0	0	90% RCA	–	59	V 5	RCA — Indo
15	56	0	90% LAD / 80% CX	–	–	69	V 5	LCA
16	41	0	85% LAD	–	–	71	V 5	LCA
17	54	0	90% LAD	–	–	68	V 5	LCA

α-Adrenergic coronary constriction in effort angina

18	58	0	90% LAD 85% CX	70% RCA	–	73	V 5	LCA
19	61	TMI (INF)	75% LAD 75% CX	100% RCA	–	60	V 4	LCA
20	65	0	90% LAD	70% RCA	–	73	V 5	LCA
21	67	SMI (INF)	90% LAD 100% CX	–	RCA to CX	65	V 5	LCA Placebo
22	55	TMI (INF)	80% LAD	100% RCA	–	59	V 6	LCA
23	42	0	90% LAD 70% CX	–	–	59	V 6	LCA

SMI = Subendocardial myocardial infarction; TMI = transmural myocardial infarction; INF = inferior; LAD = left anterior descending artery; CX = left circumflex artery; LCA = left coronary artery; RCA = right coronary artery; phento = phentolamine; indo = indoramin.

Table 2. Results at rest and end-exercise before (exercise 1) and after (exercise 2) intracoronary injection

		HR (bpm)	Mean BP (mm Hg)	HR · systolic BP (mm Hg·min^{-1})	ST↓ (mm)	Cardiac index (l/min/m^2)	PA wedge (mm Hg)
Phento	Rest	80 ± 3	102 ± 5	10,950 ± 664	0	3.0 ± 0.2	9 ± 1
	End — exercise 1	121 ± 5	126 ± 7	22,533 ± 2,471	2.5 ± 0.3	4.0 ± 0.3	26 ± 2
	Rest	82 ± 3	101 ± 6	11,186 ± 790	0	3.0 ± 0.2	10 ± 1
	End — exercise 2	123 ± 5	126 ± 7	22,962 ± 2,526	1.5 ± 0.3**	4.2 ± 0.2	19 ± 2**
Indo	Rest	77 ± 3	111 ± 4	12,138 ± 750	0	3.3 ± 0.2	12 ± 1
	End — exercise 1	112 ± 3	128 ± 3	20,652 ± 951	2.0 ± 0.1	4.4 ± 0.2	27 ± 1
	Rest	75 ± 3	112 ± 4	11,777 ± 758	0	3.3 ± 0.2	11 ± 1
	End — exercise 2	113 ± 3	128 ± 3	21,339 ± 1,173	1.7 ± 0.2	4.4 ± 0.3	24 ± 2*
Placebo	Rest	74 ± 3	90 ± 3	9,474 ± 470	0	3.5 ± 0.2	10 ± 1
	End — exercise 1	114 ± 6	114 ± 3	19,052 ± 882	1.9 ± 0.2	4.3 ± 0.3	25 ± 1
	Rest	75 ± 3	90 ± 4	9,501 ± 458	0	3.5 ± 0.2	9 ± 1
	End — exercise 2	117 ± 7	115 ± 2	19,824 ± 1,274	1.8 ± 0.2	4.4 ± 0.2	25 ± 2

*: $p < 0.05$ versus end of exercise 1.
**: $p < 0.01$

HR = Heart rate; BP = blood pressure; ST↓ = ST-segment depression; PA wedge = pulmonary artery wedge pressure. Phento: Phentolamine; Indo: Indoramin.

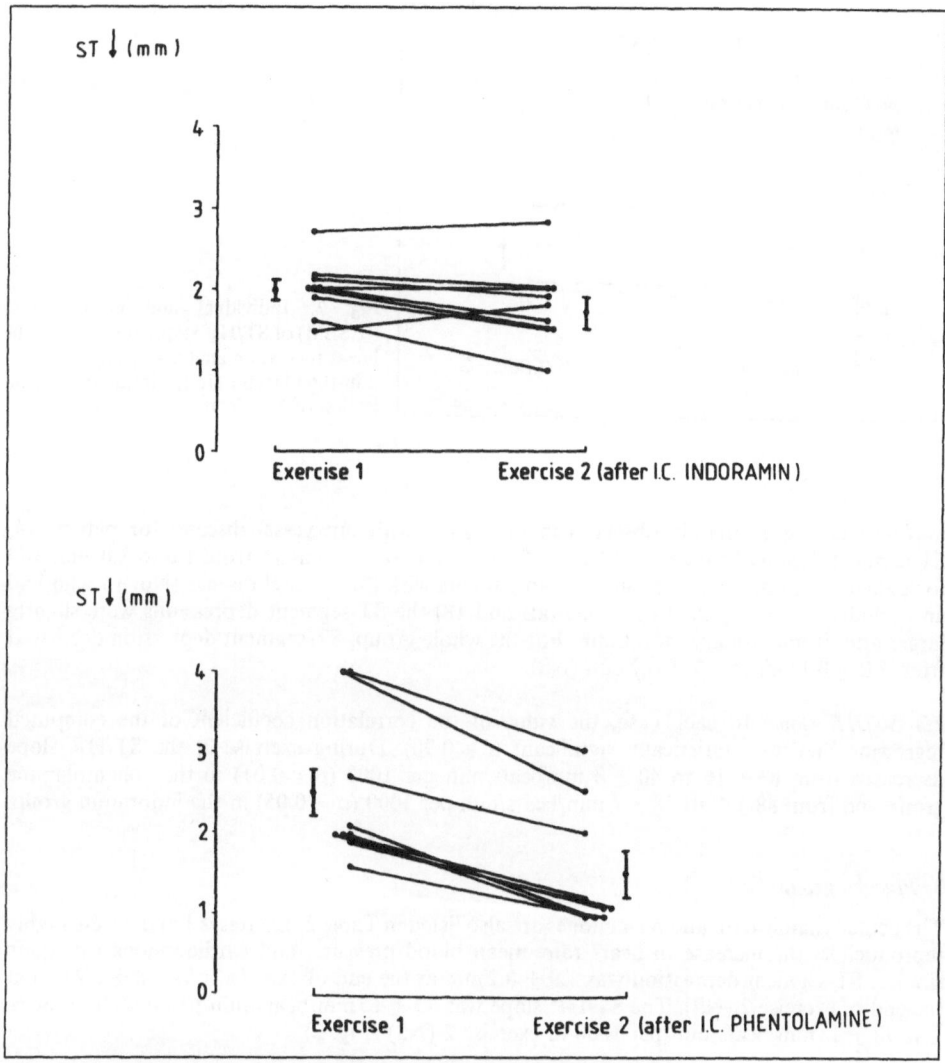

Fig. 1. Individual and mean values (± SEM) of ST-segment depression. Exercise 1 is the control exercise, exercise 2 is the exercise after intracoronary injection of phentolamine or indoramin. The changes were significant only for the phentolamine group (p < 0.01).

2.2. *ST-segment depression*: The individual and mean values of ST-segment depression are summarized in Fig. 1. In the phentolamine group, two patients with three-vessel disease (No. 1, and 3 who had an occluded vessel supplied by collaterals) exhibited the largest ST-segment depressions. After intracoronary phentolamine, ST-segment depression decreased from 4.0 to 3.1 mm in patient 1 and from 4.0 to 2.7 mm in patient 3. This improvement was similar to that observed in the other patients of the group, even those with less severe disease such as patient 9 (one-vessel disease), in whom ST-segment depression decreased from 2.0 to 1.0 mm. For the whole group, ST-segment depression decreased from 2.5 ± 0.3 mm to 1.5 ± 0.3 mm (p < 0.01).

In the indoramin group, the changes in ST-segment depression were more variable. The

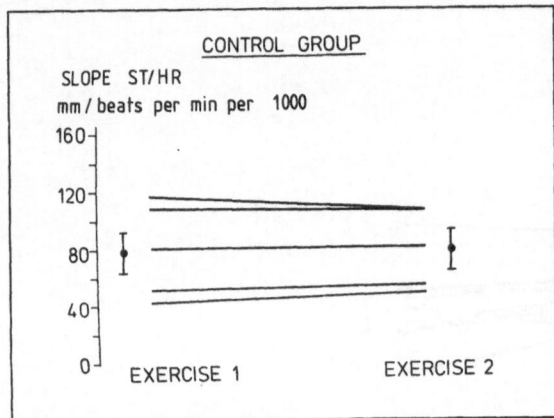

Fig. 2. Individual and mean values (± SEM) of ST/HR slopes obtained by the linear regression analysis in exercise 1 and 2 in the placebo group (intracoronary injection of 5% dextrose in water).

improvements were mainly observed in the subset with one-vessel disease: for patient 14, ST-segment depression decreased from 2.0 to 1.5 mm, for patient 16 from 1.6 to 1.0 mm, and for patient 17 from 1.9 to 1.5 mm. In two patients with three-vessel disease (No. 13 who had an occluded vessel supplied by collaterals and 18) the ST-segment depressions were slightly larger after intracoronary indoramin. For the whole group, ST-segment depression decreased from 2.0 ± 0.1 mm to 1.7 ± 0.2 mm (NS).

2.3. *ST/HR slope*: In each case, the value of the correlation coefficient of the computed regression line was statistically significant ($r \geqslant 0.90$). During exercise 2, the ST/HR slope decreased from 83 ± 16 to 40 ± 8 mm/beats/min per 1000 ($p < 0.01$) in the phentolamine group and from 88 ± 9 to 58 ± 6 mm/beats/min per 1000 ($p < 0.05$) in the indoramin group.

3. *Placebo group*

The hemodynamic data and ST-changes are also listed in Table 2. Exercises 1 and 2 were highly reproducible: the increase in heart rate, mean blood pressure, and cardiac index was quite similar. ST-segment depression was 1.9 ± 0.2 mm at the end of exercise 1 vs 1.8 ± 0.2 mm at the end of exercise 2 (NS). The ST/HR slope was 79 ± 15 mm/beats/min per 1000 in exercise 1 vs 78 ± 13 mm/beats/min per 1000 in exercise 2 (NS) (Fig. 2).

Discussion

Our results indicate that intracoronary injection of small doses of phentolamine or indoramin decreases the exercise-induced ischemia in patients with chronic stable angina. However, these changes were less pronounced after intracoronary indoramin and the attenuation of ST-segment depression did not reach a statistical significance.

These improvements seem to be mainly related to an increase in myocardial oxygen supply rather than a decrease in myocardial oxygen consumption. Indeed, we tried to produce, as far as possible, a selective coronary α-blockade. For this purpose, we injected the α-blockers locally into the most diseased coronary vessel and we excluded patients who exhibited changes of more than 5% in the peripheral hemodynamic parameters after intracoronary injection. One main argument for the absence of peripheral effects of α-blockers is the similar behavior of heart rate, blood pressure, double product, and cardiac index during both exercises. In addition, a peripheral effect of the α-adrenergic antagonists would have reduced the PA wedge, not only at the end of exercise 2, but also at control before exercise 2. Hence, the smaller increase in PA wedge during exercise 2 can probably be related to a less severe ischemia (rather than a peripheral effect of phentolamine or indoramin). This absence of peripheral effects of the α-adrenergic antagonists prompted us to stop exercise 2 when the duration of exercise 1 was reached. Thus, for each patient we were able to compare ST-segment depression in exercises 1 and 2 for similar workloads and double-products. Moreover, the selection of patients with large ST-segment depression allowed us to obtain reliable ST/HR computed regression lines.

Although PA wedge, ST-changes, and ST/HR slope are indirect parameters to assess myocardial ischemia, they are probably more reliable than the assessment of coronary blood flow by the thermodilution technique. Indeed, displacements of the coronary sinus catheter during exercise hinder this assessment. The second reason why we did not measure this parameter is that the improvement in myocardial supply during exercise 2 could be only due to a better regional distribution of coronary flow without obvious changes in total coronary blood flow. One might speculate that the less severe ischemia observed during exercise 2 is related to a warm-up phenomenon. However, in the placebo group, the severity of ischemia was identical in both exercises. We can, therefore, reasonably exclude such a mechanism in our protocol. A 30 min interval between two consecutive exercise tests seems to be sufficient to prevent this warm-up phenomenon [11]. In the present study, in addition to this fixed time interval between the two exercises, the other criterion for starting the intracoronary injection and exercise 2 was the return of all hemodynamic parameters to control values.

With regard to the site of action of α-adrenergic antagonists, we can postulate an effect on the large coronary arteries. First, the α-adrenoceptors are better distributed in the proximal coronary arteries [15]. Second, Gage et al. [10] have recently demonstrated a worsening of epicardial stenosis during dynamic exercise in patients with chronic stable angina. This finding has also been pointed out during isometric exercise [4]. This incriminated mechanism is a competition between metabolic vasodilation, such as local oxygen tension and adenosine concentration, and superimposed α-adrenergic vasoconstriction. In patients with a major obstruction, this α-adrenergic activity rapidly exceeds the metabolic vasodilation and regional coronary flow falls. In addition to this action on epicardial vessel, an effect on the small coronary vessels is also quite possible. Indeed, in dogs with severe stenosis, intracoronary injection of idazoxan (a selective α_2-antagonist) during exercise improves blood flow mainly in the subendocardial and midmyocardial ischemic areas, while the flow in the ischemic subepicardium remains unchanged [18].

In the present study, by selecting patients with large ST-segment depression, we isolated a group of patients with severe stenosis. According to Epstein and Talbot [8] it is in this type of lesion that small changes in stenosis diameter induce the most pronounced alterations in the coronary resistance.

Furthermore, endothelial dysfunction induced by atherosclerosis might reinforce this dynamic behavior of coronary stenosis. There is now compelling evidence from in vitro [2, 9] and in vivo [13, 17] studies that in atherosclerotic human coronary arteries, the endothelium-dependent vasodilation to various endogenous substances is abolished and the responsiveness to the vasoconstrictor effects of various amines is enhanced. In addition, the more severe the atherosclerotic lesions are, the more pronounced are the pharmacological alterations [2]. Hence, we probably isolated a group of patients with atherosclerotic lesions severe enough to have a dynamic behavior and to be highly responsive to the vasoconstrictor effect of the catecholamines.

The more pronounced improvement with phentolamine than with indoramin is difficult to explain. As mentioned previously, the doses used seem to be equipotent on the peripheral circulation in other studies. From a theoretical point of view, indoramin should have been more efficient than phentolamine. Indeed, phentolamine also acts on presynaptic α_2-adrenoceptors that normally participate in negative feedback control of noradrenaline release during sympathetic activity. Inhibition (by phentolamine) of these receptors increases the noradrenaline concentration in the synaptic clefts and should, therefore, enhance the adrenergic activity locally in the coronary circulation, which could have deleterious effects. Nonetheless, several recent observations might explain the higher efficiency of phentolamine. First, from pharmacological studies [3, 20] in organ baths, it has been shown that α-adrenoceptors of human epicardial vessels are partially resistant to α_1-adrenergic antagonists. Second, the α_2-adrenoceptors are more peripherally located, away from the synaptic cleft, this being the converse for the α_1-adrenoceptors [12]. According to this concept, exercise which induces a more pronounced elevation in the plasma catecholamine levels than does direct activation of cardiac sympathetic nerves should mainly stimulate the α_2-adrenoceptors. Third, in the dog it has been shown that in the ischemic inner myocardial layers, the α-adrenergic vasoconstriction is mainly mediated by α_2-adrenoceptors [7]. This finding has been ascribed to the fact that α_1-adrenoceptors are more sensitive to acidosis than the α_2-adrenoceptors [14].

Several limitations of the study require comments. Although in the placebo group, the exercises seem quite reproducible, each patient in the treated group should have undergone a second exercise without drugs to formally exclude a warm-up phenomenon and to reinforce our results. A double-blind, randomized protocol, placebo vs phentolamine vs indoramin in the same patients would have rendered the comparison between the two α-blockers much easier. However, such a protocol would have unduly prolonged the study duration and would have necessitated a second catheterization, a few days apart, to avoid interference between the two blocking drugs. The careful selection of the patients allowed us to partially circumvent this problem; indeed, both groups were relatively homogeneous, they had severe coronary lesions and a good left ventricular function.

Another methodological problem is that ST-changes and PA wedge are very indirect measurements of regional ischemia. Therefore, it might be that the changes in these parameters are influenced by the disease in other vessels than those perfused with the blocking drugs. We might also speculate that in the presence of two-vessel disease, the α-blocker is diverted into the less diseased vessel due to a preferential flow in this area. With these assumptions, one might expect that patients with one-vessel disease were more responsive to intracoronary injection of α-blockers. This seems to be the case in the indoramin group. However, the number of patients is relatively small in each group so that comparisons between subsets of patients within each group are difficult. In conclusion, this investigation indicates that an α-blockade, relatively selective in the coronary circulation, decreases the exercise-induced ischemia in patients with severe stenosis. Therefore, our results argue for a role of α-adrenergic coronary tone in exertional angina. The relatively higher efficiency of phentolamine vs indoramin suggests that α_2-adrenergic mechanisms contribute to the inappropriate coronary vasoconstriction during exercise in these patients.

Acknowledgements: We thank Antonia Alvaro for her secretarial assistance. This work was supported by a research grant from "Fonds de la Recherche Scientifique Médicale" (3,4512.87).

References

1. Berkenboom G, Abramowicz M, Vandermoten P, Degre S (1986) Role of alpha-adrenergic coronary tone in exercise-induced angina pectoris. Am J Cardiol 57:195–198
2. Berkenboom G, Depierreux M, Fontaine J (1987) The influence of atherosclerosis on the mechanical responses of isolated human coronary arteries to substance P, isoprenaline and noradrenaline. Br J Pharmacol 92:113–120
3. Berkenboom G, Fontaine J, Desmet JM, Degre S (1987) Comparison of the effect of beta-adrenergic antagonists with different ancillary properties on isolated canine and human coronary arteries. Cardiovasc Res 21:299–304
4. Brown BG, Lee AB, Bolson EL, Dodge HT (1984) Reflex constriction of significant coronary stenosis as a mechanism contributing to ischemic left ventricular dysfunction during isometric exercise. Circulation 70:18–24
5. Cohn PF (1987) Total ischemic burden: pathophysiology and prognosis. Am J Cardiol 59:3C–6C
6. De Oliveria JM (1981) Three years experience with indoramin. Brit J Clin Pharmacol 12:1115–1195
7. Deussen A, Heusch G, Thämer V (1985) Alpha$_2$-adrenoceptor-mediated coronary vasoconstriction persists after exhaustion of coronary dilator reserve. Eur J Pharmacol 115:147–153

α-Adrenergic coronary constriction in effort angina

8. Epstein SE, Talbot TL (1981) Dynamic coronary tone in precipitation, exacerbation and relief of angina pectoris. Am J Cardiol 48:797–803
9. Förstermann U, Mügge A, Alheid U, Haverich A, Frolich JC (1988) Selective attenuation of endothelium-mediated vasodilation in atherosclerotic human coronary arteries. Circ Res 62:185–190
10. Gage JE, Hess OM, Murakami T, Ritter M, Grimm J, Krayenbuehl HP (1986) Vasoconstriction of stenotic coronary arteries during dynamic exercise in patients with classic angina pectoris: reversibility by nitroglycerin. Circulation 73:865–876
11. Kaltenbach M, Schulz W, Kober G (1979) Effects of nifedipine after intravenous and intracoronary administration. Am J Cardiol 44:832–838
12. Langer SZ, Shepperson NB (1982) Recent developments in vascular smooth muscle pharmacology: the post-synaptic alpha$_2$-adrenoceptor. Trends Pharmacol Sci 3:440–444
13. Ludmer PL, Selwyn AP, Shook TL, Wayne RR, Mudge GH, Alexander RW, Ganz P (1986) Paradoxical vasoconstriction induced by acetylcholine in atherosclerotic coronary arteries. N Engl J Med 315:1046–1051
14. McGrath JC (1982) Evidence for more than one type of postjunctional alpha-adrenoceptor. Biochem Pharmacol 31:467–484
15. Moreland RS, Bohr DF (1984) Adrenergic control of coronary arteries. Federation Proc 43:2857–2861
16. Mudge GH, Grossman W, Mills RM, Lesch M, Braunwald E (1976) Reflex increase in coronary vascular resistance in patients with ischemic heart disease. N Engl J Med 295:1333–1337
17. Nabel EG, Ganz P, Gordon JB, Alexander RW, Selwyn AP (1988) Dilation of normal and constriction of atherosclerotic coronary arteries caused by the cold pressor test. Circulation 77:43–52
18. Seitelberger R, Guth BD, Heusch G, Lee JD, Katayama K, Ross J Jr (1988) Intracoronary alpha$_2$-adrenergic receptor blockade attenuates ischemia in conscious dogs during exercise. Circ Res 62:436–442
19. Taylor SH, Sutherland GR, Mackenzie GJ, Staunton HP, Donald KW (1965) The circulatory effects of intravenous phentolamine in man. Circulation 31:741–754
20. Toda N (1986) Alpha-adrenoceptor subtypes and diltiazem actions in isolated human coronary arteries. Am J Physiol 250:H718–H724

Authors' address:

Guy Berkenboom M.D., Department of Cardiology, Hôpital Erasme, Route de Lennik 808, Bruxelles 1070, Belgium.

Subject Index

A

acetylcholine 121, 127, 128

acidosis 11, 17, 215, 220, 368

acylcarnitine 31, 37, 38, 40

acyl-CoA 37

action potential 6, 24, 40, 41

adenosine 9, 12, 13, 18, 90, 92, 122, 125, 146, 167, 177, 182, 185, 188, 189, 226, 351, 352, 367

adenosine diphosphate (ADP) 121, 127, 128

adenosine triphosphate (ATP) 20, 89, 90, 267, 279, 283, 289, 290, 338

adenylyl cyclase 9, 15, 32, 41, 47, 51, 52, 55, 58, 62, 67, 68, 72-75, 86, 319

adrenergic coronary constrictor tone 117, 167, 168, 170, 171, 174, 179, 180, 183, 185-190, 203, 207, 211,

adrenochrome radicals 15

afterload 219, 220, 223, 273, 342, 353, 356

albuterol 124

ameroid constrictor 121-123, 125, 126, 128, 148, 159, 224

amiloride 16, 18, 23, 289

amiodarone 318

anaerobic glycolysis 17

anesthetics 146, 179

angina pectoris 11, 97, 141

angiotensin II 88, 124, 134, 135

anoxia 12, 13, 16, 19, 289

ansa subclavia 4

anxiety 179

arachidonic acid 34

ARL 57 267

arrhythmias 6, 9-11, 20-24, 31, 32, 41, 51, 55, 62, 67, 68, 73, 75, 125, 220, 229, 233, 250, 283, 293, 294, 298, 300, 301, 305, 306, 310, 314, 317, 318, 325, 327

arterial blood pressure 58, 59, 114, 133, 135-137, 139, 158, 169, 180, 189, 193-195, 203, 204, 208, 209, 225, 226, 231, 232, 245, 247, 248, 268, 306, 314, 316, 327, 328, 338, 339, 341, 348, 359-361, 366

asynergy 6, 278

atenolol 16, 21, 24, 47, 51-53, 55, 60-62, 98, 102, 103, 106, 107, 124, 157, 159, 160, 296, 335, 338, 340, 343, 344

atherosclerosis 128, 168, 335, 341, 342, 355, 367

atropine 5, 133, 306

automaticity 40, 294, 298

Automatic Tone and Reflexes after Myocardial Infarction (ATRAMI) 319

autoregulation 111, 112, 114, 115, 118, 119, 145, 146, 150, 170, 171, 179, 185, 194, 203, 220

Aδ-fibers 259, 261

α-adrenergic coronary constrictor tone 81, 88, 103, 111, 115, 153, 161, 167-174, 178, 180, 181, 184, 186, 187, 189, 193, 194, 204, 211, 214, 219-221, 226, 335, 341, 343, 359, 368

α-adrenoceptor 12-14, 21, 24, 31, 32, 57, 61-63, 81, 82, 84-86, 88, 91, 97, 98, 111, 112, 118, 121, 122, 124, 132, 133, 138, 139, 167, 169-174, 184, 189, 207, 211, 214, 219, 220, 367, 368

α-blockade 172, 181, 184, 359, 366, 367

α_1-agonist 105, 107, 122

α_1-adrenergic coronary constrictor tone 177, 180, 182, 204, 215, 220

α_1-adrenoceptor 12, 14, 15, 31-42, 47, 48, 50, 54, 55, 57, 62, 68, 69, 82-86, 88, 92, 97, 98, 100, 101, 104, 105, 108, 111, 112, 115, 118, 119, 121, 131, 132, 134, 135, 139, 140, 167-170, 177, 178, 183-185, 188, 190, 193, 194, 200, 201, 204, 207, 208, 210, 211, 213, 215, 219-222, 368

α_1-blockade 36, 85, 91, 98, 101, 131, 132, 138, 140, 177, 181, 187, 202, 204, 207, 221, 359

α_{1A}-adrenoceptor 31-35, 62, 84, 86, 87

α_{1B}-adrenoceptor 31-35, 62, 84, 86, 87

α_2-agonist 105, 107, 122, 123

α_2-adrenergic coronary constrictor tone 115, 118, 173, 204, 205, 207, 208, 215, 222, 223, 226

α_2-adrenoceptor 13, 16, 21, 32, 57, 62, 82-92, 97, 98, 100, 101, 104, 105, 108, 111, 112, 115, 118, 119, 121, 131, 134, 135, 139, 140, 167, 169, 170, 177, 178, 183, 185, 188, 190, 193, 194, 201, 203, 204, 207, 208, 210-216, 219-222, 253, 254, 368

α_{2A}-adrenoceptor 32, 84, 86

α_{2B}-adrenoceptor 32, 84, 86

α_{2C}-adrenoceptor 87

α_2-blockade 82, 85, 91, 98, 101, 131, 138, 140, 183, 187, 202, 207, 220, 221, 367

B

baroreflex 134, 137, 139, 167, 168, 174, 180, 189, 190, 194, 243, 244, 247, 249, 258, 314

baroreflex sensitivity (BRS)   305-315, 319
Bay K 8644   88
Beta Blocker Heart Attack Trial   300, 328
(BHAT)
betaxolol   98, 99, 103, 104, 106, 107
Bezold-Jarisch reflex   254
B-HT 920   85, 98-101, 105-107, 123
B-HT 933   82, 83, 85, 111, 113, 114, 116-118,
   133, 136, 223, 224
bisoprolol   60, 61
border zone   20, 75
bradycardia   3, 330
bradykinin   253, 268, 338
bretylium tosylate   7
β-adrenoceptor   14, 15, 21, 24, 34, 35, 41, 47-55,
   57, 60-63, 67-75, 81, 84, 85, 87, 97-100, 106,
   121, 123, 124, 131, 132, 138, 139, 141, 158, 162,
   168, 174, 178, 181, 189, 193, 203, 204, 207, 210,
   211, 215, 221, 300
β-blockade   10, 20-22, 48, 81-83, 85, 89, 98, 99,
   101, 103, 104, 114, 131, 132, 134, 136-138, 157-
   162, 165, 168, 180, 183, 184, 187, 200, 202, 203,
   207, 214, 215, 222, 226, 293-296, 298-301, 325-
   328, 330, 335, 340, 343, 344
β-oxydation   36, 37
β-receptor internalization   15
β_1-adrenoceptor   32, 48, 51, 55, 57, 58, 60-62,
   97-99, 103, 106-108, 121
β_1-blockade   89, 102, 103, 124, 343
β_2-adrenoceptor   32, 57-62, 97, 98, 102, 104, 106-
   108, 121
β_2-blockade   81, 89, 90, 103, 124, 343

C
calcium   15-17, 19, 33, 34, 39-41, 55, 73, 86, 87,
   128, 221, 267, 278, 279, 283, 284, 286, 287, 289,
   290, 294, 298, 342
calcium antagonist   86-90, 125, 190, 219-226,
   253, 254, 318, 337-340, 353
calcium channel   16, 87, 88
calcium ionophore A 23187   128
calcium overload   15, 21
calf   97, 98, 105-107, 220
carbachol   72
Cardiac Arrhythmia Suppression
Trial (CAST)   300
cardiac sympathetic afferents   167, 243-245, 254,
   268, 294, 314
cardiac sympathetic efferents   4, 5, 9, 18, 168,
   190, 258, 294, 318
cardiac sympathetic nerves   3, 111, 178, 243,
   259, 267
cardiac sympathetic nerve
stimulation   40, 85, 107, 118, 167, 177, 178, 185,
   188, 208, 211, 219-223, 226, 267, 269-273, 275-
   278, 294, 314, 317

cardiac vagal afferents   243, 245, 254
cardiac vagal efferents   4, 11, 316, 317
cardiac vagal nerves   194, 221, 243, 244, 253,
   255, 257, 258, 269, 306
cardiocardiac sympathetic reflex   189, 190, 243,
   253, 254
cardiogenic shock   10
cardiomyopathy   41, 58, 60, 62
carnitine acyltransferase   37, 38
carotid sinus reflex   82, 84, 201
cat   35, 36, 63, 112, 188, 244, 246, 314, 316-318
catecholamines   10, 12, 15, 21, 41, 48, 51, 54, 55, 57,
   58, 62, 63, 67, 68, 71, 89, 111, 132, 178, 183, 204,
   268, 278, 283, 284, 288, 289, 294, 298, 327, 368
celiprolol   335, 338, 341, 344
C-fibers   259, 261
chemoreflex   180
chlorethylclonidine   32
circadian variation   325, 327, 330
class I antiarrhythmic agents   296, 298, 299, 318
clonidine   20, 89, 90, 122, 179
cocaine   16
cold pressor test   168, 204, 211, 220
collateral flow   75
collaterals   51, 121-125, 127, 146, 157-159, 162,
   164, 214, 220, 223, 225, 229-231, 233, 234, 238,
   362, 366, 367
collateral steal   125, 158
conduction   293, 295-297
conduction time   294, 295, 298
conduction velocity   7, 24, 294
contractile function   157, 165, 177, 186, 225, 254,
   267
contractility   11, 21
coronary arteries   112, 114-117, 351
coronary arteriography   335, 336, 347-350, 353,
   356, 359-361
coronary arterioles   112, 114, 115, 117-119, 351,
   356
coronary artery diameter   98, 100, 102, 105, 106,
   111-116, 119, 221, 335, 337-340, 342-344, 350-
   352, 356, 367
coronary artery occlusion   11, 20, 47, 50, 121,
   189, 196, 229-238, 245, 246, 249, 254, 267, 269,
   270, 275-277, 293-299, 314-316, 351
coronary blood flow   3, 82, 84, 97, 98, 100, 102,
   106, 108, 125, 132, 138, 146, 147, 159, 168, 177,
   179-181, 186, 193, 194, 196-204, 207, 215, 221,
   222, 223, 255, 256, 258, 260, 262, 335, 343, 351,
   352, 356, 359, 367
coronary constrictor tone   82, 92, 105, 107, 112,
   118, 131, 133, 181, 188, 194, 215, 253
coronary dilator reserve   125, 145-147, 150, 152,
   167, 194-196, 201, 207, 219, 220, 222, 224, 226,
   318, 335, 351, 353
coronary hypoperfusion   82, 114, 118, 163, 165,
   173, 187, 188, 195, 208, 211, 214-216, 220

coronary perfusion pressure (CPP) 122, 145, 146, 150, 159, 170, 179, 186, 187, 198, 199, 201, 203, 211, 215, 221, 223, 283-285

coronary resistance 97, 107, 111, 145, 146, 149, 180, 190, 193, 196, 200-204, 211, 215, 219, 222-224, 253-255, 257, 263, 342, 343, 351, 367

coronary resistance vessels 81, 83-85, 89, 90, 107, 122, 125, 178, 180, 184, 189, 220, 348, 356

coronary spasm 97, 188, 342

coronary steal 146, 158, 161

coronary stenosis 145, 146, 152, 153, 157-159, 162, 167, 170-174, 183, 186, 188-190, 193, 195, 197-203, 207, 210-216, 219, 220, 222-224, 238, 253-261, 263, 264, 336, 338, 344, 347, 348, 351, 354-356, 367, 368

coronary vasomotor tone 21, 81, 84, 111, 117, 122, 131, 193, 219, 220, 226, 330, 335-337, 342, 347, 348, 351, 353, 354

cortisol 327

creatine kinase 36, 231, 233, 234, 237

critical stenosis 146-153

cyanide 13, 16, 17, 70, 71

cyclic AMP 21, 24, 34, 35, 61, 89, 90, 92, 288

cyclic GMP 342

cyclic nucleotide phosphodiesterase 31, 34

D

delayed afterdepolarization 24, 41, 298

denervation 4-7, 9, 20, 22, 177, 184, 190, 229-238, 244, 245, 268, 278, 294

desensitization 31, 60, 68, 71

desipramine 14, 16, 17, 20, 22, 23

diabetes 330

diacylglycerol 31, 33, 34, 39, 40, 61

diatrizoate 336, 337

diazepam 335

digitalis 18

diprafenone 297-299

dipyridamole 226

dog 3, 6, 11, 20, 22, 23, 35, 36, 40, 62, 63, 75, 81-85, 88-91, 104, 106, 107, 112, 121, 122, 125, 126, 131-140, 146-152, 157-159, 161, 162, 164, 168, 171, 173, 177, 178, 179, 181-190, 193, 194, 203, 207-210, 212, 214, 215, 219-222, 224, 226, 229, 230, 233, 235-238, 243, 249, 253-255, 257, 258, 267, 268, 270, 293-295, 297, 298, 305-307, 311-313, 315, 316, 319

dopamine 12, 267, 278

down-regulation 51, 57, 58, 60, 62, 68

E

early afterdepolarization 41

ECG 48, 311, 338

ejection fraction 11, 310-312, 360

emotion 167, 174, 243

encainide 301

endothelium 81, 87, 89-91, 104-106, 121, 127, 128, 348, 353, 367

endothelium-derived relaxing factor (EDRF) 90, 125, 128, 335, 342, 353

energy deficiency 12, 17, 18, 294

energy demand 290, 294

energy depletion 16-19

energy metabolism 12, 13, 17

energyrich phosphates 21, 24, 70, 267, 288

epicardial coronary arteries 82-85, 88, 90, 91, 97, 99, 103, 106-108, 178, 189, 219-221, 335-337, 340, 341, 343, 367

epinephrine 10-12, 15, 20, 57, 58, 99, 103, 106, 107, 267, 278, 294-296

esophagus 180

excitement 6, 219

exercise 6, 58, 60, 82, 111, 131, 132, 134-136, 138, 141, 145, 146, 148-153, 157, 158, 161, 162, 164, 167, 170, 173, 174, 177, 178, 180-190, 204, 207, 208, 210-216, 219.220, 224-226, 305, 306, 308, 316, 341, 343, 344, 347-356, 359-361, 364, 365, 367, 368

exercise-induced ischemia 158, 159, 165, 219, 220, 224, 226, 366, 368

exocytotic norepinephrine release 9, 12, 13, 15

externalization 47, 50, 51, 55, 62, 70

extravascular compression 203

F

FADH$_2$ 37

fatty acid 34, 36, 37

fentanyl 253, 258, 259, 261-265

flecainide 301

flow-dependent dilation (Schretzenmayer-Effekt) 90, 91

forskolin 51, 74, 75

G

glucose 13, 16, 70

glycogen 24, 289

G-protein 32, 33, 40, 42, 47, 52, 61, 67, 68, 71, 74, 84, 85, 319

G$_i$-protein 32, 52, 67, 84, 305, 319

G$_s$-protein 47-49, 51-55, 71-73, 75, 84, 319

Göteborg Metoprolol Trial 328-330

GPT 52, 72

Gregg-phenomenon 138, 182, 186

growth factor 33, 128

guinea-pig 11, 12, 16, 35, 36, 48-51, 53, 62, 63, 220, 288

H

heart failure 41, 42, 57, 58, 60-62, 67, 97, 243, 325-327, 330

heart rate 3, 6, 11, 21, 24, 60, 114, 131, 132, 135, 137, 139, 147, 149, 151, 157-160, 162-165, 169, 180, 181, 185, 190, 194, 197-200, 209-211,

373

219, 220, 225, 226, 231, 232, 245, 249, 254, 269, 270, 273, 283, 284, 295, 306, 316, 318, 325-328, 330, 338, 339, 341, 348, 359-361, 366
heart rate variability 247-250, 305, 309-311, 313, 314, 319, 325, 327
hemoglobin 90
hexamethonium 98, 100, 105, 106, 195
Holter-monitoring 248, 311, 359
hypercholesterolemia 128
hypertension 97, 243, 246, 247
hypotension 243, 245, 246, 249
hypothyroidism 41
hypoxia 17, 31, 36-41, 288
6-hydroxydopamine 22, 294

I

ICI 118, 551 98, 99, 102-104, 106, 107, 124
idazoxan 84, 85, 131, 133-135, 138, 139, 153, 183, 204, 207, 208, 210-216, 367
infarct size 20, 21, 229, 231, 233, 234, 237, 238, 294, 327, 328
indoramin 359-361, 363-368
inositol trisphosphate (IP3) 31, 33, 34, 39-41, 61, 86
internalization 35, 47, 62, 67, 70
iodocyanopindolol (I-CYP) 48, 52, 54, 59, 61, 69, 70, 99, 103, 104, 123, 124
iopromide 337
isoflurane 159
isoproterenol 20, 51, 58-60, 62, 71, 75, 98-100, 102, 103, 106, 107, 145, 147, 148, 267, 278

L

lactate 11, 36, 190, 194, 219, 220, 222, 223, 253, 255-257, 283, 285-287, 289, 290, 328
leukocytes 270
lidocaine 318
light vesicle fractions 35, 47, 49-55
local metabolic catecholamine release 9, 15-17, 20, 22
low-flow ischemia 14
lung 99, 103, 106
lymphocytes 58-61
lysophosphatidylcholine (LPC) 34

M

membrane potential 40, 288
metabolic coronary vasomotion 16, 20, 82, 132, 158, 161, 167, 168, 178, 180, 189, 193, 207, 211, 215, 219, 367
methoxamine 85, 122, 133, 136, 179, 204
metoprolol 16, 24, 60, 61, 297, 298, 300, 328-330
mexiletine 318
MIAMI Trial 328-330
microcirculation 111, 112, 115, 118, 119, 121, 128, 226

microspheres 125, 126, 146, 148, 157, 159, 168, 172, 188, 207, 210, 224, 230
mitochondria 37, 38
monkey 220
morphine 327
mortality 10, 20, 21, 293, 300, 301, 310-312, 325-330
muscarinic receptor 32, 67-69, 71, 72, 74, 133, 305, 317, 318
myocardial blood flow 147, 149, 172, 207, 209, 214, 215, 226, 230, 261, 267
myocardial contractile function 182, 215, 234
myocardial infarction 6, 10, 11, 20, 21, 24, 62, 97, 125, 167, 168, 188, 246, 247, 249, 250, 278, 293, 294, 300, 301, 305-309, 311, 313, 325-328, 330, 363
myocardial ischemia 7, 9, 10, 12, 15, 17, 18, 20, 21, 23, 24, 31, 36-38, 41, 47-55, 57, 62, 63, 67-75, 81, 97, 125, 151, 159, 161, 163, 167, 172, 174, 177, 178, 183-186, 188-190, 194, 204, 207, 208, 211, 214, 219, 220, 222-224, 226, 229, 243, 244, 246, 247, 249, 250, 253, 254, 258, 261, 262, 267, 278, 284-289, 293, 295, 296, 301, 305, 306, 314, 315, 318, 327, 328, 342, 343, 347, 352, 367, 368
myocardial metabolism 106, 167
myocardial necrosis 20, 21, 209, 229, 231, 233, 270
myocardial oxygen balance 89
myocardial oxygen consumption 161, 162, 168-172, 180, 181, 186, 187, 193, 194, 196, 197, 201, 258, 294, 295, 327, 335, 343, 344, 366
myocardial oxygen demand 145-148, 150, 151, 158, 161, 162, 165, 177, 178, 180, 181, 186, 189, 207, 219, 220, 226, 343
myocardial oxygen supply 158, 177, 181, 188, 190, 204, 219, 220, 366
myogenic autoregulation 91

N

NADH 37
nadolol 41, 83, 85
naloxone 253, 259, 263
neuropeptide Y (NPY) 3, 12, 13, 15, 16, 89, 90
neurotransmission 6, 81, 88, 267, 268, 278
nitrates 128, 226, 249, 335, 337, 338, 340-344, 347-350, 352-354, 356
nitric oxide 128
nociceptive stimulation 253, 258-265
norepinephrine 3, 5-7, 9-12, 14-20, 23, 24, 39-41, 57, 58, 68, 71, 82, 83, 85, 87, 89-91, 98-101, 103-107, 111, 112, 114-118, 123, 131-134, 137-140, 173, 181, 184, 188-190, 207, 211, 214, 215, 267-273, 275-278, 283-285, 287, 290, 294-296, 367
Norwegian Timolol Multicenter Study 325, 328

O

osmolarity 284
ouabain 289
oxotremorine 318
oxygen 17, 37, 255
oxymetazoline 32

P

pacing 11, 157, 159-161, 269, 273, 295-297, 316, 318
pain 10, 67, 243, 247, 249, 253, 258, 262, 326-328, 360
perfusion-contraction matching 162
pertussis toxin 33, 305, 319
pH 12, 18, 24, 194, 268, 288, 289
phenol 5, 178, 184, 278
phenoxybenzamine 169, 172, 173, 185-188, 204, 214
phentolamine 24, 36, 48, 100, 132, 180, 181, 185, 188, 204, 359-361, 363-368
phenylephrine 83, 85, 89, 98-101, 105, 107, 111, 114, 116, 117, 122, 133, 178, 249, 306, 313, 314, 319
pheochromocytoma 20
phosphatidylcholine (PC) 34
phosphatidylinositol
biphosphate (PIP$_2$) 33, 34
phospholipase A$_2$ (PLA$_2$) 31, 34, 42
phospholipase C 31, 33, 34, 40, 61, 68, 86
phospholipase D 31, 34, 42
pig 11, 36, 90, 104, 105, 121, 157, 159, 162, 165
pindolol 47, 51-53, 55
pirbuterol 98, 99, 102, 103, 106, 107
platelets 84, 87, 327, 330, 348
POCA 38-40
post-ejection thickening 6, 273, 276
postextrasystolic potentiation 267, 279
postjunctional α$_2$-adrenoceptor 87, 118, 187, 201, 204, 207
potassium 12, 17, 21, 24, 33, 40, 73, 86, 283, 284, 286-290
prazosin 24, 32, 36, 40, 48, 82-85, 88, 98-101, 105, 106, 131-140, 177-185, 187-190, 193, 200, 202, 204, 207, 208, 210, 211, 213-215, 221, 222
prejunctional α$_1$adrenoceptor 12, 16, 87, 91, 118
prejunctional α$_2$-adrenoceptor 91, 132, 187, 203
preload 203, 353, 356
prenalterol 98, 99, 102, 103, 106, 107, 124
pressure-flow relationships 111, 145, 150, 171
Prinzmetal's angina 344
procaine 253
prognosis 293, 301, 325
procaterol 58-60
propranolol 16, 20, 24, 47, 48, 51, 52, 55, 60-62, 84, 85, 98-
100, 106, 114, 132, 134, 138, 153, 172, 187, 193, 195, 196, 198, 200, 202, 204, 207, 210, 213, 221, 222, 296, 300, 318, 335, 337, 340-344

protein kinase C 15, 16, 33, 34, 39, 40-42, 86, 87
protons 17, 18, 283, 286
PTCA 11, 301
pulmonary artery pressure 347-349, 351, 352, 360, 364
pulmonary edema 10
Purkinje fibers 41, 298
P$_{2x}$-receptors 90

R

rabbit 11, 34, 53
rat 11, 12, 14, 17, 22, 23, 33, 35, 36, 38, 47, 51, 53, 54, 60, 62, 63, 69, 75, 86, 89, 106, 278, 283, 284, 286, 289
rauwolscine 82, 83, 85, 88, 98-101, 105, 188, 190, 193, 195-204, 221, 222, 253, 254
Raynaud's syndrome 97
reactive hyperemia 195, 196, 201
reentry 24, 294
refractory period 6, 24, 40, 294
regional contractile dysfunction 194, 219, 220, 222
regional myocardial blood flow 147, 148, 157, 159, 161-163, 168, 169, 178, 183, 185, 188, 207, 208, 210, 211, 219, 224, 225, 230, 231, 235
regional myocardial function 7, 131-133, 136-139, 146, 152, 157-164, 181, 189, 207, 213, 215, 216, 223-226, 229-232, 234, 235, 238, 253, 254, 258, 259, 261, 265, 267, 268, 270, 272, 273, 275-278, 368, 378
reinfarction 21, 328
reperfusion 11-13, 15, 35, 41, 69-72, 74, 128, 268-270, 273, 276-279, 284-286, 288-290, 301
reserpine 20

S

saphenous vein 60, 61
sarcolemma 35, 37, 38, 47, 49, 51-55, 98, 105
sarcoplasmic reticulum 33-35, 37, 39, 41, 279
segment length 6, 147, 179, 181, 182, 185, 222
sensitization 31, 74
serotonin 32, 356
silent ischemia 327
SK & F 104078 86, 87
shock 326, 327
sodium 12, 17-19, 33, 41, 73, 74, 283, 284, 286-290
sodium/potassium pump 21
sodium/proton exchange 15, 16, 18, 23
sonomicrometry 85, 98, 131-133, 146-148, 157-159, 181, 185, 189, 207-209, 221-224, 253, 259, 267, 269
sotalol 60, 61
spinal anesthesia 190
spleen 194
stellate ganglion 4, 5
stop-flow ischemia 16

stress 6, 125, 141, 167, 168, 179, 207, 211, 248, 262, 306, 325, 328
stunned myocardium 7, 267, 268, 278
ST-segment changes 11, 48, 50, 247, 328, 348, 349, 360, 362, 364-367
substance P 338
sudden cardiac death 10, 22, 31, 42, 51, 97, 293, 300, 301, 305-311, 314, 325, 327, 330
supersensitivity 6, 9, 19, 75
sympathetic activity 248, 325, 327, 330

T
tetrodotoxin 16
thrombosis 330
thromboxane (A2) 127, 356
tilt 248-250
timolol 16
translocation 50, 51
transmural steal 145-147, 149, 152, 153, 161, 167, 170-172, 174
triggered activity 24, 41, 294
triphenyl tetrazoliumchloride (TTC) 230, 236
tyramine 85

U
UL-FS 49 157, 159, 162, 164
up-regulation 14, 49, 51, 60

uptake 67
uptake$_1$ 10, 12, 13, 15, 17, 19, 22
uptake$_2$ 10
U 46619 127

V
vagal activity 246, 248, 305, 308-311, 313-315, 325, 327, 330
vasopressin 81, 121, 124-126, 127
ventricular fibrillation 7, 9, 21-24, 41, 269, 294-299, 305, 314, 316, 317
ventricular fibrillation threshold 40

W
WB-4101 32
wall stretch 11
wall stress 164
wall thickness 131-135, 139, 145, 147, 148, 151-153, 158, 160, 162-164, 183, 204, 207-210, 212, 223-226, 230-232, 234, 253, 259, 261, 264, 267, 269, 270, 272, 273, 275-277

X
xamoterol 58-60

Y
yohimbine 16, 132, 139, 179, 183, 184, 187, 215